This bo

30 MAY 2006

#17803745

# *Ground-Water Contamination: Field Methods*

A symposium sponsored by
ASTM Committees D-19 on Water
and D-18 on Soil and Rock
Cocoa Beach, FL, 2–7 Feb. 1986

ASTM SPECIAL TECHNICAL PUBLICATION 963

A. G. Collins, National Institute for
Petroleum and Energy Research, and
A. I. Johnson, A. Ivan Johnson, Inc.,
editors

ASTM Publication Code Number (PCN)
04-963000-38

1916 Race Street, Philadelphia, PA 19103

**Library of Congress Cataloging-in-Publication Data**

Ground-water contamination: field methods: a symposium/sponsored by ASTM committees D-19 on Water and D-18 on Soil and Rock, Cocoa Beach, Fl., 2–7 Feb. 1986; A.G. Collins, A.I. Johnson, editors.
(ASTM special technical publication; 963)
"Papers . . . presented at the Symposium on Field Methods for Ground-Water Contamination Studies and Their Standardization"—Foreword.
"ASTM publication code number (PCN) 04-963000-38."
Includes bibliographies and indexes.
ISBN 0-8031-0968-7
1. Water, Underground—Pollution—Measurement—Congresses.
2. Water, Underground—United States—Management—Congresses.
I. Collins, A. Gene. II. Johnson, A. I. (Arnold Ivan), 1919-
III. American society for Testing and Materials. Committee D-19 on Water. IV. ASTM Committee D-18 on Soil and Rock. V. Symposium on Field Methods for Ground-Water Contamination Studies and Their Standardization (1986: Cocoa Beach, Fl.) VI. Series.
TD426.G7 1988 88-10303
628.1'68--dc19 CIP

Copyright © by American Society for Testing and Materials 1988

NOTE

The Society is not responsible, as a body, for the statements and opinions advanced in this publication.

Printed in Baltimore, MD
July 1988

Dmcc

# Foreword

The papers in this publication, *Ground-Water Contamination: Field Methods,* were presented at the symposium on Field Methods for Ground-Water Contamination Studies and Their Standardization held 2–7 Feb. 1986 in Cocoa Beach, Florida. The symposium was sponsored by ASTM Committees D-19 on Water and D-18 on Soil and Rock. A. G. Collins, National Institute for Petroleum and Energy Research (NIPER), presided as symposium chairman, and A. I. Johnson, A. Ivan Johnson, Inc., presided as symposium vice-chairman. Messrs. Collins and Johnson are editors of this publication.

# Related ASTM Publications

Permeability and Groundwater Contaminant Transport, STP 746 (1981), 04-746000-38
Hydraulic Barriers in Soil and Rock, STP 874 (1985). 04-874000-38

# A Note of Appreciation to Reviewers

The quality of the papers that appear in this publication reflects not only the obvious efforts of the authors but also the unheralded, though essential, work of the reviewers. On behalf of ASTM we acknowledge with appreciation their dedication to high professional standards and their sacrifice of time and effort.

*ASTM Committee on Publications*

# ASTM Editorial Staff

Helen M. Hoersch
Janet R. Schroeder
Kathleen A. Greene
Bill Benzing

ANDERSONIAN LIBRARY
WITHDRAWN
FROM
LIBRARY
STOCK
UNIVERSITY OF STRATHCLYDE

# Contents

WELL DRILLING AND COMPLETION



WATER SAMPLING (SATURATED AND UNSATURATED ZONES)

LABORATORY AND FIELD ANALYSES



CASE STUDIES

# Overview

Fifty percent of the United States' drinking water comes from ground water, 75% of the nation's cities obtain all or part of their supplies from ground water, and the rural areas are 95% dependent upon ground water. Therefore, it is imperative that every possible precaution be taken to protect the purity of the ground water.

Because of the increasing interest in prevention of ground-water contamination and the need for nationally recognized methods for investigation of contamination, a Symposium on Field Methods for Ground-Water Contamination Studies and their Standardization was held 2–7 Feb. 1986, in Cocoa Beach, Florida. The symposium was sponsored and organized by ASTM Committees D-19 on Water and D-18 on Soil and Rock. Symposium Chairman was A. Gene Collins of the National Institute for Petroleum and Energy Research (NIPER), Bartlesville, Oklahoma, and Vice Chairman was Ivan Johnson, A. Ivan Johnson, Inc., Soil and Water Consulting, Arvada, Colorado.

The purposes of the symposium were to foster interdisciplinary communication and to develop information that can be used to prepare guidelines for ground-water contamination studies and also be used to develop field methods that can eventually become ASTM standard methods or practices. To move in a direction to meet these stated purposes, 51 papers were presented on methods related to quality assurance; geophysical exploration; well-drilling; construction, monitoring, and development of monitoring wells; ground-water sampling and sampling in unsaturated soils; soil permeability; nonpoint source investigations; and a variety of actual case histories. Of the presented papers, 37 have been peer reviewed and accepted for publication in this special technical publication.

A session was dedicated to U.S. Government efforts to improve field methods for ground-water contamination studies and federal needs for standards. The session was sponsored by the Ground-Water Subcommittee of the Interagency Advisory Committee on Water Data. Speakers from federal agencies, such as the Environmental Protection Agency, Agricultural Research Service, Geological Survey, and Nuclear Regulatory Commission, held panel discussions on the following topics: (1) Standardization of ground-water monitoring techniques—is it desirable? (2) Quality assurance/quality control of ground-water monitoring—can it be accomplished? and (3) Ground-water monitoring research—does it help? These papers also are included in this STP.

During another special session, representatives from ASTM Committees D-18 on Soil and Rock, D-19 on Water, and D-34 on Waste Disposal discussed the philosophical approach and standards development progress and plans of their respective committees. This session also provided discussion about activities of an ASTM Ground-Water Coordinating Subcommittee. A third special session presented a panel discussion on quality assurance through education and certification. Opinions and programs were discussed by representatives from the Association of Engineering Geologists, American Institute of Professional Geologists, National Water Well Association, American Institute of Hydrology, and the Interagency Advisory Committee on Water Data. None of the discussion from these special sessions is included in this STP, but the organizers appreciate the efforts of the speakers at these sessions to provide auxiliary information of interest to symposium attendees.

A field trip to the NASA Kennedy Space Center was provided in mid-week to tour the center and to witness a demonstration of construction of ground-water monitoring wells at the center and to observe logging and interpretation of borehole geophysical logging of a well. In association with the symposium, but held during the week preceding the symposium, were the semiannual meetings of the 30 standards-writing subcommittees of ASTM Committee D-18 on Soil and Rock.

The International Symposium on Geotechnical Applications of Remote Sensing and Remote Data Sensing and Data Transmission of the International Association of Hydrological Sciences was held on 31 Jan. 1986. These additional activities provided discussion of techniques that could be of interest to symposium attendees.

This STP has been divided into the following chapters: (1) Quality Assurance, (2) Geophysical Methods, (3) Well Drilling and Completion, (4) Water Sampling (Saturated and Unsaturated Zones), (5) Laboratory and Field Analyses, and (6) Case Studies. The chapter on quality assurance points up that today monitoring and protecting ground water are performed by industry and federal, state, and local government agencies and that the driving forces behind these efforts are regulations in the Clean Water Act (CWA), the Safe Drinking Water Act (SDWA), the Resource Conservation and Recovery Act (RCRA), and the Comprehensive Environmental Response, Compensation, and Liability Act (CERCLA or Superfund). Data are needed that can be compared against standards in a scientifically and defensible manner; however, the standards may not be available. Development of standard ground-water contamination investigatory procedures are needed to assure quality studies.

**Barcelona** and **Gibb** identify in their paper the essential elements of a comprehensive sampling protocol used in designing a ground-water quality investigation. This excellent paper points up many of the complex problems involved in a ground-water investigation. Standardized quality assurance guidelines would be helpful in meeting regulations.

The chapter on geophysical methods indicates that these methods are useful because of their relatively low cost and that with further development, subsurface three-dimensional plumes of pollutant ion species may be identifiable. The paper by **Tinlin et al.** describes the application of controlled source audio-frequency/magnetotellurics to locate anomalies resulting from the upward movement of formation brines through improperly plugged wells.

In the well drilling and completion chapter, an excellent overview of ground-water monitoring, well design, and installation technology is given in the paper by **Riggs** and **Hatheway.** Drilling and completion methods along with discussions of why certain materials are used in monitoring well construction also are found in this chapter.

In the chapter on water sampling, **Panko** discusses well purging as a prerequisite to obtaining a representative sample. **Everett et al.** discuss the applications of suction lysimeters in vadose zones to obtain representative samples containing volatile organics. The paper by **Gerba** addresses methods of obtaining samples containing viruses. **Torstensson** and **Petsonk** discuss a sealed pressurized sampling device which is useful for obtaining samples as well as measuring *in situ* pressures and hydraulic conductivity. **Barcelona et al.** discuss sampling verification methods, including purging methods, and the effects that well casing materials can have upon samples.

**Oural et al.** note the problems in obtaining representative samples for gross-alpha radiation analysis. **Ficken** discusses new downhole devices for obtaining samples containing organics. **Unwin and Maltby** discusses the effects that well purging have upon obtaining representative samples.

The laboratory and field method chapter discusses new methods such as remote fiber spectroscopy (RFS) employed by **Klainer et al.** to detect and monitor ground-water contaminants; this new technology with sufficient development should be viable and could be standardized. **Crocker** and **Marchin** describe their use of a dynamic fluid flowthrough core apparatus to determine chemical interaction between chemical pollutants and ground-water reservoir rock.

The paper by **Lindsay** and **Baedecker** describes a field method for determining sulfide in the concentration range of 0.3 to 1500 μm. An accurate standard field method for determining concentrations of sulfide would be useful and is needed.

**Olsen et al.** utilize a laboratory system to measure the effects of permeant composition on pore fluid movement in soil where the soil sample is mounted in a conventional triaxial cell. A specially designed triaxial cell was used by **Fang** and **Evans** to determine the long-term effect of landfill leachate upon compacted clayey liner material.

A direct method using a static trapping device emplaced in the soil to concentrate organic contaminants is described by **Voorhees et al.** The organics are determined with a mass spectrometer.

In the case studies chapter, **Urban** and **Gburek** find that ground-water quality distribution is portrayed in shallow and deep fracture systems, if the sampling design is based upon correct geologic and flow system controls. Flow system controls considered are recharge, lateral flow, and discharge zones.

The paper by **Kimball** describes a comprehensive monitoring and evaluation project conducted to determine nonpoint source pollution impacts from agricultural activities. The goals of the project included strict adherence to high-quality monitoring techniques. **Waller** and **Howie** describe site selection criteria for monitor well placement, monitoring well installation and completion, sampling procedures, quality assurance protocol, and initial results pertinent to a nonpoint-source contamination from agricultural chemicals conducted in an unconfined designated sole-source limestone aquifer. Ground-water contamination in the Karst area of Northeast Iowa was studied by **Stan et al.** The study involved geologic mapping, inventory of ground-water wells, potentiometric surface mapping, dye tracing, historic water quality and land management inventories, stream gaging, spring gaging, land use mapping, and land treatment and chemical use surveys. **Morin** and **Cherry** describe an iterative approach that they employed to obtain high-quality data in a field investigation of a small-diameter contaminated ground-water plume.

According to **Melville et al.**, tracer injection and observation experiments are the best available field methods for determining advection and dispersion in aquifers. To conduct 11 experiments, they designed, constructed, and operated multilevel observation wells.

Overall, the papers emphasize that regulations related to the prevention of ground-water contamination were formulated before methods fully adequate for meeting regulation requirements had been developed to the point where they could be easily converted into consensus standards. Furthermore, in some cases research and experimentation, along with actual practice, have been inadequate to define some methods, materials, and guidelines, especially as they apply to ground-water contamination. Many of the standards obviously needed for investigations of ground-water contamination, to some people, may be just as obviously needed for even wider use in investigations of ground-water availability and associated problems such as land subsidence, artificial recharge, irrigation, and urban and rural supply. Thus, the editors believe that this volume will be helpful in pointing the way for the fine-tuning needed to modify some existing methods for investigations of ground-water contamination or in pointing to research needed for new methods and materials for such investigations, keeping in mind that many methods also are needed, possibly with some modification, for the frequent investigations of ground-water supply. Such work can and should lead to eventual, and as early as possible, production of consensus standards that can be referenced in federal and state regulations. It is hoped that this volume will stimulate action of individuals and organizations in meeting the above-stated needs. ASTM Committees D-18 and D-19 will welcome assistance from those readers who are willing to assist in development of such standards.

The papers published herein received at least three peer reviews and were reviewed by the editors following revisions of papers by the authors. The editors express appreciation to the reviewers who assisted so much in assuring the quality of papers in this special technical publication. Appreciation also is expressed to the speakers at the symposium, the authors who prepared, revised, and provided final papers for publication, and the ASTM staff and officers of Committees D-18 and D-19 for their assistance and support in organizing and publishing the results of this symposium. Thanks also go to the following people who served as the Symposium Steering Committee for the development of the program: Dennis C. Erinakes, U.S. Soil Conservation Service, Fort Worth, Texas; Jack Keeley, U.S. Environmental Protection Agency, Ada, Oklahoma; Conwell C. McCune, Chevron Oil Field Research Company, La Habra, California; Les G. McMillion, U.S. Environmental Protection Agency, Las Vegas, Nevada; A. G. Ostroff, Ostroff Associates, Dallas, Texas; John B. Robertson, Weston Designers and Constructors, Rockville,

Maryland; Don Viall, ASTM, Philadelphia, Pennsylvania; and Alex Crawley, U.S. Department of Energy, Bartlesville, Oklahoma.

*A. Gene Collins*
National Institute for
Petroleum and Energy
Research, Bartlesville,
OK 74005; co-editor.

*A. Ivan Johnson*
A. Ivan Johnson, Inc., Arvada,
CO 80003; co-editor.

# Quality Assurance

# Panel Discussion

## INTRODUCTION

The U.S. Geological Survey's (USGS) Office of Water Data Coordination (OWDC) has lead responsibility for coordinating water-data acquisition activities of the federal government, under authority of OMB Circular A-67. Having major input the OWDC is the Interagency Advisory Committee on Water Data (IACWD), chaired by the chief hydrologist of the USGS. The IACWD is composed of representatives of 31 federal agencies that acquire or use water data and oversees 6 subcommittees with numerous working groups.

One of the IACWD subcommittees is that on ground water. The general mission of the subcommittee is to disseminate information on ground-water data and associated matters between federal agencies, and between the federal and private sectors. As part of this effort, the subcommittee organized this session to give ASTM symposium participants an overview of federal efforts and needs in the area of standardization of field methods in ground-water contamination studies. The session was planned to inform symposium attendees of the current state of the art in guidelines, regulations, and research concerned with standardization. It was also intended to provide a framework of thoughts and problems in these areas via presentations by federal agency representatives concerned with guidelines and standards (U.S. Geological Survey), regulations and standards (EPA, NRC), and the user's viewpoint (SCS, BuRec).

Presentations were made by USGS, EPA, SCS, ARS, and NRC followed by a panel discussion involving these presenters and the audience. The panel discussion was chaired by Gene Hamilton, past chairman of the ASTM D-19 Committee on Water.

*William J. Gburek*
Hydrologist, Agricultural
Research Service, U.S.
Department of Agriculture,
University Park, Pennsylvania.

## FIELD METHODS FOR STUDYING CONTAMINATED GROUND WATER: A U.S. GEOLOGICAL SURVEY PERSPECTIVE

To understand the perspective of the U.S. Geological Survey (USGS) with regard to the standardization of field methods for the study of ground-water contamination, it is first necessary to understand the nature and mission of the survey. The three operating divisions within the USGS are the National Mapping Division, the Geologic Division, and the Water Resources Division. It is the Water Resources Division that is most involved with ground-water studies, although the specialized skills found in the other divisions are called upon when needed.

The USGS, through the Water Resources Division, has the principal responsibility within the federal government of providing the hydrologic information and understanding needed to achieve the best use and management of the nation's water resources. It is noteworthy that ''management'' refers to the actions of other agencies, and ''regulation'' is not within the mission of the U.S. Geological Survey. It is the survey's role to provide the hydrologic information and understanding to those agencies, both federal and state, who are the regulators and managers.

The USGS collects data for two reasons: (1) it is clearly in the public interest that data be collected in an unbiased, objective manner, particularly if there is a need for long-term continuity in the data collection program; or (2) the data are needed to help in understanding a critical water-resources issue of national or regional importance.

## Ground-Water Quality Studies in the Geological Survey

Ground-water contamination is clearly an area of critical national concern, and the USGS is addressing this problem in each of its three program areas: the Federal Program, the Federal-State Cooperative Program, and the other Federal Agencies Program. These three programs differ in the sources of their funding, in the role the U.S. Geological Survey plays, and in the objectives of the work. In all the programs, however, the USGS is involved in all scientific aspects of ground-water contamination studies—from basic research to resource appraisal.

### *Federal Program*

About one third of the operating funds of the Water Resources Division are appropriated directly by congress to support a research program and several high priority topical programs; one of these topical programs is the Toxic Chemical Waste/Ground-Water Contamination Program. To this element of its federal program and pertinent research projects, the USGS is directing approximately $10 million toward the study of ground-water contamination. These funds are being used to look at the occurrence, movement, and fate of organic and inorganic chemicals in a variety of hydrogeologic environments. A basic component of this work is the development and evaluation of both laboratory and field methods. Methods development and evaluation is particularly emphasized in the work at three interdisciplinary field studies being supported by the Toxic Chemical Waste Program: sewage effluent at Cape Cod, Massachusetts, an oil spill at Bimidji, Minnesota, and wood-treatment chemicals at Pensacola, Florida. Four of the papers in this symposium are the products of the Geological Survey's Federal Program.

### *Federal-State Cooperative Program*

Funds for the Federal-State Cooperative Program are appropriated by congress with the stipulation that they must be totally matched by state or local funds. The Cooperative Program comprises about one half of the total program of the Water Resources Division. Within this program, work elements are identified jointly by the U.S. Geological Survey and officials of the 800 or so cooperating agencies. Obviously, the work within the Cooperative Program is directed toward local, regional, or statewide problems that are also of national interest. More than 300 current projects in this program, with a budget of about $30 million, have a ground-water quality component. Two papers in this symposium are products of the Cooperative Program.

### *Other Federal Agencies Program*

In the other Federal Agencies Program, the U.S. Geological Survey is reimbursed for work performed for other federal agencies. It is within this program that the USGS may at times act as a consultant, generally as part of water-supply or environmental-impact studies on federal lands. Little or no consulting work has been done on federal lands containing contaminated ground water because most agencies are interested in ''turn-key'' operations including both identification and cleanup of contaminated sites, and such work is best left for the private sector. Nevertheless, a few site-specific studies have been done for the Department of Defense and an existing memorandum of understanding with the Environmental Protection Agency calls for the survey to review the work of private contractors at ''superfund'' sites.

## Standardization of Field Techniques

Across this broad spectrum of work, the U.S. Geological Survey faces the same problems as any other organization dealing with field investigations of contaminated ground water: where and how to sample. The USGS takes very seriously its unique role within the federal government as an unbiased, scientific, fact-finding agency. From this perspective as a scientific agency, the survey recognizes the need to address the problems associated with assuring that field methods for ground-water contamination studies yield consistent and accurate data. Whether this goal of consistency and accuracy is achieved through guidelines or standards is, from the perspective of a nonregulatory agency, only a debate over semantics. In point of fact, the USGS attempts to achieve these goals through an understanding of the site-specific hydrogeology and geochemistry, a field and laboratory quality-assurance program, and an exacting peer review of the plans of study and the reports.

Techniques for obtaining ground-water field data for contaminant studies are rapidly changing. This can be attested to by the number and titles of the presentations at this symposium, by papers published in many of the technical journals, and by the existence of a journal titled, *Ground Water Monitoring* that contains not only technical papers but also a myriad of advertisements for field equipment, supplies, and services. Because of this present state of flux, "standards" may be too rigid a word to apply to any set of existing field techniques. Certainly some type of guidelines are needed, but they must be based on scientific understanding, and they must be applied with enough flexibility to ensure that they meet the purpose and scale of the field problem.

Examples of various types of field problems include site-specific studies and general monitoring. A site study may be in an area of known contamination or in a potentially contaminated area; these are two very different types of problems and require different field techniques. Similarly, a monitoring program designed to protect a well field is quite different than one designed to collect regional water-quality data. For each of these problems, we still have to decide where and how to sample. Indeed, the questions of "where" and "how" to sample are the underlying theme of this symposium. Tentative answers will be given in the papers that follow; the remainder of this paper will expand upon the questions and present an overview of the general problem.

## Obtaining Representative Water Samples

Inherent in all ground-water sampling is the assumption that the water removed from the well is representative of the water remaining in the aquifer and, therefore, will tell us something about the chemistry or biology of the water in the aquifer at the time and place of sampling. This rather basic assumption is fundamental to all ground-water field studies. Another assumption is that if we sample in more than one location we can infer what is happening in the aquifer, not only over space, but with the proper selection of sampling points, over time as well.

### *Where to Sample*

The proper location of sampling points is essential to any field study; mislocated wells are probably the most common problem identified when evaluating contaminant-monitoring schemes. Wells are misplaced because they may be in the wrong direction with respect to the movement of a contaminant away from a source, because they may be too shallow or too deep to detect the contaminant, or because they may be too far from the source of the contamination. Other factors that result in misplaced wells are a disregard or misunderstanding of the time required for contaminants to reach the water table, and incomplete knowledge of the physical and chemical processes that retard the movement of contaminants both in the unsaturated and saturated flow regimes. Proper well location requires an understanding of the source of the contaminant, the local hydraulic gradient, the nature of the geologic materials, and the way water and contaminants move through those materials.

Accurate water levels must be obtained and referred to a common datum in order to determine the direction of ground-water flow. Often this may require wells at multiple depths to detect vertical gradients. Natural and artificial sources of discharge and recharge must be identified, and the way they function and affect the hydraulic gradient must be understood.

Some knowledge of the hydraulic characteristics of the rocks underlying the site is needed to estimate the rate of water movement, both vertically in the unsaturated zone and horizontally and vertically in the saturated zone. Similarly, knowledge of the regional hydrologic setting is needed to understand the vertical flow component. These data are required before the first monitoring well is installed. Depending on the purpose and scale of the sampling program, published descriptions of the hydrology and geology of the area may be sufficient to allow for intelligent estimates of the flow regime to be made. Alternately, a well inventory, which may include preliminary water sampling, supplemented by exploratory well drilling, may be required for some studies. Most likely, exploratory wells installed for general background data will not be suitable for inclusion in the final monitoring scheme. Geophysical methods may sometimes be appropriate to give a preliminary indication of the presence of contaminants, and, when used in conjunction with an exploratory well-drilling program, these methods may allow for rapid identification of the scope of the field problem.

Only after the field parameters pertinent to the problem are defined can the appropriate number and location of sampling points be selected. Too, no monitoring scheme can be installed exactly as designed on paper; preliminary results will doubtless suggest changes in drilling and sampling strategies.

### *How to Sample*

Planning the sampling strategy involves consideration of the purpose of the sampling. Although purposes vary greatly, all sampling programs have one common goal: the removal and preservation of water that reasonably represents the physical, chemical, and biologic properties of the water that remains in the aquifer. Keeping in mind that what is "reasonable" for one type of field activity may be totally inappropriate for another, we can look at the factors that affect the removal of a representative ground-water sample. These factors are: drilling method, casing material, well design, and sampling method. Sample preservation and laboratory analytical procedures are, of course, equally important but, for purposes of this discussion, are not considered field methods.

The physical installation of a well disturbs the natural environment of the aquifer. Many drilling methods introduce water or other fluids into the aquifer, and all methods permit, to varying degrees, the vertical movement of water in the bore hole during the drilling process and in the annular space surrounding the casing after the well is completed. The material used for the well casing may react with ground water. Cement or other material used to isolate the sampled zone from the remainder of the aquifer and from surface fluids, may itself introduce contaminants. The length and type of opening to the aquifer (the "screen" zone) will determine the usefulness of the sample; for some purposes a sample representing a small, one or two foot thick, zone of the aquifer may be most appropriate while for other purposes a sample integrating the entire saturated thickness of the aquifer may be desirable. The pump, bailer, or other device used to withdraw the sample may itself react with the sample or may allow gases to enter or escape from the sample. All these factors must be evaluated when choosing sampling methods that are appropriate to the purpose of the sampling program.

## Conclusions

One set of standard field methods cannot be developed to apply to all ground-water contaminant studies because of the high degree of variability of geologic conditions, the many different chemical and biologic constituents that may be studied, and the widely diverse purposes of the sampling programs. This does not, however, preclude the development of guidelines based upon the need to obtain representative samples that fulfill the purpose of the sampling program. Indeed, our understanding of both the problem and the solutions to obtaining representative samples is growing at an ever accelerating rate. Every field situation can be met with an appropriate set of monitoring techniques, costs not withstanding. On the other hand, the most elaborate set of monitoring techniques will not be appropriate for all field situations. Documentation of all field (and laboratory) aspects of ground-water contaminant studies will help us interpret today's samples in light of tomorrow's developments in this dynamic area of hydrology.

*Irwin H. Kantrowitz*
Chief, Florida District,
U.S. Geological Survey,
Tallahassee, Florida.

## Regulations and Standards

The Environmental Protection Agency's ground-water protection program is a collective response to a series of legislative mandates. Although there is no single encompassing federal legislation directed specifically at ground-water protection, eight separate laws address, either directly or indirectly, specific sources of ground-water contamination. The Safe Drinking Water Act, which establishes national drinking water standards; and the Resource Conservation and Recovery Act (RCRA), which provides for the safe disposal of hazardous waste, are perhaps the most important.

The laws generally require the agency to promulgate and enforce regulations. The regulations interpret and translate the intent of congress into administrative policy. In the case of ground water, primary responsibility for implementing policy is seen to rest with the states.

Regulations, in turn, often require additional clarification in the form of technical guidance documents. Such is the case with ground-water monitoring. Technical guidance typically attempts to strike a balance between the needs for consistency and flexibility. As applied to ground-water monitoring, guidance has appeared thus far in two forms:

1. Minimum standards with maximum flexibility.
2. Maximum specificity with minimum flexibility.

Both approaches have resulted in problems for those concerned about developing technically credible and cost-effective ground-water monitoring programs. An alternative approach based on establishing performance standards along with an equivalency provision may offer a workable compromise solution.

*Robert Snelling*
Acting deputy director,
Environmental Monitoring
Systems Laboratory,
Environmental Protection
Agency, Las Vegas, Nevada.

## Standards, Guidelines, and Regulations: A User's Viewpoint

### Why SCS Is Involved with Ground-Water Investigations

Soil and water conservation is the basic mission of the U.S. Department of Agriculture's Soil Conservation Service (SCS). Through Soil and Water Conservation Districts, which are units of state government, SCS advises landowners and others on the best use of the land and on the best management practices (BMPs) for agricultural and nonagricultural areas. Surface and ground water are both considered because they are inseparably linked.

One special interest in ground water arises from its use for irrigation throughout the country. More than twenty-one million acres of agricultural crops receive ground water nationwide; four million of these are located in the humid eastern part of the United States. SCS's responsibilities include evaluating quantity, quality, and efficient use of ground water. They assess all recommendations for land surface treatment in light of their impact on ground water. These responsibilities require SCS to work continuously with standards and regulations.

**Kinds of SCS Activities**

SCS specialists evaluate ground water for a wide variety of agency activities. Resource evaluations take place under river basin and watershed planning and conservation planning for individual landowners. Assessments of the potential effects of conservation measures on ground water are in the environmental impact statements required for many of the SCS's major projects. Land treatment measures to control erosion are perhaps the SCS's best known project thrust. Even these must be planned with knowledge of potential impacts on ground water. For example, the hazard of increasing the infiltration of nutrients and pesticides in the ground water is studied.

Animal-waste management is occupying an increasing proportion of the SCS workload. Land spreading, holding ponds and treatment lagoons all require due regard to protection of ground water.

Water conservation studies for multicounty or individual field-size areas look at better irrigation practices, better scheduling for pumping, and future ground-water conditions as a result of projected pumping patterns. In addition, SCS is involved in design and implementation of water-level monitoring systems and water-quality sampling.

Governmental units, private organizations, and individual landowners are advised on how to approach agriculturally related ground-water problems, how to get help, and what to have done. Once work is underway, SCS frequently provides overview on how well things are going.

SCS provides funding and guidance for field evaluations to determine the effects of agricultural nonpoint pollution on ground water and how well BMP's reduce these effects. The evaluators are other federal agencies, state agencies, universities, and consultants. SCS performs an oversight role to ensure that original objectives are attained.

SCS also performs a number of in-house evaluations of ground water for the agency and other federal agencies. Most often these involve monitoring to determine the effect of raised or lowered ground-water levels.

Landowners frequently ask SCS to provide guidance regarding the proper procedure and basic resource information needed to meet state and federal regulations intended for environmental protection.

**Concern with Standards and Regulations**

SCS has a very active interest in state and federal regulations that relate to ground water and agriculture. The agency is pretty much on the receiving end and must be able to comprehend and implement all of the requirements. The agency must be also able to cite appropriate standards and specifications, often obtained from non-SCS sources, to be used in contracts and agreements for guidance for SCS personnel.

Ground-water studies require investigation techniques and methods. SCS is concerned with obtaining reliable answers to such questions as: (1) How long does it take nitrate to travel to the water table? (2) Are there pesticides in the ground water? (3) How much pesticide? (4) How long does it last? Of course these answers hinge on how well we can sample, measure, and interpret.

As an agency that uses ground-water technology and uses information developed by others, SCS is concerned about the reliability of the information and whether data from one area to another is comparable. Should in-house specifications be written for needed techniques or are there adequate ones available elsewhere?

**The Problems with Standards and Regulations**

Standards for evaluating ground water are available from several sources such as ASTM, EPA, and USGS. Analytical techniques, especially laboratory techniques, have received the lion's share of attention in standardization. Certification of laboratories is commonplace. However, poor field procedures concerning other important facets of evaluation may create such problems that the best laboratory procedures will be unable to overcome the difficulties.

There are four facets to evaluation: (1) system design, (2) sample collection techniques, (3) analytical techniques, and (4) interpretation techniques. Chemical variations and concentrations are numerous and analytical techniques may have to be quite specific. They have to be guided by field observations that may not have been recorded. Faulty design of the field data collection cannot be remedied by the best of laboratory efforts.

The remaining three facets need more attention. Development of standards and guidelines is needed now. For example, it is well known that geology influences the movement of ground water. Where, how, and when to monitor must be based on the hydrogeology. How often have specifications in a monitoring design recognized this need? Only in a few states do regulations require hydrogeologic analysis.

It appears that the normal practice, in SCS and elsewhere, is to rely on the judgment of the geologist or engineer to integrate all the factors, create a design, and ultimately arrive at the proper interpretation. This is as it should be. However, the professionals need all the help they can get along the way from work that is performed properly and numbers that truly represent the field condition.

This example advances another problem, that of determining who is qualified. In-house, experience will tell who can do the job. But, to evaluate someone from outside the organization, a very detailed process of job, client, and proposal review is required. When in-house expertise is not available for such evaluation, there's a chance that the resulting work will be less than satisfactory.

SCS will continue to rely on its own professionals, but also will use appropriate standards and specifications that are prepared by others. Standards, promulgated by regulation or otherwise, must be in line with reality and be of such quality that they will do their intended job. A more efficient way to evaluate a person's ability to make ground-water investigations also is sought.

SCS encourages the continued development of standards and hopes the identified deficiencies will receive timely attention.

*Dennis C. Erinakes*
Ground water specialist,
South National Technical
Center, Soil Conservation
Service, Forth Worth, Texas.

## Agricultural Research Service Approach to Ground-Water Studies

The Agricultural Research Service (ARS) is the research arm of the U.S. Department of Agriculture. ARS research efforts are directed to answering the food and fiber needs of the nation in general. Research is directed toward specific problem areas suggested by USDA action agencies such as the Soil Conservation Service (SCS). The Agricultural Research Service budget is predominantly directed to the areas of human nutrition, and food and plant product improvement, shipping, storage, and handling. A subdivision within the organization is concerned with soil and water conservation. This group maintains research laboratories and field data gathering centers in the major physiographic areas of the nation. The research derived from studies are published in ARS publications, technical journals, and university publications. Typical publication outlets for soil-water studies are Agriculture Research Service Technical Reports, Water Resources Research, Journal of Environmental Quality, Journal of the Soil Science Society of America and the American Society of Agronomy, Journals of the American Society of Civil Engineers, National Water Well Association's, Ground Water, and others. Ground-water research is conducted as needed within the scope of ongoing studies investigating water yield, infiltration, and the impact of land use upon water quality and quantity. A relatively small technical staff is directly involved in ground-water research. The current focus of the ARS ground-water effort is the vadose (unsaturated) zone, with particular emphasis upon the water and nutrient management in the root zone. Attention is directed to the potential for management rather than resource documentation. Most studies are directed to perched ground water or shallow regional ground water. The ARS has pioneered much work involving chemical transport with soils, infiltration through soils, deep percolation, ground-water recharge, evapotranspiration, and water yield from upland watersheds.

Since the research is problem oriented to emerging short- and long-term agricultural water needs, the approach to research is that of using recommended procedures and equipment rather than establishing fixed materials specifications and standardized procedures of research. Researchers specify the instruments and conditions of experiments in the research publications. Where appropriate, researchers use standard methods of analysis which are published, referenced, and therefore have gained wide acceptance in the scientific community.

*James B. Urban*
Geologies, Northeast Watershed Research Center, Agricultural Research Service, University Park, Pennsylvania.

## WATER QUALITY AND RECLAMATION

The U.S. Bureau of Reclamation is a major water resources development agency working in the 17 western states. Activities include the development of irrigation water supplies, flood control, municipal and industrial water supplies, salinity control, improved fish and wildlife opportunities, and recreation. The agency developed a strong interest in water quality shortly after it was formed in 1902. This interest includes both surface and ground water and their reactions with natural and man-made materials encountered in water resources development. Through the years this interest has been directed most strongly towards agriculture but in recent years has included an ever increasing interest and concern in effects on human health and wildlife.

Some typical examples of water quality concerns include: chemical compatibility of applied irrigation water with local soils and ground water, leaching and concentration of salts in return flows, effects of both natural and man-induced salinity on river systems, and effects of trace elements.

The Bureau of Reclamation approach to water-quality investigations might best be described as scientific engineering, somewhere between the pure scientific and regulatory approaches. In general, an attempt is made to understand the mechanism which causes the observed condition. Experience has shown that it is desirable to establish a range of values rather than attempt to define a single representative value for parameters of this type. As a result of this approach and philosophy, most Bureau of Reclamation investigations exceed what would typically be required by a regulatory approach.

The Bureau of Reclamation has strong capabilities in water-quality studies. All six regions have laboratories with water-quality analysis capabilities, which are certified by either EPA or USGS. The research laboratory at the Bureau of Reclamation Engineering and Research Center in Denver is equipped and manned at the leading edge of the state of the art. The Bureau of Reclamation expects to remain a strong force in the development of water-quality expertise.

While the imposition of regulatory requirements for water-quality investigations would improve the quality of many marginal programs, the development of good scientific guidelines and standards would be of more value to agencies such as the U.S. Bureau of Reclamation. The Bureau of Reclamation is very interested in participating in the development of such standards and guidelines and is pleased to have the opportunity to participate in this conference.

*Lynn A. Johnson*
Geologist, Engineering and Research Center, Bureau of Reclamation, Denver, Colorado.

## U.S. Nuclear Regulatory Commission Standards Development for Ground-Water Related Issues[1]

The U.S. Nuclear Regulatory Commission (NRC) was established by congress under the Energy Reorganization Act of 1974. The primary goal of NRC in licensing and regulating nuclear reactors is to assure the health and safety of the public and to protect the environment. Similarly, NRC has the charge of insuring safe, permanent disposal of radioactive wastes by licensing nuclear activities and facilities, and by protecting the public against the hazards of low-level radioactive emissions and releases from the licensed nuclear activities and facilities.

To accomplish this mission, the NRC staff develops Federal Regulations (Title 10 of Code of Federal Regulations), regulatory guides (guidance documents), and branch technical positions (licensing assistance information) to codify and promulgate both licensing and technical issue guidance.

The standards development program can be subdivided into waste management (for example, low-level and high-level radioactive waste disposal) and reactor siting activities. The standards are derived from licensing review experience and confirmatory research findings. The entire standards development process is oriented towards active public comment reviews (all standards are formally noticed in the *Federal Register*), and a sound technical basis for the specified guidance.

The principal ground-water contaminant issues being studied and addressed in NRC's research program are assessment of:

1. Site characterization techniques to determine ground-water flow and transport parameters for low-permeability media including partially saturated media and fractured media.

2. Monitoring strategies for shallow land-buried facilities with emphasis on ground-water infiltration through trench covers.

3. Instrumentation used to determine unsaturated-zone conditions and properties in fractured media.

4. Ground-water flow and transport models used to determine site performance at low-level and high-level radioactive waste facilities.

5. Bore-hole sealing techniques to determine if preferential pathways could develop with time.

Specific guidance is being developed to address contaminant transport modeling issues such as mechanical dispersion, molecular diffusion, and chemical interactions between the native ground

[1] See *Ground-Water Protection Activities of the U.S. Nuclear Regulatory Commission,* NUREC-1243, February 1987, Washington, D.C.

water, leachate (radionuclides), and geologic matrix. These studies are developed jointly with field study programs on assessing methodology and instrumentation to derive the necessary model input values. Specific field tests under review are (1) down-hole testing for fracture properties and flow conditions, (2) tracer tests in fractured media, and (3) ground-water chemical sampling in the unsaturated zone.

The waste-management standards are needed to support licensing reviews, which requires that data be collected in such a fashion that errors in measurement and uncertainties of the obtained values (for both direct and indirect methods) be quantified. Therefore, ground-water characterization data derived from the various field methods should include an error and uncertainty analysis due to both the system and the measurement method.

*Thomas J. Nicholson*
Hydrogeologist, Waste Management Branch, U.S. Nuclear Regulatory Commission, Washington, D.C.

*Michael J. Barcelona*[1] *and James P. Gibb*[1]

# Development of Effective Ground-Water Sampling Protocols

**REFERENCE:** Barcelona, M. J. and Gibb, J. P., **"Development of Effective Ground-Water Sampling Protocols,"** *Ground-Water Contamination: Field Methods, ASTM STP 963*, A. G. Collins and A. I. Johnson, Eds., American Society for Testing and Materials, Philadelphia, 1988, pp. 17–26.

**ABSTRACT:** The demand for representative ground-water quality samples has been driven by the realization that our water resources are vulnerable to a variety of surface contaminant sources. The techniques by which samples are collected will materially affect the reliability of the analytical data. In this paper, the need for the progressive refinement of sampling and related techniques into sampling protocols is emphasized over the standardization of monitoring network designs or the drilling, sampling, and analytical protocols which are elements of the design. The most effective sampling protocol for a particular investigation must be tailored to the actual site conditions and the information needs of the program. Essential elements of a comprehensive sampling protocol are identified and an example which should prove useful for designing reliable ground-water quality investigations is provided. The example network design is developed by a series of decision trees which allow one to weigh various factors and conditions in a direct manner so as to document how decisions are made. The best model for an effective ground-water sampling protocol is one that is flexible and permits refinement as conditions at a site become better understood.

**KEY WORDS:** ground water, sampling procedures, hydrology, water quality, water chemistry

Efforts to monitor environmental conditions in air, soil, or natural water systems place strong demands on scientific understanding in many disciplines. Our knowledge of air or water chemistry variables has developed principally as a result of environmental quality concerns. Most of the available data have been collected because of the regulatory need to monitor conditions before or after the implementation of controls on contaminant emissions. In this respect, air quality monitoring programs have been supported by basic research in chemistry, meteorology, and engineering. The research was aided by the fact that many atmospheric contaminant sources were quantifiable. Surface water quality monitoring programs have developed along traditional sanitary engineering and waste-water treatment technology directions involving the disciplines noted above as well as that of hydrologists, biologists, and hydraulic engineers. Surface water contaminant sources also were well recognized, and direct effluent discharge restrictions were implemented.

The chemical quality of ground water has only recently become an area of research or regulatory activity. We know very little about the coupling of contaminant sources and the hydrologic, chemical, or biological forces in the subsurface which influence ground-water quality. Therefore, planning reliable ground-water sampling efforts requires a research approach regardless of the purpose of the monitoring program. In most cases the source of ground-water contamination is very difficult to characterize in detail, and it may not have to be fully characterized in order to meet program needs.

[1] Head, Aquatic Chemistry Section, and head, Ground Water Section, Illinois State Water Survey, 2204 Griffith Drive, Champaign, IL 61820.

Effective ground-water protection and monitoring activities require a very thorough understanding of the dynamics of the subsurface environment. Ground-water recharge, movement, and discharge processes occur over extended frames which will delay the observation of the effects of more strict water-use or water quality regulations. Potential sources of contamination can be brought under control eventually. However, the effects of past practices which may impact ground-water quantity or quality must be dealt with today.

Contamination detection, assessment, and remediation problems entail a great deal of expense and manpower to address effectively. To make these activities cost-effective as well, planners must identify the nature of the information need and then design monitoring networks which will yield reliable results on which to base decisions for further action. In this paper we seek to highlight approaches to the design of effective ground-water quality sampling efforts.

## Network Design and Sampling Protocols

Optimized monitoring network design problems have been treated in a number of scientific works [*1–4*]. Effective water quality network design must balance two major factors: cost and the significance of observed trends in ground-water quality. Designs based on sophisticated analytical procedures and multiple nested well installations may achieve the needs for sensitivity and spatial coverage for a particular dissolved contaminant, but operational costs may prohibit full implementation of the network. More serious problems in network design are presented by the vast selection of potential techniques for well development or purging and sample collection. Many of these "tools" have not been proven reliable in many hydrogeologic conditions for most contaminants. For example, a conventional bailer is a grab sampler which may be used reliably for nonvolatile chemical constituents at shallow sampling points. Problems could arise when bailers are used in deep installations when more prolonged sample handling may effect air-sensitive chemical constituents.

Effective monitoring network designs should fully describe what well construction, evaluation, sampling, and analytical techniques are to be used and how they are to be used in order to achieve a known degree of confidence in the results. This progression from techniques through methods to detailed procedures has been discussed in detail for analytical work [*5*]. The elements of sampling protocols provide a good basis for the discussion of overall protocol development within specific monitoring network designs. The ground-water monitoring network design should specify detailed procedures and steps to be executed in order to collect necessary data at a known level of significance or statistical confidence. Written descriptions of proven monitoring procedures are called protocols.

The development of a sampling protocol would seem to be rather straightforward, involving the selection of target chemical constituents, sampling points (such as wells), sampling mechanisms (such as pumps, bailers), and sample handling procedures (such as filtration, transfer, preservation). Once the sampling protocol is prepared, the design of companion analytical protocols may be expected to follow from existing, standardized procedures for the determination of the target chemical constituents.

The reasoned selection of ground-water sampling techniques and materials is more difficult than the foregoing procedure, for two main reasons. First, although many types of sampling points and mechanisms exist for ground-water investigations, very few have been evaluated for the errors which they can introduce into monitoring data. In most instances the optimal sampling point or the performance of a sampling mechanism will depend on the hydrologic and chemical conditions at the site of interest. Therefore, knowledge of the hydrogeologic setting is extremely important in selecting sampling techniques or materials. Second, the very low analytical detection limits, which are achievable for most chemical contaminants, encourage setting regulatory concentration thresholds at "detectable" levels. Detectable levels are limited practically by variations in method performance or by errors introduced during sampling operations. It should be obvious that an inflexible, "standardized" sampling protocol will simply not work in all monitoring situations. Any approach must be balanced by the need to assess the true extent of contamination to protect public and environmental health. It is more desirable to formalize the steps in network design so that effective protocols can be developed and refined to meet specific program goals.

Error minimization and quality assurance of the data over time are more important attributes of sampling protocols [*6,7*] than are requirements for "representative" samples. Representativeness is relative to the scale and detail of the investigation compared with the hydrogeologic and geochemical conditions in the system under study [*8*]. Simple sampling equipment and procedures which can be used in routine applications should be the central consideration for most programs. An overview of the network design and sampling protocol development process is given in Table 1.

One can see from the table that the development of an efficient monitoring network can be divided into four stages. These are detective work, preliminary and working network designs, followed by continuing refinement of both the network design and the sampling protocol. The principal activities undertaken during the various stages have been treated in detail by several authors [*9–12*]. Most of these references cite the importance of careful planning of the sample collection effort from the beginning of a study. Decisions which will effect the usefulness of the sampling protocol are made in the early stages of network design as the hydrogeologic setting is being investigated. These stages parallel those of the basic scientific research method, where model

TABLE 1—*Ground-water quality monitoring network design activities.*

| Stage | Activity |
|---|---|
| Detective work | Study site characterization<br>facility operations/land use<br>hydrogeologic<br>geochemical |
| Preliminary network design | Scope of network purpose and parameter selection<br>quality assurance/quality control<br>detection<br>assessment |
| | Sampling points<br>well placement and construction<br>well development and performance evaluation |
| Working network design | Preliminary sampling protocol<br>sampling mechanism and material selections<br>water level measurements<br>well purging<br>sample collection<br>sample filtration/preservation<br>field determinations, blanks, standards<br>sample storage/transport |
| Refine network design and sampling protocol | Analytical operations<br>Interpret chemical and hydrologic results |

or hypothesis testing is followed by refinement of the model and further experimentation. It is important to keep in mind that the subsurface is a hydrogeologic environment and physical factors may exert control over chemical or biological processes.

Many ground-water investigations do not go through stages of systematically increased network complexity or the need for more specific chemical constituent information. A recent reference should be particularly helpful in preparing a sample protocol which can meet the purposes of most monitoring efforts [*13*]. The approach described in the example below can guide the cost-effective collection of hydrogeologic and chemical data of known quality, particularly when it is used to continually reevaluate the information base being developed. The reader is cautioned that the example is simplified and the complexity of the design process will vary for specific sites. In this example the formalization of the process by which decision are made is emphasized over the decisions themselves. As more research evidence on validation of field methods is reported, it will be possible to more completely evaluate the consequences of these decisions.

## Sampling Protocol Development Example

Let us presume that the situation under investigation is an abandoned waste-handling site for which good background information exists on the facility's operations, as well as on the site geology and hydrogeology. Water level histories, hydraulic head differences, and hydraulic conductivities of water-bearing formations and detailed piezometric surface maps are available for the site under varied production well pumping conditions. Sampling for chemical analysis will be necessary, since virtually no background data exist on either the chemical constituents of concern or background ground-water quality. Therefore, the detective work that can be done without the construction of monitoring wells is complete and the next stage of the investigation proceeds.

The preliminary network design calls for an array of upgradient and downgradient wells which should intersect major flow paths through the most likely area of contaminant release. These wells must provide information on background water quality and the capability to detect the type and degree of potential contamination. Therefore, in addition to samples for general contamination indicator parameters [for example, pH, $\Omega^{-1}$, total organic carbon (TOC) and total organic halogen (TOX)], samples will also be taken for determinations of major dissolved cations and anions and some trace metals. In this case, the quality assurance/quality control program should be designed for the set of chemical constituents noted above. Also, since surface soils at the site are thought to be contaminated with waste oils, it is necessary to thoroughly steam-clean all drilling tools to minimize contamination of subsurface core or water samples. Drilling should also begin at the upgradient well locations to reduce the chances of cross-contamination. Though drilling is not strictly part of a sampling protocol, improper drilling methods may bias subsequent sampling steps regardless of the care taken to avoid contamination in sample collection.

Prior to beginning the drilling of chemical sampling points, the background information on the site should be reviewed with particular emphasis on the geologic setting and the chemical constituents of interest. Figures 1*a* and 1*b* depict an example selection process for a drilling method and a suitable well casing material for the situation, respectively. In this case, the decision is made to use hollow steam auger drilling methods and initially either polytetrafluoroethylene (PTFE) or stainless steel well casing and screens in the saturated zone. Based on the information gathered from the initial well completions, these decisions may be modified if additional sampling points are needed.

Once the initial wells are constructed, they can be developed to remove fines by mechanical pumping or air-lifting. If gross contamination of the water with volatile organics is discovered in preliminary sampling during the drilling or coring operations, air development methods should be used with caution [*13*].

The hydraulic performance of the monitoring well should be evaluated. Then an adequate purge

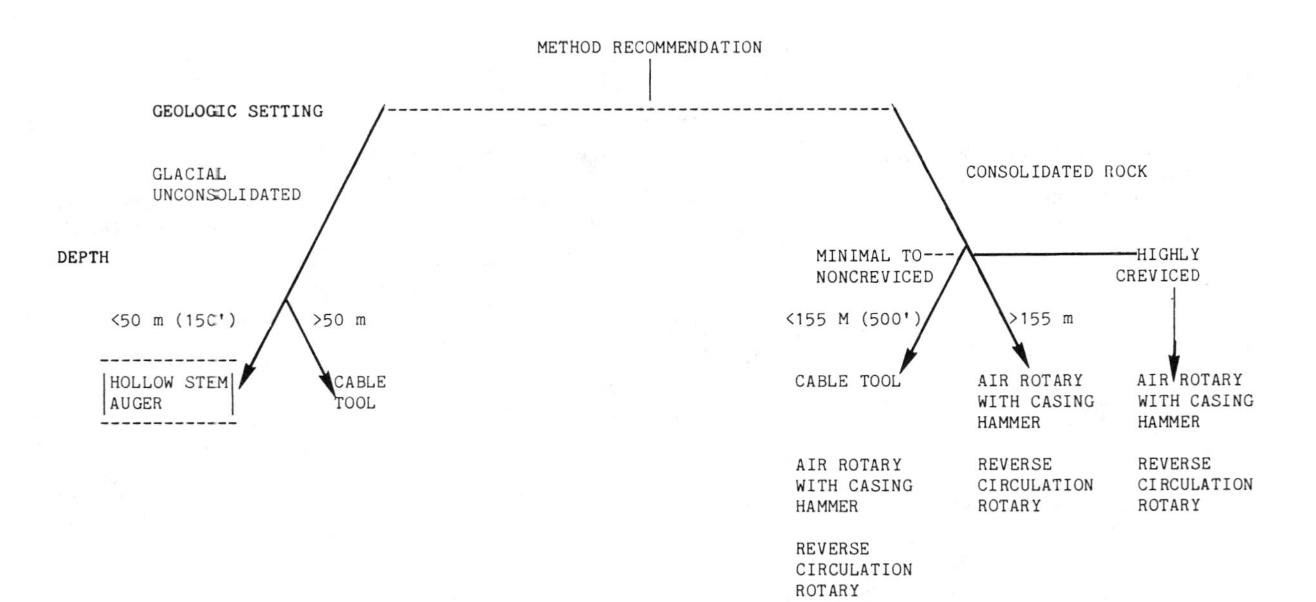


FIG. 1*a*—*Example decision tree for recommended drilling methods (adapted from Ref 13).*

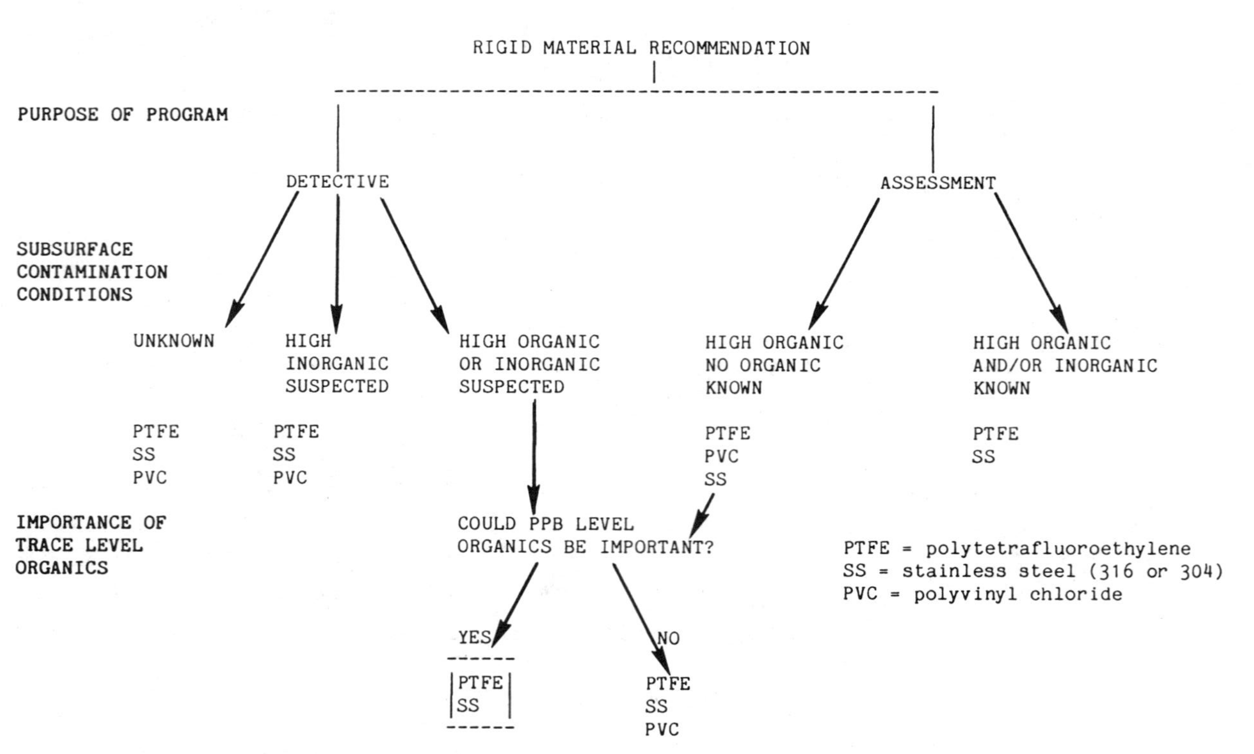


FIG. 1b—*Example decision tree for recommended well-casing/screen materials (adapted from Ref 13).*

TABLE 2—*Well purging strategy based on hydraulic conductivity (see also Fig. 2).*

Given

48-ft deep, 2-in.-diameter well[a]
2-ft-long screen
3-ft-thick aquifer
static water level about 15 ft below land surface
hydraulic conductivity = $10^{-2}$ cm/s

Assumptions

A desired purge rate of 500 mL/min and sampling rate of 100 mL/min will be used

Calculations

One well volume = (48 ft − 15 ft) × 613 mL/ft (2-in.-diameter well)
= 20.2 L
Aquifer transmissivity = (hydraulic conductivity) × (aquifer thickness)
= $10^{-4}$ m/s × 1 m
= $10^{-4}$ $m^2/s$ or 8.64 $m^2$/day

From Fig. 2
at 5 min: ~95% aquifer water and
(5 min × 0.5 L/min)/20.2 L
0.12 well volumes
at 10 min: ~100% aquifer water and
(10 min × 0.5 L/min)/20.2 L
= 0.24 well volumes

It appears that a high percentage of aquifer water can be obtained within a relatively short time of pumping at 500 mL $min^{-1}$. This pumping rate is below that used during well development.

[a] 1 ft = 0.3048 m; 1 in. = 25.4 mm.

volume can be calculated and formation water rather than stagnant water will be sampled. The evaluation of hydraulic performance allows one to relate the results from an individual well to the hydrogeologic setting at the site and provides a consistent basis for representative sampling. An example hydraulic evaluation procedure is outlined in Table 2 and Fig. 2 for one well within the hypothetical network [*13*].

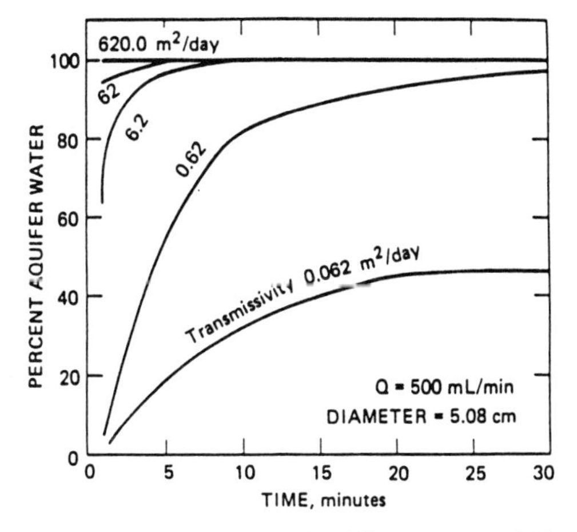


FIG. 2—*Percentage of aquifer water versus time for different transmissivities (see also Table 2).*

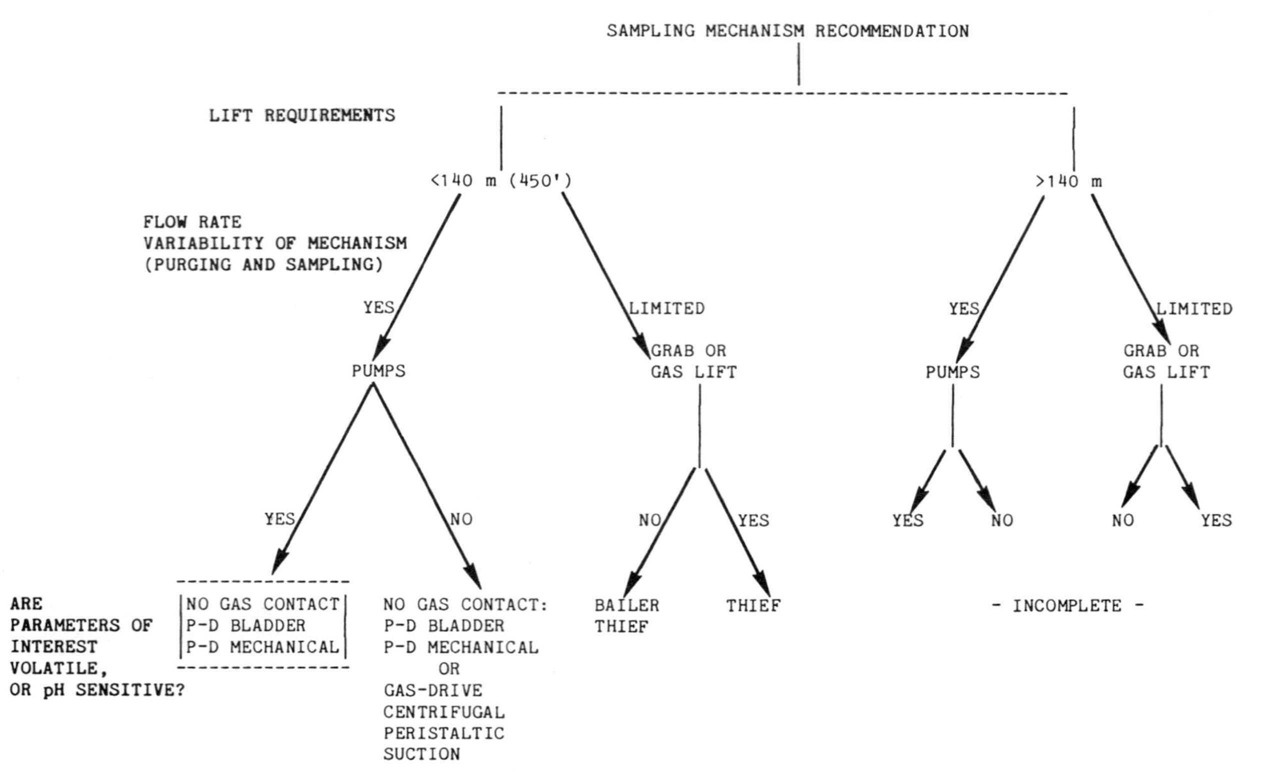


FIG. 3—*Example decision tree for recommended purge and sampling mechanism (adapted from Ref 13).*

TABLE 3—*Generalized ground-water sampling protocol.*

| Step | Goal | Recommendations |
|---|---|---|
| Hydrologic measurements | establish nonpumping water level | measure the water level to ±0.3 cm (±0.01 ft[a]) |
| Well purging | removal or isolation of stagnant $H_2O$ which would otherwise bias representative sample | pump water until well purging parameters (such as pH, T, $\Omega^{-1}$, Eh) stabilize to ±10% over at least two successive well volumes pumped |
| Sample collection | collection of samples at land surface or in well-bore with minimal disturbance of sample chemistry | pumping rates should be limited to ~100 mL/min for volatile organics and gas-sensitive parameters |
| Filtration/preservation | filtration permits determination of soluble constituents and is a form of preservation. It should be done in the field as soon as possible after collection | *filter*: trace metals, inorganic anions/cations, alkalinity *do not filter:* TOC, TOX, volatile organic compound samples; other organic compound samples only when required |
| Field determinations | field analyses of samples will effectively avoid bias in determinations of parameters/constituents which do not store well: for example, gases, alkalinity, pH | samples for determinations of gases, alkalinity and pH should be analyzed in the field if at all possible |
| Field blanks/standards | these blanks and standards will permit the correction of analytical results for changes which may occur after sample collection: preservation, storage, and transport | at least one blank and one standard for each sensitive parameter should be made up in the field on each day of sampling. Spiked samples are also recommended for good QA/QC |
| Sampling storage/transport | refrigeration and protection of samples should minimize the chemical alteration of samples prior to analysis | observe maximum sample holding or storage periods recommended by the Agency. Documentation of actual holding periods should be carefully performed |

[a] 1 ft = 0.3048 m.

Once the wells are constructed, developed and evaluated for hydraulic performance, the basis for a preliminary sampling protocol has been established. The range of lift and pumping rates has been identified, and decisions can be made on sampling mechanisms, materials, and handling procedures for the working network design (or the third stage noted in Table 1). These decisions can be made with the aid of decision trees similar to that shown in Fig. 3 for sampling mechanisms. In this example, the positive displacement (P-D) bladder or mechanical pumps are preferred because they provide reliable high recovery of samples for volatile organic determinations. Also, these pumps supply a stream of water during well purging which can be monitored for pH, $\Omega^{-1}$, temperature (T), and oxidation-reduction potential (Eh) to verify the calculated purge volume requirement.

Similarly, decision trees can be developed for the selection of sampling materials and other aspects of the generalized sampling protocol given in Table 3. One would then conduct an initial sampling run to establish concentration ranges for optimum selection of quality assurance/quality control (QA/QC) procedures, including field blanks, standards, and spiked samples.

It should be noted that the first set of samples from monitoring wells frequently is quite different from samples from older wells or future results from the same well. This is an inevitable consequence of the disturbance of the subsurface caused by drilling, well construction, and development. After several sampling runs have been completed, the working network design should be reexamined and refined to insure that necessary data are collected in the future. This continual reevaluation process can enrich the information return from the monitoring effort and provides data of documented quality.

## Conclusions

At this point the sampling protocol example has been taken through one iteration of refinement, and the general applicability of this development approach should be apparent. Careful consideration of the material in the references to this paper should aid network planners in their tasks. It is important to remember that the research basis for many aspects of ground-water quality monitoring is expanding rapidly. Therefore, the decisions and options for elements of the sampling protocol should become better defined. In the final analysis, common sense and experience should complement the decision-making process in effective sample protocol development.

*Acknowledgment*

The authors wish to thank their colleagues at the State Water Survey, especially Ms. Pamela Beavers for help in preparing the manuscript.

## References

[1] Sanders, T. G. et al., *Design of Networks for Monitoring Water Quality*, Water Resources Publications, Littleton, CO., 1983.

[2] Chamberlain, S. F. et al., "Quantitative Methods for Preliminary Design of Water Quality Surveillance Systems," *Water Resources Bulletin*, Vol. 10, 1974, pp. 199–219.

[3] Heidtke, T. M. and Armstrong, J. M., *Journal of the Water Pollution Control Federation*, Vol. 51, 1979, pp. 2916–2927.

[4] Moss, M. E., Lettenmaier, D. P., and Wood, E. F., "On the Design of Hydrologic Data Networks," *EOS Transactions of the American Geophysical Union*, Vol. 59, No. 18, 1978, pp. 772–775.

[5] Taylor, J. K., *Analytical Chemistry*, Vol. 55, 1983, pp. 600A–608A.

[6] Gillham, R. W., Robin, M. J. L., Barker, J. F., and Cherry, J. A., "Ground Water Monitoring and Sample Bias," American Petroleum Institute Report No. 4367, Environmental Affairs Department, Washington, D.C., June 1983.

[7] Barcelona, M. J., in *Proceedings*, National Water Well Association, 1983, pp. 263–271.

[8] Claassen, H. C., "Guidelines and Techniques for Obtaining Water Samples that Accurately Represent the Water Chemistry of an Aquifer," U.S. Geological Survey, Open File Report 82–1024, Lakewood, CO, 1982.

[9] *National Handbook of Recommended Methods for Water-Data Acquisition*, U.S. Geological Survey, Office of Water Data Coordination, Reston, VA, 1977.

[10] Scalf, M. R., McNabb, J. F., Dunlap, W. J., Cosby, R. L., and Fryberger, J., *Manual of Ground-Water Quality Sampling Procedures*, National Water Well Association, Worthington, OH, 1981.

[11] Todd, D. K., Tinlin, R. M., Schmidt, K. D., and Everett, L. G., "Monitoring Ground-Water Quality: Monitoring Methodology, EPA-600/4-76-026, U.S. Environmental Protection Agency, Las Vegas, NV, 1976.

[12] Barcelona, M. J., Gibb, J. P., and Miller, R. A., "A Guide to the Selection of Materials for Monitoring Well Construction and Ground-Water Sampling," Illinois State Water Survey Contract Report No. 327, USEPA-RSKERL, EPA-600/52-84-024, Champaign, IL, 1983.

[13] Barcelona, M. J., Gibb, J. P., Helfrich, J. A., and Garske, E. E., "Practical Guide for Ground-Water Sampling," State Water Survey Contract Report No. 374 (EPA 600/S2-85/104) prepared for USEPA-RSKERL, Ada, OK, Aug. 1985.

*J. Jeffrey van Ee*[1] *and Leslie G. McMillion*[1]

# Quality Assurance Guidelines for Ground-Water Investigations: The Requirements

**REFERENCE:** van Ee, J. J. and McMillion, L. G., "**Quality Assurance Guidelines for Ground-Water Investigations: The Requirements,**" *Ground-Water Contamination: Field Methods, ASTM STP 963*, A. G. Collins and A. I. Johnson, Eds., American Society for Testing and Materials, Philadelphia, 1988, pp. 27–34.

**ABSTRACT:** The U.S. Environmental Protection Agency (EPA) is required to assess the quality of data that are collected in its research and monitoring programs, and this requirement is meant to apply to established areas such as air and water pollution monitoring as well as to the emerging use of geophysics in ground-water investigations. The Agency's quality assurance requirements are outlined, and the ability to meet the requirements in ground-water investigations is discussed. Suggestions are offered on how standards and standard methods can be developed, based upon the experiences obtained in the implementation of monitoring programs in air pollution in the 1970's. The concept of reference and equivalent methods is compared in the monitoring of the air quality with the assessment of ground-water quality for the EPA. The applicability of quality assurance to the many and varied steps in a ground-water investigation is discussed with emphasis given to geophysics. The lack of standard procedures and quality assurance guidelines for geophysical investigations of hazardous waste sites poses serious questions to the acceptance of emerging disciplines for ground-water quality studies.

**KEY WORDS:** quality assurance, ground water, geophysics, hazardous waste, site investigations, standards, standard methods, reference and equivalent methods

Although the science of hydrology has been in existence for years, standard procedures for the sampling of ground water, especially for organic contaminants, are not well developed. In the past, the quantity and movement of the ground water has been of most concern. Today, with the passage of the Clean Water Act (CWA), the Safe Drinking Water Act (SDWA), the Resource Conservation and Recovery Act (RCRA), and the Comprehensive Environmental Response, Compensation, and Liability Act (Superfund), the monitoring and protection of the quality of ground water is of intense concern. Private industry and Federal, state, and local agencies are making a variety of chemical, physical, and bacteriological measurements of ground water, for example, water table level and total dissolved solids (TDS) levels. Whether or not these measurements are providing data that can be intracompared and compared against standards in a scientifically and defensible way is debatable. Data that are inaccurate, imprecise, and not representative of the monitoring zone can lead to conclusions that may be costly and hazardous to the public and private sectors. Poor quality or improperly analyzed data at Superfund or RCRA sites are particularly troublesome because of the potentially large hazards and financial risks that are involved with these hazardous waste sites. With the large amount of data being collected by a great number of organizations and individuals, there are many potential sources of error. Unfortunately, little attention has been paid to determining the extent of those errors. Documented,

[1] Electronics engineer and hydrologist, respectively, Environmental Monitoring Systems Laboratory, Office of Research and Development, U.S. Environmental Protection Agency, Las Vegas, Nevada 89114.

uniform quality assurance practices which are aimed at reducing those errors are not well developed in the ground-water industry at this time.

The Environmental Protection Agency's (EPA) Office of Ground Water Protection has recently drafted a strategy for the Agency to follow in the implementation of the numerous laws that Congress has passed to protect the nation's ground water. The strategy emphasizes the importance of quality assurance in the sampling of ground water. The role of the United States Geological Survey is also recognized by the EPA as being important to the widespread characterization of ground-water resources. Other Federal agencies, such as the Nuclear Regulatory Commission and the Department of Agriculture, are involved in the study of ground water, and the data that they collect are important in providing an understanding of the ground water in the United States. State and local agencies conduct ground-water studies within their regions. Professional organizations, such as the American Society for Testing and Materials (ASTM), the National Water Well Association (NWWA), and the Society of Exploration Geophysicists (SEG), also participate in the study of ground water. Universities, colleges, and research institutes conduct research. Finally, private industry often conducts, or contracts, ground-water studies. The degree of quality assurance exercised by these various organizations varies, and the comparability of the data they collect may be poor, given the lack of industry-wide standards in the monitoring of ground water.

## Application of Quality Assurance Guidelines to Ground-Water Investigations

The EPA Quality Assurance (QA) and Monitoring Staff recognized in 1980 the need for quality assurance in the many studies that EPA conducts and finances in its environmental studies. The EPA developed a list of 16 points (Table 1) that should be addressed prior to the conduct of those environmental studies. These points apply to in-house efforts, contracts, cooperative agreements, and interagency agreements. EPA studies involving Federal, state, and local agencies as well as private industry and academia fall under these guidelines (QAMS-001-005/80) [*1–3*]. Although all of the points may not apply to a particular study, their intent is clear. Quality assurance should be considered prior to the conduct of a study so that sound scientific practices will lead to the

TABLE 1—*Required elements (when applicable) for QA (project) plans for monitoring and measurement projects (QAMS-002/80).*

| Element |
|---|
| 1. Title page, with provision for approval signatures |
| 2. Table of contents |
| 3. Project description |
| 4. Project organization and responsibilities |
| 5. QA objectives for measurement data in terms of precision, accuracy, completeness, representativeness, and comparability |
| 6. Sampling procedures |
| 7. Sample custody |
| 8. Calibration procedures, references, and frequency |
| 9. Analytical procedures |
| 10. Data reduction, validation, and reporting |
| 11. Internal QC checks and frequency |
| 12. QA performance audits, system audits, and frequency |
| 13. QA reports to management |
| 14. Preventive maintenance procedures and schedule |
| 15. Specific procedures to be used to routinely assess and document data precision, representativeness, comparability, accuracy, and completeness of specific measurement parameters involved |
| 16. Corrective action |

collection of scientifically, and (if needed) legally, sound data. Since the EPA guidelines are written in general terms, there can be problems in applying them to specific projects, particularly when the research or environmental monitoring involves the use of innovative techniques or techniques for which there is little history in the way of established standards, procedures, or quality assurance documentation. The present increased interest and resultant studies of the quality of this nation's ground water come at a time when interest in formal quality assurance programs for ground-water investigations is just developing; thus, the application of the EPA guidelines to ground-water studies may require broad interpretation and innovation to ensure that sound data are collected.

**Air Pollution and Ground Water: The Analogy**

The present heightened interest in ground-water protection can be likened to the 1970's when air pollution was a major priority of the EPA. The present quality assurance efforts in air pollution monitoring, while not perfect, can be more readily understood because many of the struggles are now history. The lessons learned from the application of quality assurance principles to the then emerging field of air pollution study can be applied to today's problems in ground-water monitoring; however, there are limits to the analogy. Measurements of water are made more difficult by the fact that hydrological and geological investigations are conducted in an anisotropic, nonhomogeneous, relatively isolated media.

The study of air pollution can be broken down into two basic areas: the definition of the source and the measurement of the ambient air quality. Standard procedures and quality assurance guidelines have been developed for the measurement of pollutants from the source, for example, the smokestack, or the tailpipe. Standard procedures and quality assurance guidelines have also been developed for the measurement of the ambient air quality with separate procedures and QA guidelines having been developed for the location of the sampling instrumentation that is used to characterize the air quality in a large region, such as a city, or state. Standards have been developed to protect the public health and welfare from some pollutants and hazardous substances, and monitoring is performed in both the establishment of health standards for a particular pollutant and in measuring compliance with ambient standards that are based upon those health standards. Some of the same basic principles developed for air quality are just now being developed for ground water. The study of air pollution has benefited in the past from a large commitment of resources to QA. In comparison, few resources have been devoted to QA for ground-water monitoring; thus, it will be some time before methods and standards for ground-water investigations become as well defined as they are for air monitoring investigations.

This paper compares an approach EPA used to implement quality assurance guidelines for ambient air monitoring with the proposed development of procedures and criteria for the testing and certification, usage, and siting of instruments for the monitoring of the ground water and soil, for example, outside the fenceline of an RCRA or Superfund site. The EPA sanctioned the concept of "reference methods" and "equivalency" for the instrumentation used to make ambient air measurements. Guideline and quality assurance documents were developed, under different regulatory authority, for the location and usage of these instruments in measuring the quality of the ambient air. Ground-water investigations need the same structure and approaches that have been developed for air monitoring.

Two examples are provided in this paper to demonstrate how EPA certification and quality assurance requirements have been applied to the study of air pollution and how they may be subsequently applied to ground-water investigations. The first example covers a rationale for the certification of methods and instrumentation that proved to be successful in speeding the testing and certification of instrumentation for ambient air measurements. The designation of "reference methods," and a process for establishing "equivalent methods," is examined for the critical step of installing monitoring wells in a ground-water investigation. The second example applies some of the requirements for the documentation of basic quality assurance principles (contained in QAMS-002/80) to the relatively new science—as applied to ground-water quality investigations—of geophysics. Multiple disciplines and steps are required to understand the hydrogeology and ground-water quality in a ground-water investigation, and geophysics is one discipline and one step that needs to be considered in any ground-water investigation.

## QA as Applied to Ground-Water Investigations

Ground-water investigations vary. Some are strictly concerned with defining quantity and flow parameters, others involve an investigation of ground-water quality. Some ground-water investigations are directed at relatively deep aquifers while others are directed at more shallow aquifers. Some involve the measurement of organic contaminants while others involve primarily the measurement of inorganic contaminants. Despite the diversity of investigations that are made in the field of ground water, certain basic principles apply to all of the investigations.

The sampling and analysis of ground water is important to a number of regulatory programs such as: Superfund, RCRA, the Underground Injection Control program, and Underground Storage Tanks (UST). The measurement of ground water involves the same basic concepts no matter which programs are involved. Monitor wells are drilled and sampled, and the water is analyzed. Common quality assurance procedures and guidelines may be developed and applied to all of the programs. It may be advantageous to do so since all of the programs ultimately focus on the same thing—the protection of ground water; however, there are some basic procedural steps in some ground-water investigations that will be unique to a particular program. For example, the quality assurance program and documents that are being developed for the Underground Injection Control program may not be directly applicable to any other area of ground-water investigations because the depth of the ground-water zone being investigated is distinctly different from the zone of concern near a RCRA or Superfund site. The focus of this paper will be on the application of quality assurance to the shallow ground-water investigations that are typically required in the investigation of a Superfund or a RCRA site.

Some of the steps in a comprehensive site investigation are given in Table 2. Each of the steps requires, in varying degrees, quality assurance procedures and documentation to ensure the collection of quality data. The applicability of quality assurance to some research efforts and to preliminary work can be questioned, particularly if few measurements are made. The manner in which the items in Table 1 are documented is generally not important. What is particularly important is that items addressing the reproducibility, comparability, accuracy, and precision of the data are addressed so that the quality of the data will be known.

TABLE 2—*Steps in a multidisciplinary ground-water study.*

| |
|---|
| Aerial photography/surveying |
| Search of available records |
| Geologic characterization |
| Hydrologic characterization |
| Geophysical studies (Subsurface and Surface) |
| Geochemical surveys |
| Soil sampling |
| Ground-water sampling |
| Source sampling |
| Exposure assessment |
| Data interpretation |
| GIS (Geographical Information System) data displays |

The development of documented and rigorous standards and QA guidelines may produce undesirable results. The development of any QA program carries with it increased costs and paperwork, and these must be weighed against the benefits in knowing the quality of the data produced. Those in the private sector who develop a comprehensive QA program may actually be penalized in competitive procurements because the higher costs associated with the QA program would have to be passed along to the customer. If the customer had little interest or requirement for a strong QA effort, then a firm with a small QA effort and lowered costs might be chosen to perform the ground-water survey. Another disadvantage in requiring more formal QA programs is that proprietary processes may be revealed through the dissemination of the QA documents to competitors. Universal standards and QA guidelines would eliminate the possibility of proprietary processes being revealed and would presumably apply the costs of developing QA guidelines and procedures across the entire private sector; however, it is unlikely that any firm would go beyond the development of minimally required methods or standards.

*Example 1: Reference and Equivalent Methods*

The concept of equivalency and of performance standards strikes a balance between (1) the development of universal methods and standards, which might curtail the development of innovative methods and standards, and (2) the present situation of few industry or government standards, which leads to problems in comparing data and assessing the quality of that data.

Reference methods and "equivalency" have been used in the field of ambient air monitoring to ensure that basic standards are met without inhibiting the development of new methods and procedures. One such reference method was designated by the EPA for the measurement of sulfur dioxide. Important performance standards were provided by the EPA at the time the method was designated as a reference method, and test procedures were provided on how the important parameters for the reference method were obtained. When the reference method for sulfur dioxide was first established by the EPA, it consisted of a wet chemical method which required air to be passed through a series of impingers for 20 min. The response time, or performance standard, for the measurement of sulfur dioxide became 20 min. No method is perfect. It was discovered from further testing of the reference method that the measurement of sulfur dioxide by the wet chemical method had problems; however, EPA chose to continue with the method until such time as a better method could be developed or the weaknesses in the reference method could be further defined. The process of equivalency provided the opportunity that other methods, which proved to be superior to the reference method, could eventually replace the reference method; thus, there was little disruption to a nationwide monitoring program for sulfur dioxide while the search for a superior monitoring method occurred.

At approximately the same time as a reference method was established for the monitoring of an air pollutant, EPA designated test procedures and criteria for a method to be approved as an equivalent method. The user of a monitoring method, who sought approval of it as an equivalent method, could obtain from the Code of Federal Regulations the test procedures and performance specifications for the reference method. The tests could be performed by virtually anyone with the basic requirement being that the data had to be supplied to EPA for approval. Virtually every test for every reference method included an assessment of potential interferences, sensitivity to temperature, response time, drift and noise, and minimum detectable concentration. Shortly after the establishment of the reference method for sulfur dioxide, a number of instrument manufacturers began producing instruments that possessed superior characteristics to that of the wet chemical reference method. Response times were on the order of seconds. EPA approved the instruments as "equivalent" methods, and flame photometric and pulsed fluorescent instruments gradually phased out the wet chemical reference method. Through the transition period, EPA had some measure of the data quality from the reference and equivalent instruments, and private industry had design criteria that could be met or exceeded with little curtailment in the development of new technologies.

The degree to which reference and equivalent methods (and the related concept of performance standards) may be applied to various aspects of ground-water investigations varies. The reference method/equivalency process could be applied to the various disciplines that are involved in a ground-water investigation. One example where the concept may be applied is in the sampling of water from a well for volatile organics. Water from a well is evacuated for some number of well volumes or some period of time. A sample of water is drawn by some type of pump, bailer, or sampler and taken back to the laboratory for analyses. The water is then analyzed for parts-per-billion levels of a variety of organics, and an assessment is usually made on (1) whether the water is contaminated, and (2) the rate at which the water is being contaminated. The procedures for sampling a well and the composition of the well casing are interrelated. The composition of the well casing is important because it may affect the representativeness of the water sample if the casing adsorbs or desorbs organics. The method in which the well is sampled will influence the effect of the well casing material on the sample. There are several recommended approaches for the sampling of organics, but there are no "reference" methods. Likewise, there are no reference materials for the well casing.

Monitoring wells may be constructed from a variety of materials such as stainless steel, Teflon™, fiberglass, or polyvinyl chloride (PVC). If one were to apply the concept of performance standards to just one component of this sampling procedure—to the well casing material—a reference material would be selected. An example might be stainless steel or Teflon. Suggested performance standards would include mechanical strength, chemical absorption/desorption characteristics, and manufacturing tolerances on the inside dimension of the casing (this would be important for the use of logging tools and pumps). Tests for the measurement of these parameters would be prescribed. Manufacturers who sought to market casings made out of nonreference materials such as Teflon-coated polyethylene, epoxy/fiberglass, Kynar™-coated steel, or even PVC would know the minimum standards that would have to be met for the successful marketing of their product. Formal approval would come after the data were reviewed and accepted. This process of developing standards, test procedures, and an approval mechanism should be an improvement over the present situation of no standards (or test procedures) and an ill-defined—if even existing—approval process.

*Example 2: QA and Geophysics?*

The application of geophysics to ground-water investigations is relatively new. Much of the pioneering work in geophysics has been for the petroleum and minerals industries. With the decreased exploration activities in these industries, private industry is eager to use geophysics in ground-water investigations, particularly Superfund and RCRA. Geophysical measurements for the petroleum and mining industries have not lent themselves to the development of quality assurance procedures and guidelines as specified by EPA in QAMS 002/80. The consideration of geophysics by regulators would be enhanced if the basic QA requirements in QAMS 005/80, which have been applied to the more established sciences, could be applied to geophysics.

The acceptance of geophysics by regulatory agencies depends first on the ability of the geophysicist to correctly describe the subsurface environment. Unique solutions are not easily obtained through the use of geophysics alone. The experience of the practitioner in geophysics and the supporting data provided to the person responsible for interpreting the geophysical measurements will determine to a large extent the success of the method in characterizing a site. The use of geophysics in ground-water investigations has been inhibited to a certain extent by the lack of standard, or uniform, data gathering and interpretation procedures; however, the application of uniform procedures to non-uniform sites may lead to unsatisfactory results if the procedures are followed blindly. Also, there are instances in some of the physical sciences where the establishment of standards may be a costly, time-consuming, and technically difficult task. An example would be in the selection of a calibration standard or facility, or both, for a magnetometer.

The establishment of calibration standards or facilities, or both, for most instruments would be desirable if one were to compare data collected at different times with different instruments; however, the difficulty and lack of standards for the calibration of a magnetometer are not necessarily bad. The magnetometer measures the earth's magnetic field, which varies with time and location. In hazardous waste site investigations the magnetometer is often used to locate buried ferrous metal such as metal drums. Relative measurements of the magnetic field over a metal object are usually compared with readings over other areas. For example, it may not be important to know that the magnetic field over a drum was 50000 gammas or 50300 gammas. (Typically the magnetic field of the earth is on the order of 50000 gammas.) The important point is knowing the difference and rate of change of the magnetic field over an area. An absolute calibration of the magnetometer may not be required if the magnetometer was used only to locate buried drums. Thus, while it may be desirable to develop uniform standards and standard operating procedures in ground-water investigations, there is also a need for flexibility and a basic understanding of the method when those standards are considered and developed.

Perhaps the most important step in an investigation of a hazardous waste site is the documentation of how the data were acquired and interpreted. Since each site is different, it is difficult, but not impossible, to develop standard procedures for these steps. Describing a measurement process can be easy, but critiquing the process to improve the quality of the data and conclusions can be more difficult. Since the subsurface can neither be seen nor measured completely, the process of locating monitoring points or wells can be subject to a lot of uncertainty. Typically, "best judgment" is used. How does a client know, however, whether the decisions made by an investigator are indeed the best decisions? The development of standards by which data quality may be judged is important. A "standard set of data" may be submitted to the geophysical investigator for interpretation to ascertain whether geophysical data is being correctly interpreted. The "standard set of data" may represent voltage and current readings from a particular resistivity array at different electrode spacings.

The application of conventional QA principles becomes more complicated in critiquing how the data were acquired, for example, the location of the sampling points. Usually it is not known until an independent investigation has been performed whether a particular approach yielded the correct results. Monitoring wells may be drilled after a geophysical investigation to investigate the anomalies; however, even after this independent step has been taken it can be difficult to know how accurate or precise the measurements were in defining the magnitude and extent of the contamination. Data from a monitoring well may be representative for the small area near the well, but surface-based geophysical measurements may be more representative of a large volume of earth and water under the geophysical instrumentation.

## Training and Certification

If the investigation of ground water, particularly at hazardous waste sites, is fundamentally an imprecise process and the development of standards is difficult for certain aspects of an investigation, it is nevertheless important to attempt to define how imprecise the process is. The establishment of documented standards and procedures is one step. The training of individuals in the use of the methods and in the application of QA principles is another step. An important step in assuring that ground-water data is correctly collected and interpreted may be the licensing and certification of the individuals and organizations involved in those investigations. Some professional organizations have a certification program for their members, while some states have licensing programs. Minimum standards of education and experience are established for the qualification of people in these programs. The certification process may be time-consuming and meet with some initial opposition; however, the present situation is not satisfactory. Qualified, certified geologists, hydrologists, geophysicists, and chemists, who are experienced in hazardous waste site investigations, are in short supply as compared to the great number of sites that require sometimes complex and costly studies.

## The Future

These examples and arguments buttress the position that greater attention and resources need to be devoted to QA for the various disciplines that are used in a variety of ground-water investigations. There is a growing recognition by government, industry and academia of the need for greater QA; however, who will be responsible for the implementation of a QA program, and how will it be developed? The EPA has responsibility for ensuring that QA is considered in the studies it authorizes, but past and present efforts to improve QA in ground-water studies have been inadequate. Further, other government agencies and private organizations that do not fall under the EPA umbrella are conducting ground-water studies. Professional societies such as the SEG, NWWA, American Institute of Professional Geologists, American Institute of Hydrologists, and ASTM have begun to recognize the need for the development of standards, but they do not represent all those who are involved in ground-water studies. There needs to be a coordinated effort between the organizations who represent the professions that are employed in ground-water investigations because the study of ground water is an interdisciplinary science. The best approach appears to be a team approach in which all professionals recognize the need for greater QA in their professions and work towards that end. While it may take some time to form the team of government, private industry, academia and the professional organizations, the joint effort, with proper guidance, could lead to the establishment of standards and QA principles that may be characterized as a blessing rather than a curse. It has to be done if the challenges are to be met in cleaning up thousands of hazardous waste sites and monitoring the multitude of aquifers in the nation.

Reference methods, with procedures for demonstrating "equivalency" to those methods through the establishment of "performance standards," allow for innovation in the attainment of those standards. EPA could establish this concept in its ground-water programs as it has in the air programs; however, the process will not be easy, quick or universal. Guidelines could be established for how the reference/equivalent methods would be applied to a site investigation, but again the responsibility rests with the investigator to use "best judgment" at a site in the selection of the monitoring location and methods. Reference methods, equivalency, guidelines, and rigorous QA requirements proved to be useful in the 1970's in the air pollution sciences, and these concepts may prove even more useful in the 1980's in ground-water studies.

## References

[1] "Guidelines and Specifications for Implementing Quality Assurance Requirements for EPA Contracts and Interagency Agreements Involving Environmental Measurements," QAMS-002/80, Office of Research and Development, U.S. Environmental Protection Agency, 19 May 1980.

[2] "Guidelines and Specifications for Preparing Quality Assurance Program Plans," QAMS-004/80, Office of Research and Development, U.S. Environmental Protection Agency, 20 Sept. 1980.

[3] "Interim Guidelines and Specifications for Preparing Quality Assurance Project Plans," QAMS-005/80, Office of Research and Development, U.S. Environmental Protection Agency, 29 Dec. 1980.

*Carl D. Tockstein*[1] *and Brian C. Dorwart*[2]

# Availability and Access to Ground-Water-Related Information and Data

**REFERENCE:** Tockstein, C. D. and Dorwart, B. C., "**Availability and Access to Ground-Water-Related Information and Data,**" *Ground-Water Contamination: Field Methods, ASTM STP 963,* A. G. Collins and A. I. Johnson, Eds., American Society for Testing and Materials, Philadelphia, 1988, pp. 35–40.

**ABSTRACT:** For many years it has been apparent that the ever-increasing volume of information and data accumulated in all fields of human experience is far outpacing the individual's capacity to absorb even a small part of it. It is, in fact, impossible to keep current in one's own field of expertise without a systematic approach to screening, cataloging, and retrieving pertinent information. In recent years, the use of computer technology to create automated systems has provided a means to at least partially mitigate the problem.

ASTM Subcommittee D18.95 on Information Retrieval and Data Automation has, as part of its scope of responsibilities, the charge to keep current on the status of such systems and report this information to the geotechnical community. This paper is a means of fulfilling that charge by reporting the availability and accessibility of ground-water information known to the subcommittee.

Information available to the subcommittee shows that a well-coordinated program of water-data collection and dissemination is currently ongoing under the auspices of the U.S. Geological Survey's Office of Water Data Coordination. The interests of the ground-water community can best be served by encouraging the continued and expanded use of this program to provide a means for all to benefit from the collective experience and knowledge available.

**KEY WORDS:** information retrieval, ground water, data storage, data retrieval, data base

The proliferation of information and data available has far outpaced the individual's capability to grasp even a small part of it. Keeping current in one's own discipline requires an organized, intentional approach. The problem is not a new one. Over the years various individuals and groups have developed viable storage and retrieval methods, ranging from hand indexing and filing to mechanical sorting, to electronic computer technology [*1*]. In the past two decades, however, the increase in volume of information and data has been exponential. Only application of advanced computerized data storage and retrieval methods can keep pace. The American Society for Testing and Materials Committee D18 on Soil and Rock, recognizing a need for an organized approach in the rapidly expanding field, in 1969 organized Subcommittee D18.95 on Information Retrieval and Data Automation. The charge of the subcommittee was to investigate the entire gamut of information retrieval activity (that is, selecting, indexing, abstracting, storing, searching, disseminating, displaying, and transmitting information on data) and to make known to the geotechnical community the currently available methodology. More recently the emphasis has changed from strict concentration on the *how* of information retrieval to include *where* information can be found. It is in the latter mode that this paper is prepared.

[1] Principal civil engineer, Tennessee Valley Authority, Knoxville, TN 37921.
[2] Senior project manager, Goldberg Zoino & Associates, Inc., Newton, MA 02164.

## Approach to Study

ASTM Subcommittee D18.95 has used various means to fulfill its charge to accumulate and disseminate information relative to the availability and accessibility of geotechnical information and data. Information has been provided through presentation and publication of papers, articles for journals, mini-symposia at ASTM meetings, etc. In 1982 a standing Task Group on Geotechnical Data Base Inventory was established to systematically survey Federal agencies and private organizations to locate and identify sources of geotechnical data bases and catalog them according to the type of information and data available. The initial survey was completed in 1985 and cataloging is in progress.

A portion of the information obtained from the survey was related to geohydrology. This information provides a partial basis for this report.

## Discussion of Study Findings

An examination of the survey responses indicates that it is both more appropriate and beneficial to potential users to catalog data base "centers" that have access to many individual data bases on specific topics than to catalog the individual data bases. In the ground-water arena this approach has been and is being used effectively. Currently the U.S. Geological Survey (USGS) Office of Water Data Coordination (OWDC) is charged with assembling information on all available data bases. Approximately 30 Federal agencies and 500 state and county agencies provide information on their data holdings to OWDC. This information is available as printed data indexes or computer pulls from the following address:

Office of Water Data Coordination
U.S. Geological Survey
MS 417, National Center
Reston, VA 22092
Telephone 703-648-5016

Additional free publications available from the OWDC include:

1. *National Handbook of Recommended Methods for Water-data Acquisition*
2. *Index to Water Data Activities in Coal Provinces in the United States*
3. *Notes on Sedimentation Activities Calendar Year*
4. *Plans for Water Data Acquisition by Federal Agencies Through Fiscal Year*
5. *Guidelines for Determining Flood Flow Frequency*

Since 1966, the OWDC has cataloged water data in the *Catalog of Information on Water Data*. This is a catalog of information about water data collection activities and not actual data. Typical information available from the catalog is illustrated in Fig. 1.

These and other references containing water data are available from:

Water Resources Scientific Information Center
U.S. Geological Survey
MS 421 National Center
Reston, VA 22092
Telephone 703-648-5683

The major data bases and the majority of the others identified by the survey are listed by or directly accessed by the following three sources:

1. NAWDEX (National Water Data Exchange Index)
   U.S. Geological Survey
   MS 421 National Center
   Reston, VA 22092
   Telephone 703-648-5683
2. STORET (Storage and Retrieval System)
   Environmental Protection Agency
   Research Triangle Park, NC
   Telephone 800-424-9067
   or 202-382-7220
3. Found Water On-Line
   National Water Well Association
   6375 Riverside Drive
   Dublin, OH 43017
   Telephone 614-761-1711

The primary data bases identified by the survey were the USGS National Water Data Storage and Retrieval (WATSTORE) system, the U.S. Environmental Protection Agency's (EPA's) Storage and Retrieval (STORET) System, and the National Water Well Association's (NWWA's) Ground Water On-Line. Most states are compiling data for their individual data banks through the water well-permitting process. Federal agencies, while they may maintain individual data bases for specific purposes, are also tied to one or all of WATSTORE, STORET, or Ground Water On-Line.

## Conclusions

It is the opinion of the Task Group based on the results of the survey and other available information [*2,3*] that to attempt to catalog data sources related to ground water would be an unnecessary duplication of the USGS effort. A better approach is to insure that the profession is aware of the full-time efforts of the USGS, EPA, NWWA, and others to this task, and to urge expanded participation in and use of the systems so that accumulated knowledge will be readily available to all.

It is suggested that the user's most efficient initial point of contact would be the local NAWDEX Assistance Center, which is usually located in the District Office of the USGS Water Resources Division in each state. At those centers, referrals to appropriate local information can be obtained. Access is available also to the main data bases and the Water Data Sources Directory (WDSD) at USGS headquarters, where data from hundreds of organizations which collect water data are indexed. Listings of the local assistance center locations are available from:

National Water Data Exchange
(NAWDEX)
U.S. Geological Survey
421 National Center
Reston, VA 22092

| MAP | | OWDC NUMBER | AGENCY STATION NUMBER | STATION NAME | LATITUDE | LONGITUDE | STATE | COUNTY | SITE | PERIOD OF RECORD | | INTERRUPTED RECORD | STORAGE OF DATA | | | | |
|---|---|---|---|---|---|---|---|---|---|---|---|---|---|---|---|---|---|
| NUMBER | LETTER | | | | | | | | | BEGAN | DISCON-TINUED | | PUBLISHED | NOT PUBLISHED | DATA ON PUNCHCARD | DATA ON MAG STOR. DEVICE | OTHER |
| 30 | F | 57393 | | WARSING | 475010 | 0990630 | ND | 027 | RESER | 1963 | | | * | | | | |
| 30 | F | 50529 | 05055520 | BIG CL NR FT TOTTEN ND | 475257 | 0985802 | ND | C05 | STREAM | 1962 | | * | * | | | * | * |
| 30 | F | 50530 | 05056000 | SHEYENNE R NR WARWICK ND | 474820 | 0984257 | ND | 027 | STREAM | 1951 | | | * | | * | * | * |
| 30 | E | 73239 | 050561CC | MAUVAIS CL NR CANCC ND | 482653 | 0990608 | ND | 095 | STREAM | 1970 | | | | * | | | * |
| 30 | F | 63158 | | BATTLE LAKE | | | ND | 027 | LAKE | 1967 | | | * | | | | |
| 30 | F | 66C33 | WARWICK N D | SHEYENNE R - HWY 20 CROSSING | 474742 | 0983515 | ND | 027 | STREAM | 1968 | | | | * | | | |
| 30 | E | 63209 | | CAVANAUGH LAKE | | | ND | 071 | LAKE | 1956 | | * | * | | | | |
| 30 | E | 73240 | 05056200 | EDMCRE CL NR ECMCRE ND | 482010 | 0983940 | ND | 071 | STREAM | 1970 | | | | * | | | * |
| 30 | E | 50531 | 05056220 | SWEETWATER LK AT SWEETWATER N DAK | 481237 | 0985215 | ND | 071 | LAKE | 1954 | | * | * | | * | * | * |
| 30 | E | 63208 | | CARLSON POND | | | ND | 005 | OTHER | 1967 | | | * | | | | |
| 30 | E | 50532 | 05056250 | LAC AUX MORTES NR CHURCHES FERRY N | 482107 | 0990542 | ND | 071 | LAKE | 1954 | | | * | | * | * | * |
| 30 | E | 50533 | 05056260 | LAKE IRVINE NR CHURCHES FERRY N DAK | 481657 | 0991025 | ND | 071 | LAKE | 1954 | | | * | | * | * | * |
| 30 | E | 50534 | 0505640C | BIG CL NR CHURCHES FERRY NC | 481040 | 0991315 | ND | 005 | STREAM | 1949 | | | * | | * | * | * |
| 30 | E | 63206 | | SAND LAKE | | | ND | 069 | LAKE | 1954 | | * | * | | | | |
| 30 | E | 63139 | | LEEDS RESERVOIR | | | ND | 075 | RESER | 1958 | | | * | | | | |

| OWDC NUMBER | TYPES OF DATA: PHYSICAL | | | | | | | | | | CHEMICAL | | | | | | | | | | | | | | BIOLOGIC | | | SEDIMENT | | | | SUPPLEMENTARY DATA | | | | | | AGENCY REPORTING |
|---|---|---|---|---|---|---|---|---|---|---|---|---|---|---|---|---|---|---|---|---|---|---|---|---|---|---|---|---|---|---|---|---|---|---|---|---|---|---|
| | TEMPERATURE | SPECIFIC CONDUCTANCE | TURBIDITY | COLOR | ODOR | PH (FIELD) | PH (LAB) | EH | SUSPENDED SOLIDS | OTHER | DISSOLVED SOLIDS | CHLORIDE | NUTRIENTS (NITROGEN) | NUTRIENTS (PHOSPHORUS) | COMMON IONS | HARDNESS | RADIOCHEMICAL | DISSOLVED OXYGEN | OTHER GASES | MINOR ELEMENTS | PESTICIDES | DETERGENTS-MBAS | BIOCHEMICAL OXYGEN DEMAND | CARBON (TOTAL, DISSOLVED ETC.) | COLIFORMS | OTHER MICRO-ORGANISMS | OTHER | CONCENTRATION (SUSPENDED) | PARTICLE SIZE (SUSPENDED) | PARTICLE SIZE (BED MATERIAL) | OTHER | SURFACE WATER STATION | GROUND WATER STATION | WATER STAGE OR LEVEL | WATER DISCHARGE | TIME OF TRAVEL | DRAINAGE AREA | |
| 57393 | | | | | | | | | | | | | | | | | | 8 | | | | | | | | | | | | | | | | | | | | Q50 |
| 50529 | 5 | 5 | | 7 | | 6 | 8 | | | | 6 | 6 | 6 | 6 | 6 | 6 | | | | 6 | | | | | | | | | | | | * | | * | * | | * | GS |
| 50530 | 3 | 3 | | 6 | | 3 | 3 | | | | 3 | 6 | 5 | 5 | 6 | 3 | | | | 6 | | | | | | | | | | | | * | | * | * | | * | GS |
| 73239 | 5 | 5 | | | | | 8 | | | | | | | | | | | | | | | | | | | | | | | | | * | | * | * | | * | GS |
| 63158 | | | | | | | | | | | | | | | | | | 7 | | | | | | | | | | | | | | | | | | | | Q50 |
| 66033 | 5 | | | | | | 5 | | | | 5 | 5 | | 5 | 5 | 5 | | 5 | | | | | 5 | | 5 | 5 | | | | | | * | | | | | | Q51 |
| 63209 | | | | | | | | | | | | | | | | | | 7 | | | | | | | | | | | | | | * | | | | | | Q50 |
| 73240 | 5 | 5 | | | | | 8 | | | | | | | | | | | | | | | | | | | | | | | | | * | | * | * | | * | GS |
| 50531 | 6 | 6 | | 6 | | 6 | 6 | | | | 6 | 6 | 6 | 6 | 6 | 6 | | | | 6 | | | | | | | | | | | | * | | * | | | | GS |
| 63208 | | | | | | | | | | | | | | | | | | 7 | | | | | | | | | | | | | | * | | | | | | Q50 |
| 50532 | 6 | 6 | | 6 | | 6 | 6 | | | | 6 | 6 | 6 | 6 | 6 | 6 | | | | 6 | | | | | | | | | | | | * | | * | | | | GS |
| 50533 | 6 | 6 | | 6 | | | 6 | | | | 6 | 6 | 6 | 6 | 6 | 6 | | | | 6 | | | | | | | | | | | | * | | * | | | | GS |
| 50534 | 5 | 5 | | 5 | | 5 | 5 | | | | 5 | 5 | 5 | 5 | 6 | 5 | | | | 6 | | | | | | | | | | | | * | | * | * | | * | GS |
| 63206 | | | | | | | | | | | | | | | | | | 7 | | | | | | | | | | | | | | | | | | | | Q50 |
| 63139 | | | | | | | | | | | | | | | | | | 7 | | | | | | | | | | | | | | * | | | | | | Q50 |

FIG. 1—*Example of pages on surface-water quality from the* Catalog of Information on Water Data (*from Ref* 1).

## References

[1] Doyel, W. W., Johnson, A. I., and Lang, S. M., "Federal Information and Data Centers for Engineers and Scientists," *Materials Research and Standards*, American Society for Testing and Materials, Vol. 12, No. 1, pp. 17–20.

[2] *Water Data Management Systems, Proceedings of the American Society of Civil Engineers, National Capital Section, Water Resources Engineering Technical Committee*, U.S. Department of Interior, Geological Survey, Nov. 29, 1978.

[3] *Proceedings of the Speciality Conference of ASCE Computer Applications and Water Resources*, Buffalo, NY, Harry C. Torno, Ed., American Society Civil Engineers, New York, 10–12 June 1985.

# Geophysical Methods

*Donald J. Stierman*[1,2] *and Lon C. Ruedisili*[1]

# Integrating Geophysical and Hydrogeological Data: An Efficient Approach to Remedial Investigations of Contaminated Ground Water

**REFERENCE:** Stierman, D. J. and Ruedisili, L. C., **"Integrating Geophysical and Hydrogeological Data: An Efficient Approach to Remedial Investigations of Contaminated Ground Water,"** *Ground-Water Contamination: Field Methods, ASTM STP 963,* A. G. Collins and A. I. Johnson, Eds., American Society for Testing and Materials, Philadelphia, 1988, pp. 43–57.

**ABSTRACT:** Geophysical methods have been proven a cost-effective tool for investigating ground-water contamination near some landfills and hazardous waste disposal sites. Establishing standard methods or practices for geophysical investigations is made difficult, however, by the many geophysical tools available and the broad range of site conditions encountered by the geophysicist and hydrogeologist. The case histories reported here illustrate, first, that electromagnetic conductivity surveys are less cost-effective than d-c resistivity in detecting and mapping contaminated ground water. Second, they demonstrate the value of employing geophysical measurements as an integral component of a remedial investigation. Geophysical investigations must be planned and data must be interpreted in light of available geologic and hydrologic data. Geophysical, geological, and hydrologic observations must all be used to develop the overall conceptual model. Preliminary analysis of geophysical measurements should be made in the field so that the investigator can adjust his strategy to site-specific conditions. New data that are collected should be compared with predictions made based on the overall conceptual model, and the degree to which new data conform to such predictions serves as an indicator of the reliability of the model and the adequacy of the data. Finally, there is a need to standardize units and data presentations so that information developed by different investigators can be easily compared and fully exploited.

**KEY WORDS:** electrical prospecting, electromagnetic prospecting, hydrogeology

Geological and hydrological settings for ground-water contamination investigations are widely varied. Each site has unique characteristics which can inhibit successful application of practices and standards that were useful at a different site. This is particularly true in regard to geophysical techniques as they are used to detect and map ground-water contamination. Geophysical exploration can be a cost-effective method for reducing the number of expensive monitoring wells needed to characterize a site, but geophysical studies are often inconclusive. The purpose of this paper is, first, to report on successful application of geophysical exploration for contaminated ground water at three sites, located in California, Ohio, and Michigan. At two of these sites, d-c resistivity was found far superior to electromagnetic (EM) conductivity measurements in detecting contaminated ground water and in providing a general image of the subsurface. We shall discuss basic limitations of EM methods as well as suggest proper procedures for d-c resistivity. Second, we shall discuss

[1] Associate professor and professor, respectively, Geology Department, The University of Toledo, 2801 W. Bancroft St., Toledo, OH 43606.

[2] Also associate geophysicist, Applied Geomechanics, Inc., 1336 Brommer St., Santa Cruz, CA 95062.

the importance of formulating and testing an overall conceptual model in the field as well as during the formal interpretation phase of a geophysical program. Finally, we will show how this overall conceptual model serves as a test for determining the adequacy of a hydrogeology field investigation.

Most ground-water contamination occurs in shallow aquifers where the effects of natural and cultural heterogeneities on geophysical signals are most severe. Seasonal patterns can also influence geophysical measurements focused on shallow targets. The impact of such noise can be minimized by field methods designed to isolate or attenuate noise. Geological information, hydrological data, and geophysical measurements must be integrated into an overall conceptual model not only in the interpretation phase but also in the planning of an investigation. For example, well logs and water quality data are needed to interpret geophysical measurements, so that the geophysical measurements can be used to interpolate between and extrapolate beyond wells used to sample the subsurface. A firm understanding of the geological factors influencing geophysical signatures of a specific study area is also fundamental to a successful and efficient geophysical exploration campaign.

Development of an overall conceptual model for a site not only enhances the reliability of geophysical interpretations, but also provides a standard for deciding if sufficient knowledge has been generated to begin design and implementation of a restoration program. At the point where additional data confirm predictions based on an overall conceptual model, the model can be judged sufficient to be used as a basics for the feasibility study.

**Resistivity and Electromagnetic Measurements at the Stringfellow Site Riverside County, California**

Approximately 140 million litres (L) of liquid industrial wastes, including organic solvents and acids containing heavy metals, were discharged into surface lagoons at the Stringfellow site between 1956 and 1972. In addition to infiltration during episodes of surface discharge from the site in 1969, 1978, and 1980, percolation of liquid wastes through alluvial materials and through fractures in the crystalline basement has contaminated an alluvial aquifer downgradient from the site. A brief history of Stringfellow and discussion of earlier electrical exploration in the vicinity of the site was reported by Stierman [*1*].

The Stringfellow site was given high priority for remedial action under the Comprehensive Environmental Response, Compensation and Liability Act, PL 96-510 (known as CERCLA, or "Superfund"). Prior to Federal involvement, the State of California had secured the site against further surface discharge and had installed a system of monitoring and extraction wells. Once designated a Superfund site, further studies were conducted to provide (1) information needed for "fast-track" abatement action to attenuate the flow of contaminated ground water beyond existing extraction wells and into the downgradient basin, and (2) information needed to design a final solution to the threat posed to ground water by this site.

The "fast-track" investigation included EM measurements used to aid in selecting the locations of four additional monitoring wells. Locations of two EM traverses (A-A′, B-B′) are shown in Fig. 1, and the subsurface images constructed from these measurements are shown in Figs. 2 and 3. Fifteen months after completion of the EM measurements, as part of the remedial investigation/feasibility study conducted by Science Applications International Corp., d-c resistivity measurements were made to, among other things, locate the front and determine the width of the plume. Profile locations (A-A′, C-C′, D-D′, E-E′) are shown in Fig. 1 and the interpretations shown in Figs. 2 and 4. The locations of numerous other geophysical measurements not discussed in this paper have been omitted.

Figure 2 compares the results of a dipole-dipole d-c resistivity profile with the resistivity pattern determined using EM methods. Both the EM and dipole-dipole cross sections are only semiquantitative; that is, they represent apparent, not true, earth resistivities, but these can be interpreted

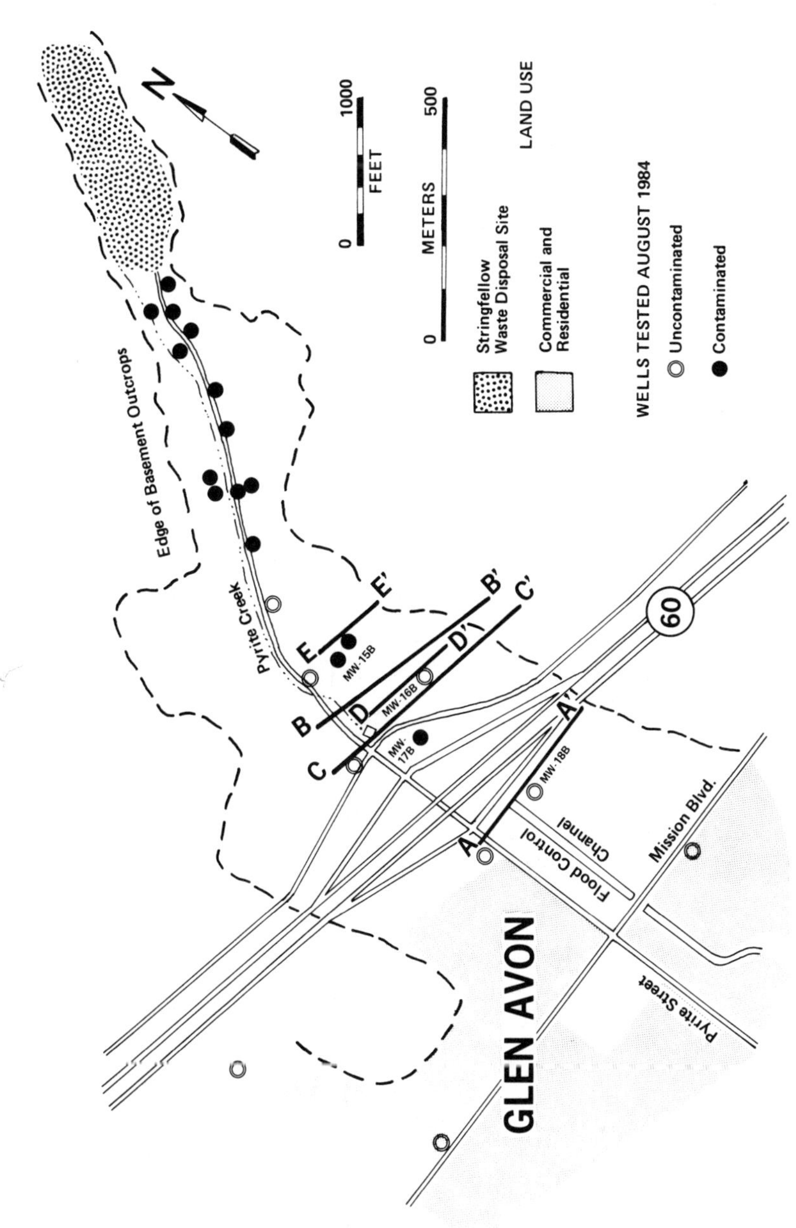


FIG. 1—*Map of Pyrite Canyon showing the location of the EM profiles (AA' and BB'), dipole-dipole profile (AA'), and Schlumberger resistivity profiles (CC', DD' and EE') discussed in this paper, with respect to monitoring wells and natural and cultural features of Pyrite Canyon.*

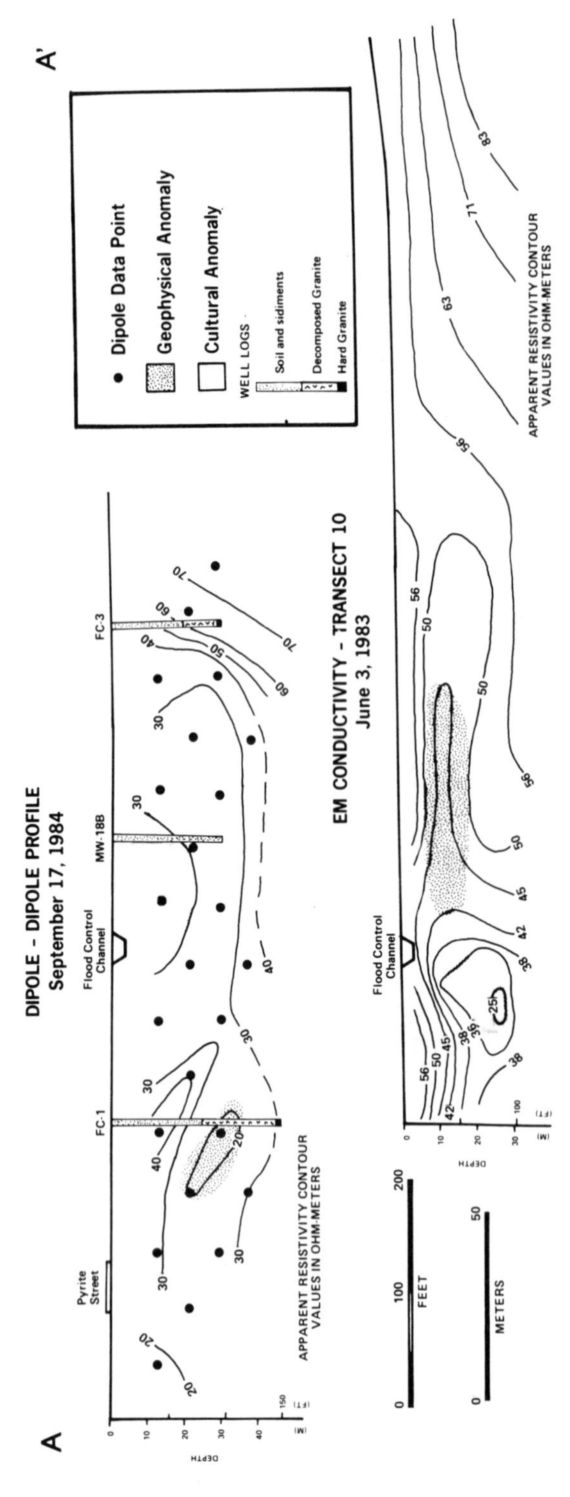


FIG. 2—*Comparison of the dipole-dipole pseudosection modified after Edwards [2] with the electromagnetic conductivity cross-sectional representation [3], modified here to show resistivity rather than conductivity.*

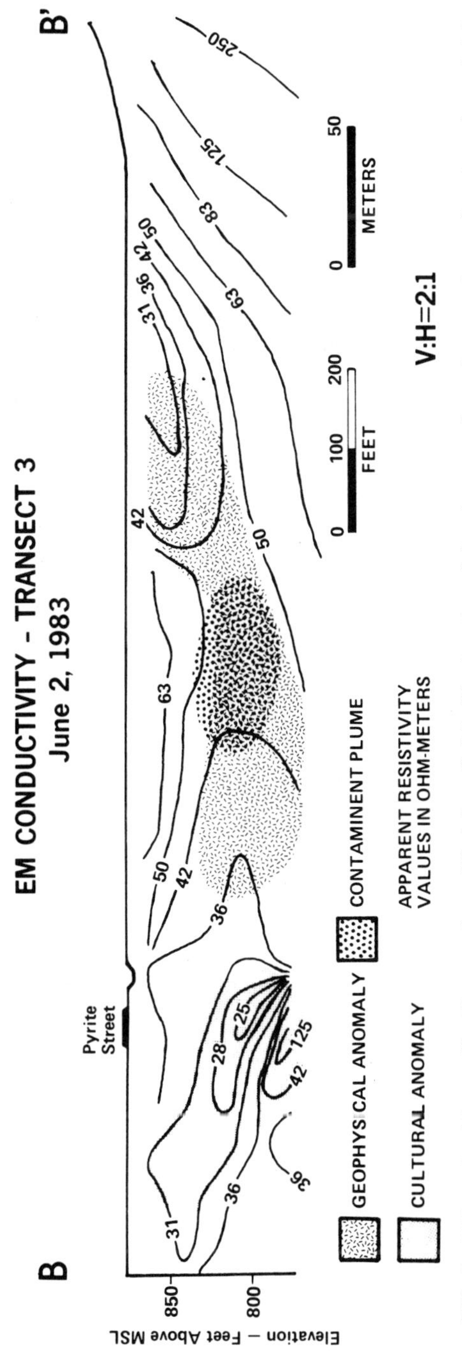


FIG. 3—*EM conductivity profile, Transect 3, modified from Ref 3. The location of the plume as shown is approximate, based on d-c resistivity interpretations, water quality data, and soil-gas surveys.*

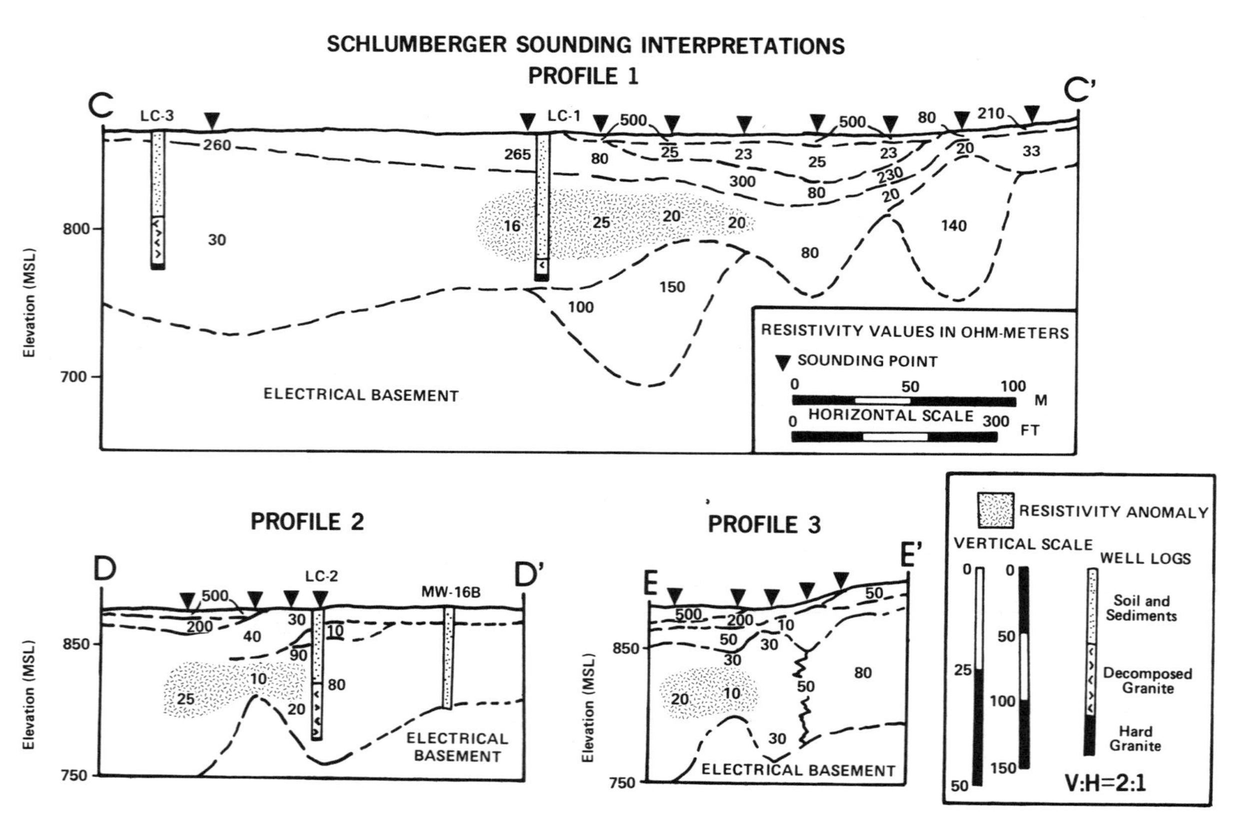


FIG. 4—*Interpretations of electrical resistivity from Schlumberger profiles CC', DD', and EE' (Fig. 1). Numbers indicate electrical resistivity in Ω-m of plane-layered models consistent with the sounding curves* [4]. *Dashed lines are interpretive representations of electrical stratigraphy.*

as an image of subsurface geology. Based on the interpretation of this EM profile, Well MW-18B was drilled about 20 m east of the concrete flood-control channel. No contamination was detected in water from MW-18B. EM measurements made in the vicinity of Pyrite Street, west of the flood-control channel, were so overwhelmed by cultural noise that nothing meaningful could be detected in the subsurface and, thus, values were not plotted [*3*].

The dipole-dipole survey was not hampered by this high noise level, and a narrow low-resistivity anomaly was defined about 25 m east of the centerline of Pyrite Street. Monitoring Well FC-1, drilled to investigate this anomaly (Fig. 2), encountered contaminated water [*5*]. This well marked the discovery of contaminated ground water south of Highway 60 (Fig. 5) which, until that time, had served as a psychological barrier separating the site and contaminated areas of Pyrite Canyon from the community of Glen Avon, a residential area largely dependent on private wells as a source of drinking water. In addition to detecting the contaminant plume, the dipole-dipole data better reflect the rapid decrease in low-resistivity sediment cover on the eastern limit of the profile, near Well FC-3.

North of Highway 60, Schlumberger soundings were employed as part of the d-c resistivity investigation. The chief advantage of the Schlumberger array over dipole-dipole profiles is that data can be matched with earth models in which true resistivities and layer thicknesses are specified. This can also be done for EM measurements but in a much more limited sense. In the course of this d-c resistivity interpretational phase, as many as five distinct layers overlying an insulating "electrical basement" were needed to match the observed Schlumberger sounding curves. EM measurements as they are routinely interpreted are limited to two-layer-over-basement models.

Profile 1 (Fig. 4) shows a 16 Ω-m resistivity low extending from 15 to 30 m below the surface near the center of the profile. Monitoring Wells LC-1, LC-2, and LC-3 were drilled after these d-c resistivity measurements were completed. LC-1, drilled into the resistivity low, encountered contaminated water. The eastern lobe of the EM anomaly in this area (Fig. 3) is probably a shallow, conductive clay layer that is shown at depths of less than 15 m on Profile 1, CC′ (Fig. 4). Well MW-16B, drilled to investigate this eastern lobe of the EM anomaly shown in Fig. 3, did not encounter contaminated water. Schlumberger Profiles 2 and 3, a short distance to the north of and running parallel to Profile 1 (Fig. 1), also show a narrow resistivity low between 15 and 30 m below the surface (Fig. 4). The interpretation that these anomalies indicated a narrow plume of contaminated ground water has been confirmed by the pattern of contaminated monitoring wells (Fig. 5) and by soil-gas measurements, which detected anomalous concentrations of trichloroethylene directly over the resistivity lows in April 1985 [*6*].

**Novaco Industries—Monroe County, Michigan**

Novaco Industries of Temperance, Michigan, is an active tool and die design, manufacturing, and repair facility. On June 13, 1979, personnel noticed leakage of a chromic and sulfuric acid solution from a joint in a 7930-L storage tank buried near one corner of the facility (Fig. 6). It is estimated that 300 to 400 L of this acid escaped into the subsurface. Although pumping removed some quantity of these contaminants from the ground water, this site remains under study. Chromium has contaminated residential water wells nearby [*7*] and the site appears on the National Priority List for CERCLA Remedial Actions [*8*].

EM measurements made by a consultant for the State of Michigan in late January and early February 1984 failed to detect any electrical resistivity pattern indicative of ground-water contamination (Fig. 7). It was concluded that severe cultural noise limited the resolution of EM measurements at this site [*9*]. D-c resistivity, however, appears promising. Five Wenner soundings completed one afternoon in May 1985 detected a resistivity low associated with contaminated wells. In addition, sounding data document the highly variable resistivity of the surficial soil and

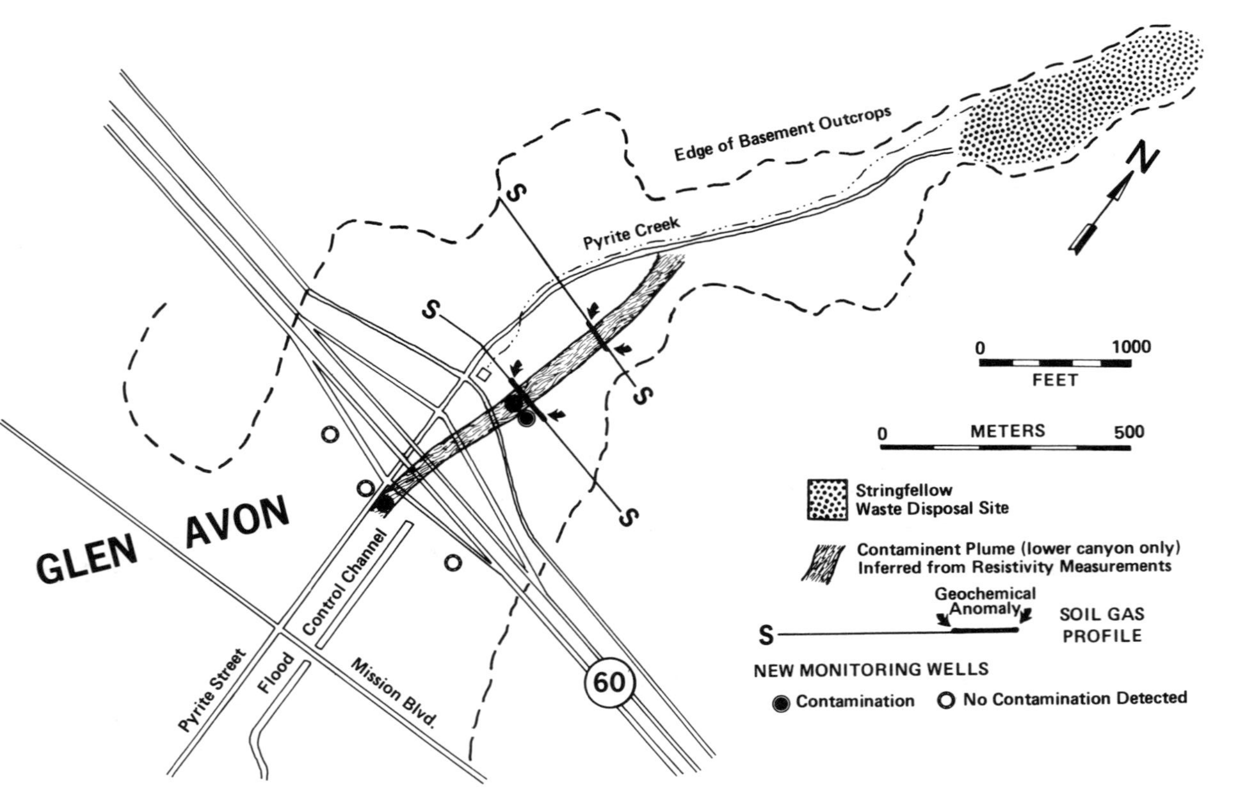


FIG. 5—*Map of Pyrite Canyon showing location of contaminant plume as deduced from Schlumberger profiles (Fig. 4) and dipole-dipole data (Fig. 2). Also shown are locations of six new monitoring wells* [5] *and selected soil-gas profiles* [6].

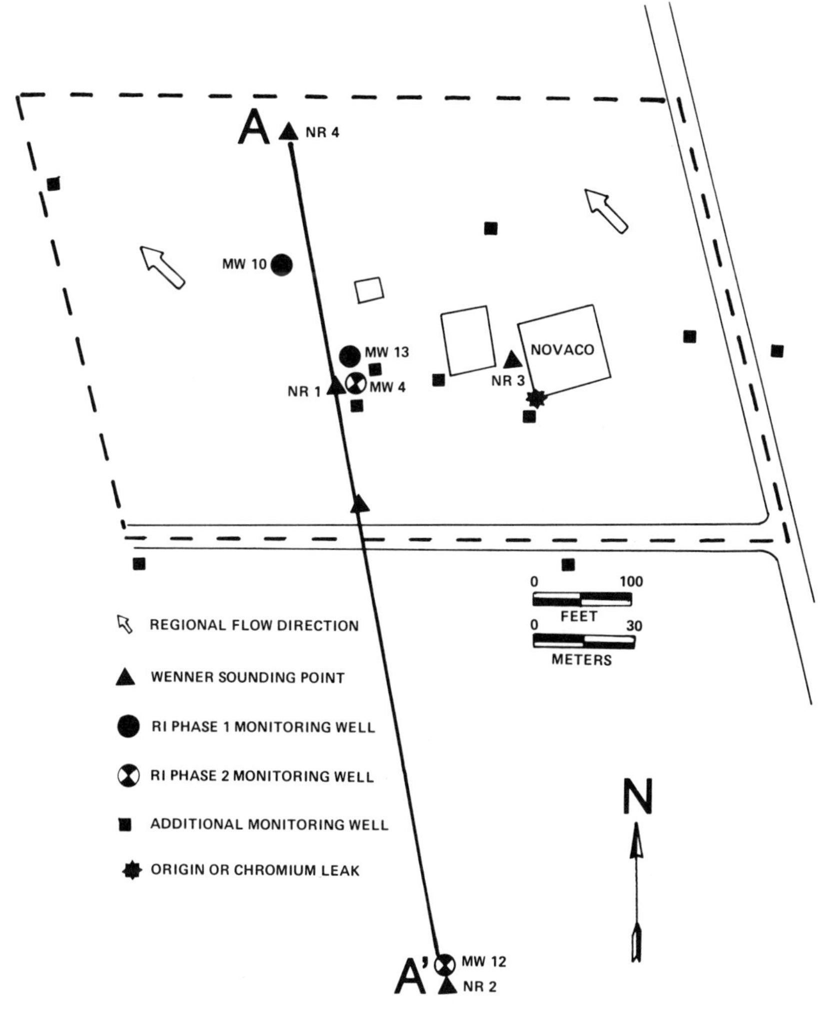


FIG. 6—*Map showing location of monitoring wells, Wenner electrical soundings, and chromic acid spill at the Novaco site in southeast Michigan* [7].

sediment layer. Resistivity of the uppermost material varied from 70 Ω-m in marshy ground near Wenner sounding point NR4 (Fig. 8) to 790 Ω-m in the dry sand near NR2. Surface resistivity near NR1 (Fig. 6) was so high (3000 Ω-m) that the resistivity of the underlying saturated zone could not be accurately determined with the equipment available. Although these shallow resistivity variations influence both d-c soundings and EM measurements, the influence of such heterogeneous materials on a d-c sounding curve is easily appraised and can, except where resistivity contrasts are high and the layers thin, be isolated from resistivity variations in the underlying ground-water zone where such variations reflect water quality.

## King Road Landfill—Lucas County, Ohio

The King Road Landfill operated as a dump for residential and commercial customers from 1954 through 1976. The 42-ha (104 acre) site was covered by sand and the top of the resulting

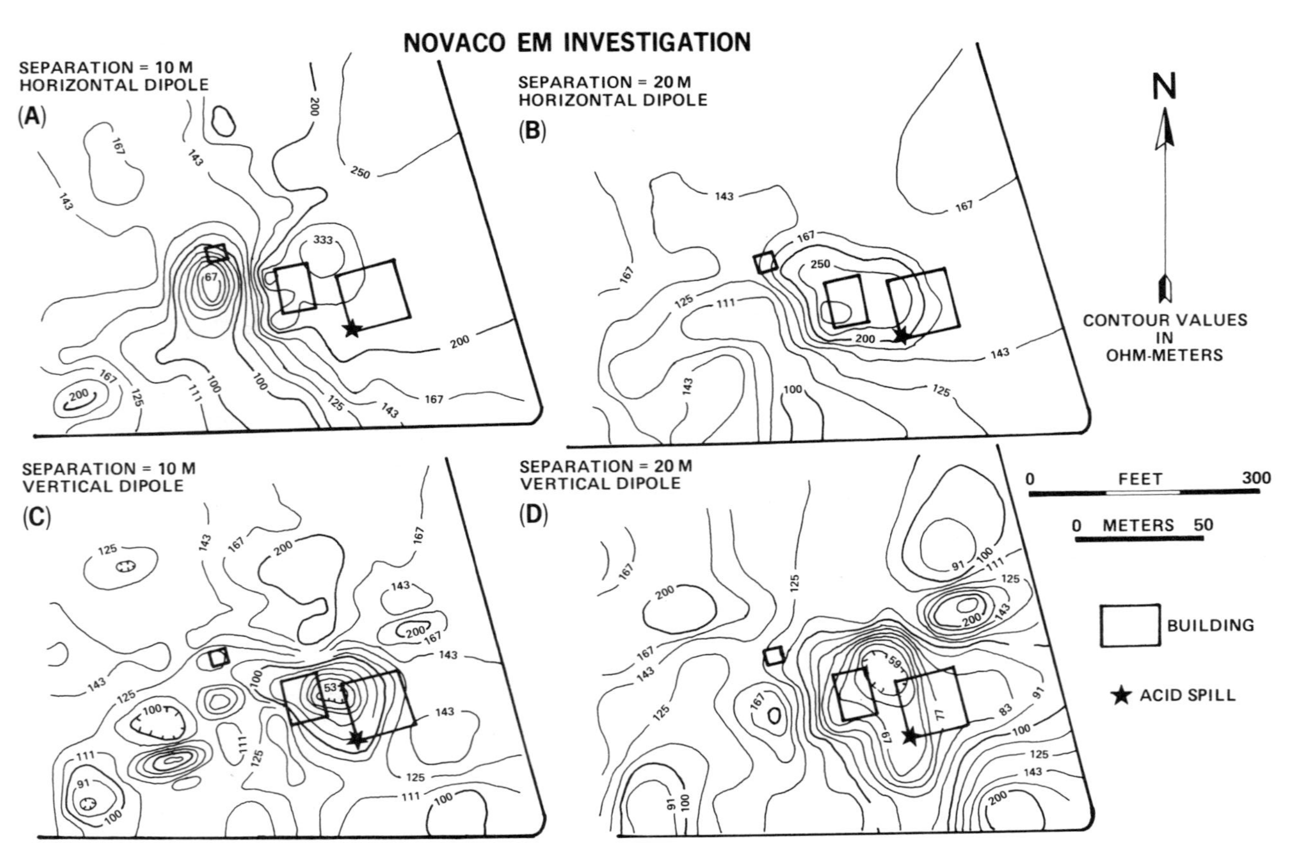


FIG. 7—*Resistivity contours representing eletromagnetic conductivity measurements made at the Novaco site in late January and early February of 1984* [9]. *Conductivity values originally reported have been converted to apparent resistivity.*

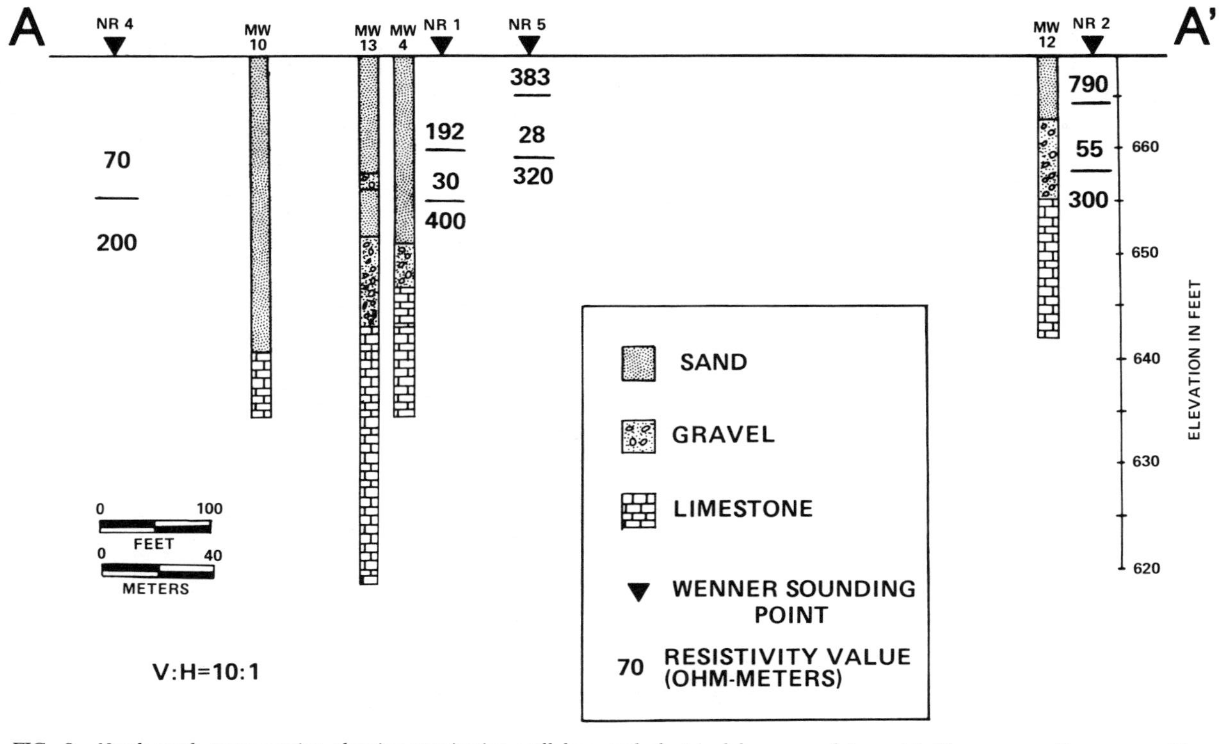


FIG. 8—*North-south cross section showing monitoring well logs and electrical layers used to match Wenner sounding curves. The location and orientation of profile AA' is shown in Fig. 6.*

mound leveled off. In addition to leachate seeps noted on the flanks of the site following periods of high precipitation, contaminated water was encountered when monitoring wells were drilled adjacent to the site in 1983 [*10*]. However, large gaps between monitoring wells have made it difficult to develop a coherent model for water flow patterns and contaminant pathways. Geophysical methods are contributing to development of an overall conceptual framework. Electrical resistivity again proved useful in mapping contaminated ground water along the west flank of the site, and additional electrical measurements will be made when suitable equipment becomes available. Seismic refraction proved more reliable than d-c resistivity in defining the water table.

Preliminary results suggest that bedrock, encountered at between 7 and 25 m below the surface, plays an important role in the details of horizontal ground-water flow. Seismic refraction detected, for example, a water table dipping away from a bedrock high and in a direction inconsistent with the potentiometric surface derived from monitoring well information. This bedrock high had also been discovered as the result of geophysical measurements. As at Stringfellow and Novaco, d-c measurements documented a broad range (30 to 800 Ω-m) of shallow electrical resistivities. This heterogeneous nature of the surface was, however, not clearly indicated during an EM soil conductivity test [*11*].

## Discussion

In each of the three case histories outlined above, a flexible geophysical strategy was employed regarding d-c resistivity. That is, although there were specific targets for electrical exploration, the details of instrument deployment in the field were left to the discretion and judgment of the supervising geophysicist, providing him with the freedom to respond to the site-specific conditions. This strategy is in accordance with the published policies of the U.S. Environmental Protection Agency (EPA) [*12*]. However there are procedures and standards that can be set which do not constrain the field program but, rather, guarantee that site-specific conditions will be properly evaluated.

The first stage of a geophysical exploration program of contaminated ground water should consist of measurements adjacent to existing wells. The geophysical signature of both background and contaminated zones must be determined. This aids in developing efficient plans for investigating the rest of the site and is essential for interpreting results. If ground-water contamination is suspected but discovery wells not yet drilled, there is generally an uncontaminated well where a geophysical signature for the area can be obtained. A second recommended field practice is the preliminary reduction, plotting, and interpretation of data as they are collected. Not only does this permit a geophysicist to gage the quality of his/her data while still on site, but it also enables the investigator to modify field strategy in response to site-specific conditions. In addition, this procedure requires the supervising geophysicist and chief hydrogeologist to develop an overall conceptual model for the area under investigation. Formulating and testing hypotheses while in the field is part of the interpretation process, although it cannot replace the more rigorous data reduction and modeling of geophysical data possible in the office. At the point where additional geophysical measurements and drilling results confirm predictions based on the overall conceptual model and no major questions remain unresolved, the site can be considered well-characterized and sufficiently well understood to phase out exploration and proceed with selecting appropriate restoration methodologies.

Preliminary reports on well logs and water conductivities should be routinely routed to the interpreting geophysicists. In this way, the significance of geophysical signals can be recognized and the interpretations modified. At Stringfellow, for example, electrical basement (determined geophysically) is consistently deeper than hard crystalline rock as determined by drilling (Fig. 4). At Novaco, it appears that electrical basement underestimates the depth to bedrock (Fig. 8). The geophysical interpreter must adjust depth-to-basement estimates accordingly in those areas lacking well control. Finally, geophysical measurements should remain open for reinterpretation throughout the remedial investigation phase. The significance of an observation might not be appreciated except in light of data collected at some later time.

Based on these case histories, we conclude that d-c resistivity is superior to EM techniques in detecting contaminated ground water. First, d-c resistivity is far less susceptible to cultural noise. Second, it is easier to isolate the effect of heterogeneous surficial resistivities from signals originating in the saturated zone when using d-c resistivity than with EM. Third, d-c resistivity is superior in resolving lateral and vertical variations in subsurface electrical properties. EM measurements appear, at Stringfellow, to sample (and thus average) a much larger volume than do d-c resistivity measurements. Although it is recognized that an EM measurement provides a composite of the various subsurface properties at various depths and is less sensitive to vertical structure than is d-c resistivity [*12,13*], the relatively poor lateral resolution at Stringfellow suggests EM is less sensitive to horizontal changes as well. EM was ineffective in discriminating between the plume and surrounding materials near the center of Transect 3 (Fig. 3), a zone where d-c resistivity methods detected a resistivity change of a factor of 2 (Fig. 4) in each of three profiles.

Isopeth plots (Fig. 7) of either EM or d-c resistivity data are unlikely to provide useful information if surficial resistivities vary significantly. Measurements alone are insufficient; they must be properly interpreted in light of geologic and hydrologic data. The basic attractiveness of EM over d-c resistivity is based on speed and costs; however, at Novaco and Stringfellow, the effectiveness of d-c resistivity has been shown superior. We think this is a result of limitations inherent to the nature of the EM signal. The large subsurface volume sampled and averaged by the EM signal, coupled with its susceptibility to cultural interference, renders EM surveys less sensitive than d-c resistivity to subtle changes in ground-water resistivity.

In developing the figures needed to compare d-c resistivity with EM conductivity results, we encountered two factors which inhibited rapid evaluation. First, some of the EM data were plotted on east-west profiles with east on the left, rather than the standard north-looking view. The EM patterns shown in Figs. 2 and 3 have been transposed (and the original vertical exaggeration reduced) so that comparison with the d-c resistivity patterns is straightforward. We suggest that cross sections follow geologic [*14*] and engineering tradition, placing east (for east-west profiles) or north (for profiles striking due north) on the right of a cross sectional representation of geologic, hydrologic, or geophysical data.

Second, EM results are typically reported in terms of electrical conductivity, the reciprocal of electrical resistivity. Electrical properties of fluids are traditionally specified in terms of conductivity, while properties of solids are listed in terms of resistivity [*15*]. This presents a problem regarding which should be used for ground-water studies, since what is being probed is a porous solid whose electrical properties are largely determined by the properties of the liquids contained in the pores [*16*]. Virtually all geophysical textbooks and editors of scholarly geophysical publications, as well as the U.S. Geological Survey [*17*], favor the use of electrical resistivity rather than conductivity. We therefore suggest that resistivity, rather than conductivity, be established as the standard unit for characterizing the electrical properties of geological structures probed during ground-water investigations. This standard may become increasingly important as other electrical measurements such as induced potential are developed as methods for mapping contaminated ground water.

## Conclusions and Recommendations

Electromagnetic conductivity measurements, although somewhat less costly than d-c resistivity, are significantly less successful when used in areas where cultural noise or large variations in the electrical properties of shallow materials are present. This lesser effectiveness of EM methods is probably a fundamental weakness that cannot be improved by modifying field techniques or interpretive analysis. Isopethic maps, on which only raw EM or d-c data are presented, will not reflect ground-water quality except where geologic and cultural conditions are particularly favorable. In most cases, variable surface resistivities or cultural noise will obscure signals of interest originating in the saturated zone unless proper analytical steps are taken as part of an integrated interpretation.

Prior to attempting the geophysical mapping of a contaminant plume, the geophysical signature of both background (uncontaminated) and contaminated zones must be carefully examined. This is the fundamental procedure that can, first, serve as a test as to the applicability of a geophysical survey in a specific situation and, second, provide the investigators with data necessary for designing a survey and for interpreting the resulting measurements. During the course of a study, geophysical and hydrogeologic investigators should develop an overall conceptual model of the site and test this model by shrewd selection of additional measurements or monitoring wells. At the point where the investigators are successful in predicting these additional measurements, assuming no major questions remain unresolved, the site can be considered well-characterized and restoration plans begun.

Because most investigations involve a diverse team of experts whose data must ultimately be woven into a comprehensive picture, cross-sectional views should be standardized with respect to their orientation. Cross sections should be oriented so that the right-hand sides are either the most easterly end or else due north.

*Acknowledgments*

We wish to thank Dr. Robert Shokes of Science Applications International Corp., project manager of the Stringfellow Remedial Action/Feasibility Study, for permission to utilize data collected as part of that project. We also thank Ecology and Environment, Inc., and the Michigan Department of Natural Resources for providing reports on electromagnetic conductivity measurements.

*Legal Notice*

Portions of this report were prepared as the result of work sponsored by the California Department of Health Services. It does not necessarily represent the views of the Department, its employees, or the State of California. The Department, the State of California, its employees, contractors, and subcontractors make no warranty, express or implied, and assume no legal liability for the information in this report; nor does any party represent that the use of this information will not infringe upon privately owned rights.

## References

[1] Stierman, D. J., *Environmental Geology and Water Science,* Vol. 6, 1985, pp. 11–20.
[2] Edwards, L. S., *Geophysics,* Vol. 42, 1977, pp. 1020–1036.
[3] "Electromagnetic Conductivity Survey—Stringfellow Hazardous Waste Site, Glen Avon Heights, CA," unpublished report, Ecology and Environment, Inc., San Francisco, CA, 1983.
[4] Stierman, D. J. and Hozhausen, G. R., "Results of Electrical Exploration of Pyrite Canyon," unpublished report, Applied Geomechanics, Inc., Santa Cruz, CA, 1985.
[5] *Stringfellow Update,* Office of Public Information, California State Department Health Services, Sacramento, CA, Jan. 1985.
[6] "Draft Interim Report on Development and Screening of Remedial Technologies and Alternatives," Science Applications International Corp., Riverside, CA, Aug. 1985, pp. 3–32; Figs. 3–7.
[7] de Venecia, K. Wolberg, "Processes Controlling the Migration of a Contaminate Plume Through the Application of a Finite-Difference Solute-Transport Models," M.S. thesis, Geology Department, The University of Toledo, Toledo, OH, Dec. 1987.
[8] "Michigan Sites of Environmental Contamination Proposed Priority List," Michigan Department of Natural Resources, Lansing, MI, Nov. 1985, p. 189.

[9] "Draft Remedial Investigation Report," Michigan Department of Natural Resources, Lansing, MI, Oct. 1984, pp. 71–79.
[10] Stangl, J. M., "A Preliminary Hydrogeologic Investigation of the King Road Landfill, Lucas County, Ohio," unpublished manuscript, Geology Department, The University of Toledo, Toledo, OH, May 1985.
[11] Kier, D. E., "Conductivity Survey at the King Road Landfill, Toledo, Ohio," unpublished report, Geology Department, The University of Toledo, Toledo, OH, May 1985.
[12] Benson, R. C., Glaccum, R. A., and Noel, M. R., "Geophysical Techniques for Sensing Buried Wastes and Waste Migration," Report submitted to Environmental Monitoring Systems Laboratory, Office of Research and Development, U.S. Environmental Protection Agency, Las Vegas, NV, 1982.
[13] McNeill, J. D., "Electromagnetic Terrain Conductivity Measurement at Low Induction Numbers," Technical Note TN-6, Geonics, Ltd., Mississauga, ON, Canada, 1980.
[14] Compton, R. R., *Manual of Field Geology,* Wiley, New York, 1962, p. 45.
[15] *Handbook of Chemistry and Physics,* 51st ed., R. C. Weast, Ed., Chemical Rubber Co., Cleveland, OH, 1970.
[16] Archie, G. E., *Transactions,* American Institute of Mining, Metallurgical, and Petroleum Engineers, Vol. 146, 1942, pp. 54–62.
[17] Zohdy, A. A. R., Eaton, G. P., and Mabey, D. R., "Application of Surface Geophysics to Ground-Water Investigations," Chapter D1 of *Techniques of Water-Resources Investigations of the United States Geological Survey,* U.S. Government Printing Office, Washington, DC, 1974.

*Richard C. Benson,[1] Matthew Turner,[1] Paula Turner,[1] and William Vogelsong[1]*

# *In Situ*, Time-Series Measurements for Long-Term Ground-Water Monitoring

**REFERENCE:** Benson, R. C., Turner, M., Turner, P., and Vogelsong, W., "***In Situ*, Time-Series Measurements for Long-Term Ground-Water Monitoring,**" *Ground-Water Contamination: Field Methods, ASTM STP 963*, A. G. Collins and A. I. Johnson, Eds., American Society for Testing and Materials, Philadelphia, 1988, pp. 58–72.

**ABSTRACT:** Because recent legislation has made long-term ground-water monitoring an important issue, the methods by which data at a site are collected are of utmost importance. The traditional method of site assessment, which consists of drilling a number of monitor wells from which water samples can be taken and tested, does not provide enough information on which to base a long-term ground-water monitoring program. *In situ*, time-series measurements using a variety of geophysical techniques, on the other hand, can provide a wealth of information not available through traditional measurement methods.

Over the past ten years, geophysical measurements made from the surface have been used extensively at waste-disposal sites to assess natural hydrogeologic conditions and map the lateral and vertical distribution of contaminants. In recent years, downhole measurements have been used to produce more-detailed information about vertical contaminant distribution through the use of vertical profiles. Examples illustrating these applications are presented in the text. Made at appropriate intervals over a period of time, surface and downhole measurements can be used to provide highly accurate, three-dimensional maps of contaminant flow in ground water and provide a means of early leak detection at waste-disposal sites.

Because the values measured by geophysical methods are dominated by the specific conductance of pore fluids, they can be related to the specific conductance of ground-water samples, thus correlating geophysics with traditional assessment methods. Correlation between *in situ*, geophysical measurements and inorganic, ground-water sample analyses have been as good as 0.96 at the 95% confidence level. At one landfill, the correlation with total organic carbon was 0.85 at the 95% confidence level.

Three examples of sites where the spatial extents of contaminant plumes have been mapped using geophysical, time-series measurements are presented in this paper. They consist of a localized spill, a large landfill, and a flowing, abandoned well. Measurements taken over a period of time at each of these sites clearly illustrate how the actual dynamics of a contaminant plume can be assessed as it interacts with natural and man-induced variables.

**KEY WORDS:** *in situ* time series, ground-water monitoring, geophysics, electrical methods, resistivity, electromagnetic conductivity, ground-water contamination, hydrogeologic variability

Stringent new legislation regulating the discharge of effluents to ground water has been recently adopted by many states, making long-term ground-water monitoring a critical issue. Implementation of these regulations on a state-to-state basis not only reflects growing concern for the quality of the nation's ground-water supply, but also points to the need for more accurate, cost-effective, long-term monitoring programs.

The purposes of long-term ground-water monitoring are to assess the hydrogeologic setting of

[1] Technos, Inc., Miami, Florida, 33142.

facilities discharging effluents to ground water, provide long-term monitoring of discharge zones, and develop three-dimensional descriptions of contaminant distribution which may occur within the zones of discharge. All this must be accomplished with sufficient accuracy so as to confidently describe the current hydrogeologic setting as well as detect and predict future changes in contaminant migration.

The most critical part of development any long-term ground-water monitoring program is in eliminating, or at least minimizing, the errors made in the assessment of overall site conditions. If the earth were a uniform, layered media with isotropic properties, accurate site description would be easy; a single boring and sample analysis would be sufficient. Rarely is this the case, however. Site conditions are always more complex than they are initially thought to be. Many times what has been reported as a "simple hydrogeologic setting" has turned out to be a "hydrogeologic nightmare" five to ten years later due to improper assumptions based upon too little information and oversimplification. Only recently have errors in drilling, sampling, sample analyses, data interpretation, and the impact of not understanding complex hydrogeologic settings become recognized as significant. Numerous papers dealing with these topics have been written over the past few years [*1–7*].

Traditionally, the approach to long-term ground-water monitoring has been to drill a limited number of holes from which localized water samples were tested. In terms of accuracy and acceptable confidence levels, this approach is inadequate. In an attempt to improve the accuracy of site assessments, the United States Environmental Protection Agency (EPA) [*8*] has proposed that borings be placed at approximately 46-m (150-ft) intervals. There are several drawbacks to this proposal, however. One is higher program cost and the other is the possibility that vital information between borings may still be lacking due to the presence of permeable sand channels or vertical fracture zones which may be much less than 46 m (150 ft) in width. A real and more cost-effective alternative to the EPA's proposal exists when geophysical measurement techniques are combined with traditional drilling methods to produce data representative of true site conditions.

## Geophysical Measurement Techniques

Geophysical surveys can be useful in the study of most subsurface geologic problems, especially where there is a good contrast (at least 1 to 1.5) between what is considered background and what is considered the anomaly. Geophysical measurements can be made using a variety of techniques either from the surface or down a borehole to produce two- and three-dimensional descriptions of site conditions. Although measurements made from the surface are limited by decreasing resolution with increasing depth, this limitation can be overcome by taking measurements down a borehole. Downhole measurement provides excellent resolution within 0.3 m (1 ft) regardless of depth.

Two of the most commonly used geophysical measurement techniques are the resistivity and electromagnetic conductivity methods. The resistivity method includes the use of profiling (lateral measurements made at a constant sample depth) and sounding (vertical measurements made at a single location at increasing depths). Resistivity measurements can be made by using a variety of available equipment and result in a value of electrical resistivity, expressed in ohm-metres (Ω-m), for a volume of the subsurface. There are many publications which describe in detail the theory of the resistivity method [*9–16*].

The electromagnetic conductivity method includes the use of measurements made with instrumentation such as the EM38, EM31, and EM34-3, which are surface measurement systems, and the EM39, which is a downhole measurement system manufactured by Geonics, Ltd. Although surface-measurement systems can be used for sounding, they are generally used for profiling because they do not require ground contact as resistivity measuring instruments do. Because the surface electromagnetic conductivity method can make continuous measurements to a depth of 15 m (50 ft) and provide total site coverage in a quicker, more cost-effective manner, electromagnetic profile surveys are generally the preferred approach over resistivity surveys. Measurements made using the electromagnetic conductivity method result in a value of electrical conductivity, expressed in millimhos per metre (mmhos/m), for a volume of the subsurface. It is important to note that electromagnetic conductivity values are the reciprocal of resistivity values. The electromagnetic method is discussed in detail in numerous publications [*10,13,15,16*].

Both resistivity and electromagnetic conductivity methods can be used to assess variations in natural site conditions, locate and map lateral and vertical variations in soil and rock, identify subsurface fractures and permeable zones, and detect and map subsurface resistivity and conductivity contrasts which may reflect the distribution of contaminants in the unsaturated and saturated zones. This type of information is not always obtainable through the traditional method of drilling and water sampling because important details are often missed. One of the most obvious benefits of geophysical measurement is the significant improvement in sampling density which yields more detailed results. This and other benefits have been described in numerous periodicals and conferences on the subject [*17–27*]. Extensive listings of additional publications are given in Zohdy et al [*16*], Merrick [*20*], and Mooney [*14*].

Because resistivity and electromagnetic conductivity values are dominated by the specific conductance of pore fluids, they can be correlated to the specific conductance of ground-water samples. In turn, these measurements can be used to indicate the presence of contamination in ground water through the conductivity contrast between natural, background, values, and contaminated area values. Resistivity and electromagnetic conductivity are primarily used to map inorganic contaminant distribution through correlation with specific conductance (a key indicator of the presence of inorganics); however, they may be also used to estimate the extent of organic contamination at some sites. Quite often there is good correlation between inorganic and organic contaminant distribution at sites such as landfills. Correlation between *in situ*, geophysical measurements and ground-water chemical analyses has been as good as 0.96 at the 95% confidence level for organics at landfills [*18,28*]. However, organic contaminants may not move at the same rate or in the same direction as inorganics because of several mechanisms such as chemical and biological degradation, volatilization, and adsorption [*29*].

**Using Geophysics to Assess Natural Site Conditions and Map Contaminant Distribution**

Figures 1 through 5 illustrate the use of geophysical measurement techniques to assess natural site conditions and map contaminant distribution. In Fig. 1, electromagnetic conductivity data from 11 station measurements are compared with those of a continuous electromagnetic profile run over the same area of gypsum rock to detect vertical fractures. Here, the advantages of continuous profiling over station measurements are clearly seen. Data from the continuous profile have sufficient resolution to detect and map vertical fractures in the gypsum rock, indicated by the high peaks in the data, whereas the limited amount of data available from station measurements is not sufficient to detect the localized fractures. This ability, coupled with the fact that the cost of acquiring data for both sets of measurements is approximately the same, shows that the benefits of continuous profiling far outweigh those of station measurements.

In Fig. 2, shallow 3-m (10 ft) and deep 9-m (30 ft) resistivity measurements were taken to map the distribution of contaminants originating from a large [1 mile$^2$ (2.59 km$^2$)] landfill. As Fig. 2 indicates, the data obtained at shallow depth [3 m (10 ft) reveal a much more complex flow behavior than the data obtained at greater depth [9 m (30 ft)]. In the final analysis, the complex behavior of the shallow contaminant plume was due to the unsaturated zone [approximately 2 m (6.6 ft) thick] interacting with the upper portion of the saturated zone. Shallow contaminant flow was controlled by the effects of the Hialeah-Miami Springs well field within 6.5 km (4 miles) southeast of the landfill and the interbedded sands and marls. Deeper contaminant flow was

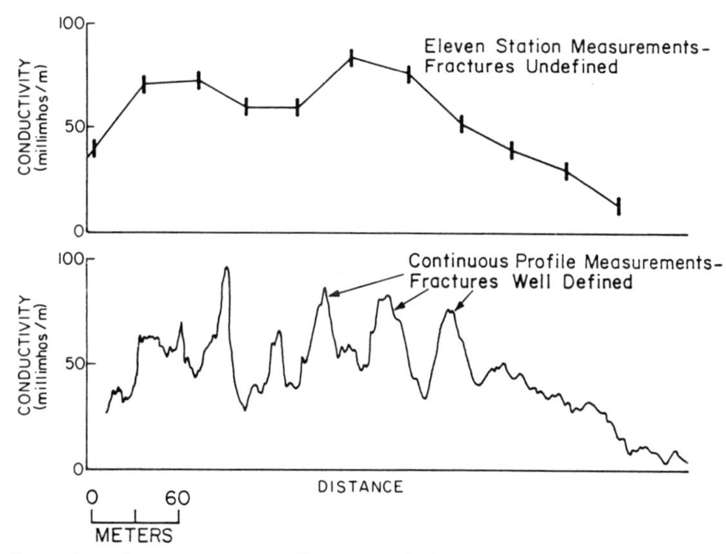


FIG. 1—*Comparison of electromagnetic conductivity results from station profile measurements and continuous profile measurements.*

controlled by the effects of regional geology and regional hydrologic gradients, both of which had a more uniform effect than conditions encountered in the shallow portion of the flow system.

Whereas Fig. 2 shows the effect of the unsaturated zone (interbedded sands and marls) and cultural factors (well field) on a very large area, Fig. 3 illustrates the complexities of localized

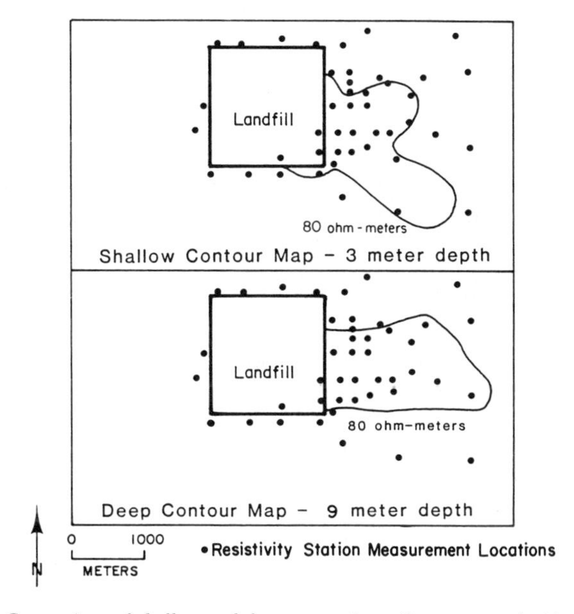


FIG. 2—*Comparison of shallow and deep contaminant flow measured with resistivity.*

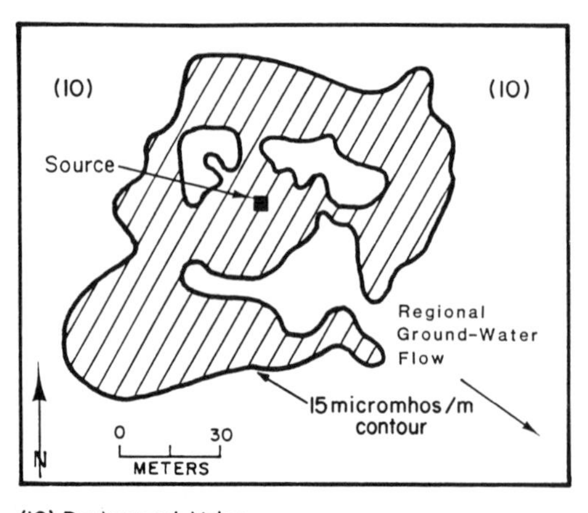


FIG. 3—*Electromagnetic conductivity map of the complex distribution of an inorganic chemical spill.*

flow over an area of 10 000 m$^2$ (2.5 acres). At this site, electromagnetic conductivity was used to map a point-source inorganic contaminant spill where there was approximately 2 m (6.6 ft) of unsaturated zone within a bryozoan facies of a limestone riddled with cavities, which contribute to its high horizontal and vertical permeability.

The electromagnetic conductivity survey lines were run from north to south, approximately 3 m (10 ft) apart over the area within the border in Fig. 3. Background values were about 10 μmhos/m. Regional ground-water flow at the site was toward the southeast. Within the unsaturated zone and the upper 3 m (10 ft) of the saturated zone, however, ground-water flow was dominated by the local subsurface conditions. At this site, it is thought that most of the contaminant flow is controlled by the unsaturated zone where interconnected cavities and localized fractures affect the radial spread of contaminants.

Vertical distribution of contaminants can also vary widely due to changes in permeability, which can range as high as a factor of ten within 0.5 m (1.6 ft). Figure 4 shows the results from a continuous, downhole electromagnetic induction log used to map the vertical distribution of inorganic contaminants through rock of varying permeability. The rock was an oolitic limestone from the ground surface to the bottom of the well [8.23 m (27 ft)], with numerous cavities filled with sand below a depth of 4 m (13 ft). No clay is present. The figure also shows the test results if discrete measurements were to be taken at standard 1.5-m (5-ft) intervals. Examination of both profiles quickly indicates the difference in the quality of data gathered. Although results from the discrete measurements indicate that inorganic contaminants are present, results from the continuous electromagnetic induction log shows that the conductivity values fluctuate below a depth of 4 m (13 ft). The fluctuations indicate zones of higher inorganic contamination because of increased permeability where the limestone has a cavity filled with sand. Variations in permeability can significantly affect the results obtained from a monitor well. It is obvious that by screening at different depths and varying the length of the screen, a wide range of chemical concentrations can be obtained at a single location.

Another common problem associated with accurately describing contaminant flow is the highly variable conditions encountered in glacial till. Thin, permeable zones can allow contaminants to flow over considerable distances in an otherwise uniform, impermeable till. In Fig. 5, a natural gamma log taken in glacial till indicates the presence of clay, and a gamma-gamma log shows variations in density. These downhole logs clearly show a relatively uniform clay content, except

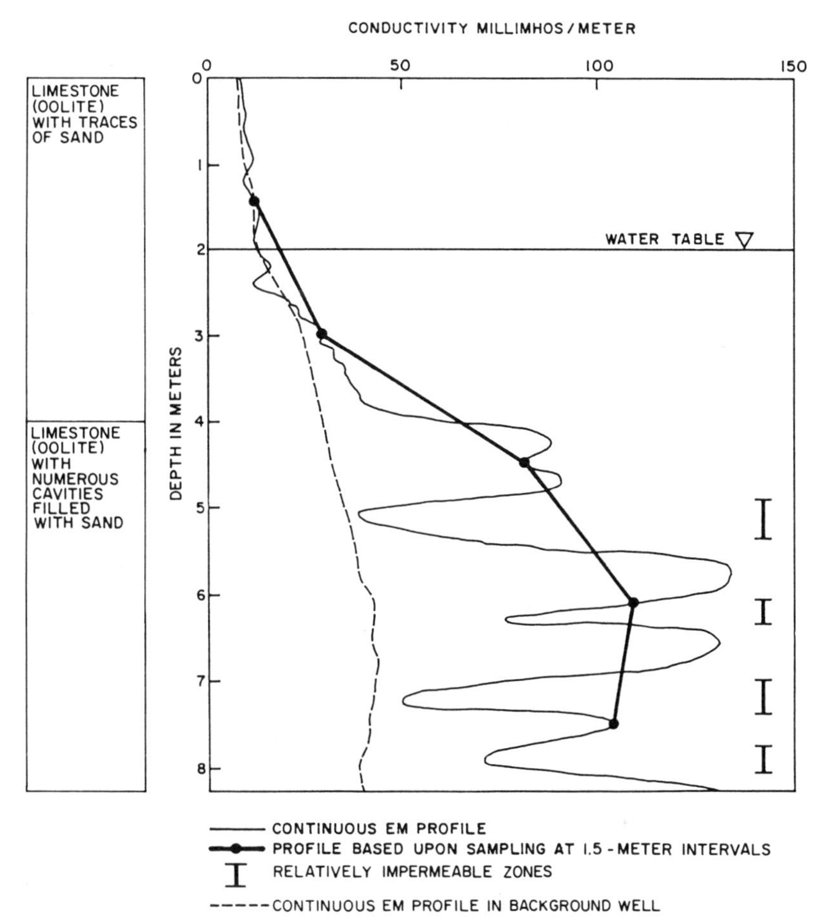


FIG. 4—*Continuous electromagnetic downhole induction log showing vertical distribution of inorganic contaminants.*

near the top of bedrock, where a relatively clean, sandy silt with greatly increased permeability occurs. The density log shown in Fig. 5 clearly indicates a high-density zone about halfway down the record, resulting in lower permeability. The clean, sandy silt overlying the bedrock shows up as a low-density material in the density log, indicating a higher permeability. These variations may be often missed by a traditional drilling program alone. Without such data, the detection and assessment of contaminant distribution tends to become a guessing game.

## Geophysics and Time-Series Measurements

The preceding examples demonstrate how geophysical-measurement techniques can be combined with traditional drilling programs to more accurately assess natural site conditions and map contaminant distribution. These same methods can be also repeated at regular intervals to map the movement of contaminant plumes and monitor for leaks at waste disposal sites. Traditional time-series water quality sampling from monitoring wells is discussed in other publications [*30–32*]. Time-series measurements should be taken at short, regular intervals over long periods of time for

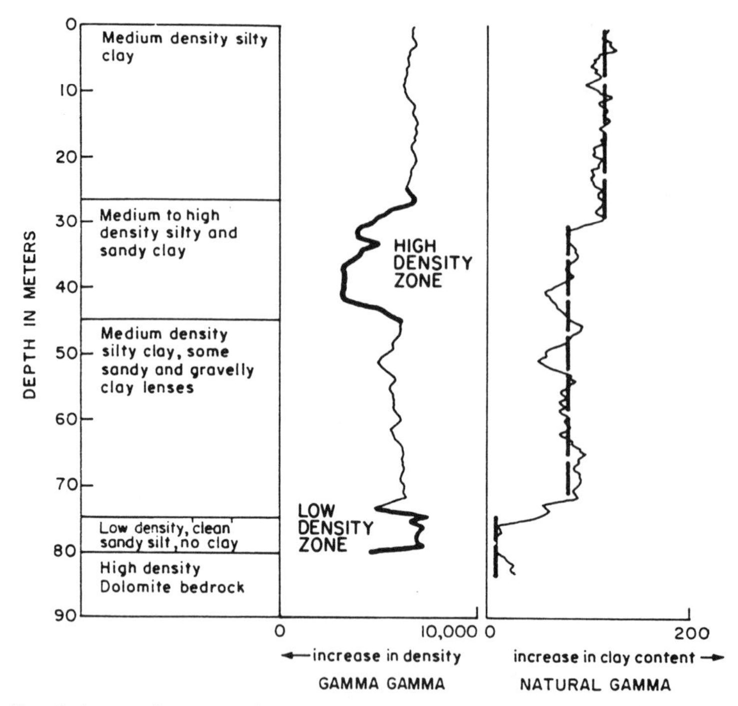


FIG. 5—*Downhole natural gamma and gamma-gamma density logs showing vertical changes of permeability in glacial till.*

long-term ground-water monitoring. However, the following three site examples are given to illustrate how geophysical measurements combined with drilling and sampling can be used to begin a program which will provide *in situ*, time-series measurements for long-term ground-water monitoring. These examples include:

1. Three time-series measurements of a localized spill taken over a period of 20 days.
2. Three time-series measurements of a leachate plume from a large landfill taken over a period of eight years.
3. Two time-series measurements mapping the brine flowing from an abandoned well over a period of five years.

*Example 1*

Example 1 consists of a small site where an inorganic contaminant spill spread over an area of approximately 6000 m$^2$ (1.5 acres). The effects of surface topography and natural drainage controlled the initial spread of the contaminants which eventually seeped down through about 2 m (6.6 ft) of unsaturated limestone. There, they came into contact with the water table. Electromagnetic conductivity measurements, made north to south over a 0.02-km$^2$ (0.000772 mile$^2$) area centered around the spill at 1, 10, and 20 days after the initial spill, are illustrated in Fig. 6. They show a complex distribution of contaminants and localized flow in all directions, following fractures and cavities in the limestone. This example shows that regional ground-water flow does not necessarily dominate local or short-term contaminant flow. When dealing with long-

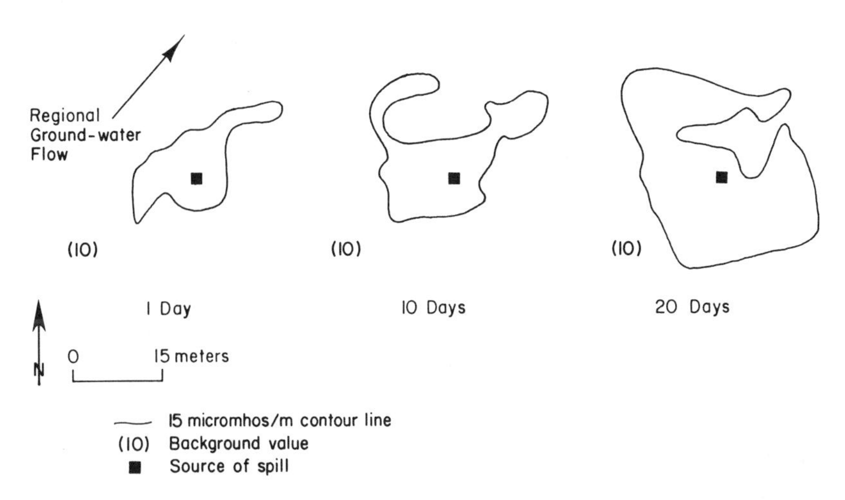


FIG. 6—*Short-term time-series monitoring of the extent of an inorganic contaminant spill using continuous electromagnetic conductivity profiles.*

term monitoring of spills of this nature, more frequent time-series measurements over a longer period of time should be made.

*Example 2*

Example 2 consists of a large, 2.59 km$^2$ (1 mile$^2$) landfill, the 58th Street Landfill, located in Miami, Dade County, Florida. In operation for nearly three decades, this landfill is situated less than 4.8 km (3 miles) upgradient of the largest well field in Dade County, Florida. Withdrawals from the well field range from 3280 L/s (52 000 gal/min) to nearly 5700 L/s (90 000 gal/min). The 210-day travel time contour for this well field extends within 1.6 km (1 mile) from the eastern edge of the landfill.

The Biscayne aquifer, a surficial aquifer, is the only source of fresh ground water in Dade County. Near the landfill, the Biscayne aquifer has an unsaturated zone of about 2 m (6.6 ft), is 24 m (79 ft) thick, and consists of highly porous and permeable limestone and sand. Geology throughout the area is reasonably uniform.

From 1973 through 1975, the United States Geological Survey conducted an initial evaluation of ground-water flow and water quality by using a series of well clusters located on a line extending through the landfill. Their water-quality analysis included various inorganic physical parameters, including specific conductivity. A cross section of specific conductivity values through these wells is shown in Fig. 7. The results of their evaluation indicated the presence of a contaminant plume, which, at distances much greater than 0.8 km (0.5 mile) from the eastern edge of the landfill, became dispersed and diluted to concentrations close to background levels of about 500 μmhos/cm [*33*]. They had clearly identified a major plume, but its lateral extent was undetermined.

In 1977, the United States Geological Survey (USGS) contracted Technos to conduct a resistivity survey that would map the lateral extent of the plume. In all, a total of 102 resistivity profiling stations were used to develop the data. The results of this study are shown in Fig. 7. By means of the resistivity survey, Technos, Inc. was able to provide the third dimension of the contaminant plume with a high degree of accuracy. The maximum ratio of specific conductivity at the disposal site to background values was approximately 4:1, and the same was true for the resistivity contrast

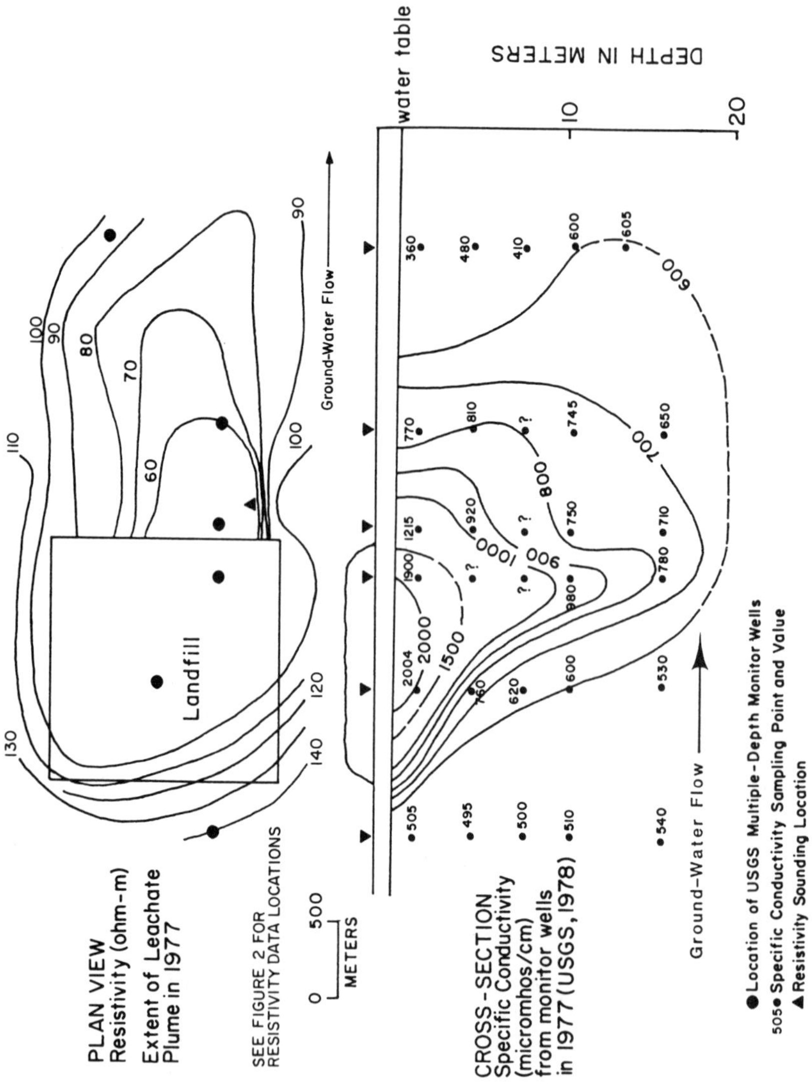


FIG. 7—*Plan view and cross section of landfill leachate plume (1977).*

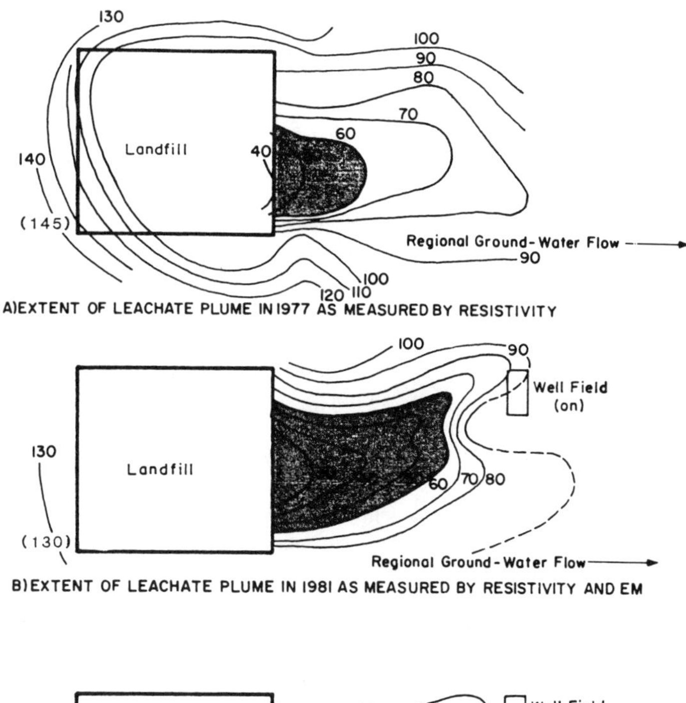


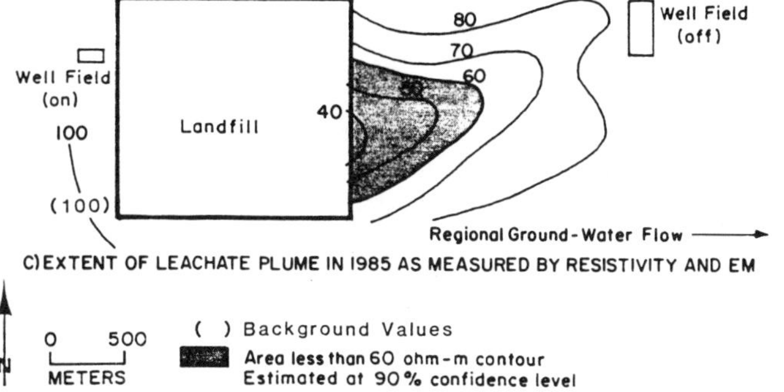


N
0 500
METERS
( ) Background Values
Area less than 60 ohm-m contour
Estimated at 90% confidence level

FIG. 8—*Three time-series measurements over an eight-year period at the 58th Street landfill, Dade County, Florida.*

(with a background value of 145 Ω-m and lowest resistivity value of 40 Ω-m) providing a good signal-to-noise ratio for the measurements.

Four years later, in 1981, litigation between the State of Florida and Dade County prompted that the survey be repeated. This time, a total of 130 electromagnetic conductivity stations and 50 resistivity stations located over approximately 15.5 km$^2$ (6 mile$^2$) were measured. The extent of the leachate plume in 1981 can be seen in Fig. 8. Background values are about 130 Ω-m. One of the most obvious differences between the 1977 and 1981 surveys is the enormous increase in the spatial extent of the leachate plume. Based on the 60-Ω-m contour in the resistivity survey, the contaminant plume's spatial coverage more than doubled in the four-year period. In addition, the plume migrated approximately 0.8 km (0.5 mile), which is equivalent to approximately 0.5 m/day (1.8 ft/day), or roughly one half as fast as the local ground-water flow calculated by the

United States Geological Survey. Another difference between the 1977 and 1981 surveys is that contaminant flow shifted from the east toward the northeast. This directional shift appears to be in direct response to pumping activities, which began in 1979, at the new Medley well field. This well field is located 2.4 km (1.5 miles) downgradient from the northeastern edge of the landfill.

In 1985, a reconnaissance investigation showed that the plume had reduced in size. This was probably due to three factors:

1. Reduction in pumping of the Medley well field 2.4 km (1.5 mile) to the northeast.
2. Closing and partial capping of the landfill.
3. Pumping of 44 L/s (700 gal/min) of water used for utility and cooling tower operations at a new resource recovery facility located upgradient of the landfill.

These three sets of *in situ* time-series measurements, shown in Fig. 8, illustrate the dynamic changes in the spatial extent of the plume in direct response to site conditions. They also provide an estimate of the *in situ* ground-water flow rate. Results from the studies reveal that background resistivity values have gradually decreased (specific conductance of the ground water increased) by about 30% over the past eight years due to development in the area.

Correlation of resistivity survey data with the physical and chemical parameters of specific conductivity, ammonium nitrogen, sodium, and total organic carbon (TOC) of water samples collected at the site ranged from 0.756 to 0.885 at the 95% confidence level. As might be expected, specific conductivity and sodium showed the greatest correlation. A correlation of 0.851 for TOC indicated that at this site little separation of the organic and inorganic chemical parameters occurred within the contaminant plume and that inorganics were a good indicator of the distribution of organics.

*Example 3*

Example 3 consists of a flowing artesian well tapping the Floridan aquifer at Chekika Hammock State Park, Dade County, Florida. This well has been contaminating the sole-source Biscayne aquifer with brackish water since 1944. The flow rate decreased from an initial 145 L/s (2300 gal/min) to about 63 L/s (1000 gal/min) over the years. The primary contaminating chemical constituents are chloride, sodium, and sulfate ions. High-chloride-content water flows from the well onto the surface and forms several ponds and a marshy area. As this water percolates into the ground, it becomes a major source of contamination to the surficial aquifer.

Data collected from eight USGS monitor-well clusters indicated that ground-water contamination was occurring. In 1979, Technos, Inc. mapped the brackish plume by means of an electromagnetic conductivity survey consisting of 133 station measurements distributed over approximately 31 km$^2$ (12 mile$^2$) and concentrated over the major portions of the contaminant plume. The plume was easy to identify because the specific conductivity of water samples from the monitor wells was more than an order of magnitude (4500 to 5000 μmhos/cm) greater than the background values (300 to 400 μmhos/cm) of local ground water. In addition, the geohydrologic setting was relatively uniform. The survey showed that the plume extended more than 11 km (7 miles) downgradient and covered an area of more than 80 km$^2$ (31 mile$^2$) [*34*]. The survey was repeated in 1984. Results showed the plume extending more than 16 km (10 miles) downgradient and covering an area of more than 41 km$^2$ (16 mile$^2$). The extent of the plume in the 1979 and 1984 surveys is shown in Fig. 9. Correlation of the electromagnetic conductivity survey data with the primary physical and chemical parameters of specific conductivity, chloride, sulfate, and sodium ranged from 0.959 to 0.969 at the 95% confidence level.

Profiles of electromagnetic conductivity measurements, obtained both downgradient, along the axis of the plume, and nearly perpendicular to the axis of the plume, were drawn. Figure 10 shows the gradient of contaminant concentrations along the axis and along a cross section of the plume.

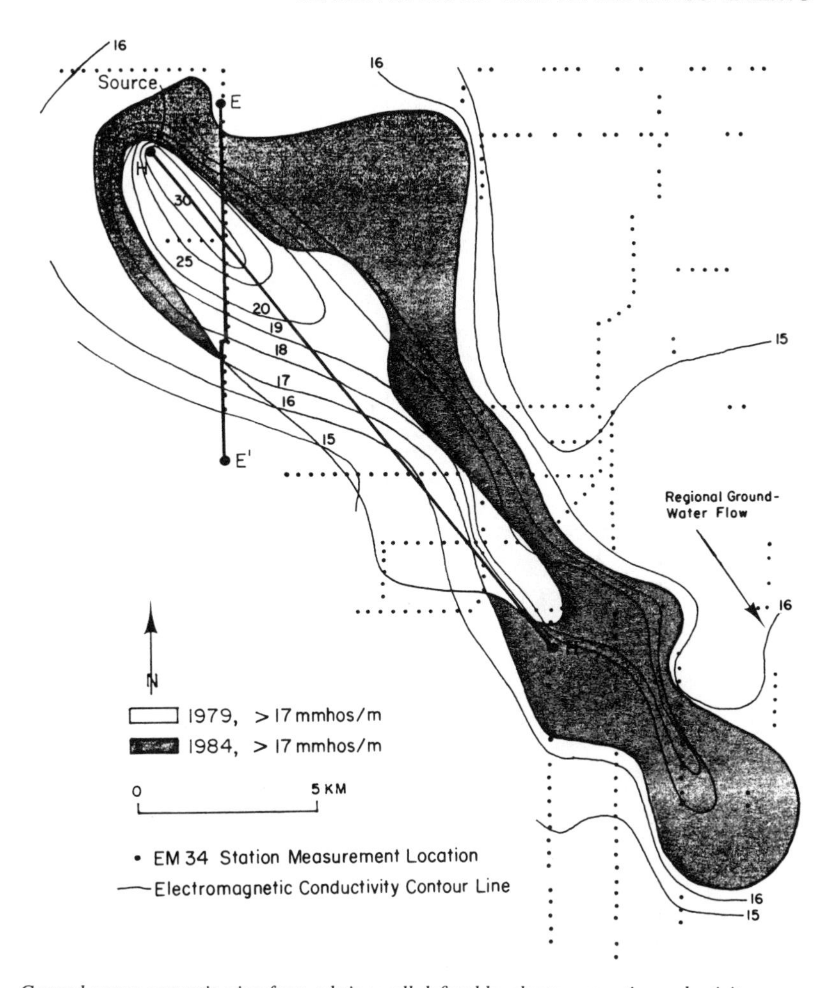


FIG. 9—*Ground-water contamination from a brine well defined by electromagnetic conductivity measurements.*

The map of spatial extent (Fig. 9) and the change of concentration along the axis and cross section (Fig. 10) are excellent indicators of the flow dynamics and mixing of the plume. In the electrical cross section E′-E of Fig. 10, changes in conductivity of 250% within 1 km (0.6 mile) in 1977 and of 300% within 1.5 km (0.9 mile) in 1981 indicate that very little mixing is occurring perpendicular to the flow direction of the plume. Armed with this information, a very meaningful and accurate ground-water flow and transport model can be constructed for predicting future conditions. Time-series measurements of electromagnetic conductivity around this plume every four years would confirm or disprove the predictions.

## Conclusion

Surface and downhole geophysical measurement methods can be an excellent means of assessing hydrogeologic conditions and mapping contaminant distribution at a site. Combined with traditional methods and implemented over regular periods of time, this approach can provide the additional information necessary to make accurate site-condition assessments and plan long-term ground-water monitoring programs. Surveys conducted at a number of sites have already shown that the

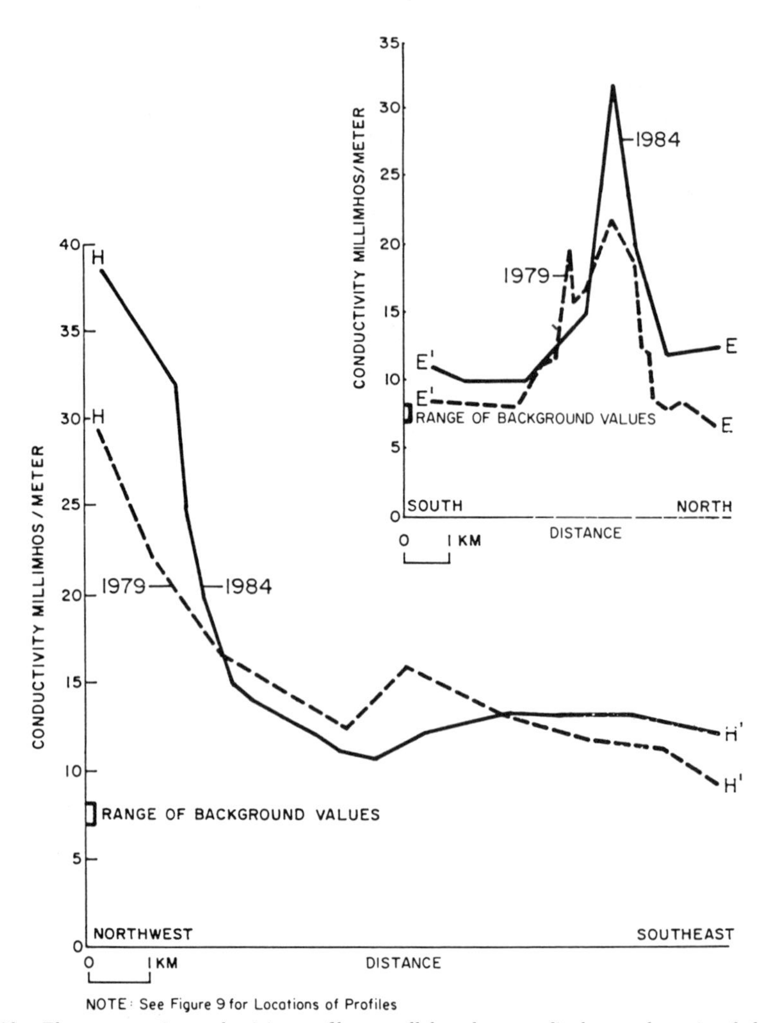


FIG. 10—*Electromagnetic conductivity profiles parallel and perpendicular to the axis of the brackish ground-water plume.*

actual dynamics of contaminant plumes, as they interact with natural and man-induced variables, can be determined using *in situ*, time-series measurements. Furthermore, when predictive modeling for risk assessment and remedial design is required, it can be done with a much higher degree of confidence. The technology is available to do a quality job. No longer is the traditional method of merely drilling a limited number of holes from which localized water samples can be taken an adequate approach.

## References

[1] Perazzo, J. A., Dorrler, R. C., and Mack, J. P., "Long-Term Confidence in Ground Water Monitoring Systems," *Ground Water Monitoring Review*, Vol. 4, No. 4, 1984, pp. 119–123.

[2] Hileman, B., "Water Quality Uncertainties," *Environmental Science and Technology*, Vol. 18, No. 4, 1984, pp. 124–126.

[3] Walker, S. E. and Allen D. C., "Background Ground Water Quality Monitoring: Temporal Variations,"

Fourth National Symposium and Exposition on Aquifer Restoration and Ground Water Monitoring, National Water Well Association, 1984.
[4] Dunbar, D., Tuchfeld, H., Siegel, R., and Sterbentz, R., "Ground Water Quality Anomalies Encountered During Well Construction, Sampling and Analysis in the Environs of a Hazardous Waste Management Facility," *Ground Water Monitoring Review*, Vol. 5, No. 3, 1985, pp. 70–74.
[5] LeGrand, H. E., "A Standardized System for Evaluating Waste-Disposal Sites," National Water Well Association, 1980.
[6] Nacht, S. J., "Ground-Water Monitoring System Considerations," *Ground Water Monitoring Review*, Vol. 3, No. 2, 1983, pp. 33–39.
[7] Nacht, S. J., "Monitoring Sample Protocol Considerations," *Ground Water Monitoring Review*, Vol. 3, No. 3, 1983, pp. 23–29.
[8] "RCRA Ground-Water Monitoring Technical Enforcement Guidance Document," United States Environmental Protection Agency, 1985.
[9] Ginzburg, A., "Resistivity Surveying: Geophysical Surveys," Vol. 1, 1974, p. 325.
[10] Griffith, D. H. and King, R. F., *Applied Geophysics for Engineers and Geologists*, Pergamon Press, New York, 1969.
[11] Orellana, E. and Mooney, H. M., *Master Tables and Curves for Vertical Electrical Sounding Over Layered Structures*, Interciencia, Madrid, Spain, 1966.
[12] Orellana, E. and Mooney, H. M., *Wenner Master Curves for Geoelectrical Sounding*, Intercencia, Madrid, Spain, 1970.
[13] Heiland, C.A., Sc. D., *Geophysical Exploration*, Hafner, New York, 1968.
[14] Mooney, H. M., *Handbook of Engineering Geophysics, Volume 2: Electrical Resistivity*, Bison Instruments, Inc., 1980
[15] Telford, W. M., Geldart, L. P., Sheriff, R. E., and Keys, D. A., *Applied Geophysics*, Cambridge University Press, U.K., 1982.
[16] Zohdy, A. A. R., Eaton, G. P., and Mabey, D. R., "Application of Surface Geophysics to Ground-Water Investigations," *Techniques of Water-Resources Investigations of the United States Geological Survey*, Chapter D1, 1974.
[17] Benson, R. C., Glaccum, R. A., and Noel, M. R., "Geophysical Techniques for Sensing Buried Wastes and Waste Migration," Environmental Protection Agency—Environmental Monitoring Systems Laboratory, Las Vegas, NV, 1983.
[18] Benson, R. C., Turner, M., Volgelson, W., and Turner, P., "Correlation Between Field Geophysical Measurements and Laboratory Water Sample Analysis," *Proceedings*, National Water Well Association/Environmental Protection Agency Conference on Surface and Borehole Geophysical Methods in Ground Water Investigations, National Water Well Association, 1985.
[19] Benson, R. C., and Pasley, D. C., "Ground Water Monitoring: A Practical Approach for a Major Utility Company," Fourth National Symposium and Exposition on Aquifer Restoration and Ground Water Monitoring, National Water Well Association, 1984.
[20] Merrick, N., "Resistivity Sounding for Groundwater," M. S. thesis, University of Sydney, New South Wales, Australia, 1977.
[21] Cartwright, K. and McComas, M., "Geophysical Surveys in the Vicinity of Sanitary Landfills in Northeastern Illinois," *Ground Water*, Vol. 6, 1968, pp. 23–30.
[22] Pasley, D., Jr., and Benson, R. C., "Salt Water Intrusion—Measurement in Developing Coastal Zone Communities Adjacent to Biscayne Bay, Dade County, Florida," presented to the 45th Annual Meeting of the Society of Exploration Geophysicists, Denver, CO, 1975.
[23] Keys, W. S. and MacCary, L. M., "Application of Borehole Geophysics to Water-Resources Investigations," *Techniques of Water-Resources Investigations of the United States Geological Survey*, Chapter E1, 1976.
[24] "Surface and Borehole Geophysical Methods in Ground Water Investigations," National Water Well Association (NWWA) Second National Conference and Exposition, Americana Hotel, Fort Worth, TX, 12–14 Feb. 1985.
[25] Greenhouse, J. P. and Monier-Williams, M., "Geophysical Monitoring of Ground Water Contamination Around Waste Disposal Sites," *Ground Water Monitoring Review*, Vol. 5, No. 4, 1985, pp. 63–69.
[26] Ringstad, C. A., and Bugenig, D. C., "Electrical Resistivity Studies to Delimit Zones of Acceptable Ground Water Quality," *Ground Water Monitoring Review*, Vol. 4, No. 4, 1984, pp. 66–69.
[27] Weber, D. D., Scholl, J. F., LaBrecque, D. J., Walther, E. G., and Evans, R. B., "Spatial Mapping of Conductive Ground Water Contamination with Electromagnetic Induction," *Ground Water Monitoring Review*, Vol. 4, No. 4, 1984, pp. 70–77.
[28] Lysyj, I., "Indicator Methods for Post-Closure Monitoring of Ground Waters" *Proceedings*, National Conference on Management of Uncontrolled Hazardous Waste Sites, Hazardous Materials Control Research Institute, 1983, pp. 446–448.

[29] Freeze, R. A. and Cherry, J. A., *Groundwater*, Prentice-Hall, Englewood Cliffs, NJ, 1979, p. 425.
[30] Zar, J. H., "Power and Statistical Significance in Impact Evaluation," *Ground Water Monitoring Review*, Vol. 2, No. 3, 1982, pp. 33–35.
[31] Keely, J. F., "Chemical Time-Series Sampling," *Ground Water Monitoring Review*, Vol. 2, No. 4, 1982, pp. 29–39.
[32] McBean, E. A. and Rovers, F. A., "Alternatives for Assessing Significance of Changes in Concentration Levels," *Ground Water Monitoring Review*, Vol. 4, No. 3, 1984, pp. 39–42.
[33] Mattraw, H. C., Hull, J. E., and Klein, H., "Groundwater Quality Near the Northwest 58th Street Solid Waste Disposal Facility, Dade County, FL," United States Department of the Interior Geological Survey, Water-Resources Investigation 78-45, 1978.
[34] Waller, B. G., "Areal Extent of a Plume of Mineralized Water from a Flowing Artesian Well in Dade County, FL," United States Geological Survey, Water-Resources Investigations 82-20, 1982.

*W. E. Kelly,[1] I. Bogardi,[1] M. Nicklin,[1] and A. Bardossy[2]*

# Combining Surface Geoelectrics and Geostatistics for Estimating the Degree and Extent of Ground-Water Pollution

**REFERENCE:** Kelly, W. E., Bogardi, I., Nicklin, M., and Bardossy, A., "**Combining Surface Geoelectrics and Geostatistics for Estimating the Degree and Extent of Ground-Water Pollution,**" *Ground-Water Contamination: Field Methods, ASTM STP 963,* A. G. Collins and A. I. Johnson, Eds., American Society for Testing and Materials, Philadelphia, 1988, pp. 73–85.

**ABSTRACT:** Geostatistical techniques are presented for estimating the areal distribution of specific conductance as an indicator of the degree and extent of ground-water pollution. Combinations of measurement types including surface geoelectrics and monitoring wells are evaluated in terms of the prediction error. A realistic example using several combinations of surface resistivity and borehole measurements of specific conductance is presented. Results of the example are analyzed and show the increase in prediction accuracy possible by combining the two measurement types.

**KEY WORDS:** ground-water pollution, geoelectrics, geostatistics, observation network, error analysis, areal mapping

The purpose of this paper is to present a methodology, which combines surface geoelectrics and geostatistics, for quantitatively evaluating ground-water pollution.

Surface geoelectrical methods are attractive because of their relatively low cost, and because they can be used to estimate ground-water specific conductance, which is a common indicator of ground-water pollution.

Ground-water pollution is one of the nation's most serious environmental problems. Efficient methods to detect and quantitatively evaluate ground-water pollution are needed.

Surface electrical measurements cannot completely replace monitoring wells, for a number of reasons. First and foremost, they do not directly measure ground-water specific conductance, but rather involve the solution of an inverse problem to determine the appropriate geoelectric-geologic model and a material level relation between aquifer resistivity and specific conductance. In applications to pollution, the first step is to determine an appropriate model and then the degree of variability of the model over the study area. Except for the simplest geologic conditions, this is difficult to do with geoelectrical measurements alone. As a consequence, pollution predictions based on surface geoelectrics alone are either not possible or would be very uncertain. The usefulness of geoelectrics depends ultimately on the extent to which they quantitatively improve the accuracy of pollution estimates and decrease the consequence of pollution [*1*].

Surface geophysical techniques can be used in ground-water pollution studies to determine soil type and depth to water table and bedrock, and to estimate the degree and extent of ground-water pollution [2]. Electrical methods respond directly to ground-water specific conductance, which can

[1] Civil Engineering Department, University of Nebraska-Lincoln, Lincoln, NB 68598-0531.
[2] Tiszadata Consulting Engineers, Kikelet u. 19, Budapest, Hungary.

frequently be directly related to the degree of ground-water pollution. As a consequence, electrical methods have been used to delineate the extent of ground-water pollution, track movement, and even to make rough estimates of the degree of contamination. Early applications of surface geoelectrics employed direct current (d-c) methods. Recently, electromagnetic methods (EM) have become popular for surveying or mapping applications since they are considerably faster and less labor-intensive than conventional resistivity surveys.

Since both the soil material and the saturating fluid control aquifer resistivity, it is convenient to characterize soil electrical properties by their formation factor defined as

$$\mathrm{FF} = \frac{\underset{\sim}{\rho_e}}{\rho_w} \tag{1}$$

where $\rho_e$ is the resistivity of the saturated soil-water mixture, and $\rho_w$ is the water resistivity. Since soil properties exhibit spatial variability (both stochastic and deterministic), the formation factor cannot be constant even under constant aquifer geometric conditions. As a consequence, Eq 1 is used to calculate ground-water resistivity (or specific conductance SC = constant/$\rho_w$) from electrically measured resistivity and statistically known formation factors in the form

$$\underset{\sim}{\rho_w} = \frac{\underset{\sim}{\rho_e}}{\underset{\sim}{\mathrm{FF}}} \tag{2}$$

where ~ means random variable.

If geologic conditions are relatively uniform, treatment of the formation factor as a random variable should be satisfactory. However, the methodology must ultimately allow for variation in the geological/geoelectrical model. Variations in the geologic model could mean variations in the number of layers, layer thicknesses, and resistivities for layers other than the aquifer layer itself. Variations in thicknesses and layer resistivities of the nonaquifer layers could mask pollution variations and at some point would make profiling alone ineffective.

Spatial parameters such as aquifer hydraulic and electric parameters generally exhibit geostatistical properties [*3–5*]; that is, they can be assumed to have a deterministic component (drift), and a spatially correlated stochastic component.

Geostatistics in its simplest form applies a first-order analysis based on the following intrinsic hypothesis [*6*]:

1. The expectation of $Z(x)$, a single realization of the parameter $z$, exists and is independent of the point $x$

$$E[Z(x)] = m \tag{3}$$

or, if the expectation of $Y(x)$ is a known function, then

$$E[Y(x)] = m(x)$$

that is, the drift is known. Then, let

$$Z(x) = Y(x) - m(x)$$

and

$$E[Z(x)] = 0 \text{ for all } x$$

2. For all vectors $h$, the increment $[Z(x+h) - Z(x)]$ has a finite variance independent of $x$

$$\mathrm{Var}[Z(x+h) - Z(x)] = 2\gamma(h) \tag{4}$$

The function $\gamma(h)$ is called the variogram.

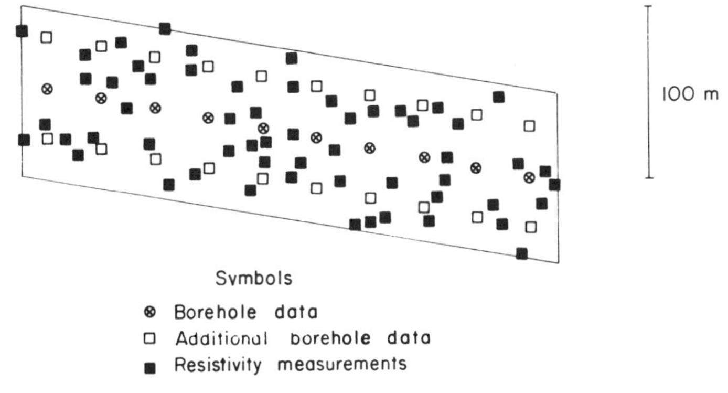


FIG. 1—*Investigated area with measurement locations.*

In the following sections the methodology is presented using a realistic numerical example based on conditions at a landfill in southern Rhode Island as described in detail by Kelly [*7*].

## Primary Analysis of Observation Data

Figure 1 shows that over the area the following field observations are available:

(*a*) vertical average specific conductance in 10 wells along the centerline of the area,

(*b*) vertical average specific conductance in 30 wells (including the above 10 wells) located in a regular network, and

(*c*) aquifer resistivity determined from profiling measurements at the 30 wells and at an additional 57 randomly located points for a total of 87 measured resistivities.

Based on the above data availability, the areal prediction and error analysis of the distribution of specific conductance will use four combinations of data:

(*a*) 87 resistivity data,

(*b*) 87 resistivity data and 10 specific conductance data,

(*c*) 30 specific conductance data, and

(*d*) 87 resistivity data and 30 specific conductance data.

Trend surfaces, quadratic in distance from the landfill, were fitted to both the 30 specific conductance data and the 87 resistivity data; this was also done in evaluating the actual field data [*7*]. Note that the selection of the trend surface attempt is not directed at a true least-squares fitting but rather attempts to define the physical characteristics of the plume.

Residuals from the trends of both specific conductance and resistivity were statistically analyzed. Figures 2 and 3 show the corresponding histograms. Formation factors were computed for the 30 borehole data, and the histogram is given in Fig. 4. In addition, prior statistics for formation factor were assessed based on the known variability of the aquifer [*8*]. Table 1 gives statistics for formation factor. Note that we are assuming that a single geological/geophysical model applies over the entire site as did Kelly in his original analysis [*7*]. The geoelectrical model is used to convert apparent resistivities, which are measured in profiling, to layer resistivities, which must be used to estimate formation factors or specific conductances.

## Structural Analysis

Structural analysis is the study of spatial correlation of the parameters in question, and here variograms are used for this purpose. Empirical variograms were fitted with spherical, polynomial, and Gaussian schemes [*6*]. Thus, it is possible to study the sensitivity of the prediction and its

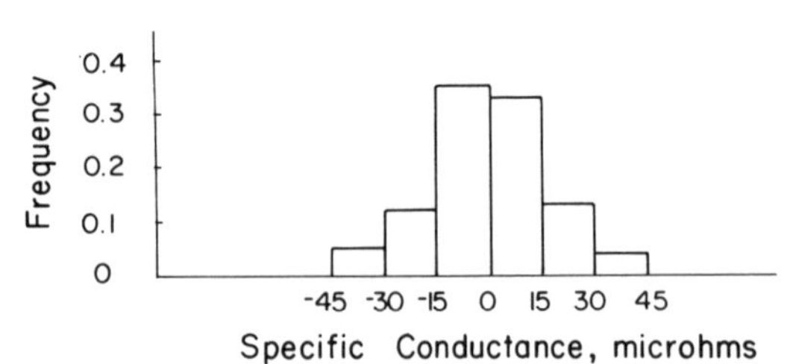


FIG. 2—*Histogram of residual specific conductance SC.*

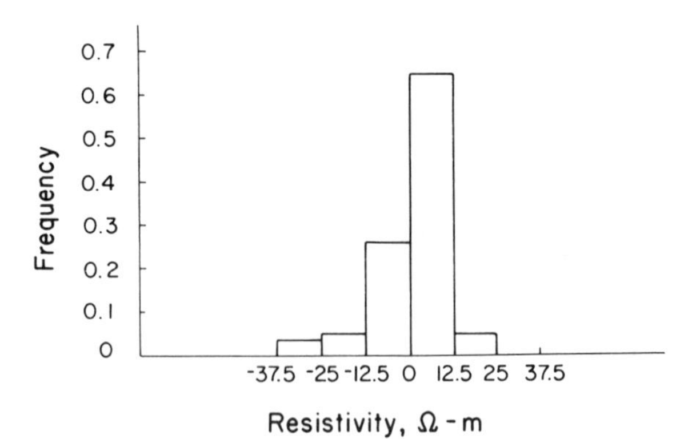


FIG. 3—*Histogram of residual resistivity RE.*

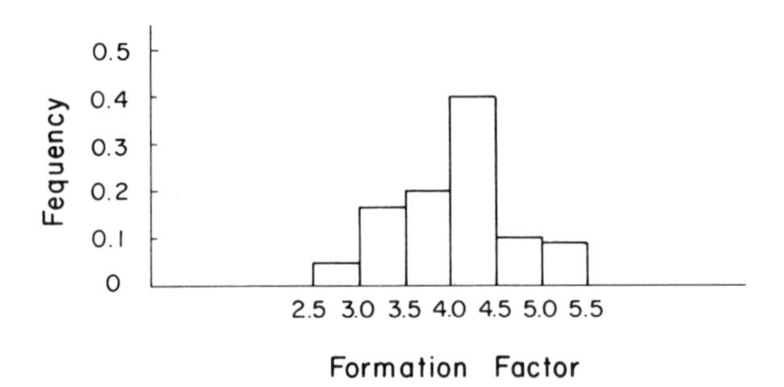


FIG. 4—*Histogram of formation factor FF.*

error in view of the three types of variogram. Figure 5 gives the variogram for resistivity from 87 data, while Fig. 6 shows the variogram for specific conductance calculated from the 87 resistivity data using the mean formation factor. The variogram for specific conductance from the 30 direct measurements is illustrated in Fig. 7. Finally, the variogram for the formation factor calculated from 30 pairs of corresponding values of resistivity and specific conductance is given in Fig. 8. The structural analysis leads to the following results:

1. Both the residual variograms of resistivity and specific conductance, and the formation factor variogram, reflect spatial dependency.
2. Numerical values for variogram parameters (range, sill, and nugget effect) are obtained.

TABLE 1—*Formation factor statistics.*

| Type of Data | Mean | Standard Deviation |
|---|---|---|
| 30 wells | 4.1 | 0.62 |
| 10 wells | 4.0 | 0.58 |
| Prior | 5.0 | 1.00 |

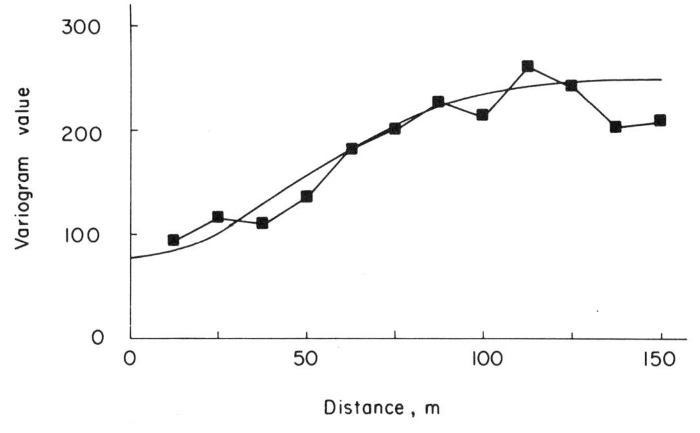


FIG. 5—*Gaussian residual variogram of RE.*

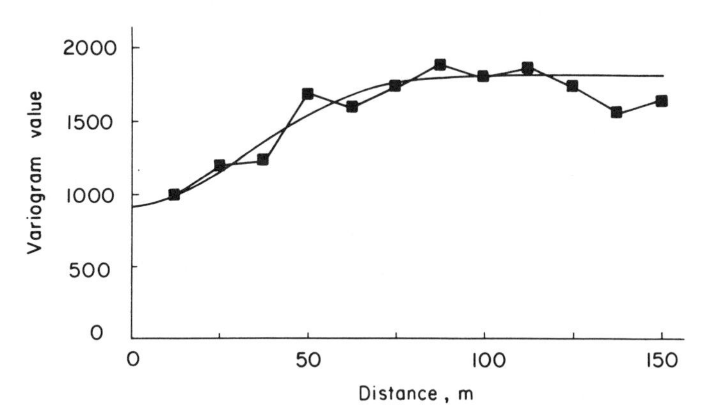


FIG. 6—*Gaussian residual variogram of SC calculated from RE.*

3. The variogram of specific conductance obtained from resistivity measurements and formation factor (Fig. 6) shows a much larger nugget and sill than the one determined from direct specific conductance measurements (Fig. 7).

## Mapping

The variograms obtained from the structural analysis are used with kriging techniques to estimate the areal distribution of specific conductance characterizing the extent and intensity of pollution.

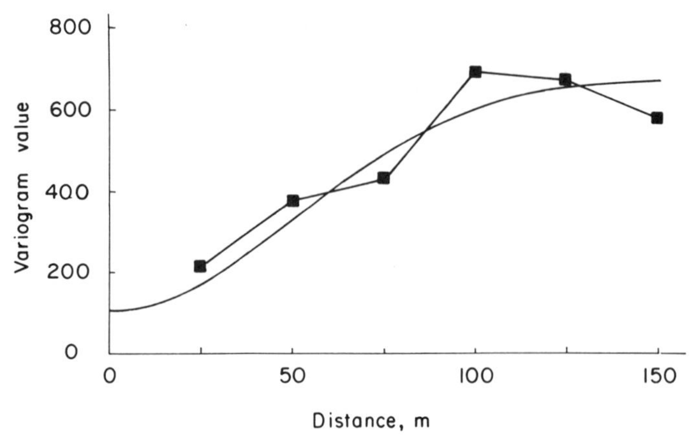


FIG. 7—*Gaussian residual variogram of directly measured SC.*

Kriging as a minimum variance interpolation technique starts with calculation of the kriging matrix $C$

$$C = [c_{ij}] = [\gamma(x_i - x_j)] \tag{5}$$

where $x_i$ and $x_j$ are observation points. Next, the variance vector $b$ to the estimated point $x$ is defined as

$$b = \gamma(x - x_i) \qquad i = 1, \ldots, n \tag{6}$$

where $n$ is the number of observation points. Now the matrix equation

$$C\lambda = b \tag{7}$$

is solved for $\lambda$, and the best linear unbiased estimator $\tilde{f}$ for point $x$ is found to be

$$\tilde{f}(x) = \sum_{i=1}^{n} \lambda_i f(x_i) \tag{8}$$

The areal distributions of kriged points were determined for the four data combinations given

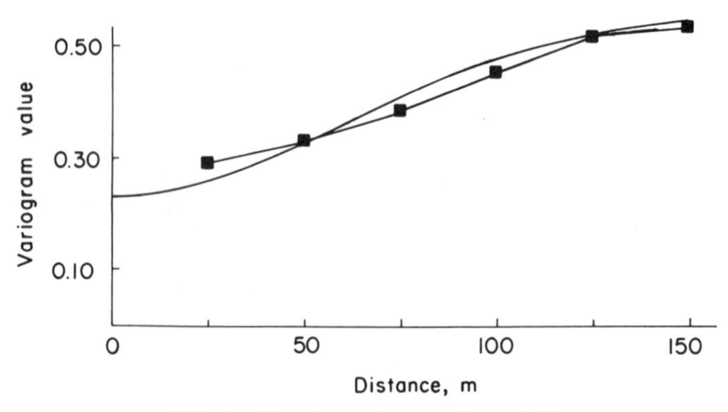


FIG. 8—*Gaussian variogram of actual FF.*

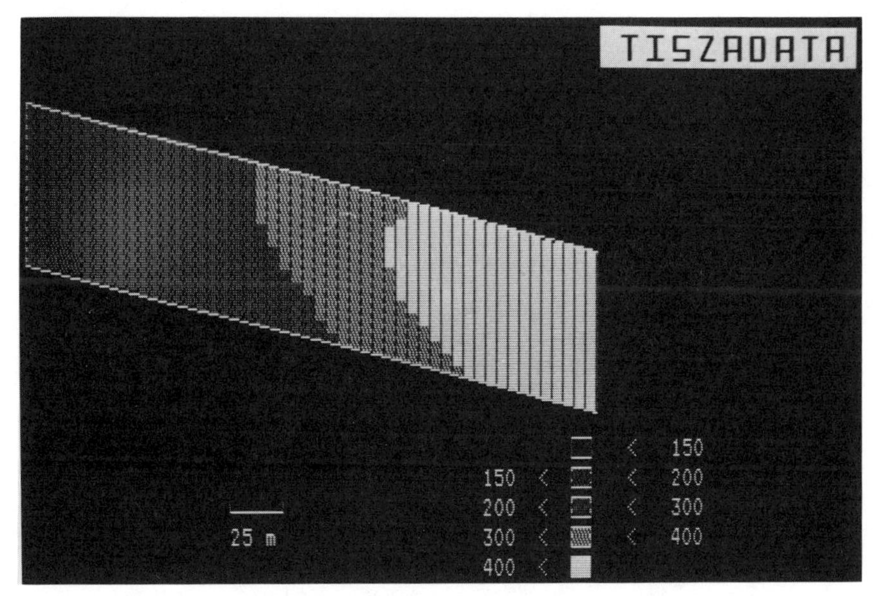


FIG. 9—*Map of SC distribution from 87 RE data and prior FF statistics.*

in the previous section. Consequently, Figs. 9 and 10 refer to Case (*a*); that is, 87 resistivity data are available and the formation factor can be characterized either by prior statistics (Fig. 9) or the actual statistics (Fig. 10). Maps for Cases (*b*), (*c*), and (*d*) are given in Figs. 11, 12, and 13, respectively. All specific conductance values are in micromhos.

Now the logical questions are: How accurate are these predictions? And can one accept the

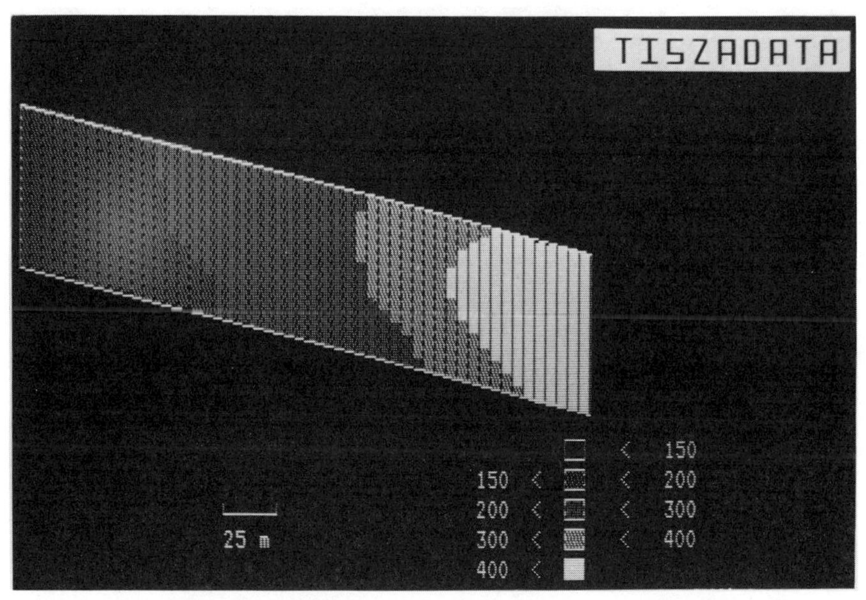


FIG. 10—*Map of SC distribution from 87 RE data and actual FF statistics.*

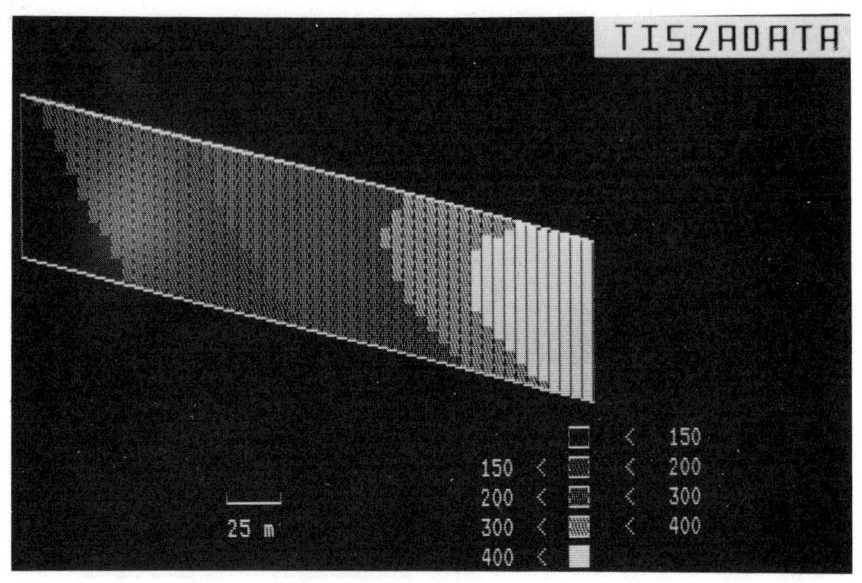


FIG. 11—*Map of SC distribution from 87 RE data and 10 direct SC data.*

results based on geophysics and the limited well measurements or are additional wells necessary? Such questions can be answered by the following error analysis.

**Error Analysis**

The estimation variance of kriging for point $x$ can be calculated as

$$\sigma^2(x) = \sum_{i=1}^{n} \lambda_i b_i = b^* C^{-1} b \tag{9}$$

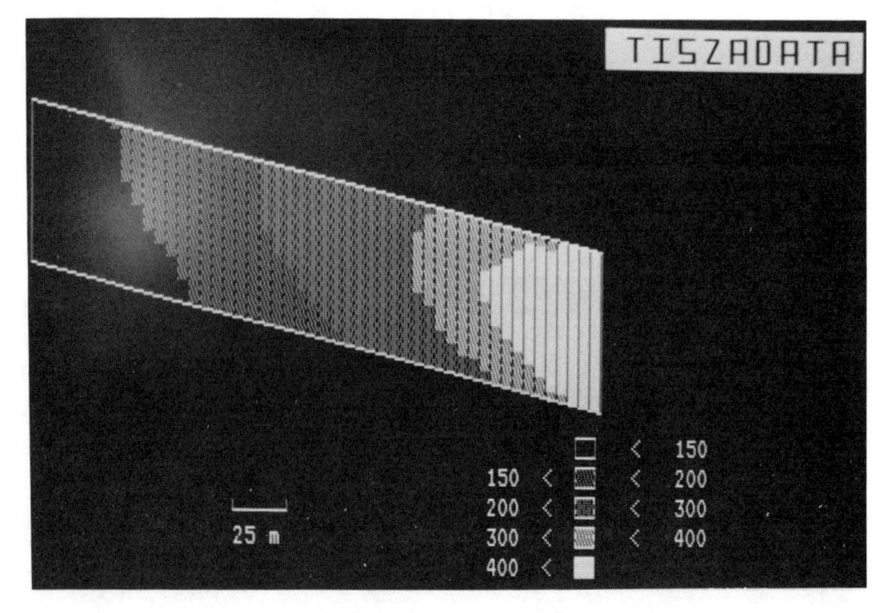


FIG. 12—*Map of SC distribution from 30 direct SC data.*

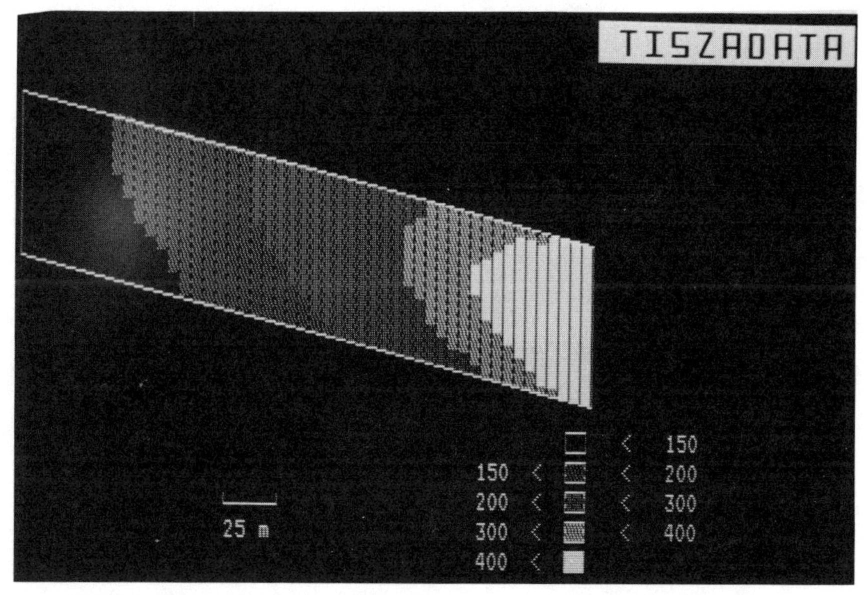


FIG. 13—*Map of SC distribution from 87 RE data and 30 direct SC data.*

The areal distribution of estimation standard deviation as calculated by Eq 9 is illustrated for the four data combinations in Figs. 14–18. Note that the numerical values pertaining to the four error categories are different in each figure.

An important part of the methodology is the combination of information stemming from different types of measurements. Cases (*b*) and (*d*) are for combined information. In Case (*b*), the following procedure can be used:

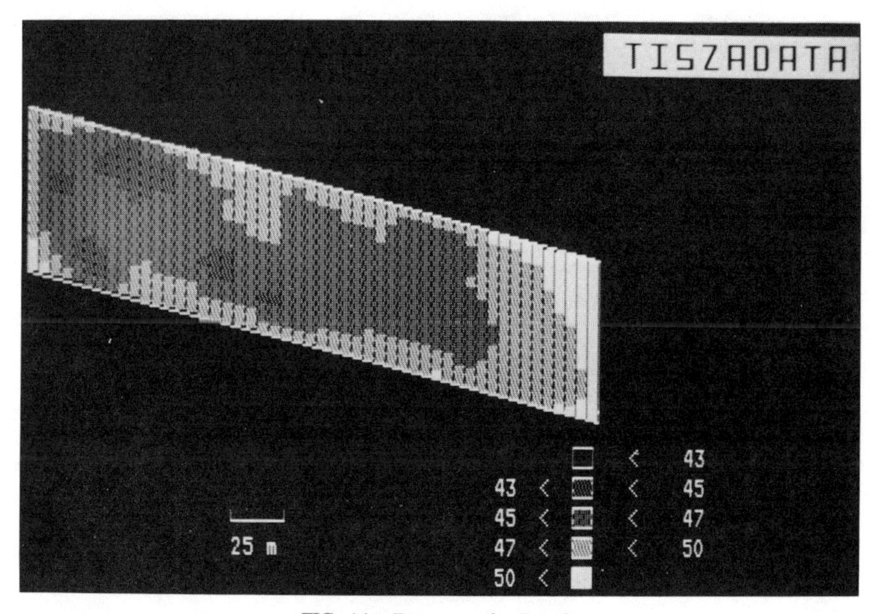


FIG. 14—*Error map for Fig. 9.*

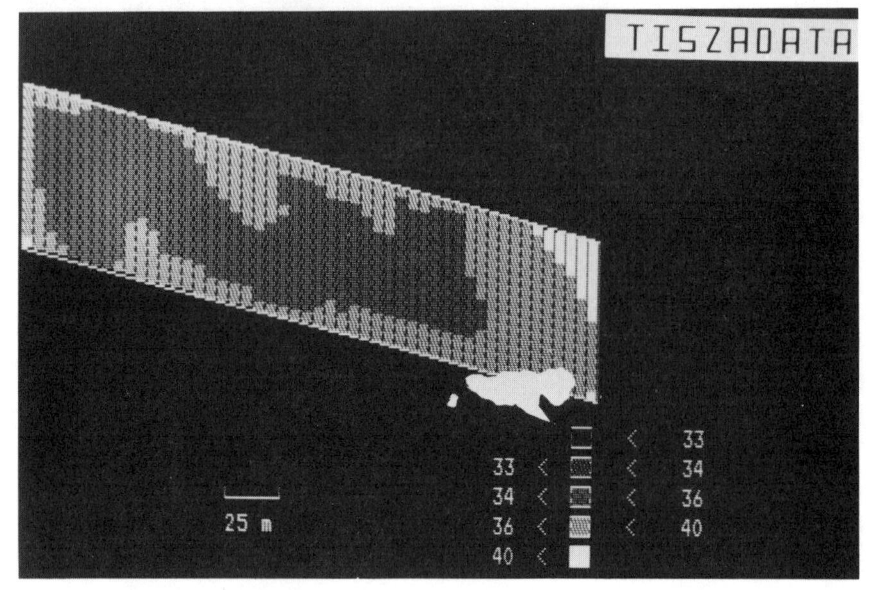


FIG. 15—*Error map for Fig. 10.*

1. Calculate 87 specific conductance values from 87 resistivity values using the mean formation factor from 10 borehole data.
2. Fit a trend surface to the 77 specific conductance values and the 10 measured specific conductance values.
3. Determine the residual specific conductance variogram from the total 87 points.

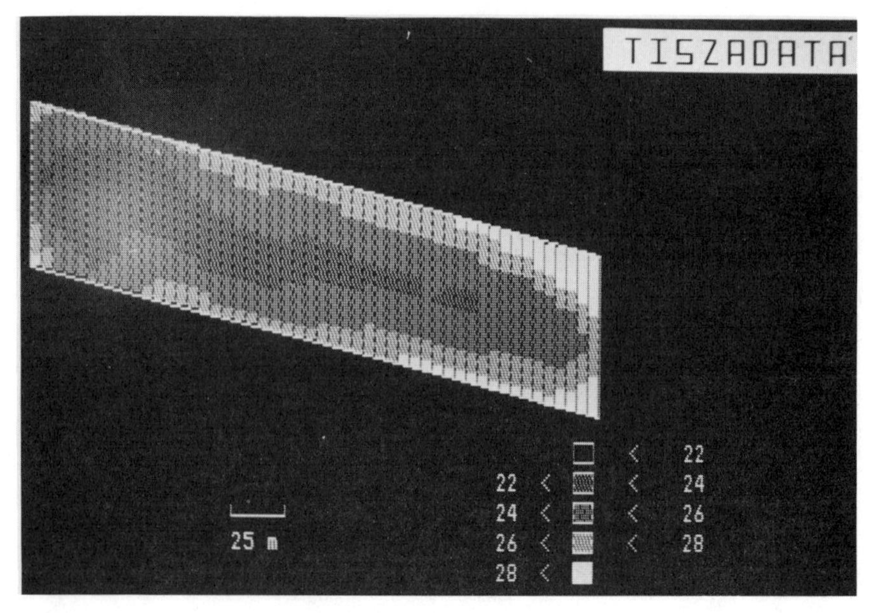


FIG. 16—*Error map for Fig. 11.*

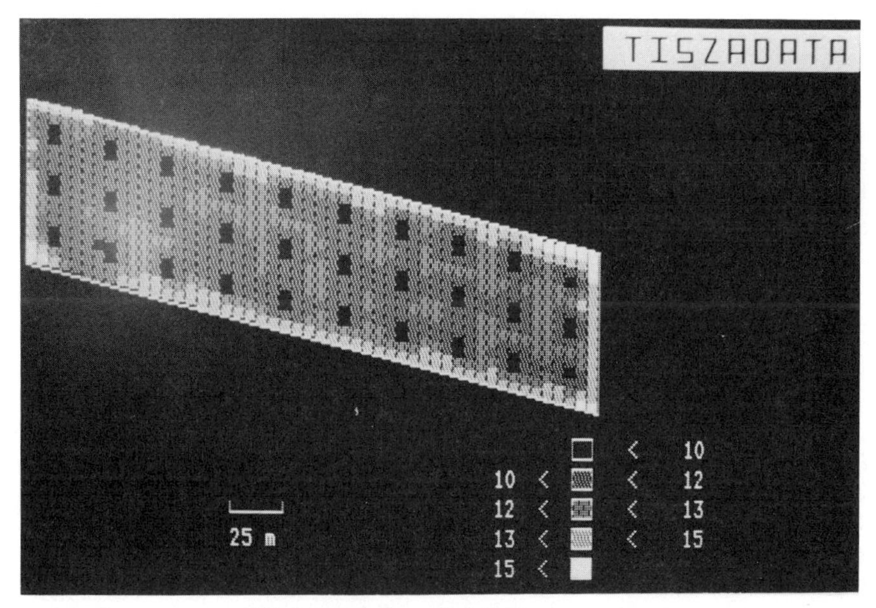


FIG. 17—*Error map for Fig. 12.*

4. Calculate standard deviations of specific conductance from the 87 calculated values (*A*), and from the 10 measured values (*B*).

5. Calculate an updated specific conductance variogram as the original variogram multiplied by *B*/*A*.

6. Perform kriging "with noisy data" [*5*], that is, zero-error in the 10 measured points and the calculated errors in the other 77 points.

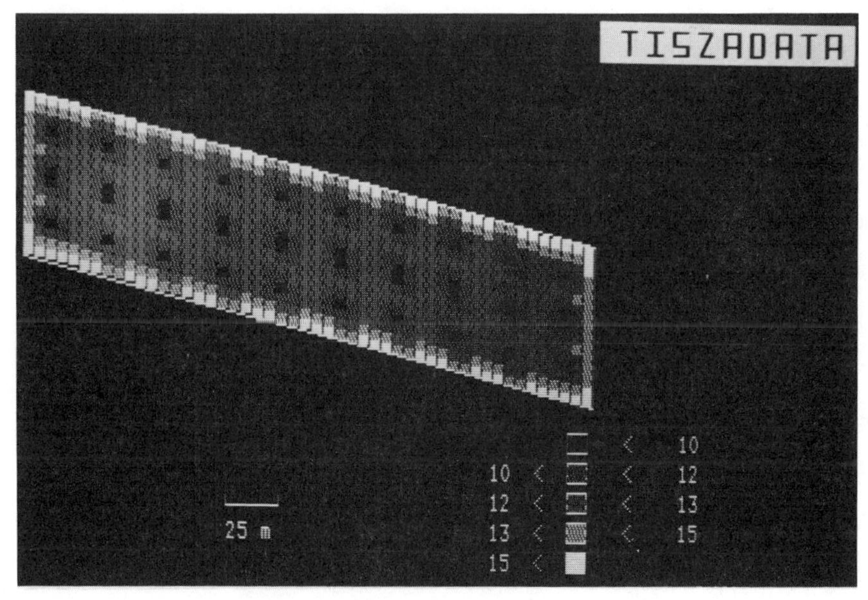


FIG. 18—*Error map for Fig. 13.*

TABLE 2—*Average estimation error of specific conductance prediction in micromhos.*

| Information Cases | Variogram | | |
|---|---|---|---|
| | Spherical | Linear | Gaussian |
| (*a*) 87 RE (prior FF) | 47.2 | . . . | 45.8 |
| 87 RE (calculated FF) | 36.3 | . . . | 35.3 |
| (*b*) 87 RE + 10 SC | 25.3 | . . . | . . . |
| (*c*) 30 SC | 12.6 | 13.4 | 12.1 |
| (*d*) 87 RE + 30 SC (statistical FF) | 12.1 | 13.1 | 11.9 |
| 87 RE + 30 SC (geostatistical FF) | 11.9 | . . . | 11.7 |

For Case (*d*) (57 resistivity and 30 specific conductance) at least two procedures—geostatistically considered formation factor and statistically considered formation factor—can be applied.

Alternative methods, such as universal kriging, intrinsic random functions of order k (IRF-k), or cokriging [*6*], could also be applied. However, the above approach appears preferable since often more is known about the properties of the trend surface than universal kriging would require. On the other hand, the nonlinear relationship between resistivity and specific conductance does not favor the use of cokriging, which is normally recommended for linear relationships [*6*]. However, cokriging could certainly be applied to combinations of measured and calculated specific conductance values. A further possibility for an overall judgment of the prediction accuracy could be the average estimation error (kriging standard deviation) over all kriged points. To this end, Table 2 gives values of average estimation error for each case and variogram type.

## Conclusions

The following conclusions derive from the results of the example presented herein:

1. All the four cases result in a similar shape for the estimated pollution; however, considerable differences appear in the actual extent of the plume.
2. There is almost no difference between the plumes estimated for Cases (*c*) and (*d*) (Figs. 12 and 13).
3. Cases (*a*) and (*b*), in decreasing order, result in a greater plume extent compared with Cases (*c*) and (*d*).
4. For the example presented, even resistivity measurements alone are sufficient for a relatively good estimation of the plume [Case (*a*), Fig. 9].
5. A better knowledge of formation factor would result in a distinct error reduction [Case (*a*)].
6. Even a relatively small number of specific conductance measurements combined with surface geoelectrical measurements lead to further error reduction [Case (*b*) in Table 2].
7. A larger number of specific conductance measurements would further decrease prediction error (Case (*c*) in Table 2).
8. Though a measurement density corresponding to Case *c* is rarely achieved in practice, additional resistivity data could still contribute to additional error reduction [Case (*d*) in Table 2].
9. The type of variogram has only a minor effect on estimation error.

## References

[*1*] Keeney, R. and Raiffa, H., *Decisions with Multiple Objectives: Preferences and Value Tradeoffs*, Wiley, New York, 1976.

[2] *Procedures Manual for Groundwater Monitoring of Solid Waste Disposal Facilities*, EPA/530/SW-616, U.S. Environmental Protection Agency, Washington, D.C., 1977.
[3] Matheron, G., *The Theory of Regionalized Variables and Its Applications,* Ecole des Mines, Fontainebleau, France, 1971.
[4] Neuman, S. P., "Role of Geostatistics in Subsurface Hydrology" in *Geostatistics for Natural Resources Characterization,* 2nd NATO Advanced Study Institute, Stanford Sierra Lodge, CA, 16–17, Sept. 1983.
[5] de Marsily, G., "Spatial Variability of Properties in Porous Media: A Stochastic Approach" in *Fundamentals of Transport Phenomena in Porous Media,* J. Bear, Ed., The Hague, The Netherlands, 1984.
[6] Journel, A. G. and Huijbregts, Ch. J., *Mining Geostatistics,* Academic Press, New York, 1978.
[7] Kelly, W. E., "Geoelectric Sounding for Estimating Aquifer Hydraulic Conductivity," *Ground Water,* Vol. 15, No. 6, Nov.–Dec. 1977, pp. 420–425.
[8] Kosinski, W. K. and Kelly, W. E. "Geoelectric Sounding for Predicting Groundwater Properties," *Ground Water,* Vol. 19, No. 2, 1981, pp. 163–171.

*A. Tamburi,*[1] *U. Roeper,*[2] *and A. Wexler*[3]

# An Application of Impedance-Computed Tomography to Subsurface Imaging of Pollution Plumes

**REFERENCE:** Tamburi, A., Roeper, U., and Wexler, A., "**An Application of Impedance-Computed Tomography to Subsurface Imaging of Pollution Plumes,**" *Ground-Water Contamination: Field Methods, ASTM STP 963,* A. G. Collins and A. I. Johnson, Eds., American Society for Testing and Materials, Philadelphia, 1987, pp. 86–100.

**ABSTRACT:** An impedance-computed tomography (ICT) algorithm and system, known as the Electroscan[4] system, for impedance imaging with surface measurements is described. The algorithm determines the current flow paths iteratively and employs sparse-matrix techniques. The measurement system uses a grid of electrodes over the ground surface and introduces controlled currents into a subset of electrodes in a prescribed sequence. Voltage measurements (using high-impedance equipment) are taken at passive electrodes only. Tests are reported for controlled saline plumes injected into a porous media flow field in a laboratory plume, and an uncontrolled leachate plume coming from a sanitary landfill located in floodplain alluvium.

**KEY WORDS:** tomography, geophysics, impedance, imaging, graphics, ground water, pollution

A summary of top-surface resistivity is given in Chapter 8 of Telford et al. [*1*]. Basic electric field theory and various electrode spreads (for example, Wenner, Schlumberger, dipole-dipole) are discussed in some detail along with practical measurement experiences. Section VI of Benson et al. [2] discusses resistivity methods within the context of problems related to buried waste. Telford et al. (page 686) express the view that the resistivity "sounding technique requires that site conditions be relatively homogeneous laterally."

In their Abstract, Benson et al. point out that "Electromagnetic and resistivity methods can help define plumes of contaminants in ground water. Resistivity and seismic techniques are useful in determining geological stratigraphy. . . . The resistivity method measures the electrical resistivity of the geohydrologic section . . . and provides a tool to evaluate contaminant plumes and locate buried wastes." Sendlein and Yazicil [*3*] state unequivocally: "It is clear that the only method that can be used for direct measurement of contaminated ground water is the electrical resistivity method."

Therefore, it appears that the resistivity method has quite a lot to recommend it for purposes of detection of contaminated ground water and buried hazardous wastes as long as these regions are basically layered and broad in extent. However, conventional resistivity methods produce results that are difficult to interpret at discontinuities in subsurface conductivity.

[1] Head, Hydrotechnical Division, E. J. Faraci and Associates Ltd., Winnipeg, Manitoba, Canada, at the time of this work. He was employed in the same position by Wardrop Engineering Inc., Winnipeg, at the time of his accidental demise on 27 February 1988.

[2] Hydrogeologist, Saskatchewan Water Corp., Regina, Saskatchewan, Canada.

[3] Professor of electrical engineering, University of Manitoba, Winnipeg, Manitoba, Canada.

[4] Electroscan is a trademark of Quantic Electroscan Inc., Winnipeg, Manitoba, Canada.

This paper describes an impedance-computed tomography (ICT) algorithm and system, called the Electroscan system, for impedance imaging or inversion. Examples of the application of ICT to ground-water pollution plumes are presented.

## The ICT Algorithm

This algorithm was presented at an Optical Society of America topical meeting, Industrial Applications of Computed Tomography and NMR Imaging, at Hecla Island, Manitoba, on August 13–14, 1984, and is published in the proceedings of that meeting [*4*]. The main principles are outlined in the following are fully described in several patents [*5*].

In many respects, when applied to subterranean imaging through top-surface measurements, the ICT measurement procedure is a generalization of the various electrode spreads commonly used. While a 16 by 16 array is being developed, electrodes are introduced into the ground for this work over an 8 by 8 orthogonal grid with variable interelectrode spacing. One measurement set (described as an excitation) is obtained by using any pair of electrodes as current electrodes and a selection of the remaining ones as potential measurement electrodes. Clearly, this measurement set includes the Wenner, Schlumberger, and dipole-dipole spread configurations as special cases.

Because a unique interpretation is not possible with the results of a single excitation, a number of other linearly independent measurement sets are taken using various current electrode-pair configurations.

In theory, a gradient optimization scheme could be used to adjust an assumed subterranean conductivity distribution (described by finite-difference or finite-element nodes, probably) in order to minimize the difference between the calculated and the measured voltages over the surface. Dines and Lytle [*6*] describe a scheme of this nature. This, however, produces dense matrices of order corresponding to the number of nodes employed. For problems with more than a few dozen nodes, this optimization procedure becomes impossibly lengthy. Clearly, fine definition cannot be obtained in this way.

The following algorithm has the advantage that it employs only sparse matrices and, consequently, has the capability of yielding fine definition.

The algorithm employs the Poisson equation

$$-\nabla \cdot \kappa \nabla \phi = f \tag{1}$$

where

$\kappa$ = conductivity,
$\phi$ = electrical potential, and
$f$ = any impressed current source distribution within the region.

Equation 1 holds for d-c and for low-frequency excitations for which reactive effects are negligible.

Figure 1 depicts a portion of a subterranean region and the top surface. The cube is divided into finite elements. One particular excitation pair, through which current I passes, is shown at points $a$ and $b$. Node $r$ is taken as the reference potential.

At the outset, a conductivity distribution $\kappa(\bar{r})$ must be assumed. It is most convenient to take $\kappa(\bar{r})$ as a constant value throughout initially. Then the current distribution pattern, for each excitation, is estimated by solving Eq 1 subject to the Neumann boundary condition

$$\kappa(s) \frac{\partial \phi}{\partial n} \bigg|_s = h(s) \tag{2}$$

At active electrode sites, $h(s) \neq 0$. This states that a current density $h(s)$ crosses the surface.

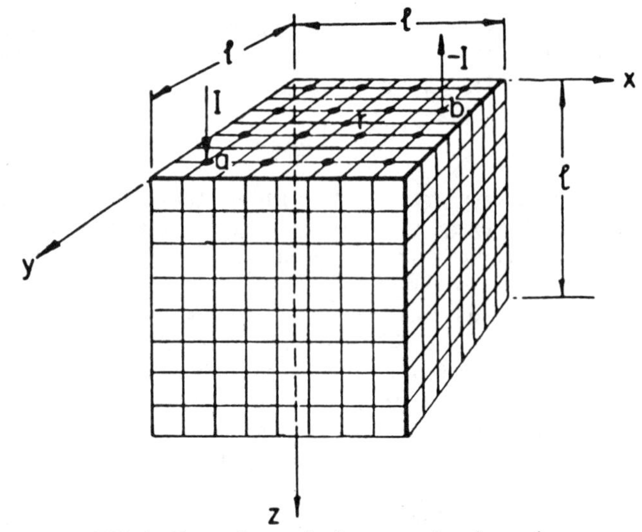


FIG. 1—*Top surface and subterranean imaging region.*

Therefore, the surface integral of $h(s)$ in the region of an electrode equals the current I. Over the remainder of the top surface, $h(s) = 0$. This states the fact that no current crosses the top surface (other than at the active-electrode sites). As far as the remaining five walls are concerned, the assumption is that negligible current crosses them. This is valid if the active electrodes are reasonably remote from those walls.

Using the finite-element method (FEM) or another appropriate numerical procedure, Eq 1 is transformed to the discrete form

$$S\underline{\phi} = \underline{b} \tag{3}$$

where

$S$ = large sparse matrix,
$\underline{b}$ = column matrix resulting from the sources, and
$\underline{\phi}$ = column matrix containing the voltages at node points $\phi_i$ throughout the cube.

Equation 3 is solved for each excitation with the other number of excitations dependent on the array size. Other than for the reference potential, this field solution uses only Neumann boundary conditions.

Calculated top-surface potentials, for all excitations, are compared with measured values at all surface electrodes. If the agreement is acceptable, then one may assume that the originally assumed κ-distribution is correct. The computational sequence then passes to the final image processing and output preparation stage (Fig. 2). If agreement is unacceptable, it is necessary to improve the κ-estimate.

Because

$$\overline{E} = -\nabla\phi \tag{4}$$

gives the electric field intensity $\bar{E}$, and as an approximation to κ is available, the current flux density is

$$\begin{aligned}\overline{J} &= \kappa\overline{E} \\ &= -\kappa\nabla\phi\end{aligned} \tag{5}$$

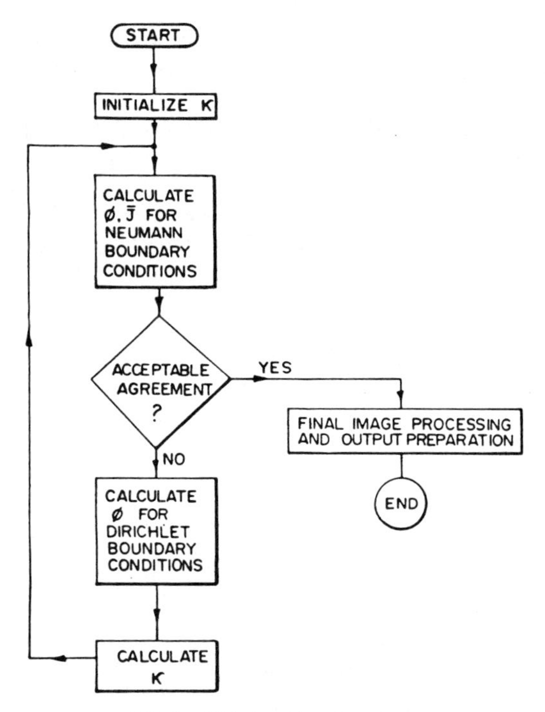


FIG. 2—*ICT algorithm flow chart.*

as a first approximation. This is Ohm's law in point form. Clearly, whatever realistic (for example, positive and finite) approximation to $\kappa(\bar{r})$ is used, the general current flux density distribution will turn out to be reasonable.

Using FEM, the electric potential distribution is expressed in terms of element node-point potentials $\phi_i$ and shape functions $\alpha_i(\bar{r})$ by

$$\phi(\bar{r}) = \sum_i \phi_i \alpha_i(\bar{r}) \tag{6}$$

within each element. Thus

$$\nabla\phi = \sum_i \phi_i \nabla\alpha_i(\bar{r}) \tag{7}$$

and so $\bar{J}(\bar{r})$, in Eq 5, may be computed throughout the region (see, for example, Ref 7).

In summary, an estimate to $\bar{J}(\bar{r})$ is the outcome of this first step of the iterative procedure.

The second step of the algorithm uses the measured potentials in order to influence a subsequent estimate of $\kappa(\bar{r})$. This is done, in the first instance, by applying the measured potentials (that is, Dirichlet boundary conditions) to the appropriate locations in the computer model for each excitation. An influence is then felt throughout the region in terms of a perturbation to $\phi(\bar{r})$. The Dirichlet boundary condition

$$\phi(s) = g(s) \tag{8}$$

is applied at voltage-measurement sites, thus replacing the previous Neumann boundary conditions. Neumann boundary conditions remain unchanged at all other locations.

Notice that with this scheme there is no need to measure voltages at active current sites. As a practical consideration this is most important as such measurements are unreliable due to contact and spreading resistances.

With the inclusion of Dirichlet (that is, voltage) boundary conditions, Eq 3 is suitably amended and solved for each excitation.

The third step involves the calculation of a new $\kappa(\bar{r})$ estimate such that the measured potentials are more closely approximated by the computed surface potentials. This is done indirectly by adjusting $\kappa(\bar{r})$ in order to better satisfy Ohm's law (Eq 5) throughout the region, for all excitations, by minimization of

$$R = \sum_X \iiint_V (J + \kappa\nabla\phi)\cdot(\bar{J} + \kappa\nabla\phi)\, dv \tag{9}$$

where

$R$ = squared residual sum,
$V$ = imaging region (that is a cube in this case), and
$X$ = excitations.

Expanding Eq 9

$$R = \sum_X \sum_j \iiint_{Vj} (\bar{J}\cdot\bar{J} + 2\kappa_j\bar{J}\cdot\nabla\phi + \kappa^2_j\nabla\phi\cdot\nabla\phi)\, dv \tag{10}$$

where the volume integration has been broken into the sum of volume integrals over individual elements $j$. The terms $\phi$ and $\bar{J}$ are held constant, at this stage of the iteration. Minimization of the residual then yields

$$\frac{\partial R}{\partial \kappa_i} = \sum_X \iiint_{Vi} (2\bar{J}\cdot\nabla\phi + 2\kappa_i\nabla\phi\cdot\nabla\phi)\, dv = 0 \tag{11}$$

where (for simplicity) it is assumed that $\kappa(\bar{r})$ is constant over each element.

Rearranging Eq 11

$$\kappa_i = \frac{-\sum_X \iiint_{Vi} J\cdot\nabla\phi\, dv}{\sum_X \iiint_{Vi} \nabla\phi\cdot\nabla\phi\, dv} \tag{12}$$

which yields a revised estimate (that is, a new iteration) of the conductivity distribution in a least-squares sense.

Note that Eq 12 does not involve matrix operations at all. If the conductivity is assumed to vary within each element, then Eq 12 involves only very sparse matrix operations.

With the new conductivity estimate obtained from Eq 12, the Neumann problem is again solved and the calculated potentials are compared with measured values. At this stage it is decided whether or not to continue the iterative process.

## Examples

While further refinement of the algorithm driving the ICT or Electroscan system is expected and automatic data acquisition is presently feasible only at laboratory scale, it was felt that a preliminary assessment of the method's ability to image ground-water pollution plumes could still be undertaken.

The laboratory assessment was undertaken in a porous media test-bed consisting of sand with a D50 of 0.7 mm in a layer 0.4-m thick, 15-m long, and 2.5-m wide. A background flow field was provided by a constant-head tank, and the elevation of the phreatic surface was controlled with a tilting weir at the downstream end of the facility. The test-bed was retained at its upstream and downstream ends by a geotextile. The phreatic profile was monitored using 25-mm-diameter point gage wells along the edge of the facility. The pollutant consisted of either a 3% or 5% solution of sodium chloride. The pollutant was injected in both slug modes or continuously by a small positive-displacement pump. For both cases the pollutant was injected for a surface test case and a subsurface test case. The facility plan is shown in Fig. 3.

Both in the laboratory study and the subsequent field study an 8-by-8 electrode matrix provided 64 measurement or excitation nodes. Data were collected only on the nodes not used for excitation. Excitation was provided by a Bison 2350 while the potentials were measured by a Keithley multimeter. Fourteen excitations were used to construct each set of images, with the excitation electrodes being varied over the grid. As data acquisition is not yet automatic, four to six hours are required to collect the data for an image set. Multiplexing may be random, but the spacing of the excitation electrodes must be varied to obtain a reasonably constant information density from each level.

## Laboratory Results

The laboratory results were encouraging in that both the Electroscan and conventional Wenner methods generally identified the plume. For four cases, including surface slug, surface continuous, subsurface slug, and subsurface continuous, the Wenner results were good in two cases, fair in one case, and poor in the fourth. Only in the poor case was the plume not imaged. In contrast, the Electroscan results were good in three cases and fair in one.

In the discussion to follow, "a" spacing refers to the electrode-to-electrode distance of the Wenner array with the depth of penetration (image depth) *approximately* equal to "a." The Electroscan method generates a level-to-level increment of depth equal to the electrode spacing, with the number of levels restricted to two-thirds $N$ where $N$ is the number of elements in an array row or column. As Level 1 is the ground surface, these results are not usually reproduced. The Electroscan results occasionally generated minor symmetrical structures along the lateral boundary. It was thought that this might relate to the assumption of no current passage beyond the boundary. As a consequence, a fifth nonsymmetrical case was tested. The Wenner and Electroscan results for this case are presented as Figs. 4–7. The Wenner results equivalent to Electroscan Level 2 nicely identified a slug which had been injected at the southwest corner of the array (Fig. 3). The Electroscan results are significantly superior as less information is lost with depth. For example, at a depth of one "a" spacing, 25 points were available to contour Wenner results whereas 49 points were available to contour Electroscan results for the same number of electrodes (Fig. 5). At a depth of two "a" spacings, only four Wenner results could be obtained for the array, whereas the Electroscan results still provided 49 contour points (Fig. 6). At a depth of three "a" spacings, the Electroscan technique still provided 49 points for contouring, whereas no points for the Wenner technique were available (Fig. 7).

The Electroscan results demonstrate that the denser saline plume had an asymmetric radial spread powered both by density differences and the original ground-water mound created during

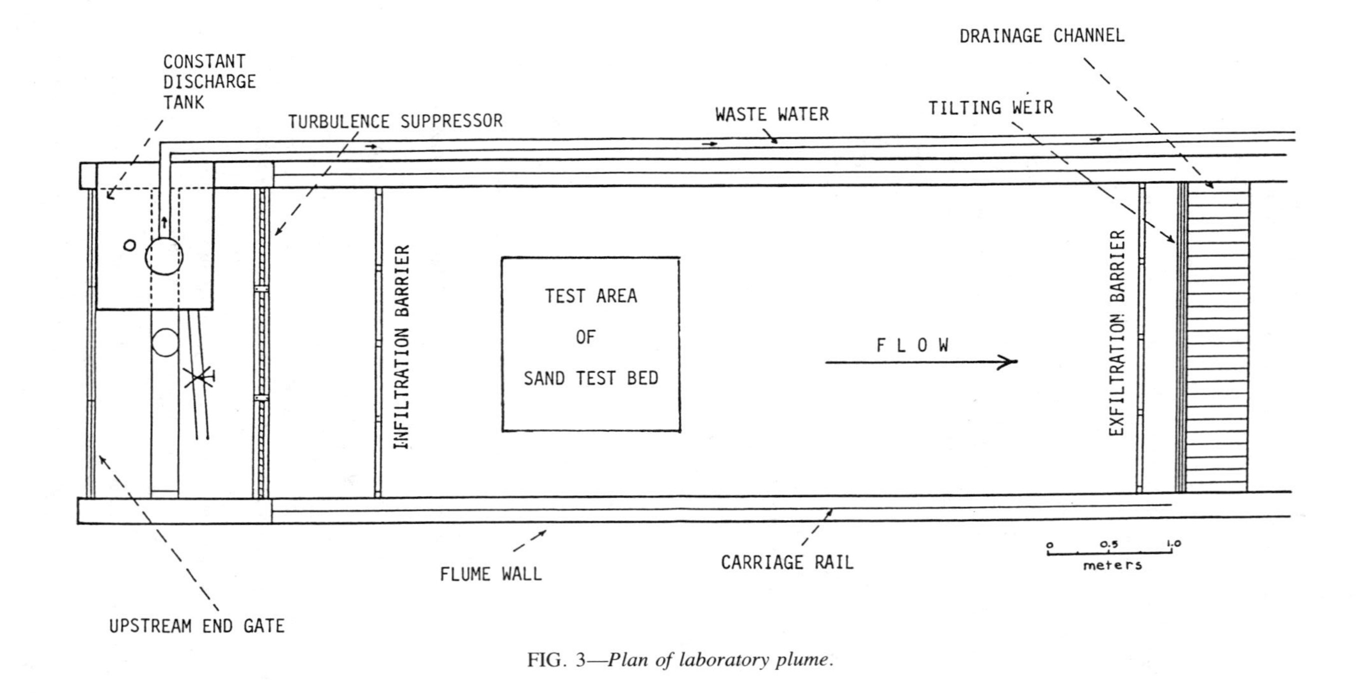


FIG. 3—*Plan of laboratory plume.*

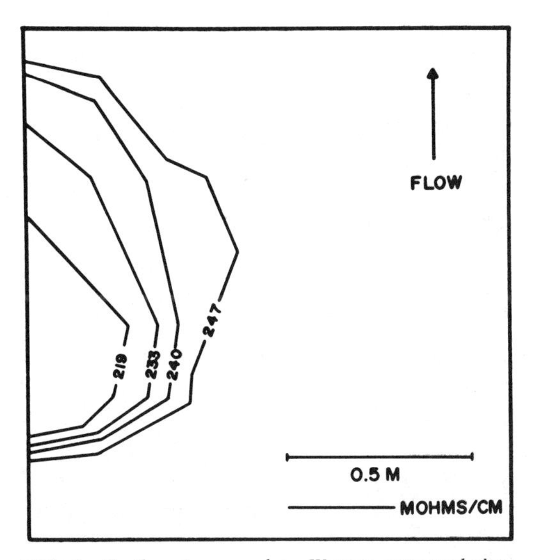


FIG. 4—*Southwest corner slug—Wenner contoured plot.*

injection. Level 3, which approached the concrete floor of the laboratory, showed significantly greater spread than would be indicated by the trend of Levels 1 and 2. This is attributed to the density plume encountering the floor of the laboratory and spreading radially. An asymmetry at Depths 2 and 3 is observed (Figs. 6 and 7). The test-bed material was placed by wheelbarrow and a potential for local size segregation existed as coarser particles rolled off adjacent piles. It is suspected that the northeast protrusion is a reflection of flow through a segregated coarse phase with higher permeability.

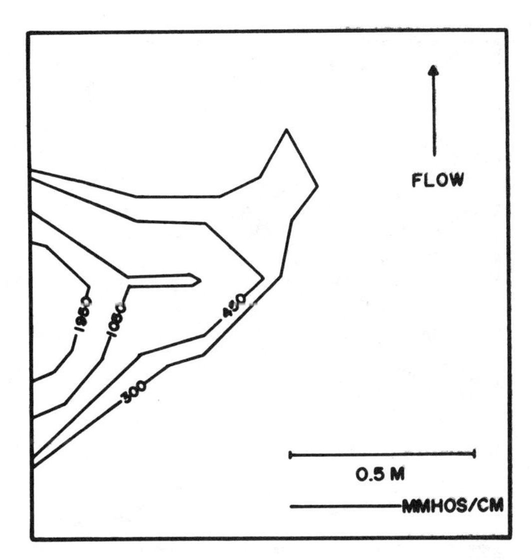


FIG. 5—*Southwest corner slug—Electroscan contoured plot—Level 2.*

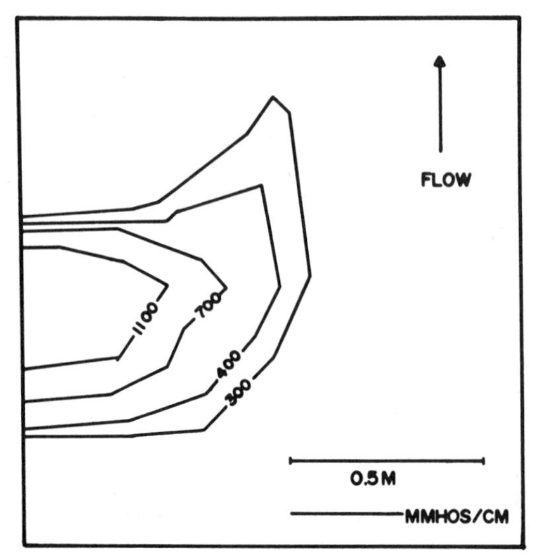


FIG. 6—*Southwest corner slug—Electroscan contoured plot—Level 3.*

## Field Results

Results from a field site near Portage la Prairie, Manitoba were also analyzed. The site was located in the floodplain of the Assiniboine River on the Portage la Prairie alluvial fan. The fluvial stratigraphy at the site is complex, both due to a geologically recent shift in the meander belt and progressive deposition in a meander which was slowly cut off to form an oxbow lake.

The postwar landfill was established within the oxbow lake while it still contained standing water. Sedimentation continued after establishment of the dump until the channel south of the dump was completely infilled.

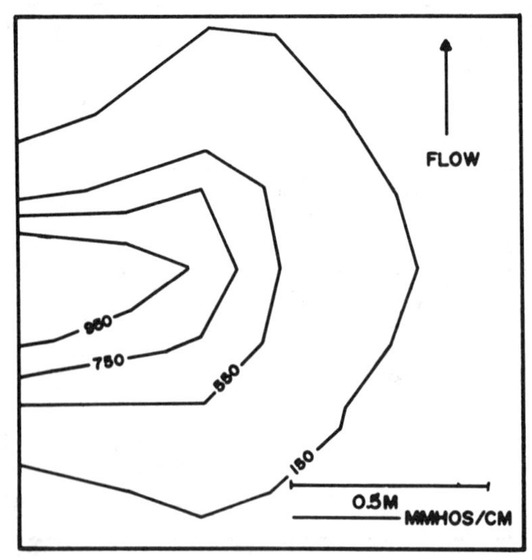


FIG. 7—*Southwest corner slug—Electroscan contoured plot—Level 4.*

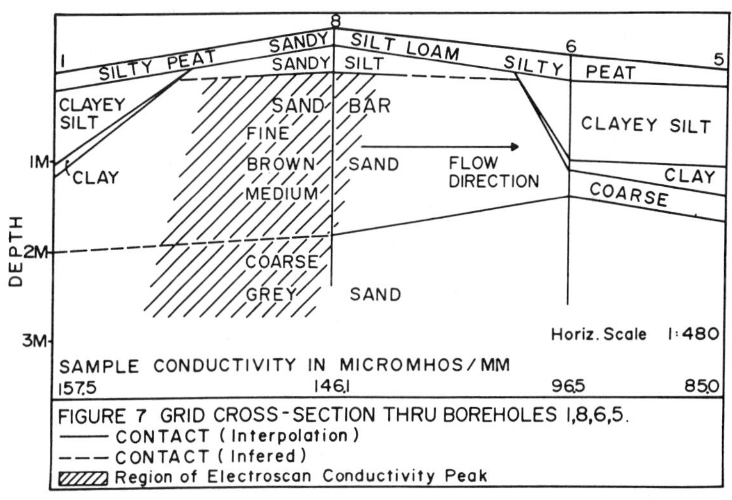


FIG. 8—*Grid cross section through Boreholes 1, 8, 6, and 5.*

The regional flow field at the site is from north to south, extending from the dump toward the Assiniboine River. Consequently it was anticipated that the contaminant plume would have its greatest extension on the south side of the dump. In order to locate an appropriate imaging site, conventional soundings and profiles were run in the infilled channel immediately south of the dump. The conventional profile revealed that substantially higher conductivity existed within 130 m of the dump than farther south. A steep conductivity gradient existed 100 to 120 m south of the dump and it was decided to attempt to image this steep gradient three dimensionally.

The 64-node grid which was established at the site also contained a fluvial sandbar deposited during the late stages of fluvial infilling. The sanbar had a surface expression in the southwest quadrant of the array (Fig. 8). However, the subsurface expression of the sandbar, as revealed in boreholes, extended virtually to the edge of the landfill (Fig. 8). Consequently, the site provided two significant impedance structures suitable for imaging within the 64 electrode array.

The electrode spacing established for the array was approximately 12 m based on conventional Wenner sounding results.

The Electroscan technique was used to determine the distribution of subsurface conductivities required to generate the observed pattern of voltages at the surface. The conductivity distribution was then contoured for each level, using the sterile aqueous suspension (SAS) contouring package (Fig. 9). As seen in Figs. 8 and 9, the Electroscan technique generated a local conductivity maximum over the northern half of the sandbar. Figure 8 shows the Electroscan results for Level 2. The sandbar in this region does not have a significant surface expression but is known to extend to the vicinity of the landfill as a result of the borehole investigation. The southern half of the bar appears to contain relatively clean water and both the Electroscan and conventional Wenner techniques revealed low conductivities in this region (Figs. 9 and 10).

It is felt that the Electroscan results are believable in this region as Boreholes 1 and 8 were sampled for conductivity and revealed very high conductivities of 1575 and 1461 μs/cm in a depth zone corresponding to Level 2. In contrast, the Wenner technique did not successfully identify the high-conductivity region due to its limited extent. While the Wenner technique requires a length of four "a" spacings to obtain a reading, the Electroscan technique, in contrast, established conductivity on an "a"-spacing-by-"a"-spacing basis. Consequently, for the same grid spacing, measurement density with the Electroscan technique is four times that obtained with the conventional

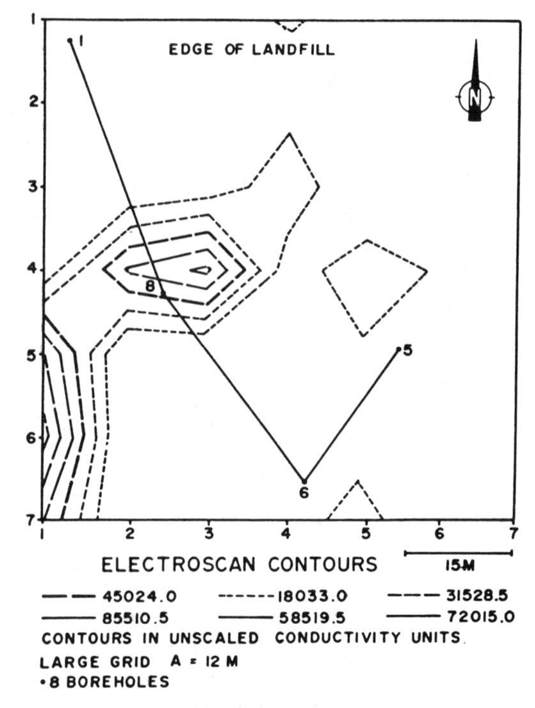


FIG. 9—*Electroscan contours.*

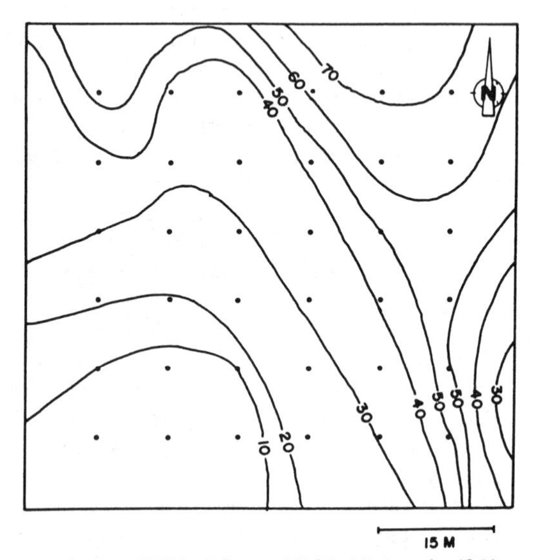


FIG. 10—*Wenner contours.*

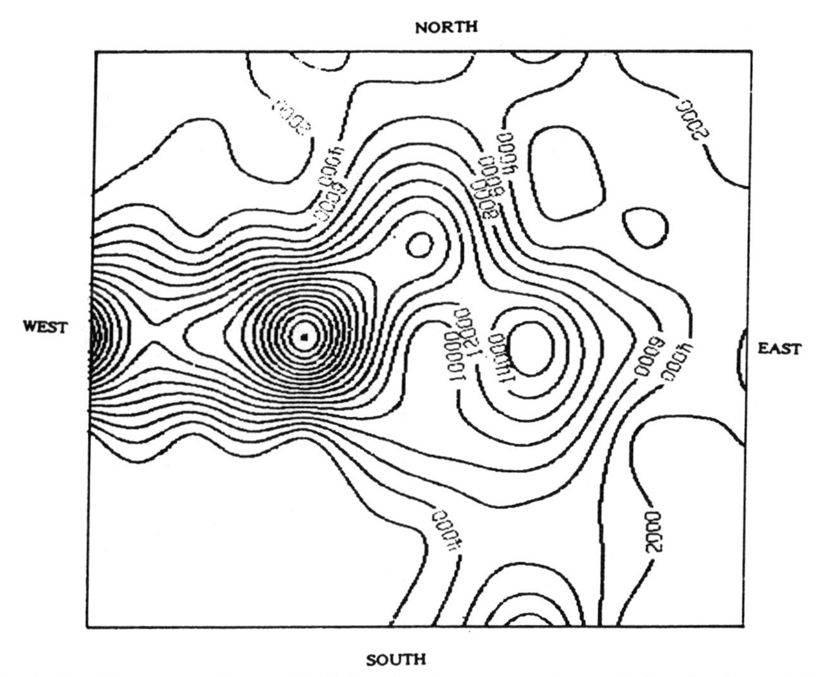


FIG. 11—*Level 2 map view, large grid, 80 iterations (contours are in unscaled conductivity units).*

Wenner technique. And so, one should expect improved image quality with the Electroscan method.

In essence it is believed that the Wenner technique did not image this local conductivity maximum because of its limited extent. The Wenner technique therefore averaged the local maximum out and merely generated the regional conductivity slope and perhaps the inflection points on the 300-μs/cm contour (Fig. 10).

The second significant impedance structure within the grid was the high-conductivity region adjacent to the landfill itself (Fig. 10). In this case, due to the extent of the high-conductivity region, the Wenner technique successfully imaged the structure. In contrast, the Electroscan technique apparently did not image the structure using SAS graph for plotting. The Level 2 results were then recontoured using PLOT 88[5] contouring software (Fig. 11). The results were substantially more interpretable, but did not eliminate the discrepancy in the northerly contours. The Level 1 Electroscan results were also assessed (Fig. 12), and showed a high-conductivity region along the north edge of the plot, thereby reducing the significance of the discrepancy. Wenner results are averaged over three horizontal "a" spacings, and lateral shifting of peaks is therefore a possibility. In contrast, the Electroscan method cannot generate a lateral shift as the data are calculated on an "a"-spacing basis.

Alternately, it is known that the Electroscan technique currently suffers from edge effects due to the assumption of no current flowing beyond the edge of the grid system. This assumption is particularly violated when a high-conductivity region with steep conductivity gradients, such as an adjacent landfill, lies just outside the grid boundaries.

Subsurface conditions, both for the field site and in the laboratory test-bed, were verified with conductivity samples. These samples verified the comparative results obtained for the Wenner and

[5] Trademark of Plotworks Inc., La Jolla, CA.

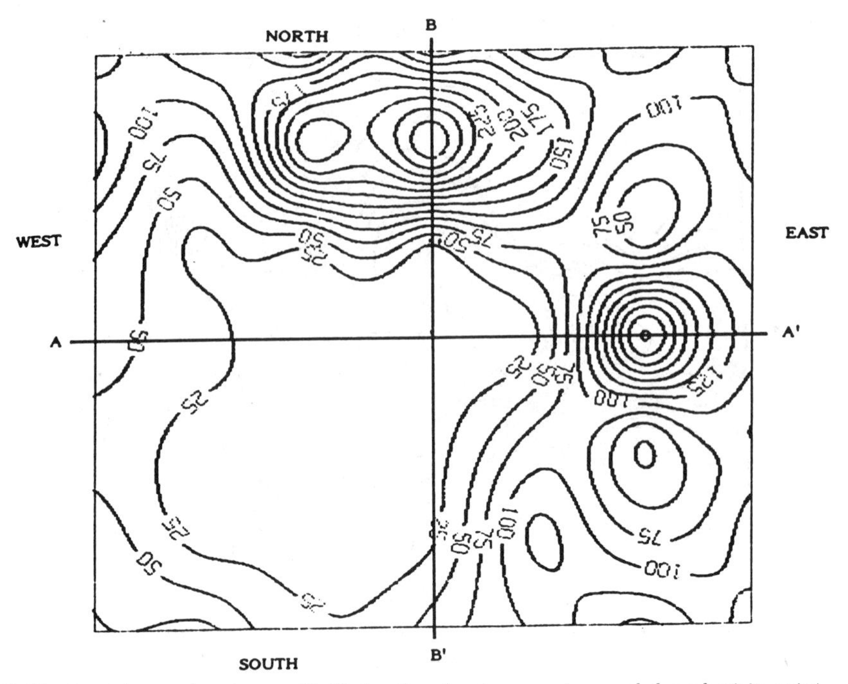


FIG. 12—*Level 1 map view, large grid, 80 iterations (contours are in unscaled conductivity units).*

Electroscan techniques. More detail was also obtained in the field for more complex field structures using a backhoe which excavated trenches on the structures, typically to depths of 2 m. A stainless steel sand point was then driven from the base of each backhoe trench, and water samples were collected for conductivity from total depths as great as 6 m.

## Conclusions

As a result of the laboratory and field studies reported herein, it is apparent that the Electroscan algorithm as presently constituted shows great promise as an imaging tool for ground-water studies.

The Electroscan technique is clearly a new geophysical tool rather than a mere rearrangement of the traditional resistivity electrode techniques. It is the first truly three-dimensional impedance geophysical tool. While layer contours can be generated for both the Electroscan and Wenner techniques, it must be remembered that contours generated from Wenner data are generated on independent data sets. That is, the bulk resistivity or conductivity generated by a Wenner array in one region is largely independent of the bulk resistivity or conductivity generated by a Wenner array some distance away. In contrast, the Electroscan results from various electrode locations are not independent and conductivities are solved simultaneously for the entire three-dimensional region beneath the electrode grid. Consequently, errors of interpretation are far less likely. A second major advantage of the Electroscan technique is that for a given electrode spacing, definition of conductivity or impedance is greater than that provided by the Wenner technique. Consequently, smaller targets can be much more readily imaged with the Electroscan technique.

As previously noted, contouring graphics affect the interpretability of the image. The Electroscan method has another major advantage when used with a graphics package having fishnet surfaces (Fig. 13). Pseudo four-dimensional surfaces can now be generated to improve interpretation of

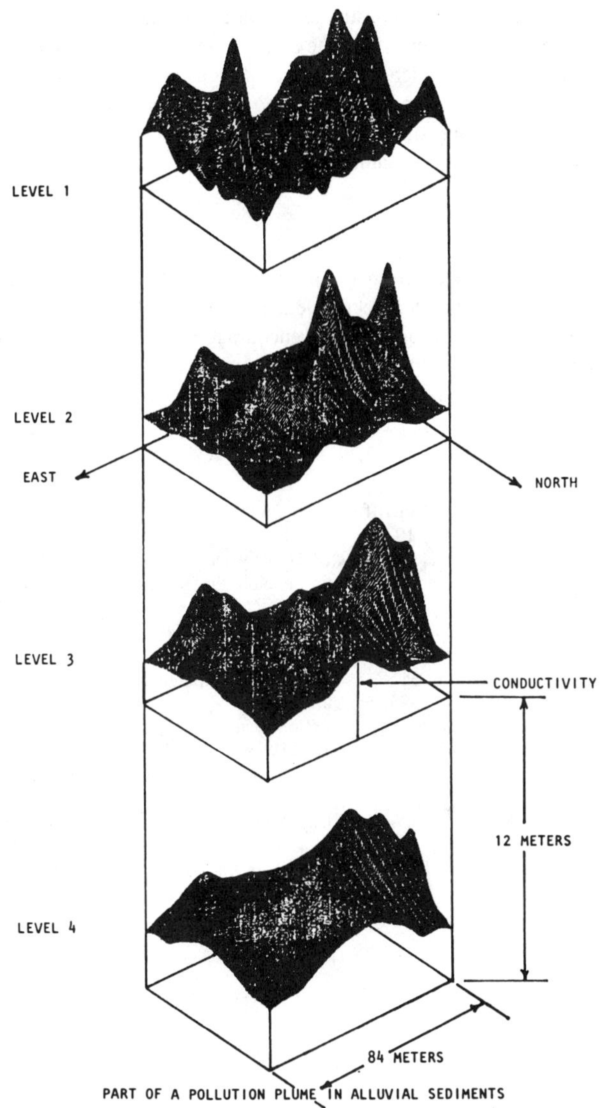


FIG. 13—*A saline plume in a laboratory sandbed.*

pollution plume geometry. Figure 13 contains the information in Figs. 11 and 12, but is considerably easier to interpret.

It is felt that currently the two most significant limitations of the Electroscan system in the field ground-water setting consist of (1) the assumption of no electric current flow outside the test volume, and (2) measurement quality in a low signal-to-noise-ratio setting. The latter limitation will certainly be overcome as funding becomes available for further hardware development. The former limitation is less tractable but can be rendered less of a problem by using large arrays and extending the geographic array coverage beyond the area of interest.

Field and laboratory results therefore reveal that the Electroscan technique requires further development but has exciting and demonstrable potential for imaging of ground-water problems, particularly pollution plumes. What is most exciting about the technique is that it is the first subsurface impedance technique which is truly three-dimensional. Furthermore, the technique has the potential, using various excitation frequencies, to image specific pollutant ion species (even if nonconducting) using techniques similar to those used in remote sensing for imaging surface conditions on the basis of spectral signatures.

*Acknowledgments*

Permission to publish these results from a contract, performed for the Canadian Defence Research Establishment, Suffield, is gratefully acknowledged. Dr. J. McFee and Dr. Y. Das, of that Establishment, are thanked for discussions and encouragement. Mr. M. R. Neuman and Mr. A. Fedorkiw are thanked for their help. The laboratory work of Mr. R. Allard and computational assistance of Mr. C. J. Mandel are especially appreciated.

## References

[1] Telford, W. M., Geldart, L. P., Sheriff, R. E., and Keys, D. A., *Applied Geophysics,* Chapter 8, Cambridge University Press, U.K., 1976.
[2] Benson, R. C., Glaccum, R. A., Noel, M. R., "Geophysical Techniques for Sensing Buried Wastes and Waste Migration," Technos, Inc., U.S. Environmental Protection Agency Contract No. 68-03-3050, 1982.
[3] Sendlein, L. V. A. and Yazicigil, H., "Surface Geophysical Techniques in Ground-Water Monitoring. Part II," *Ground Water Monitoring Review,* winter 1982.
[4] Wexler, A., Fry, B., and Neuman, M. R., "Impedance Computed Tomography Algorithm and System," *Applied Optics,* Vol. 24, No. 23, 1985, pp. 3985–3992.
[5] Wexler, A. and Fry, B. 1982 Patents: Reconstruction system and methods for impedance imaging, Canada (Filing priority 393977, 12 January 1982) No. 1196691, 12 November 1985 and United States No. 4539640, 3 September 1985: Method and apparatus for imaging the interior of a structure, Europe No. 0085490, 12 January 1983.
[6] Dines, K. A. and Lytle, R. J., "Analysis of Electrical Conductivity Imaging," *Geophysics,* Vol. 47, No. 7, 1981 pp. 1025–1036.
[7] Chari, M. V. K. and Silvester, P. P., *Finite Elements in Electrical and Magnetic Field Problems,* Wiley, New York, 1980.

*Richard M. Tinlin*,[1] *Larry J. Hughes*,[2] *and A. Roger Anzzolin*[3]

# The Use of Controlled Source Audio Magnetotellurics (CSAMT) to Delineate Zones of Ground-Water Contamination—A Case History

**REFERENCE:** Tinlin, R. M., Hughes, L. J., and Anzzolin, A. R., "**The Use of Controlled Source Audio Magnetotellurics (CSAMT) to Delineate Zones of Ground-Water Contamination—A Case History,**" *Ground-Water Contamination: Field Methods, ASTM STP 963,* A. G. Collins and A. J. Johnson, Eds., American Society for Testing and Materials, Philadelphia, 1988, pp. 101–118.

**ABSTRACT:** A significant potential for the pollution of fresh-water aquifers exists due to oilfield waterflood operations. The sources of potential pollution are surface spills, a lack of mechanical integrity of injection wells, and improperly plugged wells which are in communication with the injection zone. Surface spills are relatively easy to detect and control. Procedures for checking the mechanical integrity of a properly constructed injection well are available. Making a determination in the absence of good records as to whether or not a well is improperly plugged, providing a conduit for the vertical migration of formation brines from the production zone to shallower fresh-water aquifers, is very difficult.

Electrical surface geophysical methods offer considerable promise in detecting the movement of formation brines into fresh-water aquifers, through improperly abandoned or plugged wells.

An electrical surface geophysical technique, Controlled Source Audio-Frequency Magnetotellurics (CSAMT) has been applied to locate the presence of anomalies resulting from the upward movement of formation brines through improperly plugged wells. The primary objective in a CSAMT survey is to provide apparent resistivity and the phase angle between the electric and magnetic fields over a prospect area. CSAMT has the advantages of excellent lateral resolution, good depth penetration (a kilometre or more) and is relatively inexpensive. The frequency and resistivity of the subsurface control the depth of penetration. The lower the frequency, the deeper the penetration.

A CSAMT survey was run in an oil-producing field in east central Oklahoma, which is currently on waterflood and has many abandoned and apparently improperly plugged wells. The water in the Vamoosa aquifer underlying the study area has a high chloride content. The objective of running the CSAMT survey was to locate suspected low-resistivity anomalies due to formation brines in the vicinity of improperly plugged wells and to attempt to map their extent.

**KEY WORDS:** ground-water pollution, surface electrical geophysics, audio-magnetotellurics, oilfield waterfloods, improperly plugged injection wells

The study area includes approximately 324 hectares (800 acres) in the Sac and Fox Reservation located in Lincoln County, east central Oklahoma (Fig. 1). The area is underlain by the Vamoosa Formation, which consists of alternating thin to massive sandstones and sandy-silty shales. The

[1] Vice president, Geraghty and Miller, Inc., Phoenix, AZ 85044; formerly, senior hydrogeologist, Engineering Enterprises, Inc., Norman, OK 73069.

[2] Geophysicist, Zonge Engineering, Tucson, AZ 85716.

[3] Environmental Protection Agency, Washington, DC.

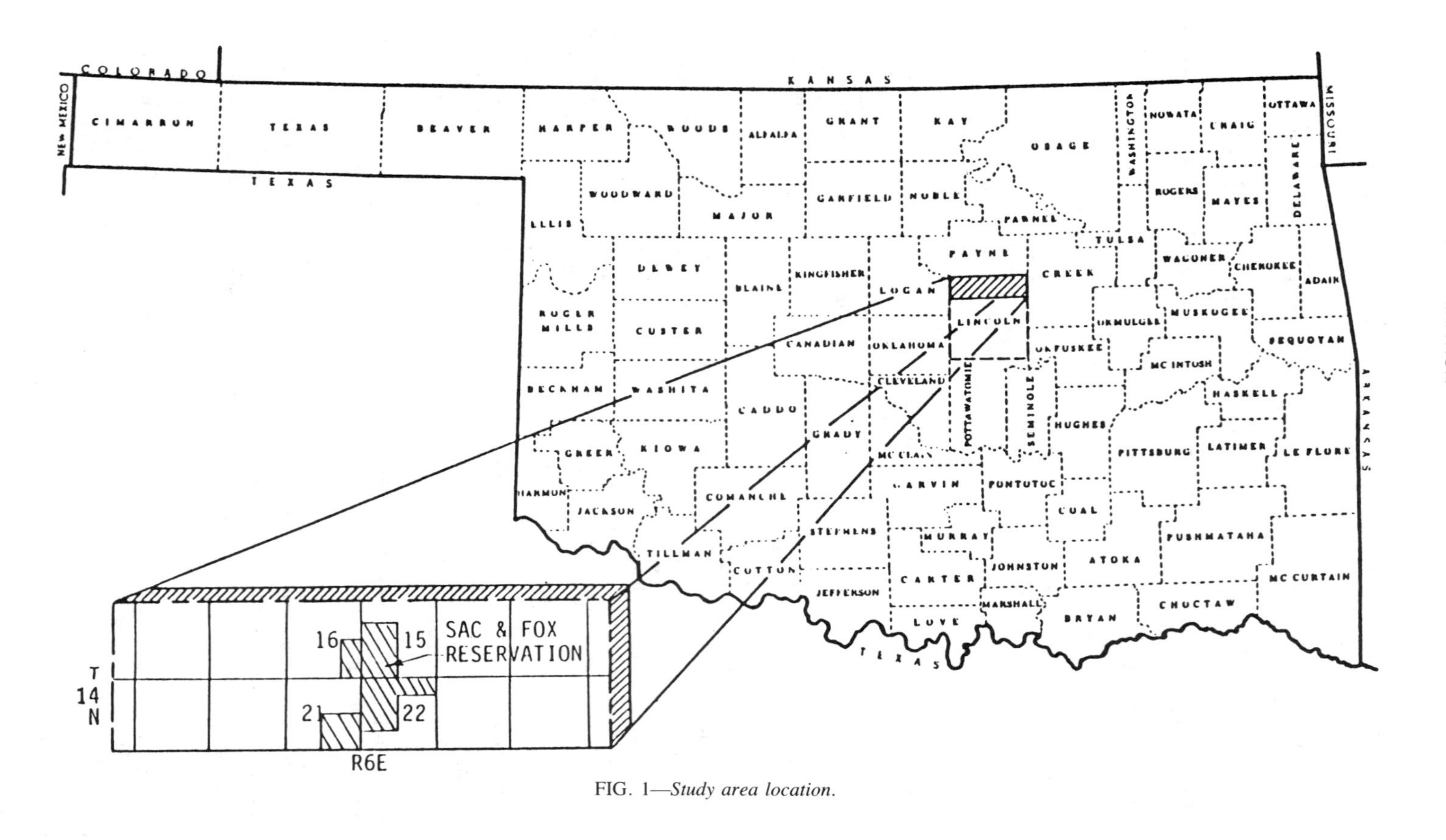


FIG. 1—*Study area location.*

sandstone layers are fine to coarse-grained and provide a reservoir source for one of the major fresh-water aquifers in Oklahoma.

Oil production in the study area began in the 1930's. The unit's cumulative production through August 1982 was approximately 28 million bbl, with an estimated current monthly production of 4500 bbl. Water injection for secondary recovery or salt-water disposal purposes or both began in the 1950's and the cumulative water injected through August 1982 was approximately 75 million bbl, with an estimated current monthly injection of 48 000 bbl.

The major objectives of the ground-water contamination study included the determination of the cause of the high chloride concentrations in the Vamoosa aquifer underlying the study area and whether this resulted directly from oilfield activities on and around the study area. A large number of well logs and plugging and abandonment records were evaluated and several tests conducted to determine the source of the high-chloride waters in the Vamoosa aquifer underlying the study area. Test holes were drilled and logged to obtain ground-water samples and to determine the ground-water quality profile at the test sites. In addition, an electrical surface geophysical survey (Controlled Source Audio Magnetotellurics—CSAMT) was run in order to locate the presence of anomalies that might result from the upward movement of formation brines high in salt content through improperly plugged wells into the overlying Vamoosa fresh-water aquifer.

## Background Studies

### *Hydrogeology*

The surface geology of the area is part of the Ada Group. Underlying the Ada Group is the Vamoosa Formation. The Vamoosa aquifer includes the Vamoosa Formation and underlying and overlying Pennsylvanian formations that are lithologically similar and hydrologically interconnected. The Vamoosa aquifer consists of a complex sequence of fine to very fine-grained sandstone, siltstone, shale, and conglomerate, with interbedded very thin limestones. The water-yielding capabilities of the aquifer are largely controlled by the lateral and vertical distribution of the sandstone beds and their physical characteristics [*1*]. Figure 2 illustrates the hydrogeology underlying the area and was prepared from driller's logs and plugging records of five abandoned wells. The orientation of the geologic cross section is approximately southwest to northeast [*2*].

Earlier investigations [*3*] reported that the base of the fresh water beneath the area could range from 50 m to more than 150 m below ground level. An evaluation of electrical resistivity logs run on oil wells in and around the area shows the base of the fresh water to be in the range of 40 to 90 m below ground level. A base of fresh-water contour map (Fig. 3) was drawn utilizing data from resistivity logs and from test wells drilled earlier in the study area [*2*]. The contour map shows the base of fresh water to be relatively shallow over a large portion of the area.

### *Ground-Water Quality*

A 1975 study [*4*] showed the quality of the ground water adjacent to the area to be generally good with a total dissolved solids (TDS) of 500 mg/L or less.

In the spring of 1979, four test holes were drilled by Engineering Enterprises, Inc. in order to determine the ground-water quality. The test well data showed an anomalous occurrence of salt water in a portion of the Vamoosa aquifer that was expected to contain only fresh water. Three possibilities as to the cause of the high chlorides in the Vamoosa were postulated: (1) natural occurrences of salt (halite) in the Vamoosa, (2) upconing of the fresh-water/salt-water interface due to overpumping of the Vamoosa aquifer, and (3) accidental introduction of salt water into the aquifer due to various phases of oilfield activities. The Vamoosa and associated rock units do not contain salt beds nor does the sedimentary environment in which the Vamoosa was deposited allow for the development of bedded salt. Overpumping of the Vamoosa aquifer was also ruled

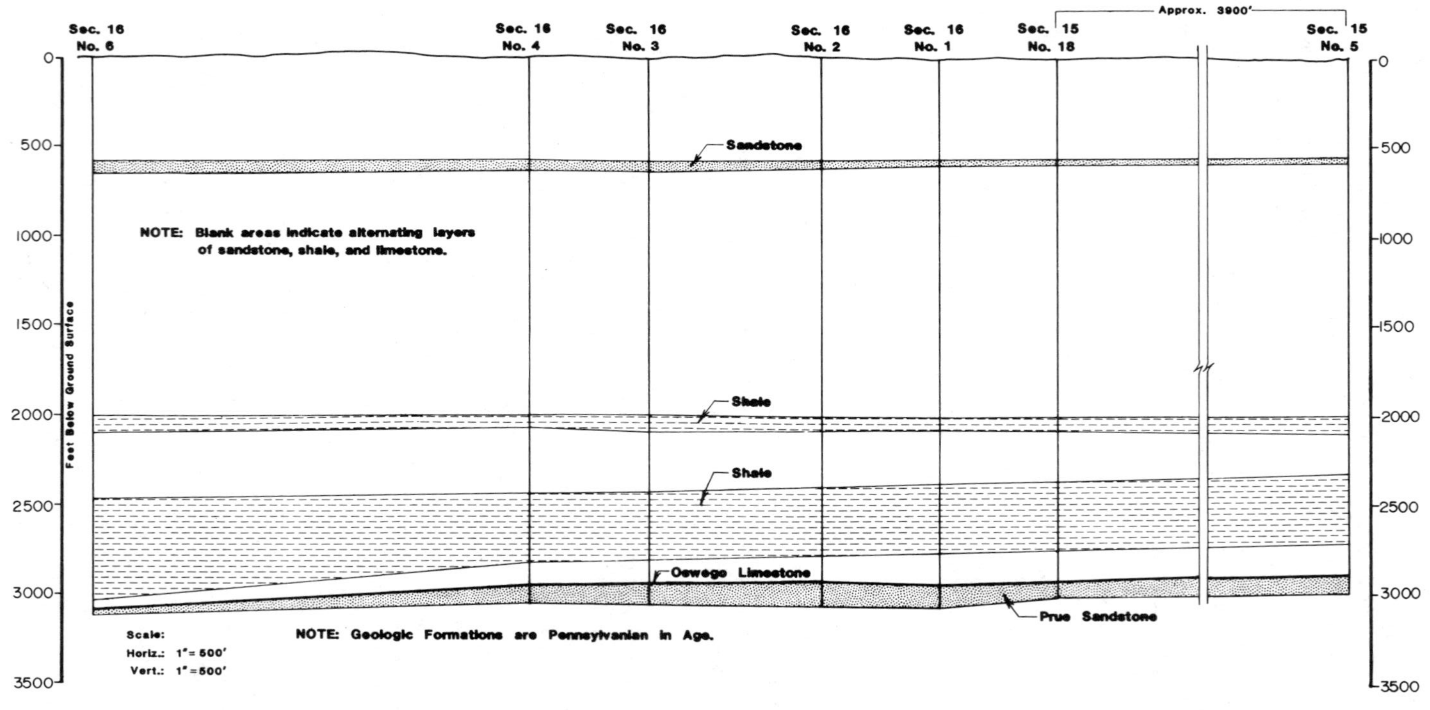


FIG. 2—*Sac and Fox ground-water contamination study, generalized geologic cross section, Lincoln County, Oklahoma, T14N-R6E.*

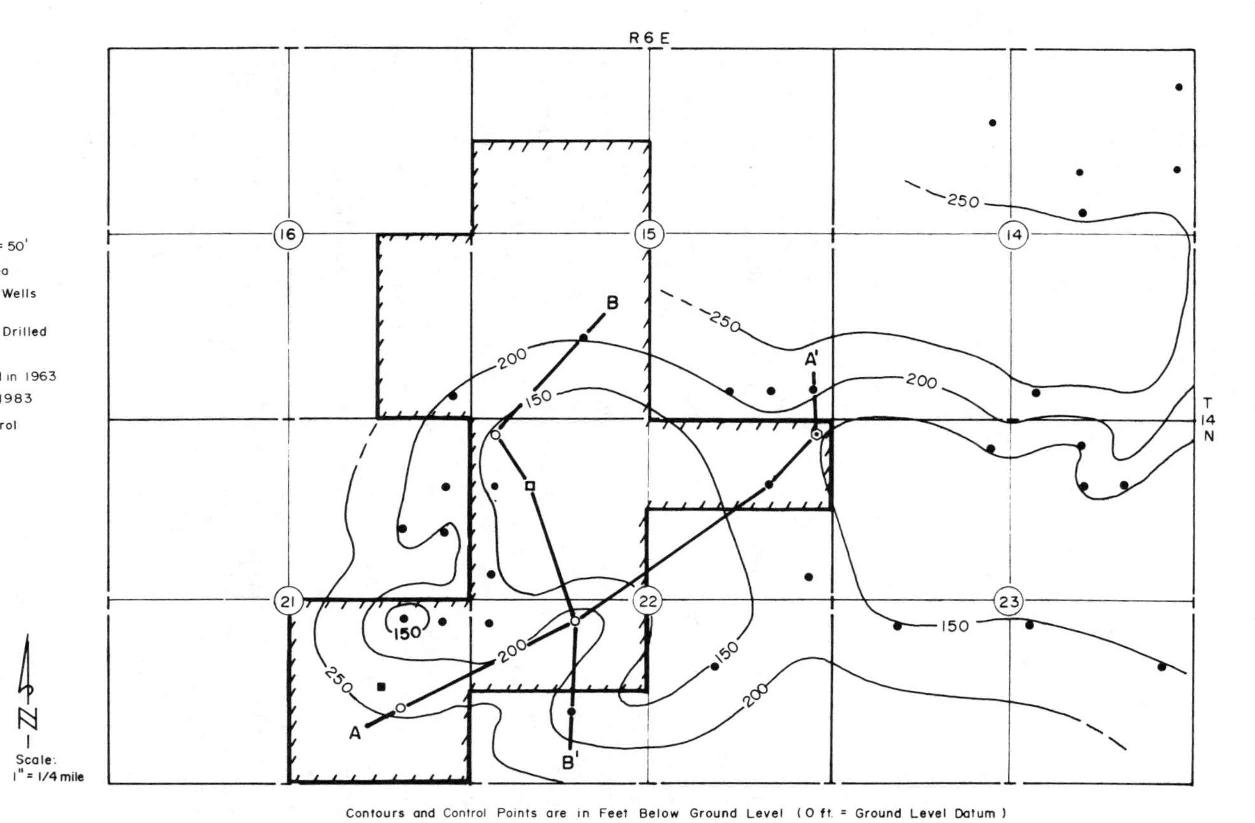


FIG. 3—*Sac and Fox ground-water contamination study, base of fresh-water contour map.*

TABLE 1—*Constituent concentrations and ratios for Vamoosa water samples.*

| Sample No. | Depth, m | Cl, mg/L | Br, mg/L | Br/Cl |
|---|---|---|---|---|
| 1 | 42 | 200 | 0.78 | 0.0039 |
| 2 | 61 | 7 520 | 33 | 0.0044 |
| 3 | 79 | 10 800 | 50 | 0.0050 |

out as the recharge rate of the Vamoosa is much higher than the pumping rates of the existing supply wells.

*Determination of Salt-Water Contamination Source in the Vamoosa by Water Sampling*

Water samples from a test well (located in Section 22) drilled under the supervision of Engineering Enterprises, Inc. in July 1983 were analyzed by Dr. Donald O. Whittamore of the Kansas Geological Survey. Whittamore's procedure [5] is very effective in distinguishing oilfield brines from halite (rock salt) solution brines and thus can determine which may be the source of ground-water contamination.

Bromide concentrations were determined by Whittamore and the chloride concentrations were determined by the Environmental Control Laboratory in Norman, Oklahoma. The results of the water analyses are presented in Table 1.

The bromide/chloride ratio is the key in determining the brine contamination source. The ratios in the saline waters are what could be expected if oilfield brine pollution had occurred. Bromide/chloride ratios for most Oklahoma oilfield brines range from 0.003 to 0.01. The bromide/chloride ratio expected for waters with chloride concentrations of 10 000 mg/L for a halite-solution source of salinity is 0.0002 + 0.0002. The bromide/chloride values are much higher than the expected values for a halite-solution source and fall within the values for most Oklahoma oilfield brines. Since this situation fits the special case where the injected fluid is the same or very similar to the Prue formation fluid, the source of salinity in the Vamoosa ground waters beneath the area was concluded to be Prue formation brine.

In December of 1983, two samples of Prue formation brine were obtained from a surface separator on site, and the bromide/chloride ratios determined by Whittamore. The purpose of this testing was to obtain a comparison of the Prue brine bromide/chloride ratios with the previous Vamoosa ground-water bromide/chloride ratios of July 1983. The bromide and chloride concentration and bromide/chloride ratios are listed in Table 2. The bromide/chloride ratios from the Vamoosa aquifer match closely the bromide/chloride ratios of the brine in the Prue Formation. Due to this close match of geochemical ratios, the Prue oilfield brine is considered to be the most probable source of brine polluting the Vamoosa aquifer.

TABLE 2—*Constituent concentrations and ratios for Prue oilfield brines.*

| Sample No. | Cl, mg/L | Br, mg/L | Br/Cl |
|---|---|---|---|
| Prue 1 | 74 200 | 377 | 0.0051 |
| Prue 2 | 74 600 | 372 | 0.0050 |

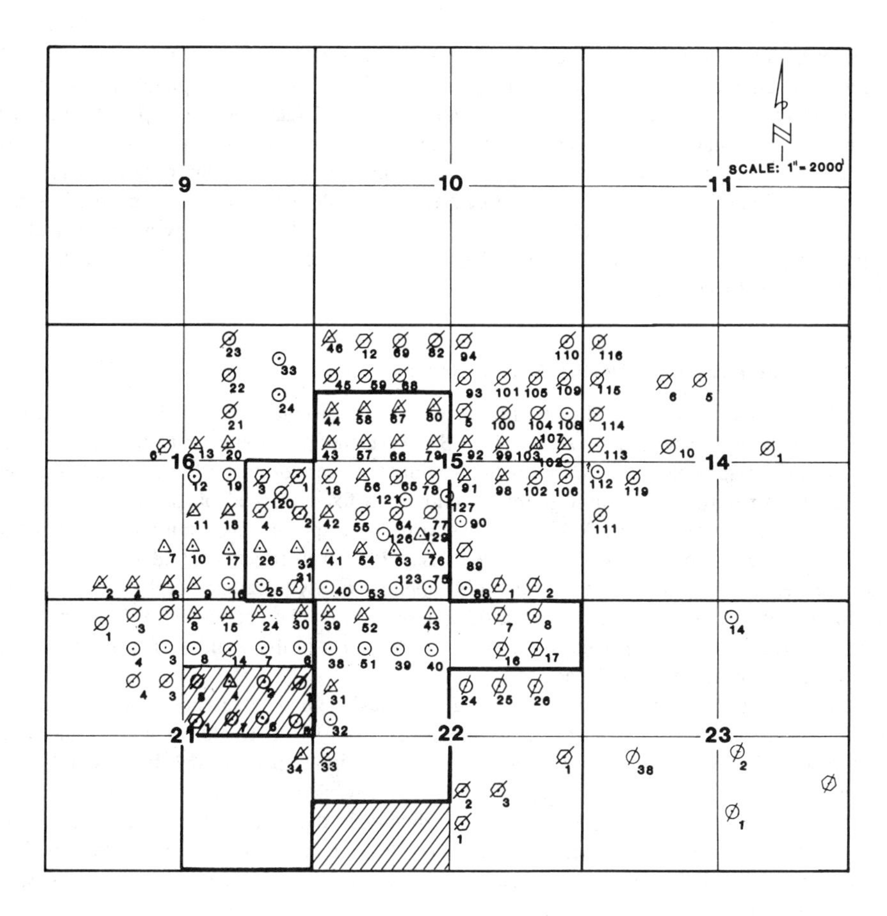


⊙ Producing Oil Well

Ø Oil Well - Properly Plugged

Ø Oil Well - Improperly Plugged

△ Injection Well

△ Injection Well - Properly Plugged

▨ Surface Rights Only

FIG. 4—*Sac and Fox unit, Lincoln County, OK.*

## *Location and Evaluation of Improperly Plugged Wells*

An evaluation of all plugged and abandoned wells in the study and surrounding areas was made from Oklahoma Corporation Commission records. A location map of these wells, including identification of the properly and improperly plugged wells, is shown in Fig. 4.

A number of the wells were determined to be improperly plugged. For a proper plugging operation in an area where protection of the fresh-water aquifer is important, at least four downhole cement plugs should be installed in every well that is to be plugged and abandoned. These four plugs are a bottom plug opposite the injection interval, an isolation plug across the top of cut casing, a surface casing protection plug, and a surface plug.

Many of the older wells were plugged by loading the hole with mud. With time mud can settle out and allow channeling of salt water through the borehole. The cement plugs in most of the older wells were determined to be inadequate. Many of the wells had only the top surface cement plug and no additional downhole cement plugs. In addition, in most of the plugged wells the surface casing was set too shallow. Surface casing should be set below the base of fresh water and cemented all the way to the surface to effectively seal fresh-water zones from deeper injection fluids. Although these plugging methods satisfied regulations at the time, they are inadequate as they provide potential flow paths for upward migration of reservoir fluids or injection fluids or both to shallower fresh-water zones.

## Controlled Source Audio-Frequency Magnetotelluric Survey

CSAMT is a relatively new technique first used on a consistent commercial basis in 1978 by Zonge Engineering of Tucson, Arizona. The CSAMT technique is similar to the conventional audio-frequency magnetotelluric (AMT) method, with the exception that a fixed current source is substituted for a natural earth-telluric source resulting in a fixed, dependable signal.

The primary objective in a CSAMT survey is to provide apparent resistivity and phase angle soundings over a prospect area. The technique is particularly effective at identifying buried, conductive features. It has been successfully applied in hydrocarbon exploration, mineral exploration, and geothermal exploration. In this particular case the application of interest was to use the CSAMT to locate suspected low-resistivity anomalies in the vicinity of wells thought to be improperly plugged and abandoned.

CSAMT has several advantages over the other geophysical methods. It has good lateral resolution, good depth penetration, is fast and relatively inexpensive, \$1000 to \$2000 per line kilometre (or approximately \$200 per station), depending on the receiver dipole spacing used. It is also relatively insensitive to "cultural" features such as pipelines, power lines, fences, well casings, etc. Disadvantages associated with CSAMT include difficulty in data interpretation due to near-field effects and difficulty in estimating depths to anomalous two- and three-dimensional features without extensive computer modeling and some geologic input. It is important to consider the specific requirements of a field project before deciding whether or not to use CSAMT.

### *CSAMT Layout in Study Area*

A typical layout for a CSAMT survey is shown in Fig. 5. The large transmitter dipole is located as far away from the receiver dipole as is practical—usually three skin depths at the lowest frequency being used. Skin depth ($\delta$) is related to the signal penetration into the ground, and is defined as:

$$\text{Skin depth } (\delta) = 503\sqrt{\rho/f}\ \text{m}$$
$$\rho = \text{ground resistivity, } \Omega\text{m}$$
$$f = \text{frequency, Hz}$$

In this study, the distance varied between 4.8 to 6.4 km.

The receiver setup comprises a grounded dipole oriented parallel to the transmitter dipole for maximum electric field pickup and a high-gain coil oriented at right angles to the dipole for magnetic field detection. A transmitter dipole length of 1500 m and a receiver dipole spacing ("a" spacing) of 60.96 m was used in this study.

The receiver system used was a Zonge Engineering GDP-12 data processor which measures both the electric (E) and magnetic (H) fields in synchronization with the transmitter output. Data

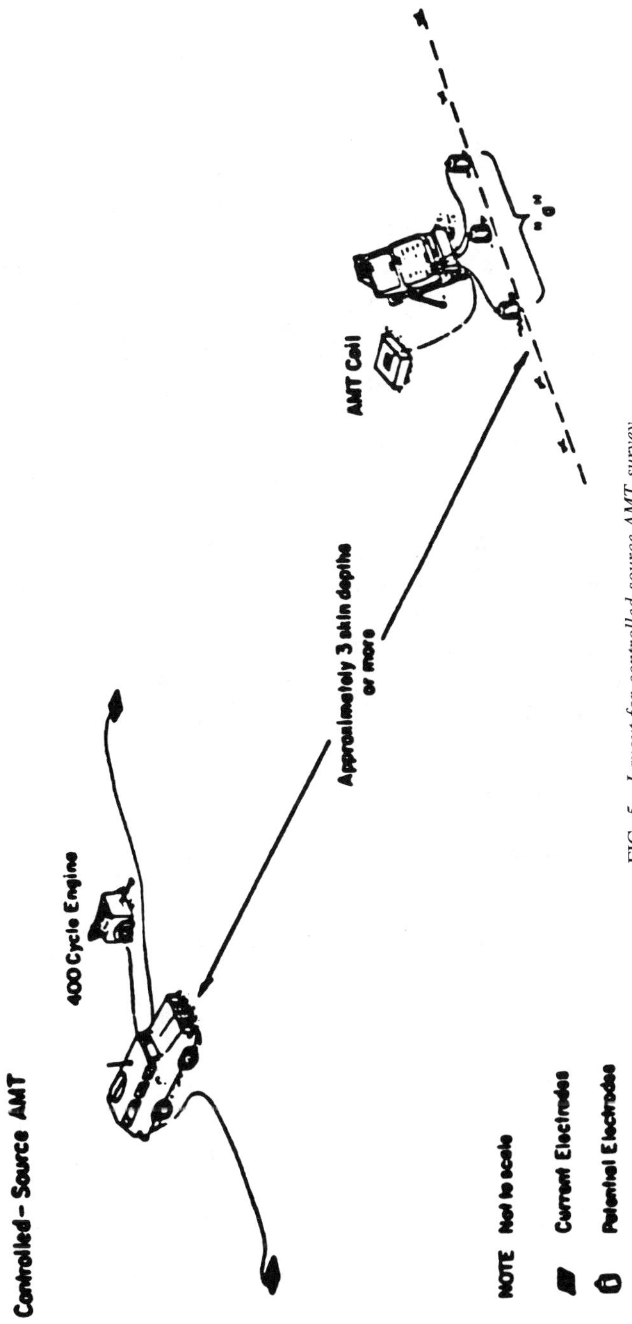


FIG. 5—*Layout for controlled source AMT survey.*

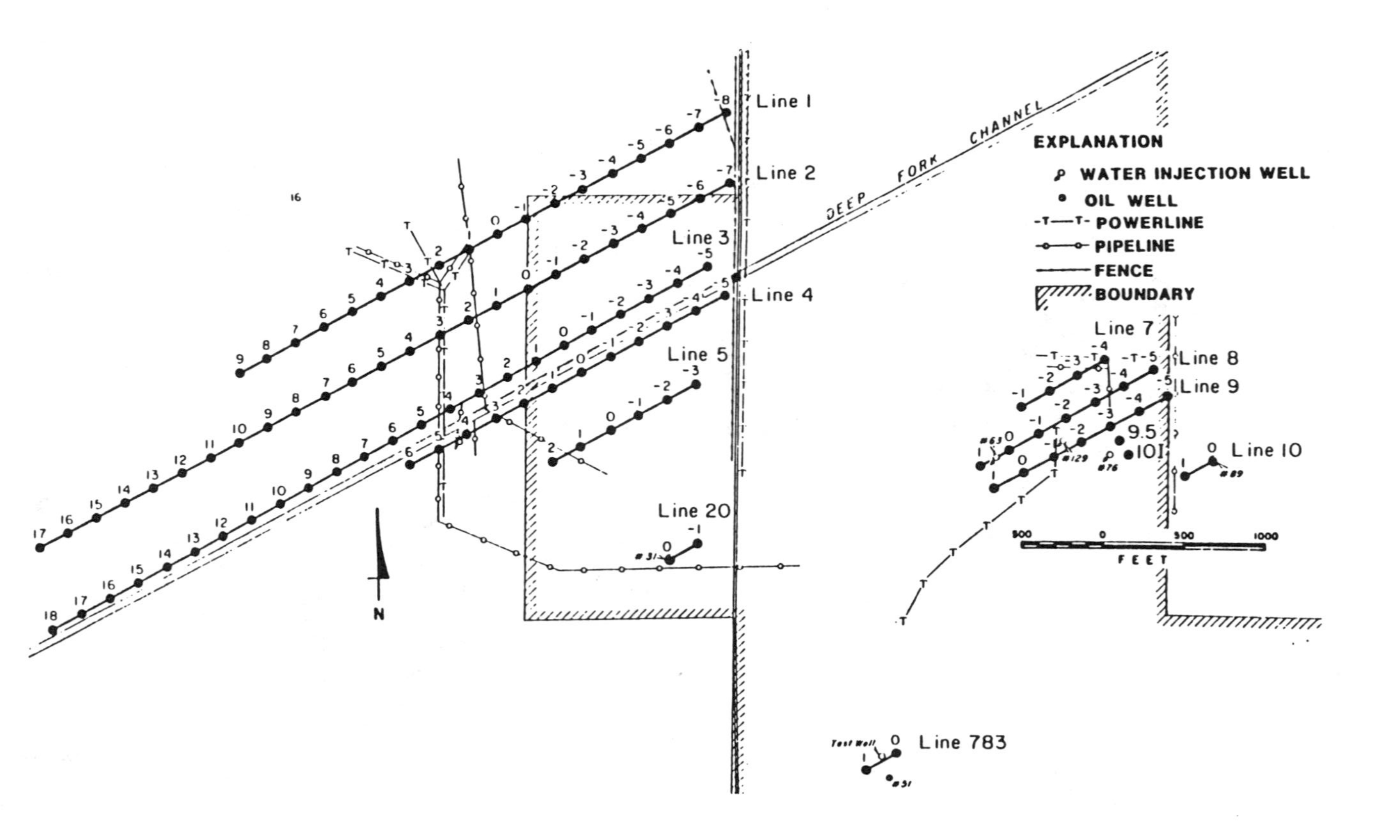


FIG. 6—*Layout for CSAMT survey.*

are obtained and recorded automatically in the field as E− and H− field magnitudes and phase, cagniard resistivity, and relative phase between the E and H fields.

Figure 6 shows the location of the survey lines. Two sets of survey lines were run in two areas, one set (Lines 1 through 5) near a cluster of wells determined to be improperly plugged in Section 16, and the second set (Lines 7 through 9) near an area of current injection activity in Section 15.

*Results of CSAMT in Area of Improperly Plugged Wells*

The survey results from Line 3 (in the area with a high density of improperly plugged wells) are presented first. Figure 7 is an electrical pseudosection showing the apparent resistivity of the ground underlying Line 3 (in ohmmetres) plotted against the frequency (in Hz) at each station on Line 3. The same data are also shown in Fig. 8 except that the apparent resistivity is plotted against depth. A pseudosection such as Fig. 8 can be viewed as showing relative depths in an approximation to a vertical slice through the ground.

The depth of penetration in a layered environment can be estimated from the following MT equation:

$$\text{Depth of penetration} = 356\sqrt{\rho/f}\ \text{m}$$
$$\rho = \text{ground resistivity, } \Omega\text{m}$$
$$f = \text{frequency, Hz}$$

Figure 7 shows the vertical low-resistivity anomaly extending laterally between Stations 10 to 16 and extending to depth below. The low-resistivity anomalies are more significant at approximate depths of 14, 40, and 80 m at Station 13 and at an approximate depth of 13 m at Station 11. The low-resistivity (high conductivity) anomalies can be identified more clearly in Fig. 8. The low anomaly extends to a depth of at least 82 m below Station 13. These significant low-resistivity anomalies are most likely caused by improperly abandoned and plugged wells. However, this will need to be verified and confirmed in future studies through detailed test drilling and water quality sampling.

Figure 9 is a horizontal view at a depth corresponding to 64 Hz. At Station 13 this frequency corresponds to a depth of approximately 63 m.

*Results of CSAMT Surveys in Area of Current Injection Activity*

In the area of current injection activity, Line 8 was selected for presentation. Figure 10 is a vertical pseudosection showing apparent resistivity versus frequency while Fig. 11 plots apparent resistivity versus depth. Note the low-resistivity anomaly extending to depth below Station −4.0, Fig. 10, and a depth of 114 m below Station −4.0, Fig. 11. It is particularly interesting because it spreads out near the surface and the apparent resistivity has the lowest value near the surface, indicating possible surface or near-surface spills or leaks.

Figure 12 is a horizontal view at a depth corresponding to a frequency of 64 Hz. This corresponds to a depth of 63 m at Station −4.0.

*Test Drilling of CSAMT Anomalies*

Two test well sites in Line 8 were selected on the basis of the CSAMT results. Test well No. 1 was located at Station −2.0, Line 8 because of the contrasting resistivities encountered with depth. Test well No. 2 was located at Station −4.0, Line 8 where the surface resistivity is the lowest and the conductive plume appears to be deep-seated.

The stratigraphic sequence of rock penetrated in both wells drilled in July 1984 is given in Table 3. Rock cuttings were logged continuously during the drilling of each well. The first good

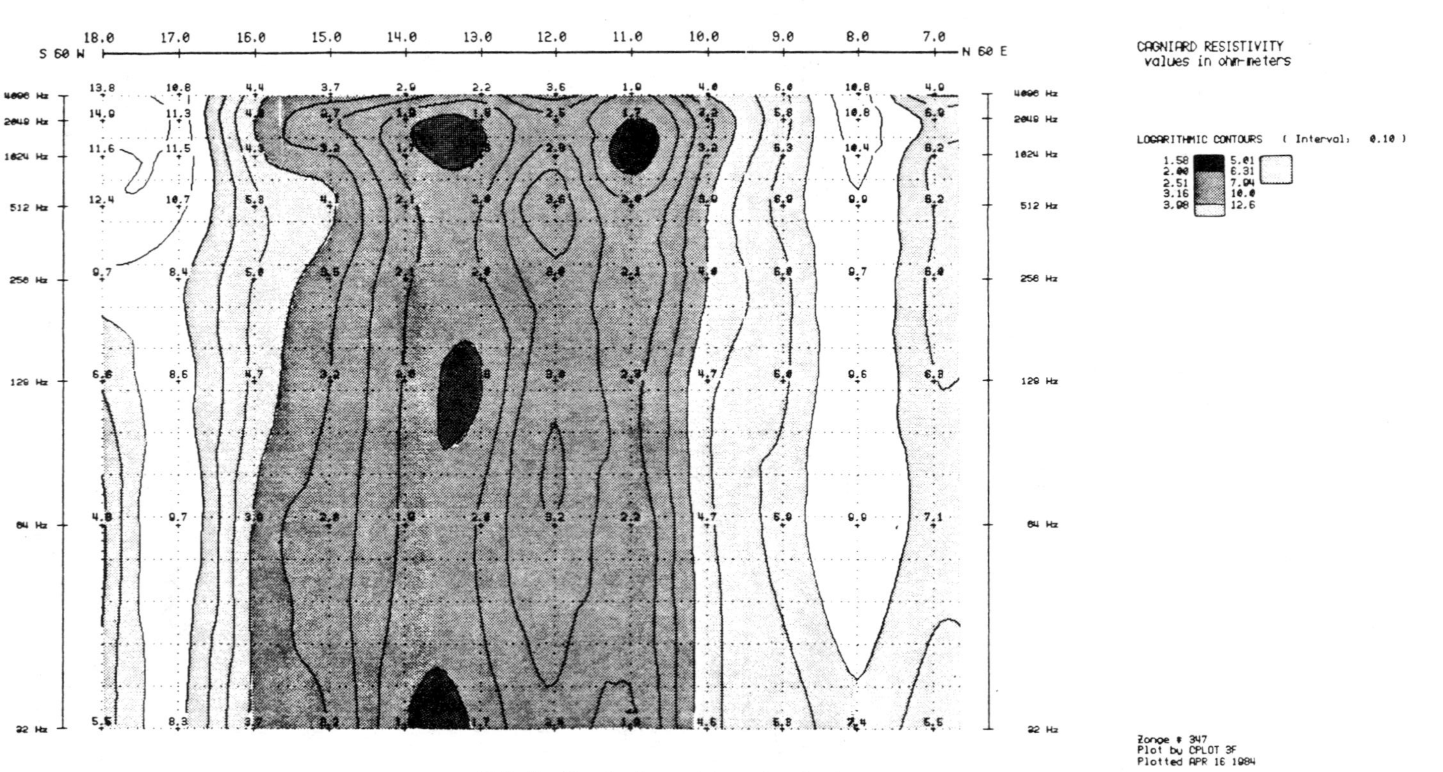


FIG. 7—*Cagniard resistivity values in ohmmetres.*

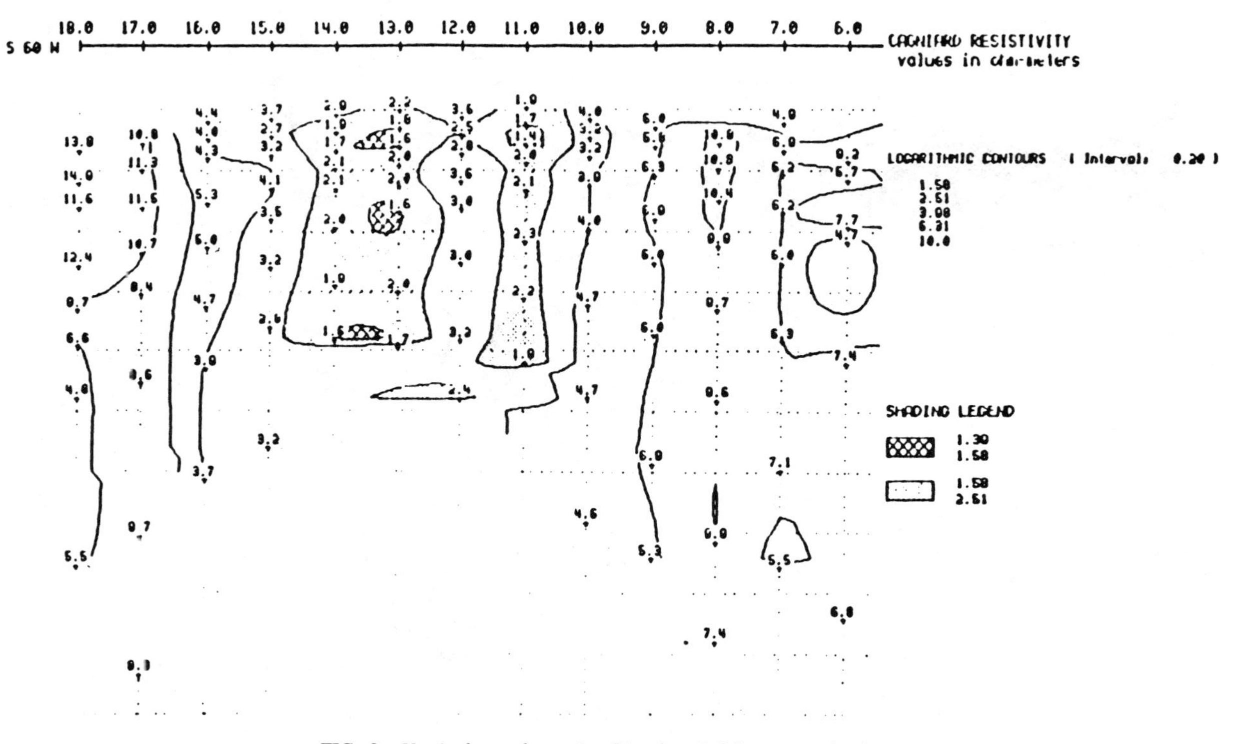


FIG. 8—*Vertical pseudo-section Line 3 resistivity versus depth.*

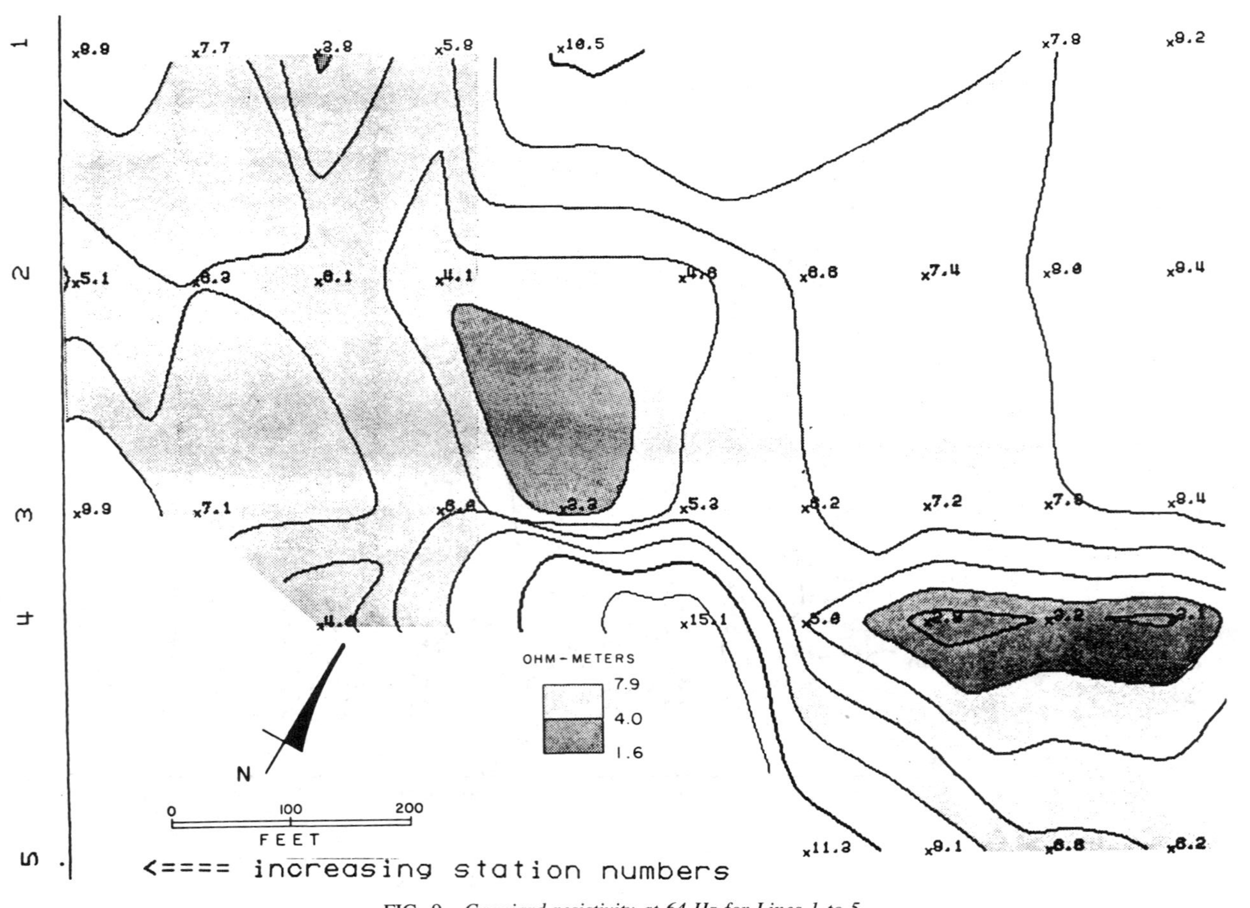


FIG. 9—*Cagniard resistivity at 64 Hz for Lines 1 to 5.*

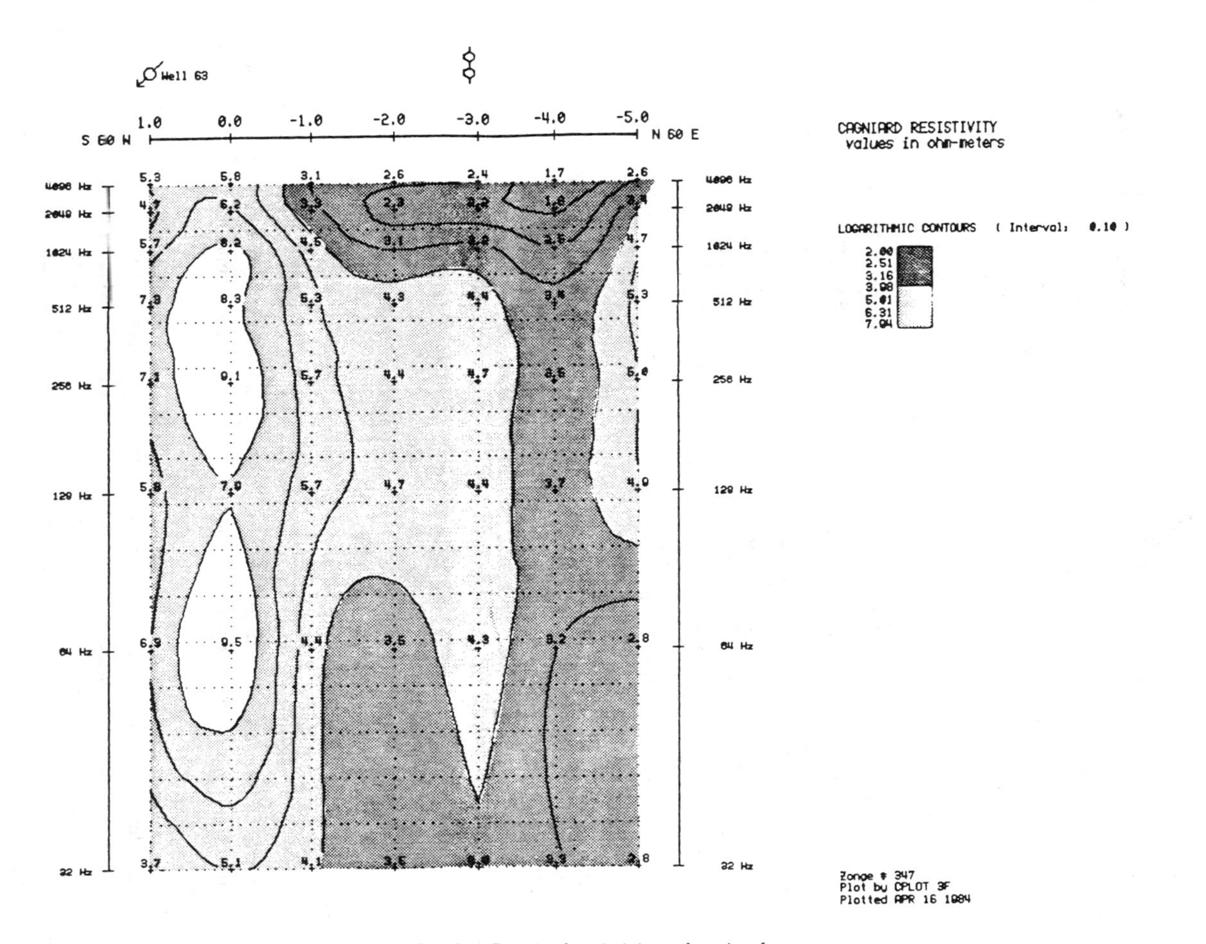


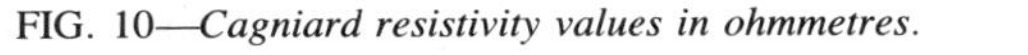
FIG. 10—*Cagniard resistivity values in ohmmetres.*

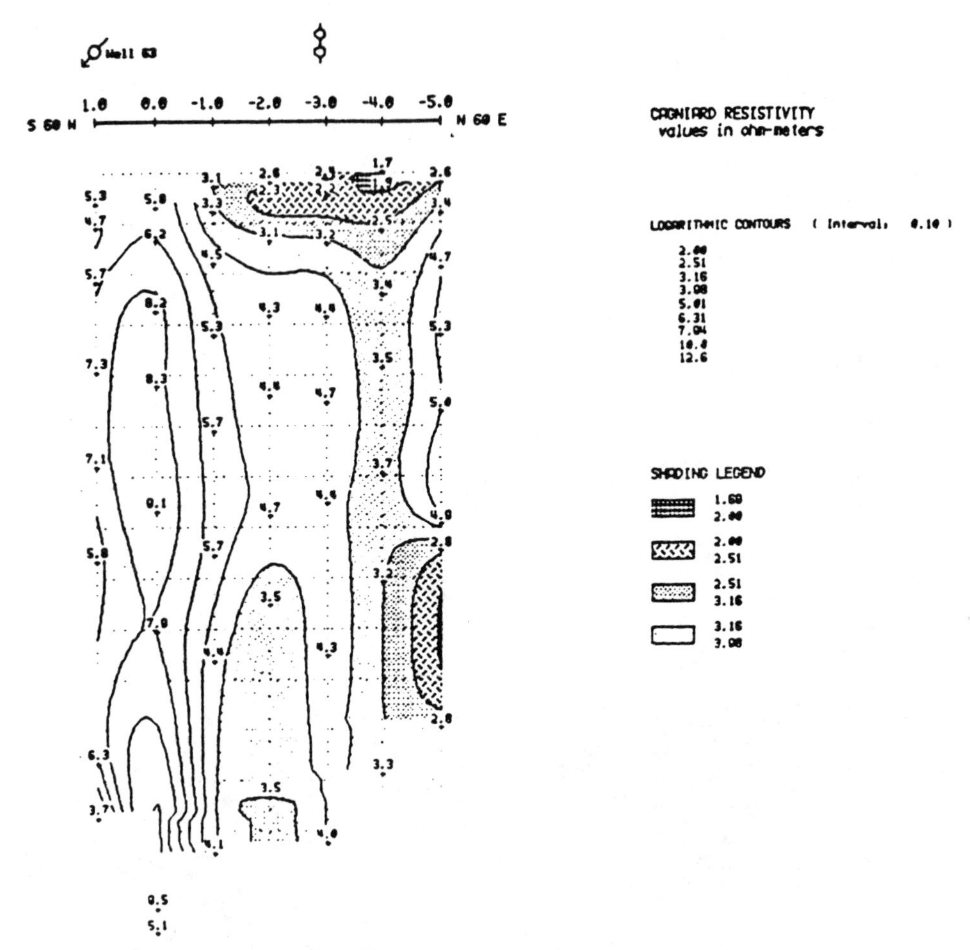


FIG. 11—*Vertical pseudo-section Line 8 resistivity versus depth.*

water-bearing zone encountered was in the Vamoosa, in the interval (36 to 46 m) in test well No. 1, and in the interval 34 to 42 m in test well No. 2. Water samples were collected from a shallow, intermediate, and deep zone in each well and the composition and bromide/chloride ratios were determined. Results of the analyses are given in Table 4.

The bromide/chloride ratios are similar to the ones obtained earlier and are within the range of typical Oklahoma oilfield brine, which is 0.003 to 0.01. This indicates that the probable source of polluted Vamoosa ground water is Prue oilfield brine regardless of whether it came from surface spills, improperly plugged wells, or mechanical integrity failure of injection wells. After logging and testing operations, both wells were completed as monitoring wells to provide future monitoring data of the ground water.

## Conclusions

The CSAMT technique has been successful in locating low-resistivity anomalies which appear to result from oilfield brines in the vicinity of improperly plugged wells and near active injection

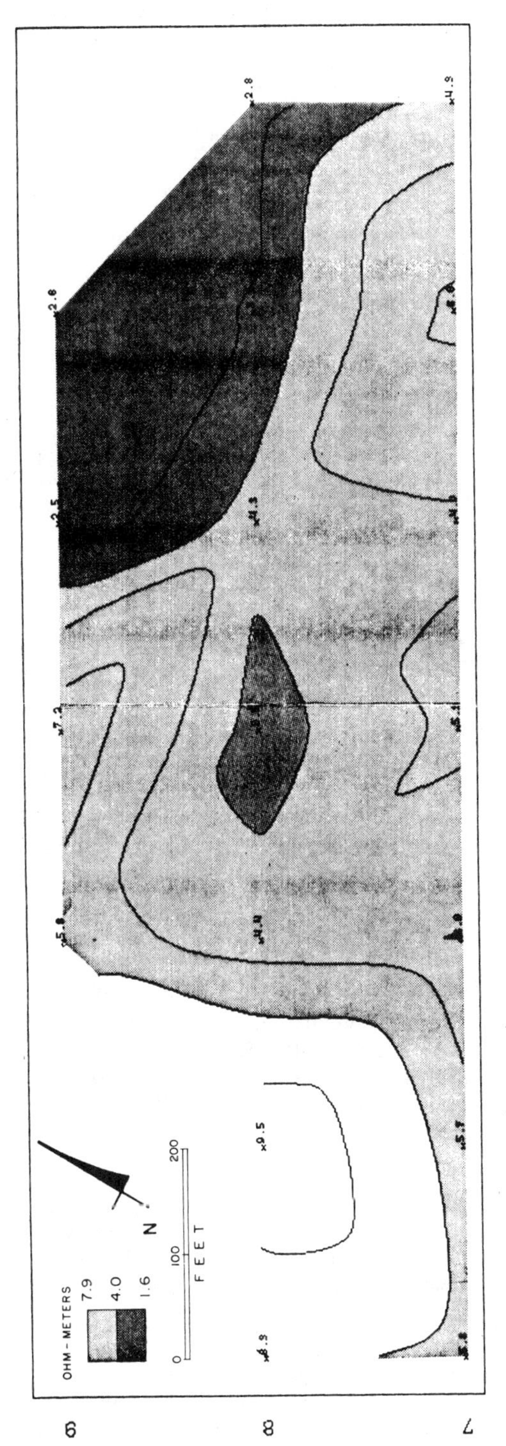


FIG. 12—*Cagniard resistivity at 64 Hz for Lines 7, 8, and 9.*

TABLE 3—*Stratigraphic sequence of rock penetrated in July 1984 test wells.*

| | Alluvium | Ada Group | Vamoosa Formation |
|---|---|---|---|
| Test well No. 1 | 0 to 9 m | 9 to 23 m | 23 to 91 m |
| Test well No. 2 | 0 to 10 m | 10 to 20 m | 20 to 91 m |

TABLE 4—*Constituent concentrations and ratios for Vamoosa water samples (July 1984 test wells).*

| Well No. | Sample No. | Depth, m | Cl, mg/L | Br, mg/L | Br/Cl |
|---|---|---|---|---|---|
| 1 | 1 | 40 | 950 | 4.9 | 0.0052 |
| | 2 | 61 | 1180 | 5.9 | 0.0050 |
| | 3 | 87 | 1200 | 6.1 | 0.0051 |
| 2 | 1 | 38 | 650 | 3.5 | 0.0053 |
| | 2 | 56 | 664 | 3.6 | 0.0054 |
| | 3 | 90 | 643 | 3.4 | 0.0053 |

wells. Two deep conductive plumes and one shallow conductive feature were detected and traced in the area of active injection wells in Section 15, while on the grid in Section 16 in the area of improperly plugged wells, one localized deep feature and two deep plumes were detected and traced. Several of these plumes ran out of the edge of the survey grids and their extent is not known.

Test drilling confirmed the low-resistivity anomalies detected by the CSAMT survey. Future studies of this type should include sufficient test drilling and logging to gather data to permit comparison and correlation of known salinity contrasts with resistivity contrasts. Once the baseline contrasts for a given area or oilfield are established, detecting and tracing of conductive plumes might be combined with estimates of water salinity to give even more useful results.

## *Acknowledgments*

Partial funding for the Vamoosa ground-water contamination study (including the CSAMT surveys) was obtained from the U.S. Environmental Protection Agency as part of Project 68-01-6389 of the Underground Injection Control Program, and is gratefully acknowledged. The cooperation of the Sac and Fox officials, including Mr. Truman Carter, is also acknowledged with thanks.

## *References*

[*1*] D'Lugosz, J. and McClaflin, "Geohydrology of the Vamoosa Aquifer, East-Central Oklahoma," U.S. Geological Survey Open-File Report 78-781, 1978.

[*2*] Fryberger, J. S. and Tinlin, R. M., "Pollution Potential from Injection Wells via Abandoned Wells," presented at First National Conference on Abandoned Wells: Problems and Solutions, Norman, OK, May 1984.

[*3*] Hart, D. L., "Base of Fresh Water in Southern Oklahoma," U.S. Geological Survey, Hydrologic Investigations Atlas HA-223, 1966.

[*4*] Bingham, R. H. and Moore, R. L., "Reconnaissance of the Water Resources of the Oklahoma City Quadrangle, Central Oklahoma, Oklahoma Geological Survey, HA-4, 1975.

[*5*] Whittamore, D. O., "Geochemical Identification of Salinity Sources" in *Proceedings*, International Symposium on State-of-the-Art Control of Salinity, Ann Arbor, MI, 1983.

# Well Drilling and Completion

*Charles O. Riggs*[1] *and Allen W. Hatheway*[2]

# Ground-Water Monitoring Field Practice—An Overview

**REFERENCE:** Riggs, C. O. and Hatheway, A. W., "**Ground-Water Monitoring Field Practice—An Overview,**" *Ground-Water Contamination: Field Methods, ASTM STP 963*, A. G. Collins and A. I. Johnson, Eds., American Society for Testing and Materials, Philadelphia, 1988, pp. 121–136.

**ABSTRACT:** The greatest difficulties in providing ground-water monitoring installations have not generally resulted from a need to develop new technologies but from the need to define and implement the administrative and communicative procedures that are required for legal and fiscal control. Many of the required field procedures that must be understood by permit writers and their inspectors have never been researched and documented but exist as evolved in day-to-day geotechnical practice and water well contracting. The field solution usually requires the selection of the appropriate available installation procedure. Selection of the well casing, intake screen, and sampling device according to material constituents is currently the most "unsettled" aspect of ground-water monitoring practice. It is emphasized that the field installation of a monitoring well system must relate to the site geology and hydrology, and that each ground-water monitoring device must be installed using the most appropriate drilling tools and techniques, and well materials, according to the suspected contaminants and the required accuracy of analysis.

**KEY WORDS:** drilling, filters, ground-water monitoring, hollow-stem augers, hollow-stem auger drilling, lysimeter, monitoring wells, monitoring well classification, monitoring well design, monitoring well installation, rotary drilling, vadose zone monitoring, well screens

For many years geologists and engineers, particularly civil and geological engineers, have been aware of the problems of surface water and ground-water contamination. Accordingly, practitioners of both professions have for many years monitored ground-water contaminants for the purpose of assuring safety of individual and municipal domestic water supplies. An accelerated need for ground-water monitoring activities was realized with the passing of the Resource Conservation and Recovery Act (RCRA; 1976) and the Comprehensive Environmental Response, Compensation and Liability Act (CERCLA; 1980). Fortunately, many of the engineering and scientific procedures needed for the implementation of ground-water monitoring activities under RCRA and CERCLA were already available from day-to-day geotechnical, geological, and other scientific practices. The performance of ground-water monitoring activities often involves a multidisciplinary, integrated team effort of geotechnical engineers, geologists, drillers, geohydrologists, geophysicists, chemists, chemical engineers, toxicologists, and microbiologists. Yet, the greatest difficulties in providing ground-water monitoring installations under RCRA and CERCLA have not generally resulted from a need to develop new technologies within the professions but from the need to define and implement the administrative and communicative procedures that are *required* for legal and fiscal control of what is a truly monumental task. The full extent and importance of the task facing the

[1] Research engineer, Central Mine Equipment Co., 6200 North Broadway, St. Louis, MO 63147.
[2] Professor of geological engineering, Department of Geological Engineering, University of Missouri-Rolla, Rolla, MO 65401.

professions is probably yet to emerge. Practicing engineers and geologists and other involved professionals should participate in the development of the administrative and communicative procedures by taking the time to study and critically review the literature, particularly the U.S. Environmental Protection Agency RCRA and CERCLA Technical Resource Documents and Guidance Manuals that are being released at a rapid rate.

The state of practice in both geotechnical exploration and water well drilling has for the most part developed and progressed in the "oral tradition." Many field techniques and procedures applicable to ground-water monitoring that must be understood by permit writers and inspectors have never been researched and documented, yet exist as day-to-day operating procedures in geotechnical exploration practice. There is a need to present and discuss those elements of geotechnical exploration and site characterization practice that are applicable to ground-water monitoring practice.

## Classification of Ground-Water Monitoring Systems

Barcelona et al. [*1*] have identified four basic types of ground-water monitoring systems according to their utilization: *ambient monitoring, source monitoring, enforcement monitoring,* and *research monitoring.*

*Ambient ground-water monitoring* has been in progress in most areas of North America for many years. Background water quality information is intermittently collected, reviewed, and placed on file by several public agencies, usually at the state or provincial level, and provides a valuable base resource for detecting and evaluating significant changes in water quality. Ambient ground-water monitoring may be also performed by an individual, a private industry, or a government to provide background data as protection against anticipated future environmental impairment claims.

*Source monitoring* differs from ambient monitoring in that it is performed within the confines or area surrounding a specific, actual, or potential source of ground-water contamination. Source monitoring systems can involve both first-phase (*detection*) monitoring wells and second-phase (*assessment*) monitoring wells. Assessment monitoring wells are installed to define the extent and concentration of a contaminant identified by one or more detection monitoring wells.

*Enforcement monitoring wells* are installed by or at the direction of a regulating agency for the purpose of collecting evidence for the prosecution of ground-water contamination cases. Enforcement monitoring wells are installed for the purpose of determining or confirming the origin, and concentration gradients at points of regulatory compliance, rather than determining the extent and concentration of individual contaminant plumes.

*Research monitoring wells* are installed for specific scientific purposes and may require a greater degree of installation care and expense than are warranted for installation of detection, assessment, or enforcement well systems.

## Physical Design of Ground-Water Monitoring Wells

Almost all monitoring wells are designed as a variation on one of three general types: (1) *saturation zone*, slotted screen-type monitoring wells in soil or weak rock; (2) *unsaturated* (vadose) *zone*, lysimeter-type monitoring wells in soil; and (3) *open hole*, often multipoint sampling wells in rock [2].

A *saturated zone, screened monitoring well* in soil is usually installed in a drilled hole (Fig. 1), with a slotted well screen below a flush-coupled riser column and with a granular filter placed between the slotted well screen and the borehole wall (Figs. 2 and 3). Well screen lengths are typically 1.5 or 3.0 m (5 or 10 ft). Since about 1984, there has been a trend toward the use of only one well screen per monitoring well in soil, in recognition of the practical difficulty in

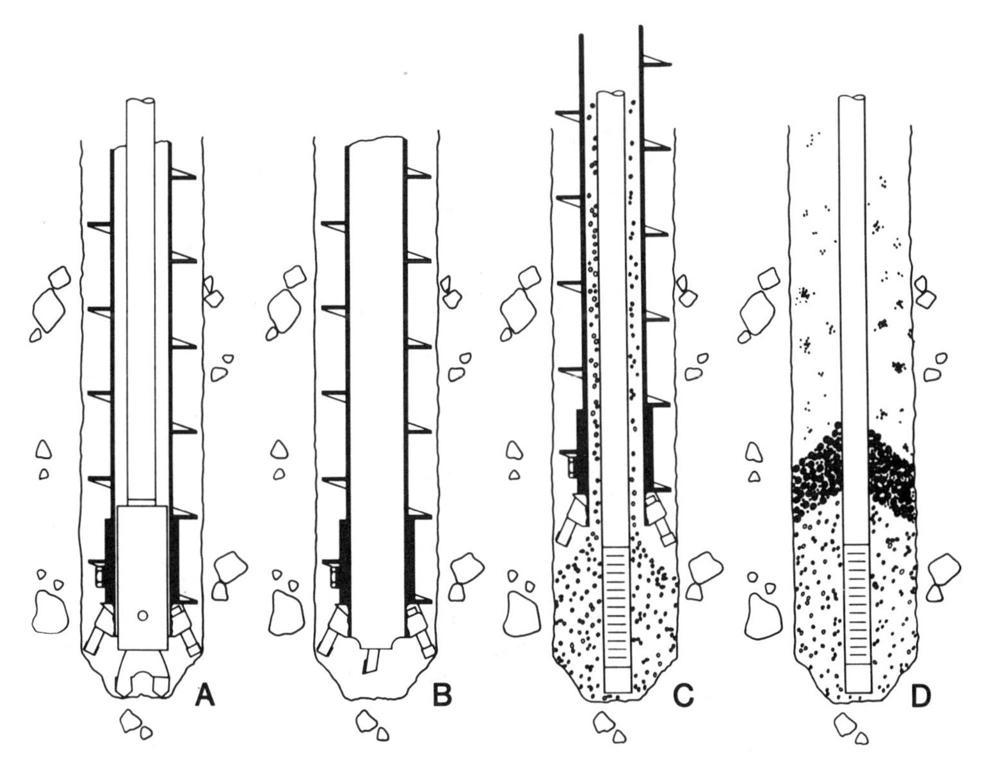


FIG. 1—*Step-by-step procedure of installing a round-water monitoring well with hollow-stem augers:* (a) *drilling with a pilot assembly;* (b) *pilot assembly removed for insertion of screen and casing;* (c) *placing the filter as augers are withdrawn; and* (d) *completed well with filter, seal, and backfill.*

constructing multiple screened wells. Such wells are difficult, if not impossible, to install with assurance that there is *no* communication along the well annulus between screened intervals. Well "clusters" are now commonly used in soil to obtain water samples at specific but different depth intervals. Most saturated zone monitoring wells are 51 mm (2 in.) in nominal diameter, to allow installation and operation of the smallest, readily available water samplers and sampling pumps [*3*]. Some 102-mm (4 in.) nominal diameter, or larger, wells are installed for operation of larger pumps, particularly when hydraulic conductivity tests are performed by pumping the well.

*Vadose zone monitoring wells* (Fig. 4) usually are installed at relatively shallow depths. The vadose zone sampling device usually consists of some design variation of a vacuum lysimeter. There have been several evaluations of using lysimeters for ground-water monitoring in the vadose zone [*4–14*]. *Vacuum lysimeters* extract soil pore water by applying a partial vacuum to the soil through a porous element of the lysimeter. The selection of the porous element of the lysimeter is important and requires consideration of the grain size distribution and degree of saturation of the soil being evaluated. The vacuum is also used for shallow installations to pump the water to a container above ground surface. Vacuum lysimeters are generally adequate for detecting the presence of a contaminant in partially saturated soil at relatively shallow depths or relative changes in contaminant concentrations. Analyses of contaminant concentrations in vadose zone soil pore water are made also by *in situ* gas chromatograph testing or by direct laboratory testing of an "undisturbed" soil sample.

*Ground-water monitoring in rock* (Fig. 5) is often accomplished with multipoint water sampling

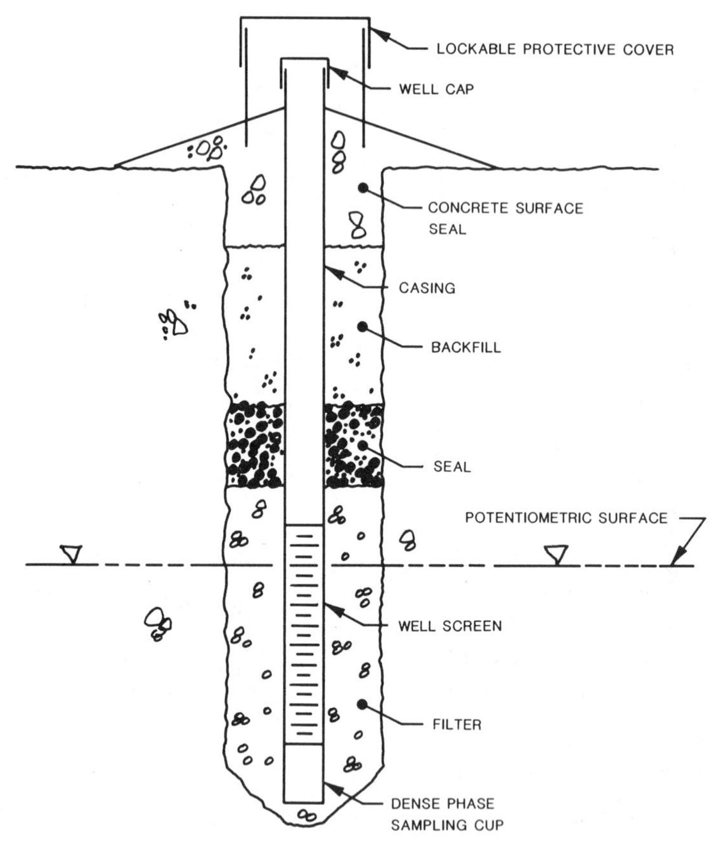


FIG. 2—*Diagram of a monitoring well with the uppermost ground-water level intersecting the slotted well screen. The positioning of the screen with respect to the potentiometric surface can be critical for sampling low-density contaminants.*

from isolated zones of an open borehole [*15–18*]. Multipoint sampling is accomplished by isolating or "packing off" a borehole interval and pumping the interval adequately to obtain a representative water sample or by using a multiport sampling device. The success of sealing and sampling can be qualitatively predicted from observations of rock core obtained from the test borehole and adjacent boreholes with a split-inner-tube core barrel.

## Site Characterization for the Design of Monitoring Well Systems

Each monitoring well system is truly unique and can be designed and installed only by following careful geologic evaluation and interpretation of each specific site. Geologic interpretation begins with the personal, site-specific knowledge of the project engineer or geologist of record, and ends with observation of the installation and development of each monitoring well. The required extent of geologic evaluation or interpretation is a qualitative decision of the engineer or geologist of record and depends upon many factors: for example, (1) the type of monitoring system being installed, that is detection or enforcement; (2) the geology of the region and of the site; (3) the quality of site-specific geologic and hydrogeologic information; (4) the experience of the engineer or the geologist of record; and (5) the possibility of secondary, off-site induced influences on

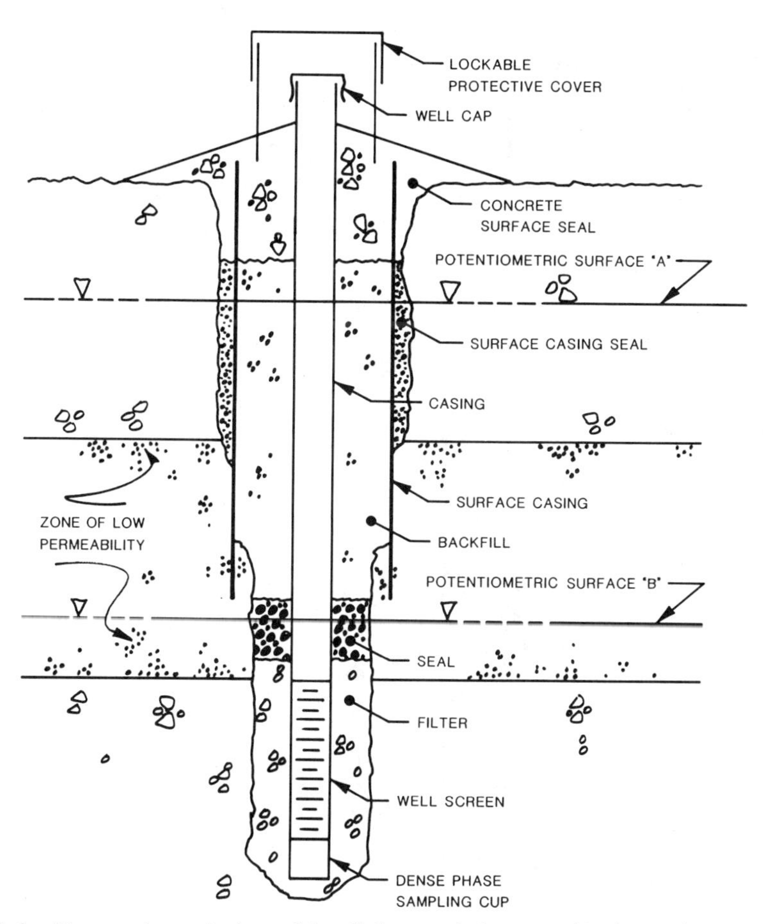


FIG. 3—*Diagram of a monitoring well installed to sample from a combined ground-water zone.*

ground-water flow or ground-water quality. In many cases, an extensive exploration drilling, and sampling program will be required to support geologic interpretation as the basis for design and installation of the monitoring well system.

## Scope of the Monitoring System

Monitoring system scope is bound to deal with one series of questions: *How many* of *what diameter* at *what locations* to *what depth* with screens of *what length*? *Any* answer (*the* answer exists but is never exactly known at the time of design or installation) depends on the purpose of the system, that is detection, assessment, or enforcement; the geologic interpretation of the site; and the personal approach of the engineer or geologist of record. Ideally, a detection system would be first installed at a site, subsequently to or concurrently with a detailed geological evaluation and interpretation of the site. Such evaluations may involve surface geophysical surveys or exploratory borings, possibly with continuous soil sampling or rock coring, downhole geophysical surveying, and in-hole hydraulic conductivity testing [*19–25*]. In all cases the "indirect" data of geophysical surveys should be substantiated with "direct" data obtained from *in situ* testing or from laboratory testing of representative samples of soil, rock, or water. It may or may not be

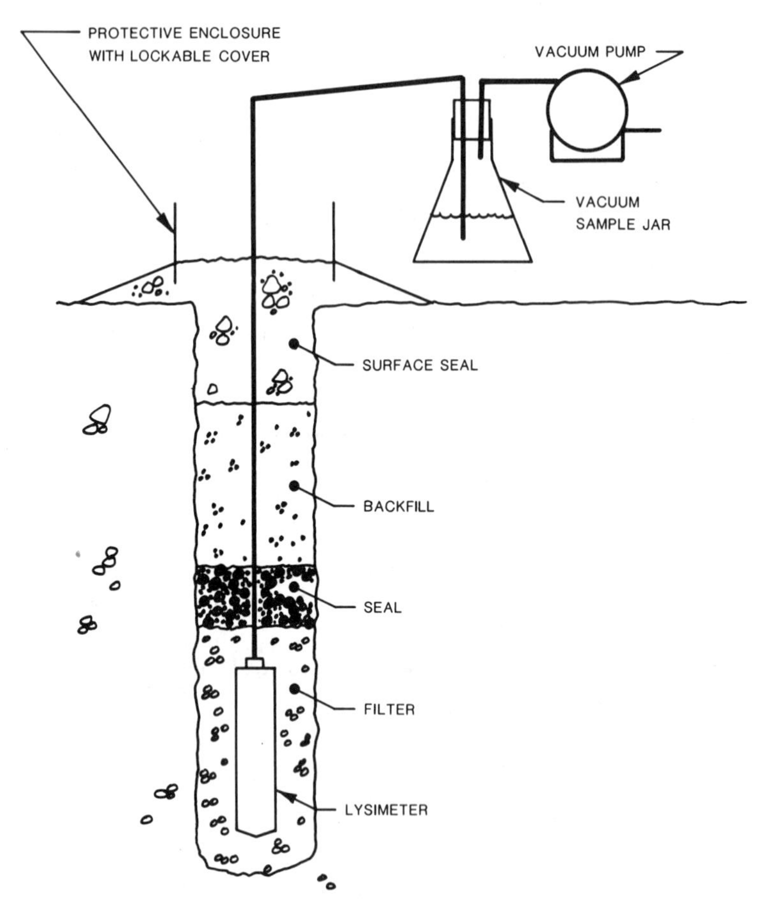


FIG. 4—*Diagram of a shallow vacuum lysimeter installation. Some alternate designs utilize two conduction tubes for pressure evacuation of the filled lysimeter.*

feasible to convert the initial exploratory borings to piezometers, or to hydraulic conductivity test wells, or to monitoring wells. Often, more exploration borings are required to characterize the geology and hydrogeology than are required for hydraulic conductivity testing and as monitoring wells.

Detection monitoring wells are usually placed to surround a suspected source of ground-water contamination. The wells are usually concentrated down-gradient from the suspected source, and within the zone of "flow paths" that pass through or below the source (Fig. 6).

Up-gradient and lateral wells are also necessary to provide a basis of interpretation of background water quality. Up-gradient and lateral wells are also used to investigate the possibility of geologically unusual, or complex, or seasonally variable seepage gradients.

Hydraulic conductivities are sometimes estimated from visual observation and mechanical identification tests of soil specimens. Falling-head "slug" tests or rising-head "slug-recovery" tests [26,27] provide design-related field conductivity values without withdrawing large volumes of potentially contaminated water such as are produced from constant-discharge pumping tests.

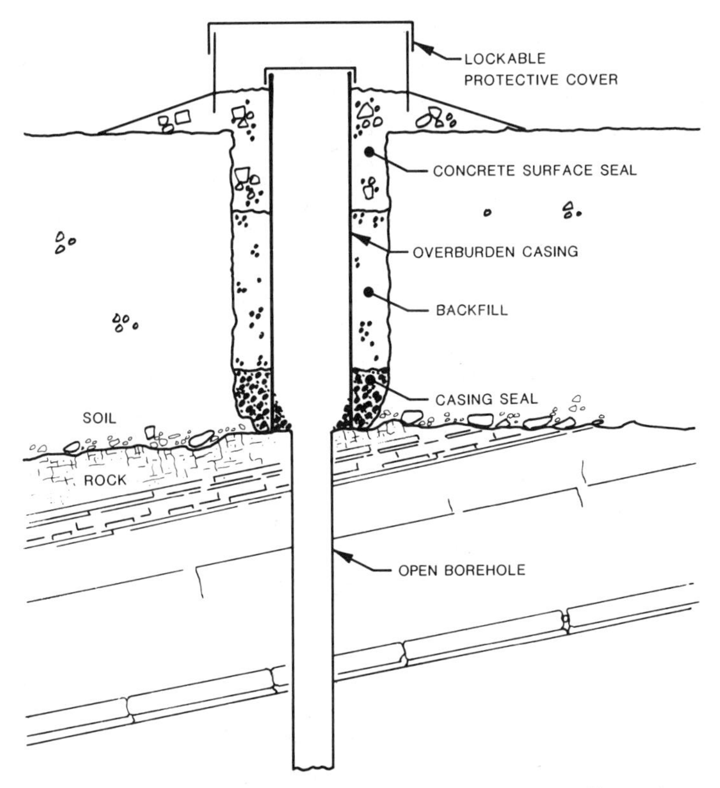


FIG. 5—*Diagram of an open-hole ground-water monitoring well in rock.*

## Monitoring Well Design Constraints

Each monitoring well requires a design that reflects site geology and hydrology, waste site operational history and anticipated contaminants to be encountered. Installation practicality must also be considered in determining the final well configuration—often to a degree greater than the theoretical performance requirements of well materials and well geometry.

For example, many ground-water monitoring wells are installed through 158-mm (6.25 in.) inside diameter hollow augers which are readily available for use on multipurpose drill rigs in North America. These hollow augers provide a well bore diameter of approximately 265 mm (10.5 in.). The designer of a monitoring well may desire and specify a 300-mm (12-in.) diameter well bore, but if a 300-mm (12-in.) well bore diameter is specified, the requirement should have definite theoretical basis because the tools to drill the 300 mm (12 in.) borehole may not be readily available and the procurement of such tools may contribute considerable time and expense to the project. Monitoring well designers should make inquiries to manufacturers and local installation practitioners as to the installation feasibility of their proposed design.

Most monitoring wells in soil are constructed using either 51-mm (2-in.) or 102-mm (4-in.) nominal diameter, flush-coupled casing and screens. The smallest hollow auger that is typically used to install a 51-mm (2-in.) nominal diameter well has an 82-mm (3.25-in.) inside diameter hollow stem. The smallest hollow auger that is typically used to install a 102-mm (4-in.) nominal diameter well has a 158-mm (6.25-in.) inside diameter hollow stem. The most difficult aspect of

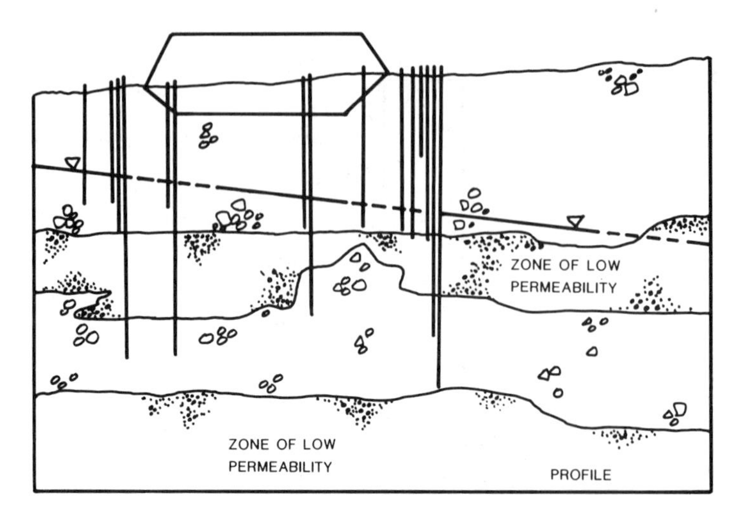


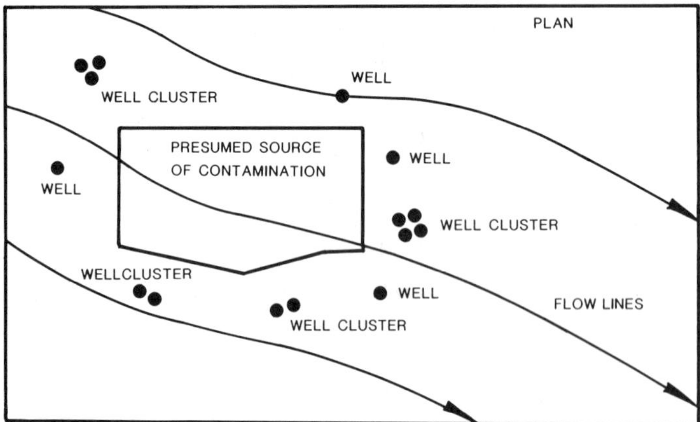


FIG. 6—*Diagram of a ground-water monitoring well system for a presumed source of contamination. The scale is purposely omitted.*

the installation procedure is the placement of the granular filter material and particularly the seal that is usually placed directly above the filter (Figs. 2 and 3). Placement of filter and sealing materials (Fig. 1*c*) for 51-mm (2-in.) nominal diameter well installations is relatively easy when 108-mm (4.25-in.) inside diameter hollow augers are used, somewhat more difficult when 95-mm (3.75-in.) inside diameter hollow augers are used, and very difficult, particularly below the ground-water level, if attempted with 82-mm (3.25-in.) inside diameter hollow augers.[3]

Filter materials must be selected not only for optimum mechanical performance, but they must be also "installable." For example, the particle size distribution of two typical well filter materials is shown in Fig. 7 (these two gradations are commonly available for use in the St. Louis, Missouri region). Material "A" will pass downward with relative ease through the annulus of a 95-mm (3.75 in.) inside diameter hollow auger and 51-mm (2-in.) nominal diameter well casing, even with a water level within the annulus, whereas Material "B" tends to "arch" or become suspended

[3] C. Harris and C. Roberts, private communication, John Mathes and Associates, Columbia, IL, Jan. 1986.

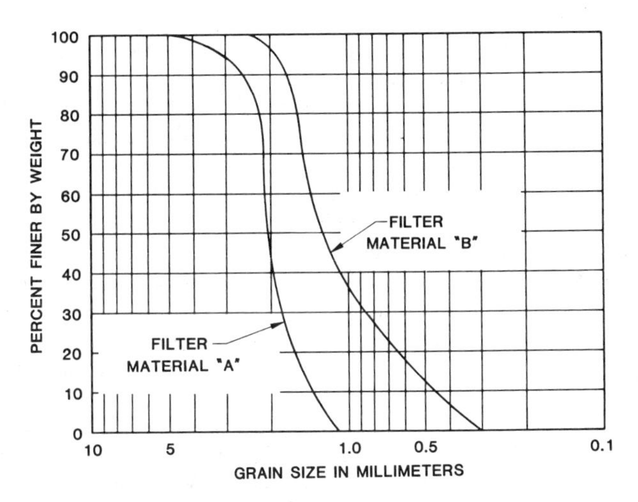


FIG. 7—*Grain size distribution of two granular filter materials.*

within the annulus, particularly in the vicinity of a water surface. Material "B" may be the better filter for the ground conditions and the selected screen openings; however, if 95-mm (3.75-in.) inside diameter (or smaller) hollow augers, rather than 108-mm (4.25-in.) inside diameter, must be used for the installation, then Material "A" will probably provide, in the end, the better installation.

The sealing materials that are placed above the filter usually consist of "pellets" of compressed, moistened, sodium montmorillonite clay powder (bentonite drilling fluid additive). Bentonite pellets are usually much more difficult to place at the desired location within the well annulus than is the granular filter. If the sealing zone is at considerable depth below the water surface, the pellets can be placed using a tremie tube, with the tube kept "dry" under gas pressure. The designer and the installer of a monitoring well *must* be assured of the chemistry of the bentonite pellets, that is, that the pellets do not contain a polymer or another chemical contaminant. Bentonite suppliers are now providing large bentonite granules which, under some circumstances, may be easier to place within the annulus than pellets. Under most circumstances the sealed zone should be at least 600 mm (2 ft) thick (vertical dimension).

The well annulus above the sealed zone should be backfilled with a low-permeability material, usually and preferably of known chemical properties. The backfill material that is often used consists of a portland cement "bentonite" water grout consisting of about 3 to 10 weight units of bentonite powder to 94 weight units of portland cement and about 110 weight units of water (usually the bentonite is first mixed with water, after which the cement is added). The bentonite-cement grout is usually mixed to a very stiff consistency and pumped to the bottom of the well annulus through a grout pipe. A 3:94 bentonite:cement ratio grout will become very rigid following installation, whereas a 10:94 bentonite:cement ratio grout will become stiff but remain somewhat "plastic." The use of cement as a component of the backfill at depths greater than about 2 m (6 ft) below ground surface is a questionable practice [*28*]. Cement does provide stability or rigidity to the backfill but also alters ground-water chemistry immediately surrounding the backfill. As a rule of good practice, drill cuttings should never be used as backfill for fear of introducing contamination.

The uppermost zone of the well annulus, about 2 m (6 ft), is usually backfilled with a concrete grout, often formulated with expanding cement. This grout is usually extended about 150 mm (6 in.) above ground surface and finished to slope (that is, to drain) away from the well axis to a radius of about 1.3 m (4 ft) (refer to Figs. 2 and 3). It is believed that this *concrete* plug and apron inhibit potential downhole infiltration of surface water.

The basic well section shown in Fig. 2 represents, schematically and practically, one of the simplest and easiest installations. The number of variations of this oversimplified well section that may be required in practice are infinite.

Figure 3 presents another oversimplified illustration of one of the difficult situations in monitoring well design and installation. This type of monitoring well is designed for placement within a ground-water zone below the uppermost ground-water zone. The intakes for these wells must be installed and isolated from the uppermost zone so that potential contaminants are not transported from the upper zone to the lower zone (or from the lower zone to the upper zone) during and following installation. This type of installation usually requires placement of sealed casing within the upper zone that is large enough to allow drilling through the casing into the lower zone without disrupting the seal of the casing. It is not unusual to use more than one drilling method to accomplish this type of installation. The design of the upper casing seal must consider liquid head differentials that may occur during the installation procedures and during the life of the well.

There is no such thing as a "standard" monitoring well. Wells must be "custom designed" according to site geology and hydrogeology, and the key parameters of the suspected contaminants, that is, ionic chemistry, solubility, miscibility, and density.

## Well Materials

Selection of well casing and intake screen materials is currently the most controversial or "unsettled" aspect of ground-water monitoring practice [*1,29–34*]. The controversy centers on the relative ability of well materials to minimize sorption, desorption, or leaching of contaminants, which can cause false indications of lower or higher ground-water contamination levels than actually exist. The same chemical sorption/desorption characteristics of sampling and sample storage devices must also be considered.

Since the early 1980's, monitoring well casings and screens have most commonly been fabricated using either *thermoplastics*, particularly high-quality, rigid polyvinyl chloride (PVC), *fluoroplastics*, the most common being tetrafluoroethylene (TFE), or *stainless steel* (SS), usually Type 304 or Type 316.

Comparative sorption or leaching characteristics should be used as a basis for choice of well materials. Stainless steel or TFE are often the preferred materials when organic contaminants are suspected. PVC or TFE are often the preferred materials when metallic contaminants are suspected. However, practice and research concerning sorption/leaching characteristics of well casing and screen material properties have not resulted in a body of *conclusive* evidence. Type 316 SS is about 40% more expensive than Type 304 SS; however, the positive performance capabilities of Type 316 are only slightly better than Type 304 for ground-water monitoring applications. Stainless steel is about 5 to 10 times more expensive than PVC, and TFE is about 15 to 30 times more expensive than PVC, depending upon the tubing wall thickness of the TFE.[4,5] The comparative qualities of TFE and PVC are the source of most disagreements among and between researchers and users; however, TFE is almost always considered to provide the higher-quality installation. The significance of the cost differences depends upon the type of system being installed and the level of detection required. If a shallow detection monitoring well system is being installed at a site where a 5-ppb level of a suspected highly toxic contaminant is significant and water quality tests are performed on a monthly basis, then the cost difference would probably be made insignificant because of the costs of water sampling and testing. However, if several assessment monitoring

[4] D. Kill, private communication, Johnson Division UOP, Minneapolis, MN, Jan. 1986.

[5] B. V. Varhol, private communication, Brainard-Kilman Drill Co., Tucker, GA, Jan. 1986.

wells are being installed at a site where 500 ppb of the same contaminant has been already detected, then the cost difference might be significant. Also, the selection of well material constituents may depend upon factors such as the "attitude" of those presumed responsible for the contamination as well as the "attitude" of those responsible for detecting and reporting the existence of a contaminant. In other words, it may be necessary to use the most chemically inert materials, which may represent the most expensive installation, to assure a court that an analysis of site-contaminant transport is as accurate as possible, whereas in the absence of a required court judgment against those causing the contamination, less expensive methods and materials might provide (at least in part) a totally adequate analysis for the design of a remedial engineering ("cleanup") program.

## Drilling Methods and Procedures

Ground-water monitoring wells must be installed with proper equipment and techniques or the wells will be of limited value or even detrimental to evaluating hydrogeology and ground-water quality. Most monitoring wells are installed in boreholes drilled with geotechnical exploration-type drill rigs. The installation of some monitoring wells requires the use of water well drilling equipment. Several drilling methods are in common use in North America but usually only one specific method or combination of methods is most economically and technically appropriate for a particular well installation. The engineer or geologist of record must understand the characteristics and applicability of these common drilling methods:

The *hollow-stem auger method* is the most used monitoring well installation method (Fig. 1). Probably greater than 90% of all monitoring wells installed in soil in North America are installed with hollow-stem augers. A hollow-stem auger column has a continuously open axial stem which allows a borehole to be simultaneously drilled and cased without the use of drilling fluids. Because the borehole is drilled "dry," there is the capability of optimum control of soil cuttings, thus minimal contamination of the ground and the air in the vicinity of the borehole. The use of hollow-stem augers also allows the option of continuous, split-barrel, core sampling of soil or weak rock simultaneously with advance of the augers. Hollow-stem augers have been used to advance soil borings to depths greater than 100 m (300 ft) but are usually used only to maximum depths of about 25 to 50 m (75 to 150 ft) according to soil density or hardness, ground-water depth, the weight of the drill rig, and the power available at the drill spindle. The typical maximum depth of drilling, in *metres*, with 82-mm (3.25-in.) inside diameter to 108-mm (4.25-in.) inside diameter hollow augers is approximately equal to one half the horsepower available at the drill spindle. Hollow-stem diameters ranging from 82-mm (3.25-in.) inside diameter to 210-mm (8.25-in.) inside diameter are in common use for monitoring well installations. The borehole diameter is usually about 100 mm (4 in.) larger than the auger stem inside diameter. When a monitoring well borehole is completed to final depth, the inner components of the auger column assembly (pilot assembly and center rods) are removed (Fig. 1*b*) and the monitoring device is inserted through the hollow axis of the auger column. The augers are then "pulled back," leaving the monitoring device in place. Filter and backfill materials are usually placed through the annular space of the auger hollow-stem and the well casing as the augers are retracted (Fig. 1*c*). When hollow augers are used to drill below the ground-water level, seepage inflow at the bottom of the auger can usually be controlled by careful drilling techniques or the use of special pilot assemblies [*35*] or by injecting a drilling fluid of known chemical quality into the hollow augers during drilling (Fig. 8).

*Continuous flight, "solid stem" augers* provide an efficient method of drilling relatively shallow borings in fine-grained soils above the ground-water level. In general, continuous flight augers are of limited use when drilling in very soft, fine-grained, caving soils, in "clean" granular soils, and in almost all soils below the ground-water level. It is important to understand that it is nearly

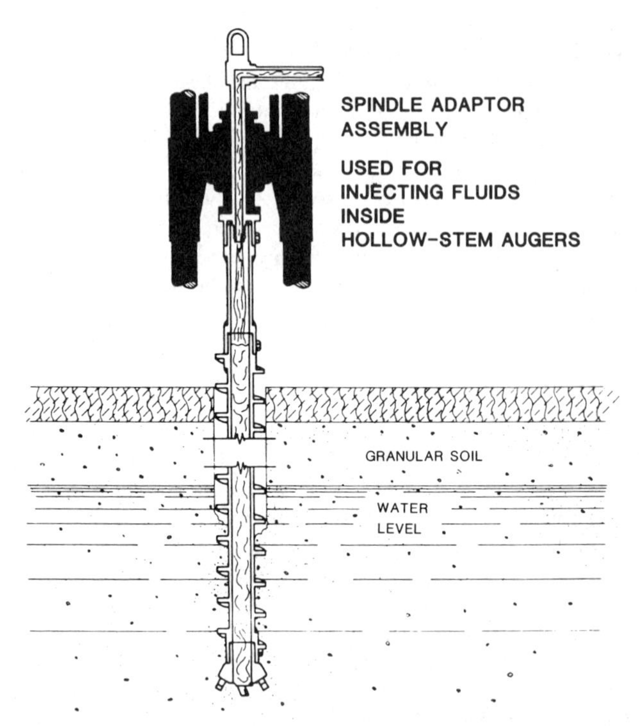


FIG. 8—*Injecting a drilling fluid through hollow-stem augers during drilling.*

impossible to drill through a contaminated soil zone with solid-stem continuous flight augers without transporting contaminants downward.

*Hydraulic rotary drilling* is also termed "straight rotary drilling" or "mud rotary drilling." Soil or rock cuttings produced by the drilling process are removed from the borehole with a liquid that is pumped downward through a drill rod column, passing through a bit and then upward within the annular space between the borehole wall and the drill rods. Drilling "mud" is the liquid circulation medium and usually consists of water and one or more *additives*. Commercially available, processed montmorillonitic clay (bentonite) powder is the most used rotary drilling fluid additive. A high-quality (that is, for the purpose of drilling) bentonite additive consists of sodium montmorillonite clay *with* polymer additives and with impurities consisting of various other naturally occurring soil particles. "Standard grade" bentonite powder is also readily available but does not contain polymer additives. Other frequently used drilling fluid additives include cellulose polymers, guar gum polymers, anionic acrylimide polymers, sodium carbonate (soda ash), potassium chloride (0-0-60 fertilizer), barium sulfate (barite), and various lost-circulation materials such as wood fibers, shredded paper, and processed mica. The use of hydraulic rotary drilling for monitoring well installations may be a requirement in some geological environments; however, the use of hydraulic rotary drilling for monitoring well installations is a matter requiring a complete understanding and a detailed analysis and reporting of the chemistry of the drilling water and of all drilling fluid additives. It is important to understand that drilling fluid circulation can easily transport contaminants along the borehole wall. At the termination of drilling of each borehole at a site of suspected contamination, the drilling fluid must be recovered, contained, and RCRA-managed as a contaminated material. The complete recirculation system of the drill rig must be thoroughly cleaned and decontaminated before moving to the next well location. When drilling mud is used to install a monitoring well, the well will usually require extensive "development" to remove "mud cake" from the borehole wall, particularly in the zone of the well screen. In some cases, development time will exceed drilling and installation time.

*Air rotary drilling* may provide the best available drilling method for some monitoring well installations. An air rotary drilling system is similar to a hydraulic rotary system. Air, possibly with foaming additives, rather than drilling mud, is the circulation medium. There are two major concerns when air is used for drilling at sites of suspected contamination: (1) injection of a contaminant such as a foaming agent or a compressor lubricant with the air, and (2) contamination of the drill site with the return air. Air filtering systems are available for both the injected air and the return air.

Under some circumstances, *core drilling* is used for installation of a monitoring device in rock. During core drilling a core barrel is rotated and axially advanced as rock is cut or abraded by the bit and forced into the core barrel. A drilling fluid, usually water without additives, is circulated to flush and cool the bit and to transport the cuttings out of the drilled hole. Core drilling with air is generally limited to coring of stiff or hard soil and sedimentary rock. Any additives to a drilling fluid, liquid or air, used for core drilling provide the same benefits to the core drilling procedure and cause the same potential contamination problems to the monitoring well installation as with other rotary drilling procedures.

*Wash borings* are advanced in soil by driving 1.5-m (5-ft) sections of threaded and coupled steel pipe or flush-coupled casing. Soil enters the inside of the pipe column or casing during driving and is removed with water applied through a small-diameter pipe column, usually with a chopping or fishtail bit attached at the bottom. Following advancement and cleaning of the casing, a monitoring device with required filter and backfill materials can be installed through the casing which is then removed, usually by back-driving.

Under some circumstances *cable-tool drilling* may be used completely or in part to install a monitoring well. A cable-tool drill is operated off of a rotating crank such that a walking beam with attached cable sheaves can be used to raise and drop a heavy string of impact drilling tools. Each drop of the tool string provides one impact of a percussion drilling bit on the rock (or soil) face at the bottom of the borehole. Drilling progress is relatively slow and usually inefficient in comparison with auger or rotary methods. Cable-tool drills can be used in ground-water monitoring practice for drilling in rock and under some circumstances to advance casing in soil, particularly on projects where a casing is required and the use of drilling fluids is not permitted or is limited. Sometimes a cable-tool drill and a drill with augering capability can be used in combination to advance and case a large-diameter borehole for installation of a ground-water sampling or monitoring device.

## Water Sampling and Water Sampling Devices

Obtaining a water sample is usually the final field "problem" of ground-water monitoring. The "problem" is to obtain a "representative" water sample, usually from within a 51-mm (2-in.) nominal diameter well screen and casing that initially contains "stagnant" water that is considered *not* representative.

Barcelona et al. [*36*] have compiled a most comprehensive study of the sampling capabilities of ground-water sampling devices. Nielson and Yeates [*3*] present an overview of the availability, advantages and disadvantages, and relative costs of several sampling devices. Several other investigators [*37–41*] have evaluated and discussed various sampling alternatives. A water sampling device must (1) fit easily within the monitoring well, (2) be fabricated of materials that will not significantly react with and alter the sample chemistry, and (3) not alter—through the physical processes of sampling—significantly the chemistry of the sample. It is economically desirable that the sampling device or pump have the capability of purging the well of stagnant water before being used to obtain a representative sample.

Several studies indicate that it is most difficult to accurately sample ground water containing volatile organic contaminants [*1,30,31,36–38*]. It is believed that some volatiles vaporize and escape when subjected to the vacuum of a vacuum pump, to a water-gas surface of a gas operated pump, to the turbulence within a pump impeller bowl, or even to atmospheric conditions. A gas operated bladder (or ''squeeze'') pump is often recommended when organic volatiles must be sampled. A relatively inexpensive, recently developed, but readily available hand-operated piston pump should also be considered for such applications.

**Reporting and Control**

Many ground-water monitoring projects are designed and installed with the anticipation of subsequent legal action. This aspect of ground-water monitoring requires project-specific quality-assurance specifications and detailed graphical and written verification of the work performed. Such items as soil or water sample location and identification, sample sealing, ''packaging'' and transport procedures, and soil or water sample accountability must be accurately reported in detail. Each well installation must be graphically recorded indicating the geometry of the drilled hole, drilling method or methods, size, location, and description of each well component, and the nature, quantities, and installation procedures for all filters, annular seals, and backfill materials.

Such documentation is better prepared with the aid of forms and checklists. Some forms and checklists can be of a general nature such as those used to report personnel presence and responsibilities, site visitors, equipment on-site (by manufacturer, model, and serial number), weather conditions, and a general description of the work completed. Other forms and checklists must be prepared on a project-specific basis, or even for individual well installations. The emphasis of detailed reporting is usually to document (1) personnel safety, (2) proper decontamination techniques, (3) verification of geology, (4) verification of soil, rock or water sample origins, and (5) verification of well construction details.

**Summary**

Ground-water monitoring well design and installation are usually managed by engineers and geologists who have been working in both the qualitative and quantitative areas of ground-water science and particularly by geotechnical engineers who have been closest to geotechnical exploration, site characterization, and ground-water engineering. However, a ''new way of thinking'' about geotechnical practice is required to document the intricate procedures and high levels of accuracy. This new level of professional practice requires an intense professional participation not previously witnessed by many engineers and geologists.

Fortunately, most of the equipment and techniques required for installation of ground-water monitoring systems were in use prior to the current upsurge in ground-water monitoring activity. All ground-water monitoring systems must be custom-designed according to site geology, the equipment and materials available for the installations, and the nature of the suspected contaminants. Monitoring well system designers must consider the costs and benefits of manufacturing custom-designed drilling tools and well materials versus the capabilities and benefits of using or adapting readily available equipment and materials. The effects of using a particular combination of well and sampler materials on sample quality is the most unsettled aspect of ground-water monitoring practice. There are indications that further research on well material constituents is required.

Engineers and geologists must be aware that equipment and installation techniques for ground-water monitoring are rapidly evolving with regard to maintaining high-quality chemical assessments. Engineers, geologists and other involved professionals must take the time to study and critically review the literature, particularly the U.S. Envrionmental Protection Agency RCRA an CERCLA Technical Resource Documents and Guidance Manuals that are being released at a rapid rate.

## References


[1] Barcelona, M. J., Gibb, J. P., and Miller, R. A., "A Guide to the Selection of Materials for Monitoring Well Construction and Ground-Water Sampling," ISWS Contract Report 327, Illinois State Water Survey, Champaign, IL, 1983.

[2] Scalf, M. R., McNabb, J. F., Dunlap, W. J., Cosby, R. L., and Fryberger, J., "Manual of Ground-Water Sampling Procedures," National Water Well Association, (NWWA/EPA Series), Worthington, OH, 1981.

[3] Nielson, D. M. and Yeates, G. L., "A Comparison of Sampling Mechanisms Available for Small-Diameter Ground-Water Monitoring Wells," *Ground-Water Monitoring Review*, National Water Well Association, Vol. 5, No. 2, 1985, pp. 83–99.

[4] Emenhiser, T. C. and Singh, U. P., "Innovative Sampling Techniques for Ground-Water Monitoring at Hazardous Waste Sites," *Ground-Water Monitoring Review*, National Water Well Association, Vol. 4, No. 4, 1984, pp. 35–37.

[5] Everett, L. G., "Monitoring in the Vadose Zone," *Ground-Water Monitoring Review*, National Water Well Association, Vol. 1, No. 2, 1981, pp. 44–51.

[6] Everett, L. G. and McMillion, L. G., "Operational Ranges for Suction Lysimeters," *Ground-Water Monitoring Review*, National Water Well Association, Vol. 5, No. 3, 1985, pp. 51–60.

[7] Everett, L. G., Wilson, L. G., Hoylman, E. W., and McMillion, L. G., "Constraints and Categories of Vadose Zone Monitoring Devices," *Ground-Water Monitoring Review*, National Water Well Association, Vol. 4, No. 1, 1984, pp. 26–32.

[8] Johnson, T. M., Cartwright, K., and Schuller, R. M., "Monitoring Leachate Migration in the Unsaturated Zone in the Vicinity of Sanitary Landfills," *Ground-Water Monitoring Review*, National Water Well Association, Vol. 1, No. 3, 1981, pp. 55–63.

[9] Morrison, R. D., "Ground-Water Monitoring Technology," Timco Manufacturing Co., Prairie Du Sac, WI, 1983.

[10] Nelson, H. B., "Use of Vacuum Tanks with Shallow Vacuum Lysimeters," *Ground Water*, Vol. 23, No. 6, 1985, pp. 802–803.

[11] Robbins, G. A. and Gemmell, M. M., "Factors Requiring Resolution in Installing Vadose Zone Monitoring Systems," *Ground-Water Monitoring Review*, National Water Well Association, Vol. 5, No. 3, 1985, pp. 75–80.

[12] Wilson, L. G., "Monitoring in the Vadose Zone Part I: Storage Changes," *Ground-Water Monitoring Review*, National Water Well Association, Vol. 1, No. 3, 1981, pp. 32–41.

[13] Wilson, L. G., "Monitoring in the Vadose Zone Part II," *Ground-Water Monitoring Review*, National Water Well Association, Vol. 2, No. 1, 1982, pp. 31–42.

[14] Wilson, L. G., "Monitoring in the Vadose Zone: Part III, *Ground-Water Monitoring Review*, National Water Well Association, Vol. 3, No. 4, 1983, pp. 155–166.

[15] Barvenik, M. J. and Cadwgan, R. M., "Multilevel Gas-Drive Sampling of Deep Fractured Rock Aquifers in Virginia," *Ground-Water Monitoring Review*, National Water Well Association, Vol. 3, No. 4, 1983, pp. 34–40.

[16] Cherry, J. A. and Johnson, P. E., "A Multilevel Device for Monitoring in Fractured Rock," *Ground-Water Monitoring Review*, National Water Well Association, Vol. 2, No. 3, 1982, pp. 41–44.

[17] Davison, C. C., "Monitoring Hydrological Conditions in Fractured Rock at the Site of Canada's Underground Research Lab," *Ground-Water Monitoring Review*, National Water Well Association, Vol. 4, No. 4, 1984, pp. 95–102.

[18] Pickens, J., Cherry, J., Coupland, R., Grisak, G., Merritt, W., and Risto, B., "A Multilevel Device for Ground-Water Sampling," *Ground-Water Monitoring Review*, National Water Well Association, Vol. 1, No. 1, 1981, pp. 48–51.

[19] Gilkeson, R. H. and Cartwright, K., "The Application of Surface Electrical and Shallow Geothermic Methods in Monitoring Network Design," *Ground-Water Monitoring Review*, National Water Well Association, Vol. 3, No. 3, 1983, 30–42.

[20] Glaccum, R. A, Benson, R. C., and Noel, M. R., "Improving Accuracy and Cost-Effectiveness of Hazardous Waste Site Investigations," *Ground-Water Monitoring Review*, National Water Well Association, Vol. 2, No. 3, 1982.

[21] Greenhouse, J. P. and Monier-Williams, M., "Geophysical Monitoring of Ground-Water Contamination Around Waste Disposal Sites," *Ground-Water Monitoring Review*, National Water Well Association, Vol. 5, No. 4, 1985, pp. 63–69.

[22] Greenhouse, J. P. and Slaine, D. D., "The Use of Reconnaissance Electromagnetic Methods to Map Contaminant Migration," *Ground-Water Monitoring Review*, National Water Well Association, Vol. 3, No. 2, 1983, pp. 47–59.
[23] Sendlein, L. V. A. and Yazicigil, H., "Surface Geophysical Methods for Ground-Water Monitoring, Part I," *Ground-Water Monitoring Review*, National Water Well Association, Vol. 1, No. 3, 1981, pp. 42–46.
[24] Sweeney, J. T., "Comparison of Electrical Resistivity Methods for Investigation of Ground-Water Conditions at a Landfill Site," *Ground-Water Monitoring Review*, National Water Well Association, Vol. 4, No. 1, 1984, pp. 52–59.
[25] Yazicigil, H. and Sendlein, V. A., "Surface Geophysical Techniques in Ground-Water Monitoring: Part II," *Ground-Water Monitoring Review*, National Water Well Association, Vol. 2, No. 1, Winter 1982, pp. 56–62.
[26] Leap, D. I., "A Simple Pneumatic Device and Technique for Performing Rising Water Level Slug Tests," *Ground-Water Monitoring Review*, National Water Well Association, Vol. 4, No. 4, 1984, pp. 141–146.
[27] Patterson, R. J. and Devlin, J. F., "An Improved Method for Slug Tests in Small-Diameter Piezometers," *Ground Water*, Vol. 23, No. 6, 1985, pp. 804–805.
[28] Dunbar, D., Tuchfeld, H., Siegel, R., and Sterbentz, R., "Ground-Water Quality Anomalies Encountered during Well Construction, Sampling and Analysis in the Environs of a Hazardous Waste Management Facility," *Ground-Water Monitoring Review*, National Water Well Association, Vol. 5, No. 3, 1985, pp. 70–74.
[29] Curran, C. M. and Tomson, M. B., "Leaching of Trace Organics into Water from Five Common Plastics," *Ground-Water Monitoring Review*, National Water Well Association, Vol. 3, No. 3, 1983, pp. 68–71.
[30] Gibb, J. P., Schuller, R. M., and Griffin, R. A., "Procedures for the Collection of Representative Water Quality Data from Monitoring Wells," Illinois State Water Survey/Geological Survey Cooperative Ground-Water Report No. 7, Champaign, IL, 1981.
[31] Miller, G., "Uptake and Release of Lead, Chromium, and Trace Level Volatile Organics Exposed to Synthetic Well Casing," *Proceedings*, Second National Symposium on Aquifer Restoration and Ground-Water Monitoring, National Water Well Association, Columbus, OH, 1982, pp. 236–245.
[32] Pettyjohn, W. A., Dunlap, W. J., Cosby, R. L., and Kelley, J. W., "Sampling Ground Water for Organic Contaminants," *Ground Water*, Vol. 19, No. 2, 1981, pp. 180–189.
[33] Reynolds, G. W. and Gillham, R. W., "Absorption of Halogenated Organic Compounds by Polymer Materials Commonly Used in Ground-Water Monitors," Second Annual Canadian/American Conference on Hydrogeology: Hazardous Wastes in Ground-Water—A Solvable Dilemma, Banff, Alberta, Canada, June 25–29, 1985.
[34] Vardy, P., "Presentation to the Ground-Water Monitoring Guidance Review Subcommittee of the Environmental Engineering Committee Science Advisory Board (USEPA)," Waste Management, Inc., Oak Brook, IL, Aug. 1985.
[35] Perry, C. A. and Hart, R. J., "Installation of Observation Wells on Hazardous Waste Sites in Kansas using a Hollow-Stem Auger," *Ground-Water Monitoring Review*, National Water Well Association, Vol. 5, No. 4, 1985, pp. 70–73.
[36] Barcelona, M. J., Helfrich, J. A., Garske, E. E., and Gibb, J. P., "A Laboratory Evaluation of Ground-Water Sampling Mechanisms," *Ground-Water Monitoring Review*, National Water Well Association, Vol. 4, No. 2, 1984, pp. 32–41.
[37] Gillham, R. W., "Syringe Devices for Ground-Water Sampling," *Ground-Water Monitoring Review*, National Water Well Association, Vol. 2, No. 2, 1982, pp. 36–39.
[38] Ho, J. S.-Y, "Effect of Sampling Variables on Recovery of Volatile Organics in Water," *Journal of the American Water Works Association*, Dec. 1983, pp. 583–586.
[39] Keith, S. J., Wilson, L. G., Fitch, H. R., and Exposito, D. M., "Source of Spatial-Temporal Variability in Ground-Water Quality Data and Methods Control," *Ground-Water Monitoring Review*, National Water Well Association, Vol. 3, No. 1, 1983, pp. 21–32.
[40] Mioduszewski, D. and Irwin, J. L., "Dedication—An Answer to Reliable and Cost-Effective Ground-Water Sampling," *Proceedings*, National Conference on Hazardous Wastes and Environmental Emergencies, HMCRI, Cincinnati, OH, 14–16, May 1985, pp. 106–110.
[41] Wilson, L. C. and Rouse, J. V., "Variations in Water Quality during Initial Pumping of Monitoring Wells," *Ground-Water Monitoring Review*, National Water Well Association, Vol. 3, No. 4, 1983, pp. 103–109.

*Drew B. Bennett,[1] Mario Busacca,[2] and Edward E. Clark[1]*

# The Modified Reverse-Circulation Air Rotary Drilling Technique for Development of Observation and Monitoring Wells at Ground-Water Contamination Sites

**REFERENCE:** Bennett, D. B., Busacca, M., and Clark, E. E., "**The Modified Reverse-Circulation Air Rotary Drilling Technique for Development of Observation and Monitoring Wells at Ground-Water Contamination Sites,**" *Ground-Water Contamination: Field Methods, ASTM STP 963,* A. G. Collins and A. I. Johnson, Eds., American Society for Testing and Materials, Philadelphia, 1988, pp. 137–145.

**ABSTRACT:** Ground-water monitoring regulations on the state and Federal level establish general performance standards for the construction of monitoring wells. Of the many drilling techniques available for the construction of monitoring and observation wells, dry techniques are recognized as superior at sites of contamination. The dry technique of modified reverse-circulation air rotary drilling (RARD) is viable and a more advantageous alternative than many traditional techniques. It requires no liquid drilling fluids or muds, and therefore no foreign fluids are added to the aquifer. Furthermore, composite soil and water samples are collected during installation specifically and exclusively from the bit depth and are uncontaminated by drilling fluids. The method utilizes compressed air to clear the borehole and a tube within a tube drill pipe system to maintain the borehole. This allows the well assembly, filter pack, and borehole sealant to be properly placed thereby satisfying the strictest regulatory requirements for short and long-term monitoring. The method accommodates various field testing such as the measurement of *in situ* hydraulic conductivity and the collection of soil cores. To insure proper construction of monitoring wells, procedures and material requirements must be designed and fully specified.

**KEY WORDS:** modified reverse-circulation air rotary drilling, monitoring well construction, ground-water contamination

Very restrictive legislation has been promulgated in the last several years regulating the discharge of effluents to ground water as well as the cleanup of past waste disposal practices. The State of Florida, where roughly 90% of the drinking water supplies come from ground waters, has been a trend-setting state in this type of legislation.

The ground-water monitoring regulations, both on the state and Federal level, establish general performance standards for a ground-water monitoring system under either a detection, compliance, or corrective action program. These standards provide guidance for the construction of monitoring wells; however, the requirements concerning well placement, well construction, and the responsibility to develop a system that yields representative samples are generally left to the permit applicant/landowner.

[1] Hydrologist and environmental engineer, respectively, Edward E. Clark Engineers—Scientists, Inc., Miami, FL 33186.

[2] Ecological analyst, DF-EMS, National Aeronautics and Space Administration, Kennedy Space Center, FL 32899.

**Objectives**

Construction of observation and monitoring wells and subsequent data collection should satisfy five major objectives:

1. Assessment of the potential for ground-water contamination from the site or operation.
2. Capability to adequately sample for suspected contaminants.
3. Collection of data to adequately evaluate contaminant migration from a given source and if so, definition of pathways of migration.
4. Collection of samples representative of depth discrete, *in situ* fluid for analysis.
5. Characterization of the site and collection of both geological, unconfined and confined head data.

The use of noncontaminating methods to construct sampling wells is critical to ensure reliable field evaluations. Today, there is little room for error in performing these field tasks because of the extremely low levels of detection in analytical laboratories, the low levels established for water quality standards and the sensitive nature of remedial action. Specific procedures have been developed to prevent contamination of the borehole from external substances and cross-borehole contamination.

Many drilling techniques are available to the professional involved in ground-water contamination investigations. Those used to drill monitoring and observation wells can be grouped into two major categories: wet methods and dry methods. Wet methods, those methods where liquid drilling fluids are added, include mud rotary, jetting, and cable tool. Dry methods, where liquid drilling fluids are not added, include solid and hollow stem augers, air rotary and reverse air rotary. Dry methods facilitate the collection of more representative formation samples and reduce the potential for ground-water contamination that could result from drilling fluids used in wet methods.

**Description**

Modified reverse-circulation air rotary drilling (RARD) is a particularly well-suited, dry method for monitoring and observation well installation. The method has the advantage of requiring no liquid drilling additives and samples specifically and exclusively from the bit depth. This minimizes complications involved with data collection (which may be encountered when traditional methods are used) by eliminating hydraulic mounds, filter cakes, flushing of the aquifer, soil and water samples tainted with additive drilling fluids, and samples that are merely a composite of the entire borehole.

RARD is commonly used in boring large diameter boreholes for water supply wells. However, small-diameter drill pipe [that is 15.24 cm (6 in.)] is available, giving these rigs the capability to be used in ground-water contamination investigations which generally require small diameter wells. The modified RARD method requires tube-within-a-tube drill pipe (see Fig. 1). Compressed air—the drilling fluid—is injected down through the annular space in the double tube drill pipe. The air exits through holes in the drill bit, from where it flows upward through the inside tube, carrying cuttings with it. After leaving the inner tube, the air is depressurized in an expansion cyclone where air escapes through a vent at the top and cuttings and formation water settle out the bottom. Drill cuttings and fluids are not recirculated in the borehole. This double-walled drill pipe with a bit attached to the bottom is rotated or allowed to settle downward. The power feed mechanism applies axial thrust to the drill pipe when the weight of the pipe can not provide adequate pressure to the bit. The well assembly is set in the inner tube while the outer tube holds the borehole open, allowing for accurate seating of the well assembly and proper installation of filter pack and borehole sealant.

The methods described herein have been used successfully by the firm of Edward E. Clark

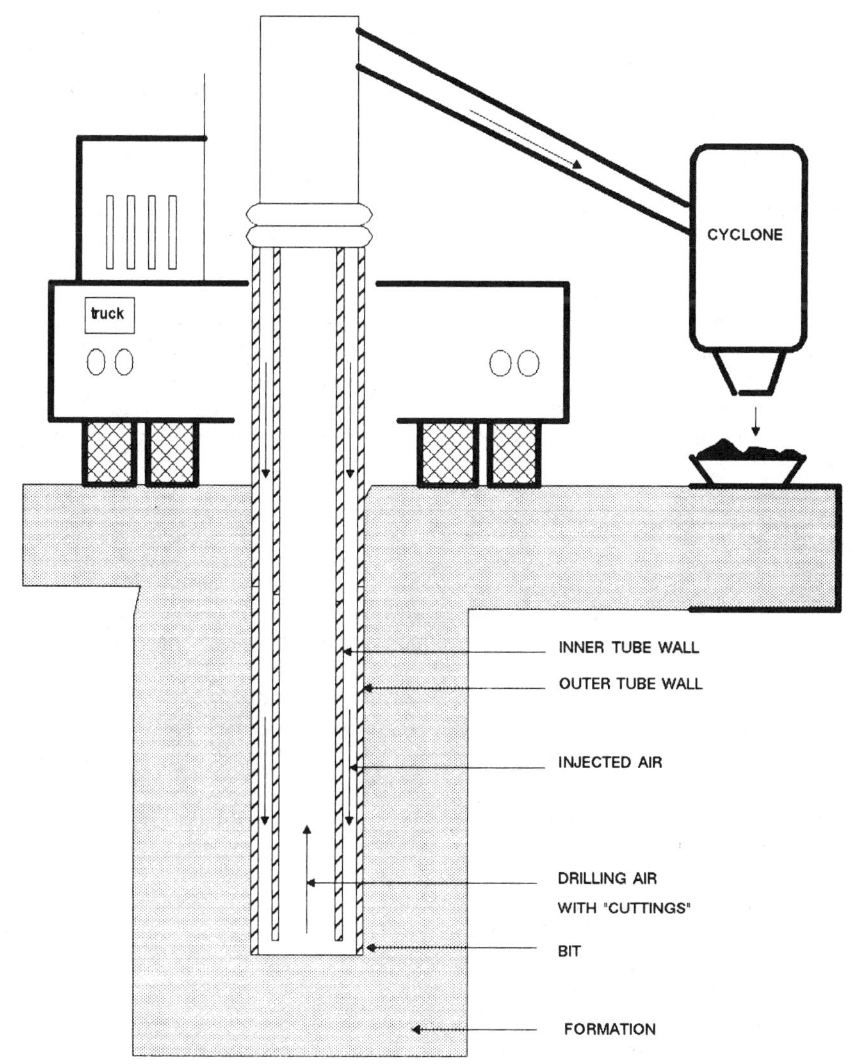


FIG. 1—*Modified reverse-circulation air rotary drilling method.*

Engineers-Scientists, Inc., to install monitoring and observation wells as part of a comprehensive ground-water study at the John F. Kennedy Space Center (KSC) located in the coastal barrier islands area of east-central Florida. A variety of unconsolidated materials, ranging from clastic sands to cohesive clays to coquina of mollusks held loosely together by calcareous cement, was encountered at depths up to 45 m (148 ft).

## Methodology

### *Predrilling Requirements*

Once a specific model of drilling equipment has been selected (for the KSC study a Drill Systems AP 1000 Center Sample Rotary Drill was used) and prior to mobilization to the investigation site, the drilling equipment, tools, and soil sampling equipment must be thoroughly cleaned. This includes a pressurized cleaning, using a potable water source, of the drill rig and drill pipe. The responsible party should be prepared to document acceptability of cleaning water. All hydraulic and oil leaks must be properly repaired prior to arriving at the site. An inspector, trained in the specifics of ground-water monitoring and drilling equipment, should be present for cleaning and repair work. All well construction materials should remain in their factory packaging until they are needed. If well construction materials require cleaning, they should be washed with a commercial-grade detergent or an approved solvent and rinsed with liberal quantities of acceptable cleaning water.

The quality of the compressed air stream used in drilling should be monitored using an approved method. Drilling air must be filtered to capture any potential compressor lubricants in the compressed air stream.

The use of air rotary drilling in badly contaminated subsurface situations must be approached carefully to minimize the exposure of drilling personnel to potentially hazardous materials. Appropriate safety equipment, apparatus, and monitors should be made available to personnel in the field.

*Drilling Operations*

Rotary drilling creates a uniform, clean hole with relatively little disturbance of the soil below the bottom of the hole [*1*]. Further preservation of the aquifer conditions can be exercised by not rotating in unconsolidated materials but by lowering the drill pipe under its own weight at a low air pressure [approximately 142 800 to 214 200 N/m$^2$ (20 to 30 psi)]. As drilling depth increases, however, air pressure must be increased as well. As in standard rotary air rotary drilling, the air compressor, hole size and drill pipe must be sized in relation to each other. The power feed mechanism will apply axial thrust to the drill pipe when the weight of the drill pipe can not provide adequate pressure to the bit. RARD has the capability to drill in a variety of conditions in excess of 300 m (984 ft) and is known to have been used in 900-m (2953 ft) exploratory holes in Puerto Rico. Regrinding and crushing of cuttings is minimized since they are removed immediately also. With a coring bit and "core catcher," lengths of limestone, coquina, and dolomite cores have been collected without crushing.

An experienced driller can detect changes in the character of the soil and rock by the rate of progress, the action of the drill rig, and by the "cuttings." The detection of these changes must be noted in the drilling logs as they provide valuable information in evaluating the hydrogeologic setting of the site.

Composite soil samples should be continuously monitored at the collection bin and collected every meter or every stratigraphic unit change. Because the outer tube of the drill pipe effectively cases off the borehole as it is drilled, soil samples can be collected specifically from the bit depth. The combination of noting drilling rate changes and the examination of representative borehole soil continuously, as well as incidental permeability tests provide the data required to delineate permeable and nonpermeable zones at great depth.

Formation water and cuttings are blown out of the hole almost immediately, making it possible to determine when a water bearing unit is encountered. After filtering or proper settling of water blown from the hole, collection and field analyses of appropriate parameters can provide preliminary information regarding changes in water quality with depth. Formation water sampling ranges from excellent in sands and cohesive silt/sand formations to marginal in fine "running" sand high in organic content. It has been reported that formation water sampling can be nonexistent when circulation is lost in formations such as cavernous limestones [*2*].

Composite water samples should also be collected every 3.05 m (10 ft), the unit length of the drill pipe, or significant water bearing unit and analyzed for suitable parameters. A log of these data must be kept, making special note of odor or color changes. Samples can be collected by appropriate methods (for example, peristalic pump or bailer) through the drill pipe or at the end of the cyclone after it has been lifted by the pressurized air, depending on the stability of the aquifer material and the parameters to be analyzed. Formation water that is obtained by the pressurized air may be more representative because it is evacuated as soon as it is encountered minimizing mixing with overlying formation water. However, mixing of water and air can lead to oxidation or carbon dioxide stripping, or both, and the attendant changes in pH and other chemical characteristics of the water.

When more than one water bearing unit is encountered and where heads are different, flow between zones will occur between the time when the drilling is completed and the time the hole is properly grouted. A great advantage of the tube within a tube drill pipe is that the borehole is cased off as it is drilled by the outer tube, which is seated against the borehole wall. The annular space between the borehole wall and the drill pipe is thus minimized [in competent rock the annular space is generally less than 1 cm (0.393 in.)] reducing the potential for inter-aquifer flow. Judgment must still be used when interpreting data collected under these situations.

At selected locations in the subsurface profile (that is, the first confining unit, major water-producing zones) incidental permeability tests should be performed following procedures outlined in several publications [*3–6*]. These techniques are applicable to completely or partially penetrating wells in unconfined aquifers. Under such conditions, these incidental tests must be considered "quick and preliminary"; however, they provide valuable data on water yielding and nonyielding characteristics of the aquifer. Slug tests should also be performed immediately after the well is completed and sealants have had time to set.

Thin-walled tube sampling of aquifer materials for laboratory analysis of horizontal and vertical hydraulic conductivity and effective porosity can be also performed with an in-the-field modification of the drill pipe. This involves fastening the thin-wall sampler to an adapter which in turn is fastened to a 3.3 cm (1.299 in.) outside diameter drill rod. The sampler and rod are lowered through the dual-tube drill pipe, where an appropriate driving technique is used. A weight can be dropped from the mast via a pulley arrangement or hydraulic downpressure from the drill head can be used.

As others have pointed up [*7*], significant margins for error in both laboratory and field tests occur. For the zone of saturation, field tests are generally preferred provided they are performed and interpreted properly since they permeate a larger volume of aquifer than laboratory tests, thus taking into account the effects of macrostructure. Thorough field investigation to delineate strata of maximum conductivity, and careful selection of zones to be tested combined with proper performing of the test, will add credibility to the results. However, a great deal of judgment of experimental technique and data collected from the field investigation remain necessary, whereas field predictions can be developed from the results of such tests [*7*].

## Well Construction

For a 5.08 cm (2 in.) inside diameter well, the borehole should be a minimum of 15.24 cm (6 in.) in diameter. While setting the well assembly, borehole caving from the bottom may be a problem. This slumping or caving of the hole is the result of various factors such as cohesionless sediments, insufficient intergranular friction, and differential hydrostatic pressures. To mitigate this, drilling air pressure can be gradually decreased allowing formation water to slowly fill the drill pipe as much as possible. This will reduce head differences between the borehole and the aquifer. If this problem of "running sand" continues, potable water should be added to the borehole to equalize the pressures. Excessive amounts of water should not be added but the head should be brought up to the approximate elevation of the water table. This water will be removed when the well is developed immediately after construction. Even after equalizing the hydrostatic pressure, some slumping may continue to occur in problem materials. However, the authors have not experienced borehole losses greater than 30 cm (11.8 in.) after mitigative measures were undertaken.

There are some design criteria for the use of filter packing around screen intakes of monitoring wells. Section 265.91(c) of the *Code of Federal Regulations* (CFR) requires that casings be screened or perforated, and packed with gravel or sand where necessary. The field investigator should determine if artificial packing is required after reviewing site conditions, composite soil samples, and any available data on grain size distribution. RARD provides a mechanism to install such filter packing properly to the intent of the regulations. Once the well assembly has been seated at the appropriate depth and centered in the borehole (since the inner tube of the drill stem is 8.47 cm (3⅓ in.) inside diameter, this is performed easily) filter packing is added. This is done by pouring uniform sand slowly down the space between the inner tube and the well assembly as diagrammatically shown in Fig. 2. This sand must be clean and should be washed quartz with a minimum of 95% by weight silicon dioxide ($SiO_2$). Combined hematite, limonite, and other iron compounds should not comprise more than 0.5%, by weight. The sand should be uniform in a gradation ranging in size from ASTM No. 6 sieve to ASTM No. 40 depending on site conditions. The sand should be applied at a slow rate in equal amounts radially in the space. This technique will avoid sand bridging in the space complicating grouting procedures later. Graded sand should be avoided because the differently sized particles will segregate and form bands of predominately fine and coarse materials in the sand pack [*4*]. This may allow the infiltration of fines (creating turbid samples which interfere in laboratory analysis) and bias water samples by drawing most water from the coarser sand material. As the drill pipe—both inner and outer tubes—is pulled up [slowly, 30 cm (11.8 in.) at a time], sand that was added to the drill pipe spreads by gravity out the bottom of the pipe, thus placing sand in the annular space between screen and hole wall. Sand pack should be run 30 cm (11.8 in.) above the screen to avoid any grout penetration into the screened area. Controlled experiments indicate that there is no significant penetration of grout in uniform sand with grain size finer than 0.06 cm (0.025 in.) [*8*]. The required amount of sand to add can be determined volumetrically and checked by depth measurements made through the drill pipe.

After the installation of the sandpack is completed, an 0.6-m (2 ft) bentonite plug is placed on top of the filter pack. Once again the annular space between the inner tube and the well casing is used as a tremmie pipe to place the pellets.

Hazardous waste regulations [Section 265.91(c) CFR] require that the annular space be sealed with a suitable material to prevent sample and ground-water contamination. Proper grouting results in a complete sheath of grout around the casing for the entire vertical distance of the borehole. The tube-within-a-tube method provides a choice of grouting procedures, gravity placement, and pumped placement via grout line, depending on drilling conditions (depth, formation, type of grout used). The tube-within-a-tube technique minimizes the channeling of slurry caused by tight places and dead spots where casing touches the wall of the hole when not properly centered.

The dual-tube drill pipe conducts grout to the bottom of the oversized borehole. The grout slurry may be allowed to flow down the stem, which acts as a tremmie pipe, by gravity as the drill pipe is retrieved a little at a time or it may be pumped down the stem via a grout pipe.

Following these procedures ensures a continuous seal which meets the CFR performance standard. The performance of this type of grouting procedure is exemplified by the three piezometer nests at KSC which have retained their head difference. Pumping one does not influence the other.

## Other Considerations

Before selecting this method, site access must be evaluated. Modified RARD equipment mobility is somewhat limited compared with a typical truck-mounted auger drill (Drill Systems AP 1000

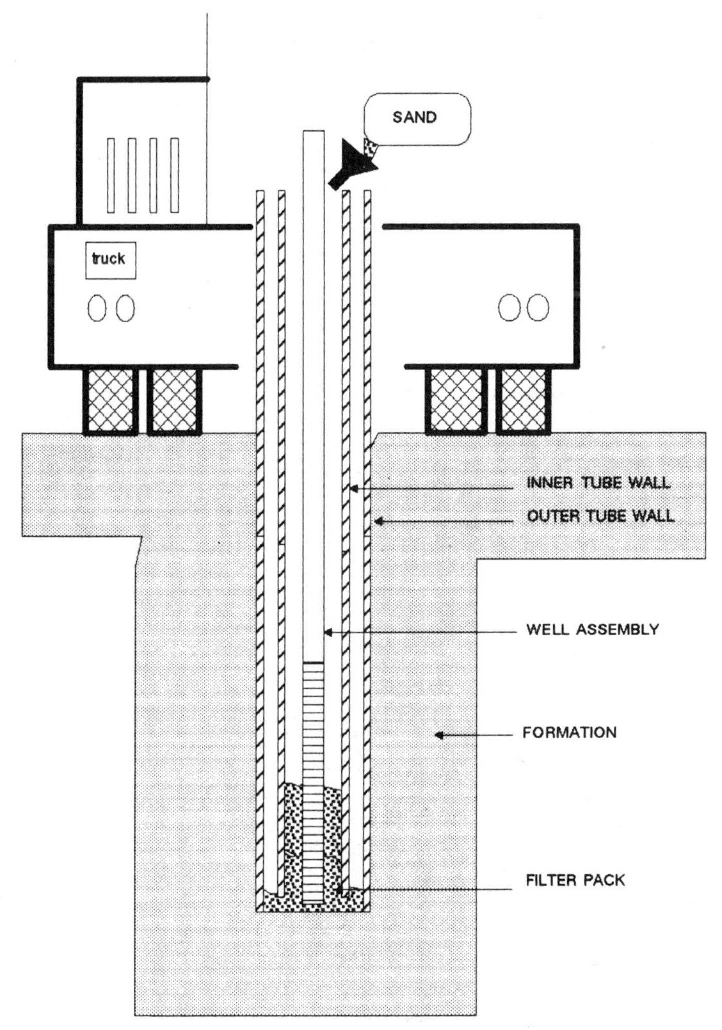


FIG. 2a—*Installation of filter pack utilizing dual-wall drill pipe.*

Center Sample Rotary Drill is commonly mounted on a 16-ton conventional tandem axle carrier). However, it has been the experience of the authors that the majority of ground-water contamination sites, generally areas where human activity has taken place, have adequate access.

Before sampling for laboratory analysis, the well should be adequately developed. More representative samples can be collected from the well in a shorter time period after well completion because modified RARD does not disturb the formation. This is particularly true when compared with wet methods, which use copious amounts of water and drilling additives.

The rig and all tools should be thoroughly cleaned and allowed to air dry before starting drilling operations at a new well site. All cleaning operations should take place in a designated area and wash water collected and disposed of properly.

Water from completed wells should not be used in drilling or in cleaning of equipment.

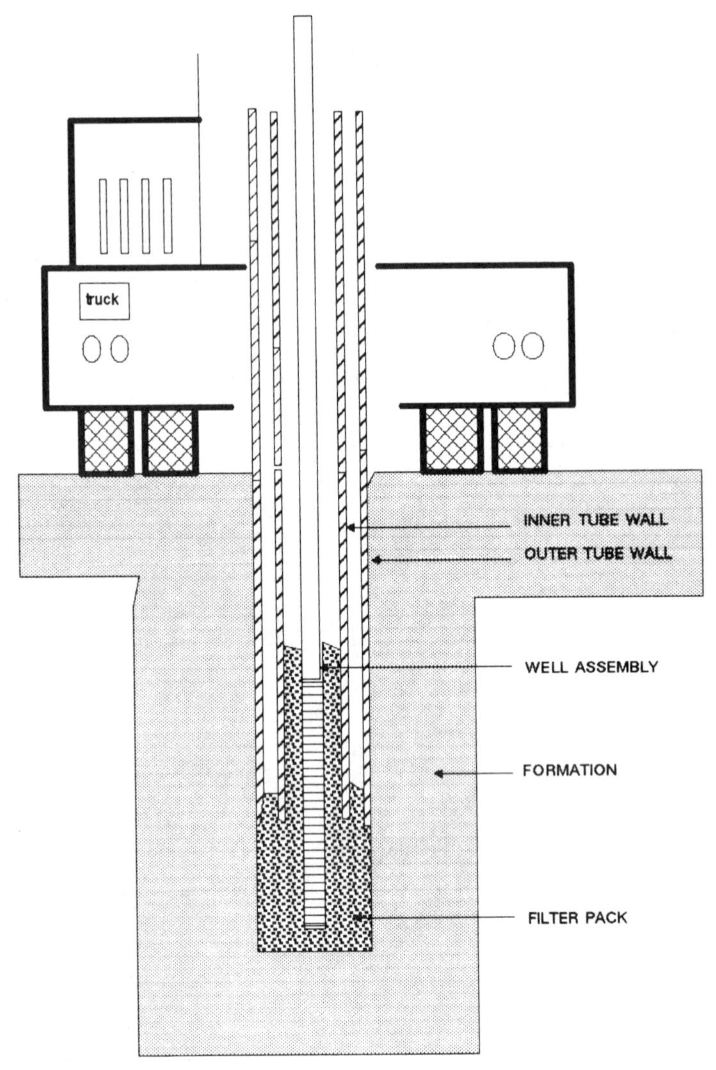


FIG. 2b—*Drill pipe is pulled back slowly, allowing filter pack to spread out of the bit surrounding the well screen.*

## Conclusions

The modified reverse-circulation air rotary drilling technique is a valid and available method for constructing monitoring wells at sites of ground-water contamination. The capabilities of this technique include the following:

1. Dry drilling method, with no drilling additions used.
2. The development of a clean borehole with relatively little disturbance of aquifer below the bottom of the hole.
3. Depth discrete aquifer and water samples, relative *in situ* hydraulic conductivity values, and core samples can be obtained during drilling.

4. By properly locating, filter packing and grouting, the most stringent well construction performance standards can be met.

## References

[1] Hvorslev, M. J., ''Subsurface Exploration and Sampling Soils for Civil Engineering Purposes,'' Committee on Sampling and Testing Soil Mechanics and Foundations Division, American Society of Civil Engineers, 1949, p. 61.

[2] Barcelona, M. J., Gibb, J. P., and Miller, G. D., ''A Guide to the Selection of Materials for Monitoring Well Construction and Groundwater Sampling,'' Illinois State Water Survey (SWS) Contract Report 327, Champaign, IL, 1983.

[3] Hvorslev, M. J., ''Time Lag and Soil Permeability in Groundwater Observations,'' Bulletin 36, U.S. Corps of Engineers, Waterways Experimental Station, Vicksburg, MS, 1951.

[4] Bouwer, H., *Groundwater Hydrology,* McGraw-Hill, New York, 1978, p. 115.

[5] Patterson, R. J. and Devlin, J. F., ''An Improved Method for Slug Tests in Small Diameter Piezometers,'' *Ground Water,* Vol. 23, No. 6, Nov.–Dec. 1985, p. 804.

[6] Bouwer, H. and Rice, R. C., ''A Slug Test for Determining Hydraulic Conductivity of Unconfined Aquifers with Completely or Partially Penetrating Wells,'' *Water Resources Research,* Vol. 12, No. 3, 1976, pp. 423–428.

[7] Olsen, R. E. and Daniel, D. E., ''Measurement of the Hydraulic Conductivity of Fine-Grained Soils, Permeability and Groundwater Contaminant Transport'' in *Permeability and Groundwater Contaminant Transport, ASTM STP 746,* T. F. Zimmie and C. O. Riggs, Eds., American Society for Testing and Materials, Philadelphia, 1981, pp. 18–64.

[8] ''Ground Water and Wells,'' Johnson Division, UOP Inc., St. Paul, MN, 1966, p. 239.

*William B. Kerfoot*[1]

# Monitoring Well Construction, and Recommended Procedures for Direct Ground-Water Flow Measurements Using a Heat-Pulsing Flowmeter

**REFERENCE:** Kerfoot, W. B., "**Monitoring Well Construction, and Recommended Procedures for Direct Ground-Water Flow Measurements Using a Heat-Pulsing Flowmeter,**" *Ground-Water Contamination: Field Methods, ASTM STP 963,* A. G. Collins and A. I. Johnson, Eds., American Society for Testing and Materials, Philadelphia, 1988, pp. 146–161.

**ABSTRACT:** Direct flow measurements using heat-pulsing techniques provide an accurate and economic method of determining direction of ground-water flow. The use of this method requires that care be taken in choice of well screen, methods of installation, measurement procedure, and method of data handling. The influence of well slot number, slot size, and screen configuration is explained. Drilling without using drilling fluid additives is recommended for installation. The use of coarse sand instead of gravel packing for the annular region also improves readings. Hydraulic conductivity and field porosity variation in annular packing versus the natural stratum is dealt with by a new calibration procedure. The use of vector analysis for this data handling is recommended. A procedure involving convergence of down-gradient direction and mean vector sum for direct flow measurement is offered to provide answers to the old question of whether enough wells have been installed for understanding flow in the natural stratum.

**KEY WORDS:** ground water, flowmeters, monitoring well, screen design, measurement, field procedures

Direct flow measurement offers the convenience of rapid simultaneous determination of ground-water flow direction and rate for subsurface engineering applications. A heat pulse is transmitted through a porous glass medium in the probe. Interstitial water modifies the heat distribution between paired thermal sensors, yielding a linear response to increase in flow rate. A more detailed discussion of the theory of the heat-pulsing technique can be found in Melville, et al. [*1*].

The ability of the heat-pulsing technique to detect and measure flow rates below $0.35 \times 10^{-7}$ m/s (0.1 ft/day) has been verified in a number of test conditions [*1–3*]. However, the accuracy of determination of the flow in the water-bearing strata depends greatly on the nature of the penetration into the strata. The choice of well screen, consistency of material placed between the well and the natural strata (annular packing), method of drilling, and centralizing of the screen play a large role in obtaining appropriate readings. The well and its surroundings become a crucial interface in channeling the natural flow across the well orifice in such a way as to be easily interpretable. The following text briefly describes preferred well conditions, flowmeter calibration, and field procedures found during five years of survey applications, principally in glacial till and outwash deposits.

The complete procedure for successful well installation and measurement can be reduced to six steps (Fig. 1):

[1] K-V Associates, Inc., Analytical Systems, 281 Main Street, P.O. Box 574, Falmouth, MA 02541.

1. Well screen selection.
2. Drilling methodology.
3. Centralizing and annular packing.
4. Calibration of the meter.
5. Vector resolution.
6. Mapping the flow network.

## Screen Selection

Initially, it is important to obtain basic field information concerning the geology of the site, the approximate depth to static water from grade, locations of any surface-water flows, such as rivers, tidal streams, or lakes, which may determine boundary conditions, and presence of withdrawal wells.

For unconsolidated deposits of sands and gravels, the nature of the soils and their particle size sets practical limits on the well slot size and probable location below grade for installation. Indirectly, the more heterogeneous the deposit and the larger the diameter of the screen, the more averaging of velocities and directions occurs because of the larger cross sectional area intercepted. Regions within known confined aquifers should be evaluated only with short screens and sealing to avoid vertical flow short-circuiting between water-bearing layers.

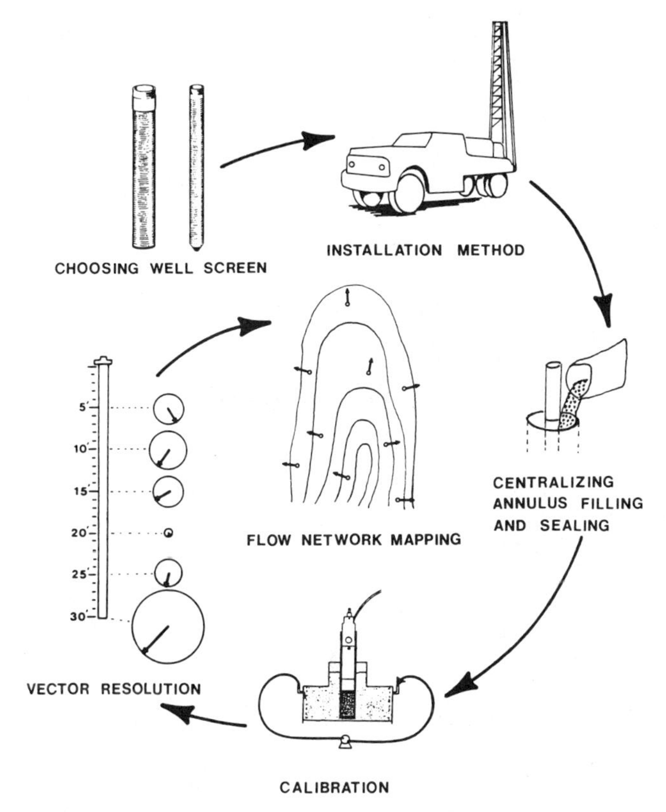


FIG. 1—*Stepwise procedure for ground-water flow measurement.*

Well screens have previously been chosen from a practical standpoint by the well driller from what was closest at hand or from the most convenient distributor in the locale with the greatest discount. For direct flow measurement, far more care must be exercised in well screen selection to avoid distortion of field flow and to assure adequate flow rates across the screened area. The type of well screen can have considerable influence on the accuracy of the ground-water flow measurements performed inside the screen [*4*]. Installation of sections of pipe consisting of random hand-drilled rows of holes or intermittent horizontal hacksawing creates distortions of flow over 40 deg and potential 100-fold rate variations, depending upon whether the thermal sensor is opposite a hole or a blank space.

A variety of commercial smooth interior wire-wound, continuous slot plastic [polyvinyl chloride (PVC)] screens and stainless steel wire-wrapped screens has been found satisfactory for sensitive measurements (Fig. 2). The preferred diameters have been 5-cm (2 in.) and 10-cm (4 in.) internal diameter screens. Although probes and well screens exist for 3.8-cm (1.5 in.) PVC, the reduction in intercepted cross section is considered undesirable for flow direction determinations.

The commonly used 5-cm (2 in.) internal diameter schedule 40 ASTM plastic screens (PVC) should possess the following characteristics:

1. Continuous slot.
2. Slot orientation: perpendicular to axis of casing.
3. Slot distance: 0.6 cm (0.25 in.) or less.
4. Slot rows: greater than 3 or continuous slot.
5. Flush jointing.

See Table 1 for some common commercial 5-cm (2 in.) PVC screens which fit these criteria.

The number of rows of slots in 5-cm (2 in.) internal diameter PVC monitoring wells has a significant influence on the accuracy of the ground-water flow direction, inferred from the interscreen direction [*4*]. Figure 3 plots the accuracy of flow direction determinations inside the well relative to the true flow outside the well against the number of rows of slots. As the number of rows of slots increases, the accuracy of measurement increases markedly.

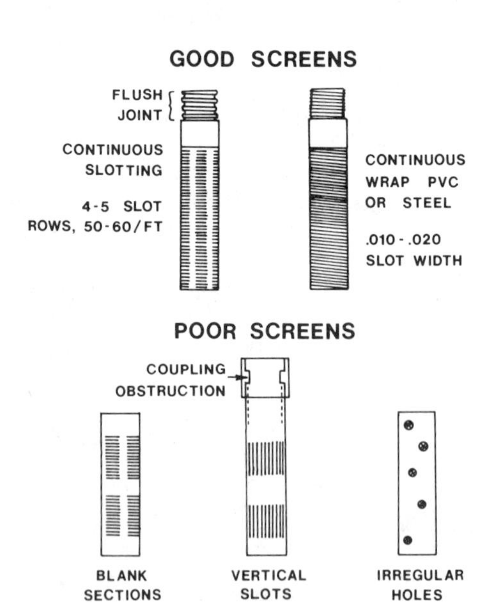


FIG. 2—*Preferred and problem screen types for direct ground-water flow measurement.*

TABLE 1—*Characteristics of selected 4-cm (2 in.) ID, plastic monitoring well screens recommended for flowmeter use.*[a]

| Manufacturer | NSF Pipe Size | Screen Gauge Size (in.) | Slots per ft | Slot Rows | Inside Diameter, in. | Outside Diameter, in. | Wall Thickness, in. | Slot Area | Resistance[b] (% flow) |
|---|---|---|---|---|---|---|---|---|---|
| Timco[c] | Schedule 40 | 0.020 | 59 | 4 | 1.89 | 2.1 | 0.210 | 5.0 | 60% |
| | Schedule 80 | 0.010 | 62 | 4 | 1.89 | 2.1 | 0.210 | 2.6 | 83% |
| Diedrich | Schedule 40 P-12 | 0.020 | 63 | 5 | 2.01 | 2.38 | 0.37 | 5.2 | 58% |
| | Schedule 40 P-12 | 0.010 | 69 | 5 | 2.01 | 2.38 | 0.37 | 2.7 | 80% |
| Johnson Well Screen[d] | continuous 2 in. | 0.020 | 79 | c | 1.939 | 2.375 | 0.218 | 5.8 | 32% |
| | continuous 2 in. | 0.010 | 84 | c | 1.939 | 2.375 | 0.218 | 2.9 | 48% |
| Johnson Screen[e] | continuous 2 ps | 0.020 | 78 | c | 1.875 | 2.375 | 0.220 | 10.2 | 29% |
| | continuous 2 ps | 0.010 | 82 | c | 1.875 | 2.375 | 0.220 | 6.8 | 45% |

[a] This is not intended as a sole endorsement of the manufacturers listed. Other comparable commercial screens exist and can be evaluated by their characteristics.

[b] Resistance is measured at 3 m/day (10 ft/day) transport velocity in 15.24-cm (6 in.) ID flow chamber with medium sand external packing and 1 mm (0.039 in.) glass bead interval packing in fuzzy packer:

$$R = \frac{\text{Unimpeded flow} - \text{flow through screen}}{\text{Unimpeded flow}}$$

[c] "Timco Geotechnical Products," Timco Manufacturing Co., Inc., P.O. Box 35, Prairie du Sac, WI 53578, 1983.

[d] "Specifications Sheet," Johnson Well Equipment, Inc., 9131 Highway 98 West, P.O. Box 3364, Pensacola, FL 32506, 1982.

[e] "Johnson PVC Plastic Water Well Screens, Specification Sheet," Johnson Division, United Oil Products, P.O. Box 43118, St. Paul, MN 55164, 1979.

[f] "Geotechnical Newsletter," K-V Associates, Inc., 281 Main St., Falmouth, MA 02540, Vol. 1, No. 1, 1980.

Conversion factors: 1 in. = 25.4 mm; 1 ft = 0.3048 m.

Interruptions in the continuous slotting create measurement and interpretation difficulties. Blank regions created by horizontal unslotted bars for strength cannot be distinguished from thin silt layers during readings in fluvial sand deposits. Similarly, vertical slotting in rows creates regions of rate change, reaching a maximum when the sensor is present at the center of the slotted region and reaching a minimum in the region of blank wall between the slots. While operationally some success has resulted by placing a nylon "tickler" or nailpoint on the end of an interval packer to "sense" the position of blank walls during readings in shallow wells [1.5 to 12 m (5 to 40 ft)], these well screens are not desirable for direct flow measurements.

Slot thickness also has a pronounced influence on accuracy of flow rates. Figure 4 shows the relationship between well screen slot size and resistance to flow. Increasing slot size dramatically increases the volume of flow through the screen. It would be advisable to use a 0.5-mm (0.020 in.) slot screen with coarse sand annular packing to keep silt out rather than use a 0.15-mm (0.006 in.) slot if low flows [less than $3.5 \times 10^{-6}$ m/s (1.0 ft/day)] are to be measured. As previously shown in Table 1, continuous slot well screen maintains the highest slot area per foot, roughly three times that of interval-slotted screen. In addition, increasing the diameter of the well screen increases the flow-through rate.

A 10-cm (4 in.) internal diameter well screen with six rows of slots and 0.5 mm (0.020 in.)

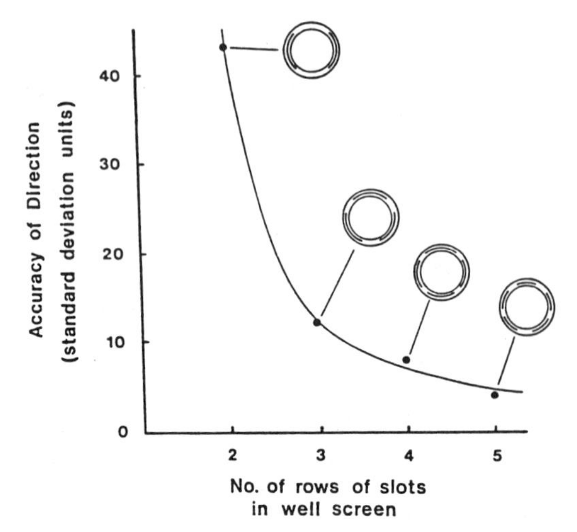


FIG. 3—*Accuracy of flow direction measured in well screen compared to actual external flow direction.*

slot width results in about a 50% increase in internal flow compared to its 5-cm (2 in.) internal diameter four-slot row counterpart.

Emplacements in hydrocarbon products should have their screens checked beforehand for swelling or dissolution if PVC is planned for use. A 5-cm (2 in.) section should be placed in a glass container with the product for 24 h, visually inspected, and poked with a pointed object to test for softness. Complete closing of 0.25-mm (0.010 in.) slots has been encountered with one

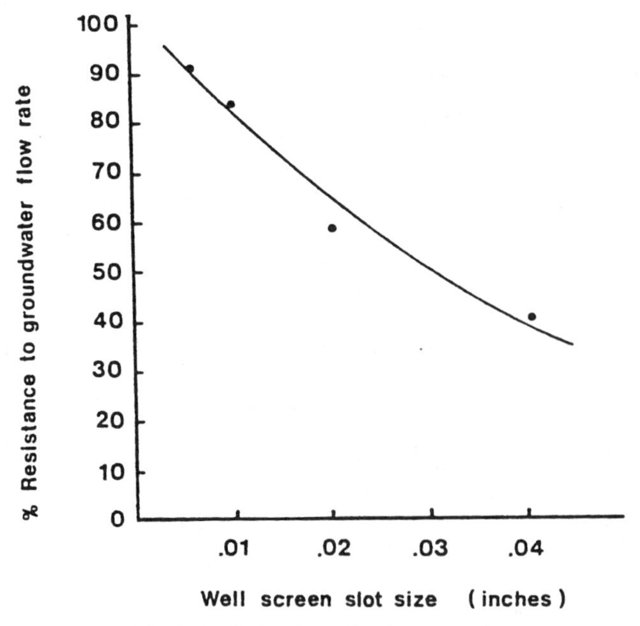


FIG. 4—*Well slot size and resistance to flow.*

commercial grade of PVC exposed to unleaded gasoline. PVC solvents such as toluene, benzene, and ketones should be tested if pure product concentrations are likely.

## Drilling Methodology

The method of drilling through the overlaying soil deposits into the geologic unit to be monitored has been found to be an important factor in obtaining valid results. The use of drilling mud to stabilize a hole often results in partial plugging of unconsolidated material opposite the screened area. Despite attempts to develop the screened region, the resulting water surge only partially removes the interstitial mud pack, resulting in an irregular patchwork of cleared regions.

The most effective methods of drilling have been hollow stem augering, air rotary, and cable tool with casing. With flush joint casings and screen commercially available, 5-cm (2 in.) internal diameter wells can be readily installed in 9-cm (3.5 in.) internal diameter, 15-cm (6 in.) outside diameter hollow stem augers. For shallow emplacements in unconsolidated sands, hand augering with a 7.6-cm (3 in.) auger inside 3-m-long (10 ft), 10-cm (4 in.) internal diameter PVC drainage pipe has proven useful.

## Centralizing and Annular Packing

Following drilling, the well screen should be inserted to avoid a space between the casing and side wall (annular space). If the well is placed into the drill hole without the addition of packing and the walls fail to collapse, the continuity of porous flow is destroyed, making any readings invalid.

The first rule of ground-water flow measurement is straightforward: *Continuity of capillary flow must be maintained across the entire screened cross section to allow measurement of transport velocities.* The location of pockets of free water induce (*a*) flow around the well screen and (*b*) localized eddies which may not bear any relationship to true flow in the geologic unit. A coarse sand packing [1.3 mm (0.051 in.) in diameter; 10–20 mesh] is commonly added down the annulus to maintain porous conditions, particularly in glacial till or weathered bedrock.

The results of use of coarse sand versus gravel packing are given in Table 2. As packing particle size increases, the direction of interstitial flow channels tends to deviate more from the mean flow direction. Large gravel [0.5 to 5 cm (0.2 to 2 in.)], like naturally rounded cobblestones, causes irregular water pressure patterns across the screen slots, distorting the flow. Table 2 shows the relationship between precision of measurement and particle size with sand and gravel packing. Flow direction was maintained at 354 deg north, determined by compass orientation in the flow chamber after cessation of measurements. For highest accuracy in direction and rate, uniform sand packing is strongly recommended.

TABLE 2—*Influence of particle size of packing material on accuracy of flow measurements.*[a]

| | Median Size, mm | Void Volume | Direction, deg | | Rate, ft/day | |
|---|---|---|---|---|---|---|
| | | | $\bar{X}$ ± SD | CV | $\bar{X}$ ± SD | CV |
| Medium sand | 0.5 | 30% | 354 ± 1.5 | 0.4% | 18.7 ± 0.6 | 3.2% |
| Mixed medium sand | 1.0 | 33% | 345.3 ± 2.1 | 0.6% | 10.3 ± 1.5 | 14.5% |
| Pea gravel | 8.4 | 42% | 357.5 ± 12.6 | 3.5% | 7.0 ± 2.3 | 32.9% |
| Medium gravel | 19 | 48% | 300 ± 59 | 19.7% | 2.6 ± 2.5 | 96.2% |

NOTE—SD = standard deviation; CV = coefficient of variation.

[a] "Geotechnical Newsletter," K-V Associates, Inc., 281 Main St., Falmouth, MA 02540, Vol. 1, No. 1, 1980.

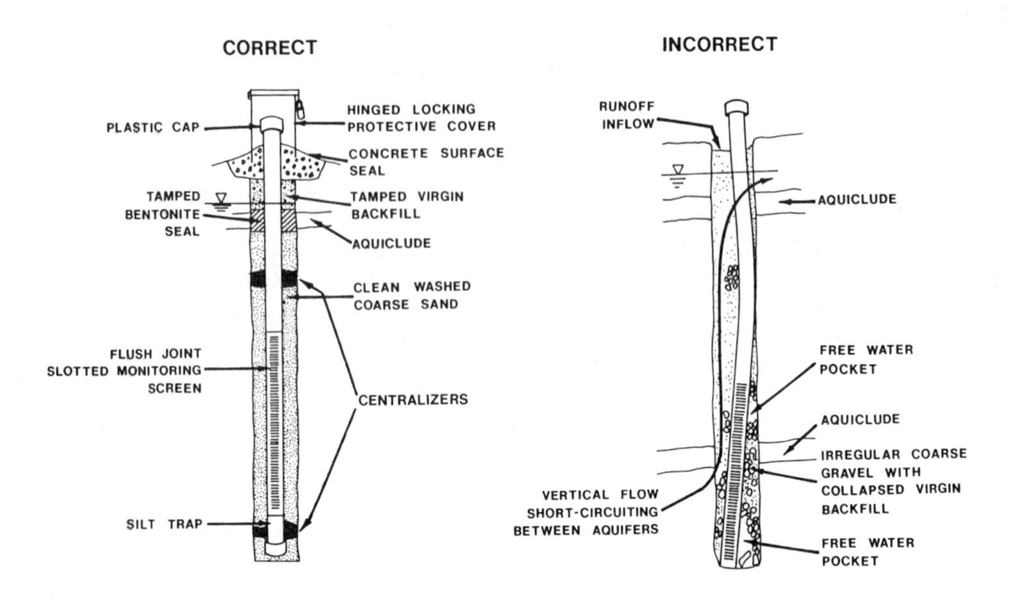


FIG. 5—*Ideal and poor well installations for direct flow measurement.*

It is also important to centralize the casing. If the casing bends and positions the screen against the wall, it can shield the immediate region from the annular packing (Fig. 5). Several commercial centralizers are available, both solvent welded and metal. Avoiding channelization around the slotted casing is extremely important. Small channels along a section of the screened area can invalidate the instrument response.

The best strategy for confined aquifers has been similar to recommended methods for emplacement of piezometers: (*a*) Use limited screen length [1.5 to 3.0 m (5 to 10 ft) at maximum] between confining layers; (*b*) seal with bentonite at penetration points of casing through any confining layers; and (*c*) use clusters of individual screens rather than multiple screens on a single casing to avoid inducing vertical movement of water up casing between confining layers under different head pressures.

## Calibration of the Meter

Ground-water flowmeters are equipped with a control box and field probe. The unit is powered by rechargeable gel-cell batteries. The field probe of the Model 30 GeoFlo meter is 4.44 cm (1.75 in.) in diameter, suitable for 5-cm (2 in.) slotted PVC monitoring wells.

Electronic ground-water flow measurement is based upon thermal transmission within a porous solid under the influence of interstitial liquid flow. The controlling factors are thermal conductance of the solid and liquid phases, surface area of the solid phase, thermal transfer coefficient of the liquid phase, and rate of movement of the liquid phase.

For example, with a porous solid phase of glass beads with distilled water, the thermal conductivity of the glass is about 0.023 cal/(s)(cm$^2$)(°C/cm) while that of the water is 0.0015 cal/(s)(cm$^2$)(°C/cm). The principal heat transfer occurs between the zones of contact of the glass beads, a small point surrounded by a thin interface of water. The slightest movement of the liquid during the period of heat transfer can profoundly influence the conductance. The GeoFlo flowmeter creates a heat pulse which is transmitted through the porous matrix. Any net movement of the interstitial water mass creates a thermal conductance bias which is linearly proportional to the rate of flow (Fig. 6).

Calibration of the flowmeter to translate the meter readouts into geological unit seepage (or transport) velocity involves the use of a flow chamber. To simplify corrections for well screen resistance and hydraulic conductivity differences between the geological unit ($K_o$), the annular packing ($K_a$), and the internal packing ($K_i$), a cross section of the well emplacement is constructed in the flow chamber, duplicating the field conditions as closely as possible. The probe packer is placed in a 0.3 m (1 ft) section of the well screen, surrounded by the annular coarse sand, with the remainder of the flow tube containing a sample of the water-bearing geologic unit (Fig. 7).

The stepwise procedure for filling the calibration flow chamber proceeds as follows:

1. Clean flow chamber of previous material.
2. Place a 25-mL (0.845 fl oz) sample of current geological material (dried) into 25 mL (0.845 fl oz) of water in a graduated cylinder, shake and let settle, and inspect to see if the fines fraction is greater than 15%.
3. Fill the chamber with the sample of the water-bearing geologic unit, assuring that the upper arch of the tube is well packed (air channels will short-circuit flow).
4. Place a 10-cm (4 in.) section of PVC drainage pipe into the neck and remove the sample material from inside the pipe.
5. Fill the drainage pipe with the annular coarse sand.
6. Insert a section of the well screen inside the center of the annular packing, using a long-handled spoon to remove the coarse sand from inside the screen.
7. Pull out the drainage pipe, leaving the annular packing in contact with the sample material.
8. Insert the flowmeter probe, with its porous-fiber packer ("fuzzy packer") filled with 1-mm (0.039 in.) glass beads, into the well screen and secure with a laboratory clamp.
9. Three sets of measurements are now taken at three different flow velocities, starting at the highest rate of flow [suggested to be 3 m/day (10 ft/day)], to produce three points on the calibration curve. The three points should lie along a straight line. See Kerfoot [*2*] for further explanation of the calibration procedure.

The sample material to fill the chamber should be obtained from *inside auger* samples during drilling or by bucket auger sampling in the field.

Melville, et al. [*1*] and Wheatcraft [*5*] have been directing attention toward hydraulic conductivity changes induced in the annular area by well emplacement. Well installations are natural magnifiers of local flow since the void region represents little resistance to flow compared to the outside strata. The use of gravel packing in the annular space between the well and the edge of the borehole can also modify local flow lines.

Distortion of flow caused by conductivity contrasts between the inside packing ($K_i$), the outside strata ($K_o$), and the annular packing ($K_a$) can be reduced by using the following guidelines:

1. Calibrate the flowmeter in the calibration chamber with the natural geological sample material and annular packing, duplicating, as closely as practical, the field well installations in the geologic unit to be monitored.
2. Ideally, match or have a higher value for the hydraulic conductivity of the internal packing material compared to the mean hydraulic conductivity of the outside strata.
3. By erring on the side of higher hydraulic conductivity for annular region and packer, if the combined packing is greater than 20 times the hydraulic conductivity of the surrounding natural strata, the magnification of flow is manageable and approaches an asymptotic maximum value of two times geologic unit flow rate.
4. Avoid channelization outside well screen borehole by washing in annular packing and using coarse sand as opposed to gravel packing, where practical.
5. Use a fuzzy packer inside well screens to avoid internal channelization. Make sure that the

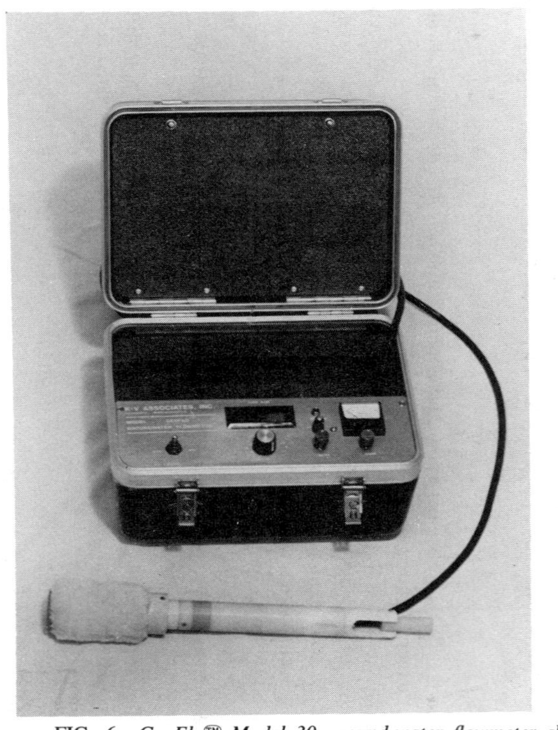

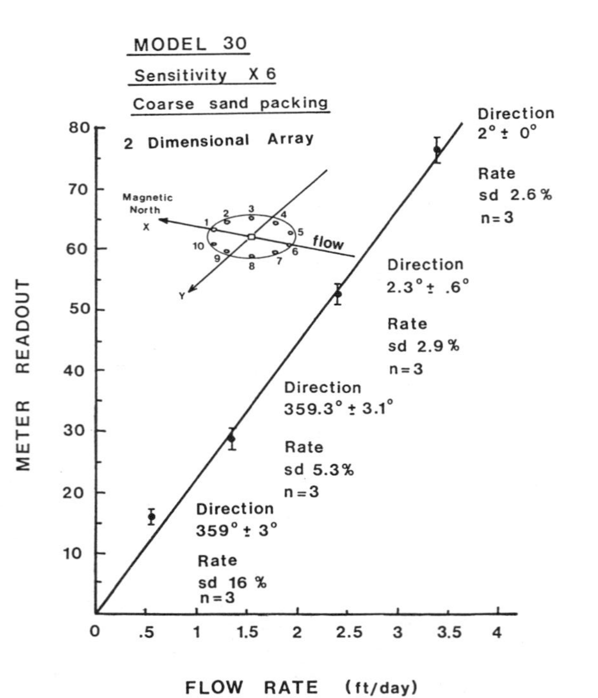


FIG. 6—*GeoFlo™ Model 30 ground-water flowmeter showing calibration curve and arrangement of thermal sensors [2,6]. Probe is inserted into porous fiber packer.*

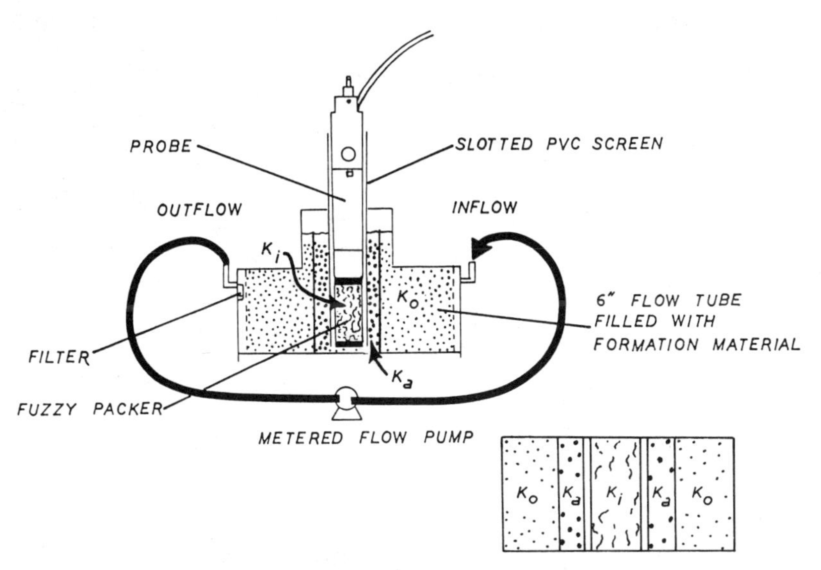


FIG. 7—*Flow chamber arrangement for calibration showing hydraulic conductivities of formation ($K_o$), annual packing ($K_a$), and internal packing ($K_i$).*

packer is suited for reasonable tightness in the well screen being used. Clean the packer after withdrawal from the well screen, particularly if silt has been encountered.

6. Avoid use of drilling fluids during installation of well screens.

While the flow system around a permeable cylinder is characterized by two factors (the location of the well in the borehole and the hydraulic conductivity contrasts between the cylinder and outside formation), a uniform ground-water flow system around a single well is defined by only one parameter, $K_r$ (the ratio between the interior packing and the outside formation). Figure 8 is a series of plots showing the streamlines passing around and through the cylinder for different values of $K_r$. Very little streamline refraction at the cylinder boundary occurs for permeability contrasts of less than about $\pm 50\%$.

If the probe is calibrated at one hydraulic conductivity (for instance, the mean condition) and corrections for magnification of flow through the screen due to hydraulic conductivity contrasts are desired at different vertical locations as the probe is moved up or down the screen, the following equation can be used:

$$V_o = n_p V_i / n_o f$$

where

$V_o$ = transport velocity in the outside strata,
$n_p$ = porosity of the packer beads,
$V_i$ = measured velocity with the probe inserted inside the monitoring well screen,
$n_o$ = observed field porosity of the outside geologic unit, and
$f$ = magnification factor of flow passing through the screen compared with flow when the hydraulic conductivity inside and outside the screen is equal.

The hydraulic conductivities listed in Table 3 can be used for approximations of contrasts.

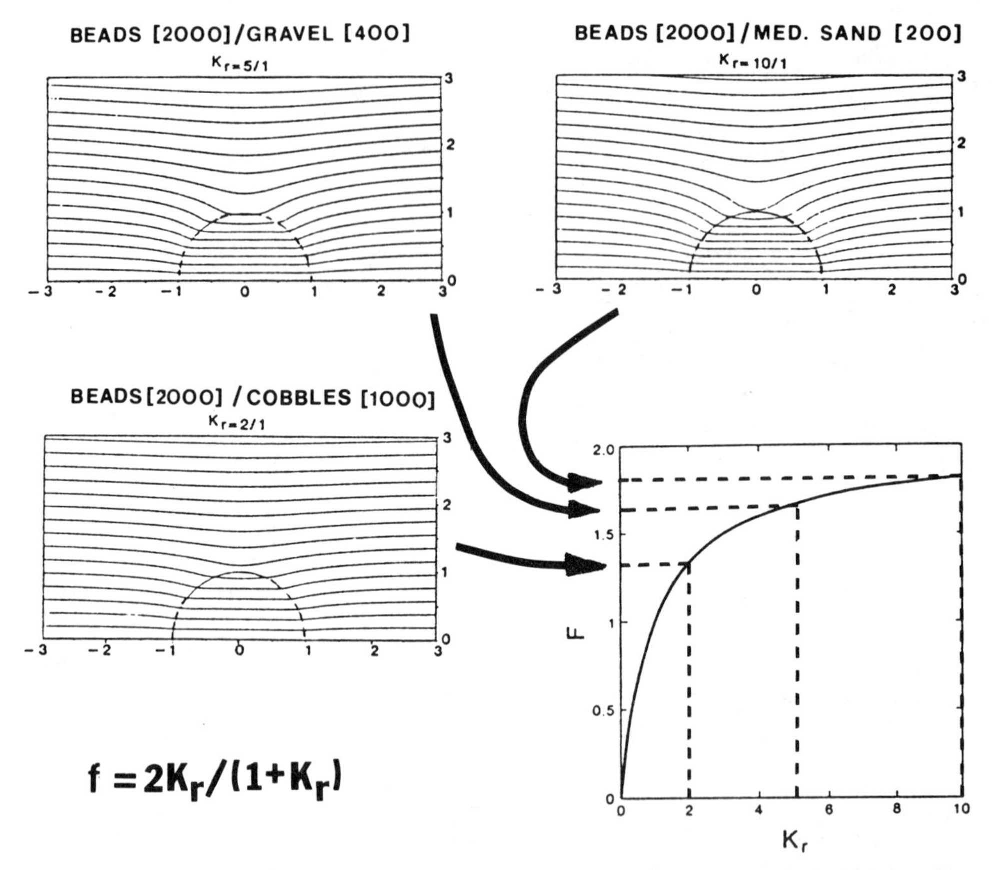


FIG. 8—*Correction factor (f) for hydraulic conductivity contrasts between internal glass bead packing [7 × $10^{-3}$ m/s (2000 ft/day)] and common geologic units.*

## Vector Resolution

The probe is inserted and its orientation controlled by snap-connect hollow aluminum rods when taking flow measurements down the well. The porous fibers of the packer surrounding the probe maintain contact with the sides of the screen. The cable should be secured to the rods by PVC tape as sections are lowered. When the desired depth is reached, the top section is secured by a

TABLE 3—*Hydraulic conductivities.*

| Geologic Unit | Hydraulic Conductivity m/day | (ft/day) |
|---|---|---|
| Fine sand | 15 | (50) |
| Medium sand | 46 | (150) |
| Coarse sand | 91 | (300) |
| Gravel | 137 | (450) |
| Cobblestones | 182 | (600) |

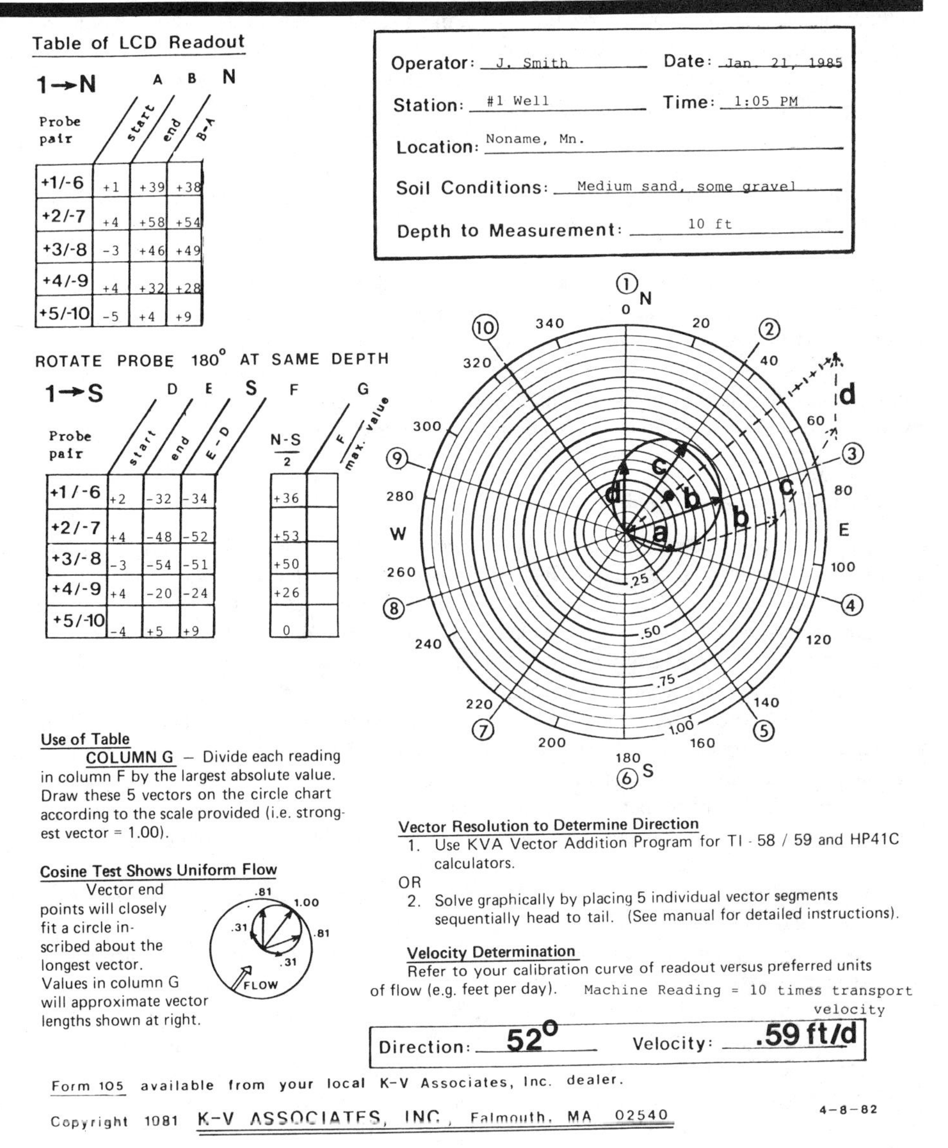

# GROUNDWATER FLOW WORKSHEET

FOR 180° ROTATIONAL DATA MEASUREMENTS

**Table of LCD Readout**

1→N

| Probe pair | A start | B end | N B-A |
|---|---|---|---|
| +1/-6 | +1 | +39 | +38 |
| +2/-7 | +4 | +58 | +54 |
| +3/-8 | -3 | +46 | +49 |
| +4/-9 | +4 | +32 | +28 |
| +5/-10 | -5 | +4 | +9 |

ROTATE PROBE 180° AT SAME DEPTH

1→S

| Probe pair | D start | E end | S E-D | F N-S/2 | G F/max. value |
|---|---|---|---|---|---|
| +1/-6 | +2 | -32 | -34 | +36 | |
| +2/-7 | +4 | -48 | -52 | +53 | |
| +3/-8 | -3 | -54 | -51 | +50 | |
| +4/-9 | +4 | -20 | -24 | +26 | |
| +5/-10 | -4 | +5 | +9 | 0 | |

Operator: J. Smith Date: Jan. 21, 1985

Station: #1 Well Time: 1:05 PM

Location: Noname, Mn.

Soil Conditions: Medium sand, some gravel

Depth to Measurement: 10 ft



**Use of Table**

COLUMN G — Divide each reading in column F by the largest absolute value. Draw these 5 vectors on the circle chart according to the scale provided (i.e. strongest vector = 1.00).

**Cosine Test Shows Uniform Flow**

Vector end points will closely fit a circle inscribed about the longest vector. Values in column G will approximate vector lengths shown at right.



**Vector Resolution to Determine Direction**

1. Use KVA Vector Addition Program for TI - 58 / 59 and HP41C calculators.

OR

2. Solve graphically by placing 5 individual vector segments sequentially head to tail. (See manual for detailed instructions).

**Velocity Determination**

Refer to your calibration curve of readout versus preferred units of flow (e.g. feet per day). Machine Reading = 10 times transport velocity

Direction: 52° Velocity: .59 ft/d

Form 105 available from your local K-V Associates, Inc. dealer.

Copyright 1081 K-V ASSOCIATES, INC., Falmouth, MA 02540

4-8-82

FIG. 9—*Ground-water flowmeter worksheet.*

special clamp which simultaneously grips the cord and rods. A magnetic compass is then secured to the top and oriented by the wrench so that Channel 1 is pointed toward magnetic north.

The machine reading results in an array of five sets of values with the Model 30 GeoFlo meter, proportional to the annular component of rate of flow in each direction (Fig. 9). If the flow across the well screen is uniform, a circular array will occur. The axis of flow can be found by two methods: (*a*) drawing a line from the origin through the center of the circle and intersecting the degrees of angle around the edge of the polar graph paper, or (*b*) adding the component vectors head to tail and drawing a line from the origin to the point of the final vector. For rate by both methods, in the first (*a*) circular array procedure, the diameter of the circle is directly proportional to flow. In the second method, the length of the line is divided by 2.7. This geometric constant results from dividing a circle into 36-deg arcs from a single point on its circumference, adding the straight-line segments retaining their original angle (vector addition), and dividing the resultant distance (sum of vectors) from the single point to the end of the segments by the diameter of the circle.

The length of the diameter of the circle is then compared with the original calibration curve to translate the proportional machine response to the rate of flow in the formation at the measured depth in the well.

The newer flow worksheets include the use of 180-deg rotation of probe to offset any bias which may have occurred due to position of the sensor pair to the heater tip. Continual use of the probe and its movement through irregularly shaped sections of well casing may move one of a pair of thermal sensors farther away from the heat source than the distance of the second pair. The rotation procedure completely compensates for any position bias. The use of porous fiber packers, which provide additional protection for the probe, also reduces bias interference.

**Common Measurement Problems**

When the probe is inserted below the static water level in the well, water is displaced upward, creating a displaced slug of water which recharges along the length of the well screen. The invading water flows into the most permeable formation, regardless of the regional flow pattern. The most pronounced impact usually occurs at the top of the screen. This phenomenon is most apparent in tighter soils, like silty sands, which have long periods of decay for the displaced water head. There are two procedures which make the user aware of the ''slug effect'' and how to compensate for it. First, take replicate measurements. The second measurement should duplicate in direction and rate the original measurement if no interference is present. Second, the user can start taking measurements near the bottom of the well screen and come up, rather than the reverse.

The probe should not be put on the bottom in any silty deposit. The compaction of silt will squeeze interstitial water outwards, causing another type of interference: compression flow. This phenomenon is also evidenced when the user is standing near a probe emplacement with 10 cm (4 in.) or less of unsaturated soil above static water. The weight of compaction will induce a local flow bearing no relationship to flow in the region.

In reviewing cases of flowmeter field use, a number of instances were found where the probe was placed down the well and rested on the bottom. This was due either to forgetting a clamp or naivete. The movement of water during measurement time (3 min with a Model 30 GeoFlo meter), if the natural transport velocity is $1.7 \times 10^{-6}$ m/s (0.5 ft/day), is only 0.13 cm (0.05 in.), below the usual capacity of the eye to perceive motion. Securing the probe to the top of the well with vinyl or rubber tape is not satisfactory since the weight of the rod and probe assembly will slowly stretch the tape, hardly visible to the eye, but significant to the probe.

In layered glacial deposits or any highly permeable deposits [hydraulic conductivity ($K$) greater than 30 m/day (100 ft/day)], the most common error during field operations comes from not taking enough measurements at different intervals along the well screen. It is not unusual to encounter a ten-fold variation in flow from one horizon to the next. Using a coarse sand annular pack can help to average the flow, but usually a minimum of three measurements are taken in 1.5 to 3 m (5 to 10 ft) well screens.

*The individual measurements must be treated as vectors* to determine the mean flow and direction across the intercepted cross section of the well screen, not as arithmetic numbers which are added and divided by the number of observations. In other words, the observation (*a*) should be taken

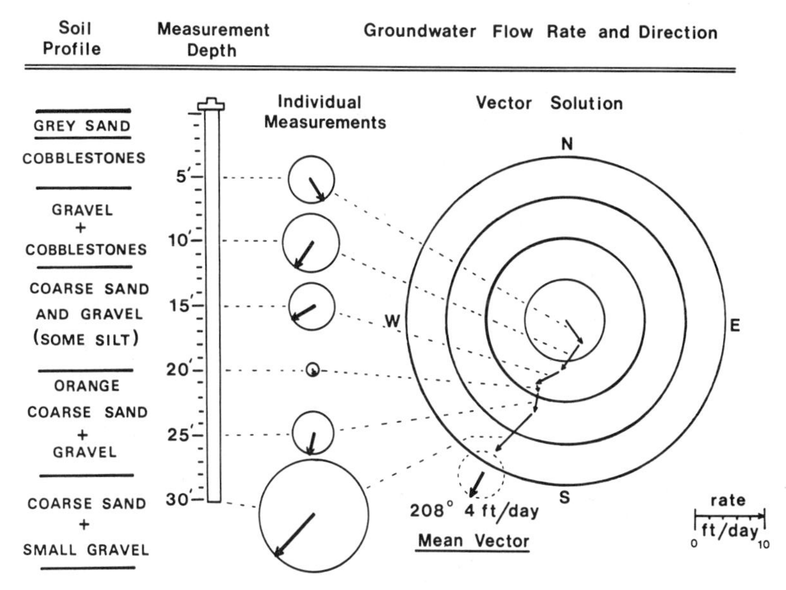


FIG. 10—*Vector resolution during profiling down the well screen.*

at equal intervals up or down the screen, (*b*) the individual vectors for each representing direction and rate of flow added head to tail, (*c*) a line connected from their origin to the point of the last to solve for mean direction, and (*d*) the length of the line divided by the number of equally spaced measurements to determine the mean rate (Fig. 10). If the rates are not treated as vectors, the arithmetic mean rate almost always overestimates the mean field velocity.

## Mapping the Flow Network

From field experience, a simple procedure has been devised to cost-effectively address the adequacy question: Are there enough wells and observation points to describe the local flow network? The answer to this question depends greatly on the geological setting. Operationally, an initial set of three wells can be tested. The mean flow direction should correspond, by surveying in the relative hydraulic head and comparing the vector set for each well, to the downgradient slope computed from triangulation between the three observation points. If not, then a set of additional wells should be placed to solve for the presence of discharge regions, recharge events, or ground water divides occurring in intervening regions. Each iteration is performed based upon an inspection of the previous corings, head elevations, and flow measurements.

Figure 11 shows an example where a recharge region lies between a set of three wells. Here a comparison of flow vectors with the gradient solution points out a clear conflict. Addition of another set of three wells and the resultant convergence of slopes and flow directions offers a more adequate depiction of local flow conditions. Field investigations have generally confirmed a 30% to 50% reduction in number of monitoring wells compared to grid or random emplacement when the iteration procedure can be employed.

## Conclusions

Direct ground-water flow measurement offers a new tool for rapid determination of local ground-water flow. However, to allow accurate field measurements to be performed, care must be exercised

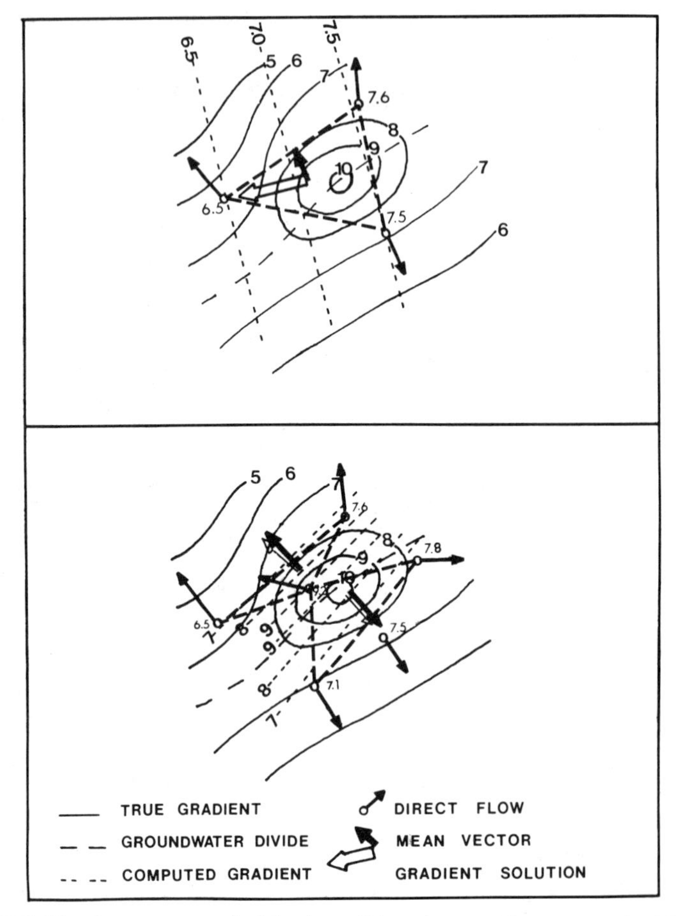


FIG. 11—*Convergence method for determining adequacy of measurement points.*

in the choice of well screens, means of installation, measurement procedure, and choice of data handling. In summary, five years' experience with direct flow measurement has produced the following helpful guidelines:

1. Optimal PVC well screens have four or greater rows of slots or continuous slotting.
2. It is better to use coarse sand annular packing to reduce silt infiltration into the screen, rather than change to a smaller slot size in the screen which would decrease internal flow.
3. Dry drilling rather than wet drilling techniques are necessary to avoid aquifer occlusion.
4. It is necessary to maintain porous flow conditions across the entire well screen. Pockets of free water interfere with measurement.
5. Hydraulic conductivity contrasts can be corrected by careful calibration.
6. Vertical currents in the well should be avoided by bentonite sealing of the annular area.
7. Flow measurements should be treated as vectors, not as arithmetic quantities.
8. Vector and static water elevations can be used to test adequacy of number of holes by convergence of downgradient and mean vector sum.

## References

[*1*] Melville, J. G., Molz, F. J., and Guven, O., "Laboratory Investigation and Analysis of a Ground Water Flowmeter," *Ground Water*, Vol. 23, No. 4, 1985, pp. 486–495.

[*2*] Kerfoot, W. B. in *Proceedings*, Second National Symposium on Aquifer Restoration and Ground Water Monitoring, Water Well Journal Publishing Co., Worthington, OH, 1982, pp. 264–268.

[*3*] Vanderlaan, G., "A Fast Track Approach to Management and Impact Assessment," *Proceedings*, National Conference on Management of Uncontrolled Hazardous Waste Sites, HMCRI, Silver Spring, MD, 1981, pp. 349–358.

[*4*] Kerfoot, W. B. and Massard, V. A., "Monitoring Well Screen Influences on Direct Flowmeter Measurements," *Ground Water Monitoring Review*, Vol. 5, No. 4, 1985, pp. 74–77.

[*5*] Wheatcraft, S. W. and Winterberg, F., "Steady State Flow Passing Through a Cylinder of Permeability Different from Surrounding Medium," *Water Resources Research*, Vol. 21, No. 12, 1985, pp. 1923–1929.

[*6*] Guthrie, M., "Use of a Geo Flowmeter for the Determination of Ground Water Flow," *Ground Water Monitoring Review*, Vol. 6, No. 1, 1986, pp. 81–86.

*Robert P. Chapuis*[1]

# Determining Whether Wells and Piezometers Give Water Levels or Piezometric Levels

**REFERENCE:** Chapuis, R. P., "**Determining Whether Wells and Piezometers Give Water Levels or Piezometric Levels,**" *Ground-Water Contamination: Field Methods,* ASTM STP 963, A. G. Collins and A. I. Johnson, Eds., American Society for Testing and Materials, Philadelphia, 1988, pp. 162–171.

**ABSTRACT:** A method is proposed to verify whether wells and piezometers have been adequately sealed into the ground to avoid vertical communication between aquifers and thus prevent misleading results of both piezometric levels and contamination degrees. The method makes use of variable-head permeability tests that are not interpreted by the common graph of ln $H$ versus $t$, but by the more fundamental graph of the water velocity versus the total head difference. Both presentations are mathematically equivalent. If there is some error in the "measured" piezometric level, it will be hidden in the common logarithmic diagram, but it can be easily detected with the proposed velocity diagram. The method is illustrated by typical examples of velocity diagrams and their interpretation. According to the author's experience, the practice of partial seals along piezometer pipes should be discouraged and a full-length seal is preferred.

**KEY WORDS:** ground water, piezometer, well, water level, contamination, permeability, reliability, proving method, evaluation

Ground-water contamination studies involve the installation of sampling wells to delineate contamination zones and hydraulic piezometers to measure hydraulic heads in aquifers and establish natural ground-water flow nets. The devices used to determine water heads (observation wells, standpipes, piezometers) may be described as watertight pipes open to water flow at their base and open to the atmosphere at their top. The name of the device seems to depend on its size (large for well, small for standpipe and piezometer) and on its location (well and standpipe for unconfined aquifers, piezometer for confined or unconfined aquifers). In practice, however, the terms are often used interchangeably even though they should not.

For clarification, the word standpipe indicates a simple pipe having its lower section slotted for water intake and placed in a borehole that may or may not be backfilled with clean sand.

To the contrary, the true sampling wells and piezometers are isolated within specific zones of soil or rock. If this isolation is not achieved:

1. There will be a vertical communication between aquifers.
2. The piezometer will indicate a water level representing some average level along the draining unsealed vertical zone around the pipe.
3. The water samples will indicate some average degree of contamination accelerated by the inadequate isolation.

[1] Research scientist, Mineral Engineering Department, Ecole Polytechnique de Montréal, C.P. 6079, Sta.A, Montréal, Québec, H3C 3A7, Canada.

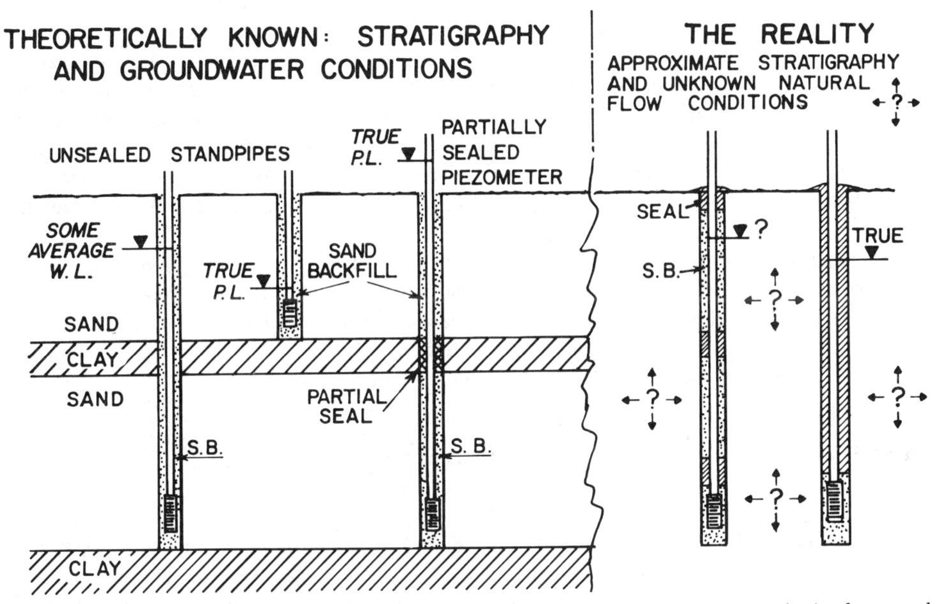


FIG. 1—*How inadequate seals result in misleading water levels. The use of partial seals, its theory and the reality (W.L. = water level; P.L. = piezometric level; S.B. = sand backfill).*

4. Consequently, piezometer readings and water analyses may be misleading and contaminants may have an easier access to high permeability layers than in the natural conditions before the introduction of the piezometer.

Several situations are given in Fig. 1; in theory, if stratigraphy and ground-water conditions are well known, it is sufficient to backfill the annular space around the pipe with materials that will have the same permeability as the adjacent natural soil. In practice, ground-water conditions remain more or less unknown as long as the piezometers are not installed. Consequently, in order to avoid meaningless readings, the author agrees with Freeze and Cherry [*1*, p. 23] that the piezometer must be sealed along its full length above its intake zone. Even if the location and depth of a piezometer are selected after adequate boring data have been obtained and analyzed, local seals above the intake or in front of identified impervious zones are inadequate; any unexpected thin layer of clay or clean sand will modify the readings. The requirement of a full-length seal may be illustrated by the analog simulation of ground-water flow by direct electrical current flow:

1. The piezometer is equivalent to an electric wire connected to a voltmeter.
2. The electrician has much more detailed drawings of the electrical installation he is testing than any drawing the geoscientific profession can obtain for natural ground and rock conditions.
3. A geoscientist using piezometers that are not sealed along their full length is equivalent to an electrician using a partly stripped electric wire to measure a voltage.

## Limitations of Common Testing Methods

Piezometers are usually tested periodically after installation to verify that they are functioning properly. Two types of tests can be performed. The simpler, common test found in most textbooks may be described as follows. If water is pumped out or poured into the piezometer, and if the water returns to its previous level after a certain time, it is assumed that the reading is representative. The second common test includes time readings to determine the *in situ* hydraulic conductivity [*1*]; a sudden change in the water level of the piezometer is obtained by the removal or introduction of a known volume of clear water and the recovery of the water level with time is then observed.

According to these two common methods for testing piezometers, the reading is assumed to be representative of the true natural total head around the intake of the piezometer if the water returns to its previous level.

The purpose of this paper is to prove that this reading is not always representative and that the results of an *in situ* variable-head permeability test may be used to detect whether the reading is a true piezometric level or not and whether the piezometer was correctly sealed or not. The paper does not discuss construction methods and problems associated with drilling techniques, filter conditions, and sealing either with bentonite pellets or grout.

## Theoretical Examination of Field Variable-Head Permeability Tests

### *Background*

For field variable-head permeability tests, Horvslev [*2*] presented formulas encompassing a wide variety of shape factors and anisotropic conditions. These formulas have been reviewed by Boersma [*3*] and Bouwer and Jackson [*4*]. For piezometers that are open over the full thickness of a confined aquifer, formulas and interpretation methods (similar to that of Theis for pumpage of wells) have been developed by Cooper et al. [*5*] and Papadopoulos et al. [*6*].

The present paper considers piezometers that are open only at their basal section.

The theory is well established and geotechnical practitioners have had extensive experience with these tests; however, they have a very limited confidence in the results (Milligan [*7*]; Canadian Foundation Engineering Manual [*8*] because of problems of execution and interpretation. Several sources of error have been studied and discussed by Horvslev [*2*], Gibson [*9*], Weber [*10*], and Bjerrum et al. [*11*]. They include:

1. Leakage by pipe fittings and between pipe and adjacent soil.
2. Hydraulic fracturing of soil by excessive water heads.
3. Soil remolding or smearing, or fine particle washing during drilling operations.
4. Clogging by sedimentation of fine particles.

In practice, many errors may be avoided by strict control of drilling operations and by applying a limited hydraulic head. This approach was adopted by the Normalization Bureau of Quebec for its two standards (BNQ 2501-135 and BNQ 2501-230) for field permeability tests performed by water addition to boreholes.

Field permeability tests in piezometers are very similar to tests in boreholes. Rat et al. [*12*] have underlined the problem of interpreting results that are supposed to give straight-line relationships but that frequently give curves. Chapuis et al. [*13*] have examined the theory of these tests and detected a previously unknown source of systematic error related to the estimation of the piezometric level. A method has been proposed [*13*] to detect and eliminate this error in the case of permeability tests performed in boreholes, a situation in which the casing is not sealed in the ground like a piezometer, and where the true piezometric level at a given depth can only be evaluated, not measured, during wash-boring operations. An analog method may be used to verify whether the water level given by a piezometer is truly representative or not. Its theoretical justification is presented below.

*Theoretical Justification*

The general equation of field variable head permeability tests is

$$\ln (H_1/H_2) = -kC(t_1 - t_2) \quad (1)$$

where:

$H_j$ = difference of total head (cm of water column) at time $t_j(s)$,
$k$ = hydraulic conductivity (cm/s), and
$C$ = shape factor ($cm^{-1}$).

This equation (or its equivalent with an exponential) is commonly used because it is practical; it seems easy to measure a total head difference $H$ at a time $t$. Equation 1 is obtained from Darcy's law

$$Q = kcH \quad (2)$$

relating the flow rate ($Q$) to the hydraulic conductivity ($k$), to the total head difference ($H$), and to a second shape factor ($c$). The use of Eq 2 does not seem practical because it requires the calculation of the flow rate ($Q$) from the water velocity ($\Delta H/\Delta t$) and the cross-sectional area ($S$) of the injection pipe

$$Q = -S(\Delta H/\Delta t) \quad (3)$$

Combining Eqs 2 and 3 gives

$$\Delta H/\Delta t = -kcH/S \quad (4)$$

In fact, the use of Eq 2 or Eq 4 simply requires the determination of the water velocity ($\Delta H/\Delta t$) in the pipe. The usual Eq 1 is obtained directly by integration of Eq 4.

If the assumed piezometric level is wrong, an error will be made on the total head difference $H$; this error will be a systematic error in Eq 2 or Eq 4 where it is included in a linear function, whereas it will be a variable error in the common Eq 1 where it is included in a logarithmic function.

Consequently, the commonly used Eq 1 may hide a variable error that cannot be easily detected. On the contrary, the equivalent error will appear as a systematic error in Eq 4 so that it may be easily detected.

## Proposed Proving Method

According to Eq 4, a plot of water velocity ($\Delta H/\Delta t$) versus the mean total head difference $H$ during the time interval $\Delta t$ should appear as a straight line if the water flow conditions are stable. The hydraulic conductivity of the soil around the intake of the piezometer will be derived from the slope of this straight line.

Again, from Eq 4, the velocity vanishes if $H = 0$, which seems rather obvious. This indicates that the straight line of an experimental velocity diagram should intersect the $H$-axis at a point where the true total head difference is zero. If the intersection point corresponds to an experimental value $H_0 \neq 0$, this means that $H_0$ is the systematic error in the assumed (experimental) total head difference calculated from an erroneous piezometric level. A velocity diagram has clear advantages:

1. It is a direct illustration of Darcy's law, whereas the common Eq 1 and its associated log-curve are an integral form of Darcy's law.
2. It shows directly whether the water flow is stable and normal, and consequently follows Darcy's law.
3. It allows for the elimination of a possible systematic error $H_0$ on the assumed piezometric level previously determined either from visual interpretations (during borehole drilling) or from piezometer readings.

The differences between a conventional interpretation (ln $H$ versus $t$) and that of the proposed testing method ($\Delta H/\Delta t$ versus $H$) are illustrated by Figs. 2 and 3 established for the same field test. These test results had received the following conventional interpretation (Fig. 2):

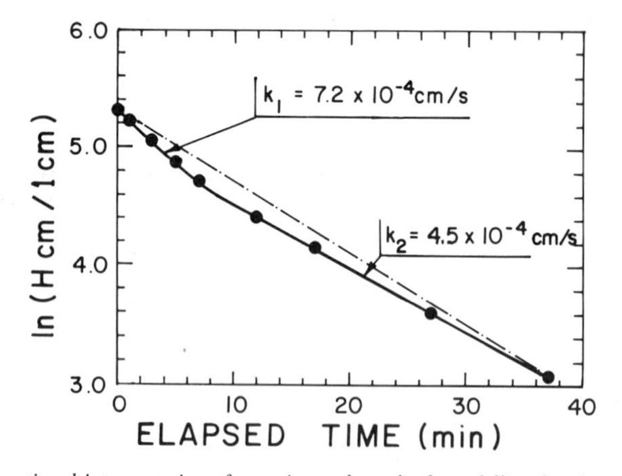


FIG. 2—*Conventional interpretation of experimental results for a falling-head permeability test.*

1. Medium-quality results.
2. A faster decrease of $H$ at the beginning of the test (corresponding to a higher value of $k$), attributed to the fact that a "certain equilibrium" had not been attained after the sudden addition of water in the pipe.
3. A slower decrease of $H$ at the end of the test (corresponding to the lower value of $k$), attributed to a progressive clogging.

This conventional interpretation of the results presented in Fig. 2 had been performed without questioning the assumption of a piezometric level at a depth of 140 cm from the natural ground level.

The interpretation of the same experimental results by the proposed testing method is illustrated in Fig. 3, which incorporates two curves: (1) the curve of the water velocity ($\Delta H/\Delta t$) versus the mean value of the total head difference during the time interval $\Delta t$, and (2) the curve of $\ln(H-H_0)$ versus elapsed time $t$, in which the true total head difference is obtained by subtracting the detected piezometric error $H_0$ from the assumed head $H$.

It appears that the test may be evaluated as follows:

1. Results are of high quality and fully respect Darcy's law.
2. The piezometer gives a water level that is 45 cm lower than the true piezometric level in natural conditions.

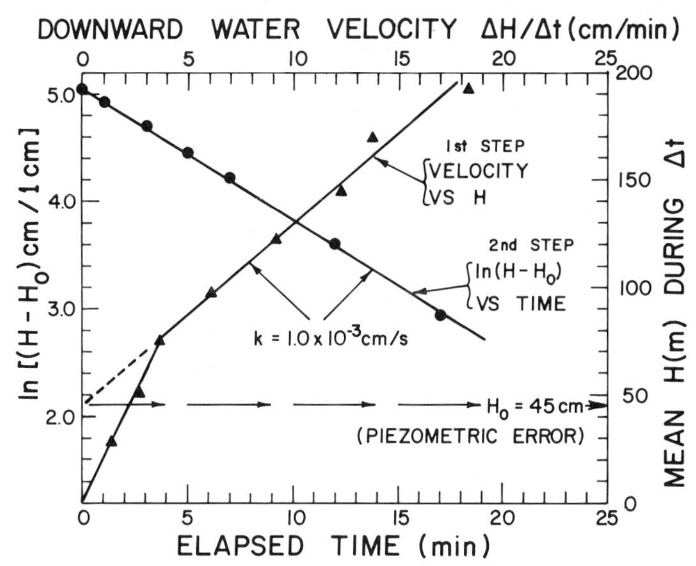


FIG. 3—*Interpretation of the experimental results of Fig. 2 by the proposed method—incorrectly sealed piezometer giving a water level, not a piezometric level (piezometric error of 45 cm). Note that the two graphs give the same value of* k.

Consequently, the proposed method may be used for the detection of incorrect seals along piezometric pipes and also for a better determination of the field hydraulic conductivity.

## Typical Experimental Results and Their Interpretation

Three typical results and their interpretation are given in Fig. 4:

1. Good test results are typical of a piezometer having a full-length seal. The fourth measurement of water depth in the pipe is erroneous, and such measuring errors can easily be detected on a velocity curve as follows: if the mean water velocity in the pipe is calculated between the 3rd and the 5th readings, the influence of the 4th is eliminated and the resulting velocity point lies on the mean straight line.
2. Such test results are typical of alluvial deposits with different aquifer layers. They are quite common for tests performed in cased boreholes and less frequent for isolated piezometers (Fig. 3). At the beginning of the test, the water-velocity curve appears as a straight line that may be extrapolated toward a value $H_0 \neq 0$. When the difference between the experimental $H$ and $H_0$ becomes small, a new water-velocity curve appears: it corresponds to the leaky flow along borehole casings or piezometer pipes that are inadequately sealed in the adjacent soil. The final water level in the piezometer will correspond to $H = 0$ and the fact that the piezometer is uncorrectly sealed would never have been detected in the common logarithmic presentation of test results.
3. Such results are characteristic of a phenomenon that may be called "hydraulic fracturing" of a soil. At the beginning of the test, the apparent hydraulic conductivity, *k*, is relatively high as indicated by the flat slope of the velocity curve. When the applied total head difference becomes smaller, the velocity curve becomes a straight line associated with a much smaller value of *k*. Such curves may be recorded for either a cohesive soil or a granular soil having no cohesion. The expression "hydraulic fracturing" is simply related to the shape of the hydraulic curve (velocity

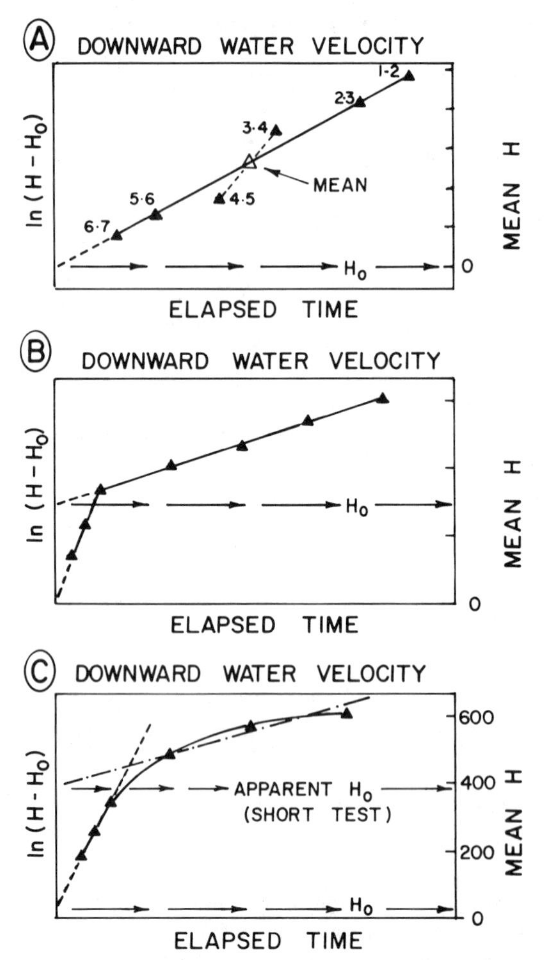


FIG. 4—*Typical results of variable-head permeability tests and their interpretations (see the text for explanations of A, B and C). Curves of* $\ln(H - H_0)$ *versus* t *are not given.*

or flow rate versus $H$) and the associated marked change in hydraulic conductivity. It does not imply a systematic mechanical fracture of the soil; instead, it simply suggests that the water flow is increased in the surrounding soil (by processes that are not fully understood in the case of granular materials for which it is difficult to imagine the opening of cracks as in rocks or in cohesive clayey materials [*14*]).

The phenomenon of "hydraulic fracturing" may be hidden in the velocity diagram of a test that has been stopped prematurely. In such cases, only the first part of the curve is registered; a straight line adjustment in Fig. 4*c* would give a hypothetical error $H_0$ = 380 cm on the piezometric level. However, the relative importance of $H_0$ as compared to an "estimated" depth of 500 cm for the piezometric level must be questioned when interpreting the test results. In a next step, the conditions for "hydraulic fracturing" should be verified with the assumption that the initially measured water level at rest is representative of the true piezometric level: the proposed verification method is presented below.

## "Hydraulic Fracturing" Conditions

When a variable-head permeability test is performed by clear water addition to a piezometer, the applied total head difference must be chosen between two extreme values. The minimum value is imposed by the accuracy of the water-level measurements. The maximum value corresponds to the appearance of unstable hydraulic conditions designated as "hydraulic fracturing."

For *in situ* tests in cased boreholes, the following rule of thumb has been suggested by Chapuis et al. [13]

$$H_{max} = H'/3 + H''/6 \tag{5}$$

where $H'$ is the thickness of the soil layer above the measured or assumed piezometric level and $H''$ is the thickness of the soil layer between this piezometric level and the water intake. This field rule corresponds to

$$H_{max} = \sigma'_{vo}/6 \quad \text{when} \quad \gamma_{sat} = \gamma_h = 2\gamma_w \tag{6}$$

where

$\sigma'_{vo}$ = initial vertical effective stress, kPa, at water intake,
$\gamma_{sat}$ = unit weight of saturated soil, kN/m$^3$,
$\gamma_h$ = unit weight of humid soil, kN/m$^3$, and
$\gamma_w$ = unit weight of water, kN/m$^3$.

If "hydraulic fracturing" is assumed to occur when the applied total head difference corresponds to an excess pore water pressure equal to the initial horizontal effective stress $\sigma'_{ho}$, then the equivalent excess water head at the beginning of the test, $H_i$, is defined as

$$H_i = \sigma'_{ho}/\gamma_w \tag{7}$$

Consequently, a safety factor (FS) against "hydraulic fracturing" may be defined from Eqs 6 and 7 as

$$\text{FS} = H_1/H_{max} = 6\,\sigma'_{ho}/\sigma'_{vo} = 6\,K_0 \tag{8}$$

where $K_0$ is the coefficient of earth pressure at rest.

If the Jaky's formula ($K_0 = 1 - \sin\phi$, where $\phi$ is the friction angle) is assumed to be valid for any soil, it appears that FS varies from 3.0 to 1.8 when $\phi$ varies from 30 to 45 deg. However, such a safety factor is not the true safety factor of a field test because the simple rule of Eq 5 that will be used in the field is derived from an assumed piezometric level and assumed unit weights.

A statistical study of "hydraulic fracturing" in three different granular soils has been presented by Chapuis et al. [13] to justify Eq 8 and its simplified version, Eq 5.

## Discussion and Conclusion

This paper proposes a method to verify whether the water level given by a piezometer is the true piezometric level and, consequently, whether a vertical communication has been established between aquifers because of incorrect seals along the pipes of the piezometer. The method makes use of variable-head permeability tests that are not interpreted by the common method (ln $H$ versus *t*) but by the more fundamental presentation of the water velocity versus the total head difference. Both presentations are mathematically equivalent. However, if there is some error in the assumed or measured piezometric level, this error will be hidden in the common logarithmic diagram whereas it will be easily detected in the proposed velocity diagram. Typical examples of velocity diagrams have been given and their interpretations discussed. The proposed method has been successfully applied to several hundred field tests performed in cased boreholes [*13*] and recently checked in scale-model tests in boreholes and piezometers performed within a large sand tank [*15*].

In the field, a casing is never correctly sealed in the ground like a piezometer, and the piezometric level around the driving shoe can only be evaluated; the errors on the assumed piezometric levels have been statistically studied by Chapuis et al. [*13*] for sites of dams and dikes. In general, for any site, successive variable-head permeability tests can be performed at different depths during field wash boring with casing; the proposed method allows for the determination of the various piezometric levels at these depths, thus enabling a better selection of the installation depths of piezometers in conjunction with stratigraphic informations.

The piezometric levels determined by the proposed method during tests in cased boreholes usually differ from those assumed in the field. However, they may be verified later by piezometric levels determined by piezometers installed at the same depth; very few disagreements have been recorded. Most of the discrepancies were associated with piezometers giving $H_0 \neq 0$ when tested, resulting either from a partial seal as requested by certain agencies, or from complex hydrogeologic situations in which different aquifers communicate along the borehole and make sealing operations difficult. According to the author's experience, the practice of only partially sealing piezometers in the adjacent ground, for example, with bentonite pellets for a distance of 60 to 150 cm, should be discouraged, and a full-length seal is preferred in all cases.

### *Acknowledgments*

This paper is based on information and experience gathered by the author from different hydrogeological studies made by Mon-Ter-Val Inc. Present research in this field is sponsored by the National Science and Engineering Research Council of Canada, Grant A-4704; this financial assistance is gratefully acknowledged.

## References

[*1*] Freeze, R. A. and Cherry, J. A., *Groundwater,* Prentice-Hall, Englewood Cliffs, NJ, 1979.
[*2*] Horvslev, M. J., "Time-lag and Soil Permeability in Ground Water Observations," Bulletin 36, U.S. Army Corps of Engineers Waterways Experimental Station, Vicksburg, MS, 1951.
[*3*] Boersma, L., "Field Measurement of Hydraulic Conductivity Below a Water Table," *Methods of Soil Analysis, Part 1,* C. A. Black, Ed., American Society of Agronomy, Madison, WI, 1965, pp. 222–233.
[*4*] Bouwer, H. and Jackson, R. D., "Determining Soil Properties," *Drainage for Agriculture,* J. van Schilfgaarde, Ed., American Society of Agronomy, Madison, WI, 1974, pp. 611–672.
[*5*] Cooper, H. H., Bredehoeft, J. D., and Papadopoulos, I. S., "Response of a Finite-diameter Well to an Instantaneous Change of Water," *Water Resource Research,* Vol. 3, 1967, pp. 263–269.
[*6*] Papadopoulos, I. S., Bredehoeft, J. D., and Cooper, H. H., "On the Analysis of Slug Test Data," *Water Resource Research,* Vol. 9, 1973, pp. 1087–1089.
[*7*] Milligan, V., "Field Measurement of Permeability in Soil and Rock," *Proceedings*, American Society of Civil Engineers, Conference on *In Situ* Measurement of Soil Properties, Raleigh, NC, Vol. 2, 1968, pp. 3–36.
[*8*] *Canadian Foundation Engineering Manual,* The Canadian Geotechnical Society, 1978.
[*9*] Gibson, R. E., "A Note on the Constant Head Test to Measure Soil Permeability *In Situ,*" *Geotechnique,* Vol. 16, 1966, pp. 256–257.
[*10*] Weber, W. G., "*In Situ* Permeabilities for Determining Rates of Consolidation," *Highway Research Record,* Vol. 243, 1968, pp. 49–61.

[11] Bjerrum, L., Nash, J. K. T. L., Kennard, R. M., and Gibson, R. E., "Hydraulic Fracturing in Field Permeability Testing," *Geotechnique,* Vol. 22, 1972, pp. 319–332.
[12] Rat, M., Laviron, F., and Jorez, J. C., *Essai Lefrance Bulletin de Liaison des Laboratoires Routiers,* No. Special N, 1970, pp. 56–66.
[13] Chapuis, R. P., Paré, J. J., and Lavallée, J. G., "*In Situ* Variable Head Permeability Tests," *Proceedings,* Tenth International Conference on Soil Mechanics and Foundation Engineering, Stockholm, Sweden, Vol. 1, 1981, pp. 401–406.
[14] Lefebvre, G., Philibert, A., Bozozuk, M., and Paré, J.-J., "Fissuring from Hydraulic Fracturing of Clay Soil," *Proceedings,* 10th International Conference on Soil Mechanics and Foundation Engineering, Stockholm, Sweden, Vol. 2, 1981, pp. 513–518.
[15] Chapuis, R. P., Soulié, M., and Sayegh, G., "Comparison de Divers Essais de Perméabilité Dans un Sable," *Proceedings,* 39th Canadian Geotechnical Conference, Ottawa, Canada, 1986.

*U. M. Cowgill*[1]

# The Chemical Composition of Leachate from a Two-Week Dwell-Time Study of PVC Well Casing and a Three-Week Dwell-Time Study of Fiberglass Reinforced Epoxy Well Casing

**REFERENCE:** Cowgill, U. M., "**The Chemical Composition of Leachate from a Two-Week Dwell-Time Study of PVC Well Casing and a Three-Week Dwell-Time Study of Fiberglass Reinforced Epoxy Well Casing,**" *Ground-Water Contamination: Field Methods,* ASTM STP 963, A. G. Collins and A. I. Johnson, Eds., American Society for Testing and Materials, Philadelphia, 1988, pp. 172–184.

**ABSTRACT:** The primary objective of this study was to determine whether or not trace organic compounds of the U.S. Environmental Protection Agency (EPA) priority pollutant list could be extracted from treated polyvinyl chloride (PVC) well casing that had been constructed with the use of PVC solvent cement and a composite epoxy-fiber reinforced casing.

Sections of PVC well casing were glued together with PVC solvent cement. In an effort to create a barrier that would reduce the leaching of various compounds from the PVC and the adhering cement, three treatments of PVC well casing were executed. PVC well casing was sulfonated with dry nitrogen gas-enriched with sulfur trioxide ($SO_3$) for 5 min, followed by a 3% calcium chloride ($CaCl_2$) wash. The remaining two treatments applied to PVC pipe were a 3% $CaCl_2$ wash and no treatment. A dwell-time study employing double-distilled water was carried out for a fortnight. Subsequently, the three leachates, the PVC solvent cement, the double-distilled water, and the 3% $CaCl_2$ wash were analyzed for their components by gas chromatograph/mass spectrometer-specific ion monitoring (GC/MS-SIM). The results indicated that both sulfonation and the $CaCl_2$ wash may act to minimize substances leaching out of the cement, but that this reduction fails to be sufficient enough to adequately protect water contained in such PVC well casing from contamination by substances leached from the cement. It has been ascertained that all detected contaminants originate from the PVC solvent cement. A field procedure is proposed for remedial action to remove PVC solvent cement contaminants from wells that have been thus contaminated.

An alternative to PVC well casing is fiberglass reinforced epoxy casing. Both powder (60 mesh, 200 mesh) and well casing were subjected to leachate tests. Of all the substances sought in these leachates, only cyclohexene oxide, bisphenol A, phenol, and di-*n*-butyl phthalate were detected, and these were encountered in the 200 mesh sample. It is hypothesized that grinding to 200 mesh exposes surfaces that would ordinarily not be exposed. A classical dwell-time study failed to confirm the presence of these compounds. As a result, a number of state agencies and EPA have approved the use of fiberglass reinforced epoxy casing for drinking water wells and monitoring wells for Superfund sites.

**KEY WORDS:** chemical composition, dwell-time study, PVC well casing, fiberglass reinforced epoxy well casing, priority pollutants

[1] Associate environmental consultant in the Health and Environmental Sciences in Department of Aquatic Toxicology, 1702 Building, Midland, MI 48674.

Polyvinyl chloride (PVC) pipe with solvent welded joints is probably the most common casing material selected for monitor well construction. It is usually the material of choice for well casings when problems of saltwater intrusion prohibit the long-term use of stainless steel. However, PVC casing is short-lived in very cold climates. In addition, if the purpose of monitoring is to detect trace quantities of organic chemicals, the use of PVC is not advisable since it may cause chemical interferences by sorption or leaching. The U.S. Environmental Protection Agency (EPA) [*1*] has found that a variety of substances leach into water supplies from PVC pipe as well as those which have solvent welded joints. The quantity of leaching chemicals can be especially large when sections of PVC casing have been connected with PVC solvent cement. Consequently, this study was undertaken to investigate whether some type of treatment could be devised that would seal off the materials that leach from PVC pipe and cement and to evaluate fiberglass reinforced pipe as an alternative well casing. Thus, several PVC well casings were prepared in various ways for evaluation. PVC casings and fiberglass casings were assembled and filled with double-distilled water, capped and maintained at room temperature (24°C $\pm$ 2°C) for two weeks. Subsequently, these double-distilled water samples were analyzed for detectable quantities of the 129 priority pollutants [*2*] chemicals in the PVC solvent cement, as well as those employed in the manufacture of fiberglass reinforced epoxy casing (FRE).

An alternative to PVC as a well construction material is a composite epoxy-fiber reinforced casing. This material is manufactured from bisphenol A type epoxy resins cured with methyltetrahydrophthalic anhydride. The result is a structurally strong material, fully reacted to the solid state. The epoxy resin used in manufacture may be either DER 331[2] or DER 325[2] epoxy resin or Shell Epon 826.[3] The final formulation of this fiberglass reinforced epoxy well casing has no inherent capability to yield halogenated hydrocarbons, and therefore would not be expected to further complicate ground-water monitoring and its subsequent interpretation. The casing tested in this study is Centron® "HP" proprietary anhydride, which, it should be noted, is far more resistant to attack by compounds of extreme pH according to an unpublished report (Centron Corrosion Studies CS-184, R. P. Johnson, 8-30-84).

Before proceeding further, it should be noted that many contracting firms have installed wells around many industrial plant sites throughout the United States employing PVC well casing with solvent welded joints. Substances originating from the cement are detectable in ground water a decade after installation. Parts of this study are devoted to the resolution of such contamination problems.

Another incident concerning PVC casing with solvent welded joints is worthy of note. In February 1983, the Midland Treatment Plant changed their water treatment. When this new water was transported to The Dow Chemical Company Aquatic Toxicology Laboratory, 402 m (¼ mile) of cast iron pipe was destroyed and replaced with PVC pipe with solvent welded joints. Eighty-four days later 1.3-mg$L^{-1}$ methylene chloride, 3-$\mu$g $L^{-1}$ chloroform, and 6-$\mu$g $L^{-1}$ tetrahydrofuran were still detected in the water despite the fact that this laboratory utilizes water at the rate of 227 L/min (60 gpm) [84 days = 27.47 million litres (7 257 600 gal) of water had passed through the 402 m (¼ mile) of PVC pipe]. It is apparent that this water at a pH 8 does not effectively remove the components that leach from PVC cement.

## Pertinent Literature

The leaching of organic substances from rigid PVC well casing and from those casings with solvent welded joints has been reviewed by Barcelona, Gibb, and Miller [*1*]. In all studies that have been reported up to the present [*1,3–9*] some substances of varying concentrations have

[2] Registered trademark of The Dow Chemical Company.

[3] Registered trademark of Shell Oil Company.

always been detected. However, none of the studies involved with primer and adhesive use report the chemical analysis of such substances beyond what has been noted on the containing can. The most detailed study of PVC cement is that of Sosebee and his co-workers [7] where they identified tetrahydrofuran, methyl ethyl ketone, cyclohexanone, and methylisobutylketone as the major constituents leaching from PVC primer and adhesive into water surrounding bonded joints. Boettner et al. [3] confirm all of the previously noted compounds with the exception of methylisobutylketone, which they did not detect in their analyses.

There are no available published data on the types of compounds that leach from composite epoxy-fiber reinforced casing.

Recently Barcelona et al. [10] published a study comparing the maximum sorption of dilute (100 ppb) halogenated hydrocarbon mixtures in water by various plastics. PVC exhibited a maximum sorption of 622 $\mu g\ m^{-2}$ while Teflon[4] sorbed only 237 $\mu g\ m^{-2}$. Unpublished data provided by Groundwater Sampling Inc. of Englewood, Colorado (personal communication, G. G. Hunkin, Sr.), following a similar but not identical procedure outlined by Barcelona et al. [10], showed that the maximum sorption of fiberglass reinforced epoxy casing was 203 $\mu g\ m^{-2}$. This number was achieved by exposing 6-g (0.21 oz) of fiberglass reinforced epoxy particles to 100 ppb concentrations of individual halogenated hydrocarbons similar to the group utilized by Barcelona et al. [10] for 72 h. The Barcelona et al. study did not employ exposure times in excess of 60 min. It is interesting to note that the use of Teflon or PVC would appear to compromise the value of total organic carbon (TOC) as an indicator of pollution. This has not been found to be the case with fiberglass reinforced epoxy casing (personal communication, G. G. Hunkin, Sr.).

## Experimental Procedure

### *Polyvinyl Chloride Casing*

Commercially available NSF-dwv 071783 2 in. Bristol-pipe PVC 1120 D-1785 280 PSI @ 73°F Sch 40 D-2665 [11] sections of PVC pipe [5 cm (2 in.) inside diameter, schedule weight 40] were purchased from a local hardware store. To simulate a typical field monitoring well, three 30.5-cm (1 ft) sections were connected to each other utilizing PVC solvent cement (PVC solvent cement clear No. 70514, Handishop Plumbing Supplies, Jacksonville, Florida 32202). The cement was allowed to harden for 48 h. Thus, each pipe or well casing was 91.4 cm (3 ft) long. A permanent cap was affixed to the bottom of each well casing. Three treatments were used in this experiment. The control well casing was untreated but rinsed with double-distilled water. A second well casing was sulfonated on the inside for 5 min with dry nitrogen ($N_2$) gas enriched with sulfur trioxide ($SO_3$). This process was followed with a 2-min rinse of 3% calcium chloride ($CaCl_2$) solution. The latter functions to neutralize the acidity brought about by the $SO_3$. A third well casing was just rinsed with 3% $CaCl_2$ solution. In essence, the third well casing functions as a control for the second well casing and the first well casing acts as a control for the other two. Top sealing caps were treated in the same fashion as their respective well casings.

Each of the three well casings [capacity approximately 1.86 L (0.49 gal) each] was filled to capacity without head space with carbon-filtered double-distilled water, capped, and allowed to remain at 24°C ± 2°C for two weeks. At the end of this period, the water in each of the three well casings was gently transferred to precleaned [soaked in 6 N hydrochloric acid (HCl) overnight, rinsed in tap water, rinsed 20 times in carbon-filtered double-distilled water, soaked in methanol for 10 min, soaked in acetone for 10 min, rinsed 20 times in double-distilled water followed by autoclaving for 20 min] glass bottles fitted with bakelite tops containing Teflon inserts and submitted

[4] Registered trademark of Du Pont.

for chemical analysis. In addition, concurrent samples of the double-distilled water, the 3% $CaCl_2$ solution, as well as the PVC solvent cement, were simultaneously submitted for chemical analysis.

## Fiberglass Reinforced Epoxy Casing (FRE) and Powder

During the manufacture of the epoxy composite tubing or casing, male-female integral coupling threads were cut with a diamond tool without using lubricants. The particles produced by this operation were removed by a suction exhaust system and subsequently were collected in filter bags. Samples from these filter bags were randomly selected during the course of routine manufacturing activities and set aside for chemical analysis. The size of the particles found in the filter bags was such that more than 90% pass through a 209-μm (0.209 mm = 60 mesh) screen. A 50-g (1.7 oz) sample thus provides an approximate surface area of 67.4 $m^2$ (725.5 $ft^2$). These particles were initially selected to obtain a high surface-to-volume ratio which on leaching would afford a greater opportunity to detect undesirable compounds in the leachate and, of course, would result in higher concentrations than would be encountered in the field.

A leaching experiment, initially carried out by the Rocky Mountain Analytical Laboratory for EPA and included here for comparison, was executed by placing 50.0 g (1.7 oz) of FRE cuttings into a precleaned 3.8 L (1 gal) glass bottle which contained a Teflon stirring bar. To this glass bottle, 2 L (0.5 gal) of carbon-filtered double-distilled water were added, and the bottle was sealed with a Teflon lined cap. The cuttings were used as received without steam cleaning or any other purification. Although further grinding was attempted, the particle size could not be reduced, employing ordinary means, beyond the 60 to 100 mesh size distribution that was submitted. The test bottle was placed in a cooler at 4°C that was sitting on a magnetic stirring unit. The cooler was packed with ice, and the test solution was agitated magnetically for 72 ± 1 h. After leaching, the solution was allowed to settle and a sample for volatile organic analysis was decanted. The remainder of the leachate was passed through a 0.45-μm filter to remove casing particles. This leachate was then submitted for chemical analysis.

Subsequent to the preceding experimental procedure, two more studies were carried out. There were two reasons for this: (1) the products involved in manufacture were not sought, and (2) a dwell-time study of 72 h may not, to some investigators, appear to be sufficient time to allow the more recalcitrant substances to leach from the particles to the contiguous water.

The first (FRE) experiment was repeated with some differences. Instead of 50 g (1.7 oz) samples, approximately 150 g (5.29 oz) were used and the typical particle size was 200 mesh obtained by use of a nonlubricated diamond tool. Two samples and a blank were employed. One sample was agitated magnetically and the other was not. The study was carried out at room temperature and lasted for three weeks. The leaching liquid was double-distilled (glass) water. At the end of the leaching experiment, the samples were filtered through a 0.22-μm millipore filter, and the filtrate was submitted for chemical analysis for the 129 priority pollutants, as well as the substances involved in manufacture.

The last (FRE) experimental procedure was a classical dwell-time study. Two well casings, 6.8 L (1.8 gal) and 6.2 L (1.6 gal) capacity, respectively, were prepared for field installation (steam cleaned). Each well casing was supplied with fitted cast-in-place bottom caps. Each of the two well casings was filled with double-distilled water, capped, and placed in a chemically clean hood at room temperature for three weeks. Simultaneously, 6 L (1.6 gal) of double-distilled water were set in the hood in 6 L (1.6 gal) glass jars fitted with bakelite caps with Teflon inserts, to serve as blanks for the period the dwell time study was taking place. At the end of three weeks, the samples were placed in precleaned (soaked in 6 N HCl overnight, rinsed in tap water, rinsed 20 times in carbon-filtered double-distilled water, soaked in methanol for 10 min, soaked in acetone for 10 min, rinsed 20 times in double-distilled water followed by autoclaving for 20 min) brown glass bottles with ground glass tops and the tops further sealed with Parafilm "M".[5] The liquid samples were submitted for chemical analysis.

## Analytical Methodology

Volatile organic compound determinations in all leachates and PVC solvent cements were performed using EPA Method 624 [2]. Aliquots of the samples were placed in a sparging device. Internal standards and deuterium-labeled surrogates were added to verify the analytical results and to provide qualitative and quantitative references for each sample. The samples were then purged with helium and the volatile organic compounds were transferred to the gas stream. They were removed from the gas stream with a Tenax[6] Silica gel trap or a Chromosorb[7] 106 trap. When purging was complete, the trap was rapidly heated and the trapped organic compounds transferred to the analytical chromatographic column of a gas chromatograph/mass spectrometer (GC/MS) with specific ion monitoring (SIM) instrumentation.

Acetone and ethyl acetate, both components of the PVC cement, and the blanks of double-distilled water and 3% $CaCl_2$ solution were concentrated using the Peters [*12*] method. Standards were run exactly in the same manner as the samples and recoveries were of the order of 88%. All other compounds encountered in the PVC leachate were injected neat onto a 0.1% SP1000 column on Carbopack C and identified by flame ionization detection (FID). All identifications were further confirmed by GC/MS-SIM.

Base/neutral and acid (BN/A) compounds were determined using EPA Method 625 [2]. Aliquots of the samples were extracted first at basic pH and then at acidic pH with methylene chloride. The extracts were concentrated and analyzed by GC/MS-SIM using the same approach as that followed for the volatile organic compounds.

Pesticides and polychlorinated biphenyls (PCBs) were determined using EPA Method 608 [2]. Aliquots of the samples were extracted with methylene chloride at a neutral pH. The extracts were then concentrated and analyzed using an electron capture/gas chromatograph (EC/GC). Identification of components was based on the retention times of the unknown, compared to those of the standards. All identifications were confirmed using a second analytical GC column.

Inorganic substances were determined according to a variety of EPA methodology [2].

## Results and Discussion

### *Polyvinyl Chloride Casing*

Tables 1 and 2 give the analytical results obtained from the leachates of the three PVC casings, the PVC solvent cement, the 3% $CaCl_2$ solution and the double-distilled water. It is apparent that sulfonation reduces the leaching of undesirable compounds. However, this barrier does not constitute a sufficient obstacle to the migration of compounds from the cement to the contained water to permit this water to pass existing water quality standards or to make PVC casing with solvent welded joints suitable for monitor wells. It should be noted that a straight 3% $CaCl_2$ wash does as well as sulfonation followed by a $CaCl_2$ wash for most compounds.

Linear correlation coefficients obtained among the nine compounds detected in each of the leachates reveal highly significant results, all in excess of $r = 0.9989$. This suggests that the rate of leaching of the nine compounds is the same for all three treatments. Partial correlation analysis [*13*] reveals some interesting results. The partial correlation coefficient between the nine compounds in the leachate obtained from the $CaCl_2$ wash treatment and those in the leachate obtained from

[5] Registered trademark of the American Can Company.
[6] Registered trademark of the Enko Institute.
[7] Registered trademark of the Johns Manville Company.

TABLE 1—*Chemical composition of the leachates from three treated PVC casings (μg $L^{-1}$) the joints of which were welded with solvent cement.*

| | Polyvinyl Chloride Casing | | |
|---|---|---|---|
| Substance | Untreated | Sulfonation Followed by 3% $CaCl_2$ | 3% $CaCl_2$ Wash |
| Methylene chloride | 4 | 3 | 3 |
| Chloroform | 8 | 1 | 1 |
| Acetone | 831 | 105 | 122 |
| Tetrahydrofuran | 360 000 | 155 000 | 150 000 |
| Methyl ethyl ketone | 975 000 | 452 500 | 437 500 |
| Ethyl acetate | 31 | 46 | 7 |
| Benzene | 1 500 | 1 000 | 2 000 |
| Cyclohexanone | 33 750 | 32 000 | 25 500 |
| Toluene | 2 000 | 1 500 | 2 500 |

the sulfonation treatment reveal that these compounds leach at the same rate when the effect of no treatment is excluded or held constant. The partial correlation coefficient in this instance is 0.969 with six degrees of freedom and is significant beyond the 0.1% level. This essentially implies that the two treatments provide a similar alteration; however, this fails to adequately protect contained water from contamination by the chemical components of the cement.

A comparison of the results depicted in Table 1 and Table 2 reveals that all the compounds detected in the three leachates originate from the use of PVC solvent cement. Unfortunately, many monitoring wells have been installed with the use of this cement and, in some cases, serious ground-water contamination may result. A procedure has been designed to remedy this situation in the field. Water must be removed from a borehole and a packer placed at the screen-casing joint. A portable steam jenny, using distilled water as the steam source, is employed to steam clean the PVC. This procedure is repeated five times. After each cleaning, the condensed water is pumped out of the well and discarded into a leak proof container. After the fifth cleaning has been completed and the condensation removed, the packer is removed. An air compressor fitted

TABLE 2—*Chemical composition of PVC solvent cement (expressed as percent), the 3% $CaCl_2$ solution (μg $L^{-1}$), and the double-distilled water (μg $L^{-1}$) used in the PVC dwell-time study.*

| Substance | PVC Solvent Cement | 3% $CaCl_2$ | Double-Distilled Water |
|---|---|---|---|
| Methylene chloride | trace[a] | 3 | ND(0.4)[b] |
| Chloroform | 0.1 | 1 | ND(0.2) |
| Acetone | trace[a] | 6 | ND(0.3) |
| Tetrahydrofuran | 22.6 | 5 | ND(2.0) |
| Methyl ethyl ketone | 51.8 | 3 | ND(1.0) |
| Ethyl acetate | <0.01 | ND(20) | ND(2.0) |
| Benzene | <0.01 | ND(20) | ND(20) |
| Cyclohexanone | 7.0 | ND(4) | ND(1.0) |
| Toluene | 0.02 | ND(8) | ND(8.0) |
| Polyvinyl chloride | 18.0 | (—)[c] | (—) |

[a] Trace = <0.004%, detection of these substances in this matrix proved to be difficult.
[b] ND = not detected at level noted in parentheses.
[c] (—) = not determined.

with a carbon filter is then employed and the bottom of the well is aerated for 72 h. Water is then removed from the well casing, a sample is collected and analyzed. Severely contaminated wells may require the preceding procedure to be repeated.

Several wells have been subjected to the aforementioned cleaning procedure. These wells were constructed by a contractor six years ago. The casing employed was PVC with solvent welded joints. The steam distillate was analyzed for the ten components of PVC cement (Table 2). The initial steam distillate contained detectable quantities of the components of PVC cement except methylene chloride, chloroform, acetone and toluene. The two most persistent substances were tetrahydrofuran (THF) and methyl ethyl ketone (MEK). By the fifth cleaning, none of the components of the PVC cement were detectable at the 0.1-μg $L^{-1}$ level. After 72 h of aeration of the ground water at the base of the wells, none of the components of the PVC cement were encountered in amounts greater than 1 μg $L^{-1}$ in that ground water. It remains to be seen how much remains detectable after a six-month rest period. These data are provisional, and this aspect of the study is still in progress. It should be noted that the analytical results create the general impression that this clean-up procedure is feasible and has a good chance of being successful.

## Fiberglass Reinforced Epoxy Casing and Powder

Tables 3 through 9 give the analytical results of the analyses of the three leachates obtained from fiberglass reinforced epoxy casing and powder.

TABLE 3—*Analytical and biological results of the two leachates (mg $L^{-1}$ or as noted) of fiberglass reinforced epoxy well casing powder.*

| | Powder Leaching Period | |
|---|---|---|
| Measurement | 73 h<br>50 g | 3 Weeks<br>150 g |
| Biochemical oxygen demand | ND(2)[a] | (—)[b] |
| Chemical oxygen demand | ND(5) | (—) |
| Total organic carbon | ND(1) | (—) |
| Total suspended solids | ND(1) | (—) |
| Surfactants | ND(0.05) | (—) |
| Oil and grease | ND(1) | (—) |
| Color, units | ND(5) | (—) |
| Fecal coliform No/100 mL | ND(0) | (—) |
| Sulfate, total | ND(10) | ND(10) |
| Sulfite, as S | ND(1.0) | ND(0.05) |
| Sulfide, as S | ND(0.05) | ND(1) |
| Total organic N | ND(0.1) | (—) |
| Ammonia, as N | ND(0.1) | (—) |
| Nitrate/nitrite, as N | ND(0.1) | 0.5/ND(0.1) |
| Gross α pCi $L^{-1}$ | ND(0.1) | (—) |
| Gross β pCi $L^{-1}$ | ND(0.5) | (—) |
| Radium, total pCi $L^{-1}$ | ND(0.1) | (—) |
| Cl, residual | ND(0.1) | ND(0.01) |
| Cl | (—) | 1.0 |
| Br | ND(1.0) | ND(2) |
| F | ND(0.1) | 2.8 |
| P, total | ND(0.1) | 0.06 |

[a] ND = not detected at levels noted in parentheses.
[b] (—) = not determined.

TABLE 4—*Elemental, cyanide and phenol results of the two leachates (mg L⁻¹) of fiberglass reinforced epoxy well casing powder.*

| Element (Total) | Powder Leaching Period 72 h 50 g | 3 Weeks 150 g |
|---|---|---|
| Al | ND(0.2)[a] | ND(0.1) |
| Ba | ND(0.1) | ND(0.1) |
| B | ND(0.1) | 3.5 |
| Co | ND(0.3) | ND(0.05) |
| Fe | ND(0.02) | ND(0.03) |
| Mg | ND(1.0) | 0.69 |
| Mo | ND(0.05) | ND(0.1) |
| Mn | ND(0.01) | ND(0.01) |
| Sn | ND(1.0) | ND(0.8) |
| Ti | ND(0.75) | ND(0.4) |
| Sb | ND(0.001) | ND(0.02) |
| As | ND(0.001) | ND(0.002) |
| Be | ND(0.010) | ND(0.005) |
| Cd | ND(0.005) | ND(0.005) |
| Cr | ND(0.035) | ND(0.05) |
| Cu | ND(0.015) | ND(0.02) |
| Pb | ND(0.035) | ND(0.1) |
| Hg | ND(0.0002) | ND(0.0002) |
| Ni | ND(0.02) | ND(0.3) |
| Se | ND(0.001) | ND(0.002) |
| Ag | ND(0.010) | ND(0.01) |
| Tl | ND(0.002) | ND(0.1) |
| Zn | ND(0.010) | 0.53 |
| Cyanide | ND(0.01) | ND(0.02) |
| Phenols | ND(0.01) | (—)[b] |

[a] ND = not detected at levels noted in parentheses.
[b] (—) = not determined.

The 72-h, 60-mesh powder leachate analysis (Tables 3–9) failed to detect the presence of any of the elements sought, any radioactivity, or any of the 129 priority pollutants of the EPA priority pollutant list.

The three-week dwell-time study involving the 200-mesh powder revealed detectable quantities of nitrates, fluorides, chlorides (non-residual), phosphorus (total), boron, magnesium, and zinc. Previous experience with this problem indicates that fluoride is a contaminant from autoclaved Teflon inserts in the bakelite caps used to seal the leachate study samples. The remainder of the detected elements were present at levels below the drinking water standards or are not regulated [*14*]. Pesticides (Table 5), PCBs (Table 5), and volatile organic compounds (Table 6) were not detected in the 200-mesh powder study. Of the base/neutral compounds (Table 7, 8) only di-*n*-butyl phthalate was detected, and furthermore the quantity found was below the EPA detection limit set at 10 μg L⁻¹ (*Federal Register,* 1984). Detectable quantities of phenol, cyclohexene oxide, and bisphenol A were encountered in the 200-mesh powder study (Table 9). Both cyclohexene oxide and phenol were contaminants contributed by the bakelite caps and their inserts. It is hypothesized that grinding the casing to 200 mesh exposes surfaces that under normal use would never become exposed, and this explanation is offered for the detectable amounts of bisphenol A, a product of manufacture of FRE. This hypothesis is further strengthened by the fact that though it was sought, it was not detected in the casing leachate study.

TABLE 5—*Pesticide and PCB determination ($\mu g\ L^{-1}$) in the two leachates from the fiberglass reinforced epoxy well casing powder.*

| | Powder Leaching Period | |
|---|---|---|
| Pesticide | 72 h 50 g | 3 Weeks 150 g |
| Aldrin | ND(0.003)[a] | ND(41) |
| α-BHC | ND(0.001) | ND(4.4) |
| β-BHC | ND(0.004) | (—)[b] |
| γ-BHC | ND(0.004) | ND(5.0) |
| δ-BHC | ND(0.002) | (—) |
| Chlordane | ND(0.04) | ND(1.9) |
| 4,4′-DDT | ND(0.012) | ND(1.3) |
| 4,4′-DDD | ND(0.016) | ND(1.0) |
| 4,4′-DDE | ND(0.012) | ND(1.8) |
| Dieldrin | ND(0.006) | ND(3.3) |
| α-endosulfan | ND(0.005) | (—) |
| β-endosulfan | ND(0.01) | (—) |
| Endosulfan sulfate | ND(0.03) | ND(110) |
| Endrin | ND(0.049) | ND(14) |
| Endrin aldehyde | ND(0.02) | (—) |
| Heptachlor | ND(0.002) | ND(2.4) |
| Heptachlor epoxide | ND(0.004) | ND(2.3) |
| Toxaphene | ND(0.40) | ND(10) |
| 2,3,7,8-TCDD | ND(0.50) | ND(1) |
| PCB-1242 | ND(0.04) | ND(1.1) |
| PCB-1254 | ND(0.50) | ND(1.1) |
| PCB-1221 | ND(0.04) | ND(1.1) |
| PCB-1232 | ND(0.04) | ND(1.1) |
| PCB-1248 | ND(0.05) | ND(1.1) |
| PCB-1260 | ND(0.50) | ND(1.1) |
| PCB-1016 | ND(0.04) | ND(1.1) |

[a] ND = not detected at levels noted in parentheses.
[b] (—) = not determined.

The three-week dwell-time study involving the analysis of casing leachate revealed the absence of detectable amounts of phenol, bis-phenol A, cyclohexene oxide, bis(2-ethylhexyl) phthalate, and di-*n*-butyl phthalate. It should be noted that since 127 of the 129 priority pollutants were not detected in either of the powder leachate studies, it was reasoned that they would not be detected in a classical dwell-time study. In addition, there is little reason to suspect their presence since they are not among the compounds of manufacture. Therefore, these substances were not sought in the analyses of the leachate from the three-week dwell-time study of the FRE casing.

## Summary

### *Polyvinyl Chloride Casing*

It is clear that both sulfonation and a $CaCl_2$ wash act to minimize the quantity of substances leaching out of PVC casing sections connected together by PVC solvent cement. However, these preventive measures fail to adequately protect water contained in such PVC well casing from contamination by substances leached from the cement. A field procedure is offered that initially appears to prove useful to remedy this contamination in wells already installed. New wells should

TABLE 6—*Volatile organic substances (mg $L^{-1}$) determined in the two leachates obtained from the fiberglass reinforced epoxy well casting powder.*

| Substance | Powder Leaching Period | |
|---|---|---|
| | 72 h 50 g | 3 weeks 150 g |
| Acrolein | ND(100)[a] | ND(130) |
| Acrylonitrile | ND(100) | ND(84) |
| Benzene | ND(1) | ND(2.0) |
| Bis(chloromethyl)ether | ND(10) | (—)[b] |
| Bromoform | ND(1) | ND(84) |
| Carbon tetrachloride | ND(1) | ND(6.6) |
| Chlorobenzene | ND(1) | ND(5.0) |
| Chlorodibromomethane | ND(1) | ND(11) |
| Chloroethane | ND(10) | (—) |
| 2-chloroethylvinyl ether | ND(1) | ND(3.8) |
| Chloroform | ND(1) | ND(3.2) |
| Dichlorobromomethane | ND(1) | ND(3.4) |
| Dichlorodifluoromethane | ND(10) | (—) |
| 1,1-dichloroethane | ND(1) | ND(17) |
| 1,2-dichloroethane | ND(1) | ND(42) |
| 1,1-dichloroethylene | ND(1) | ND(120) |
| 1,2-dichloropropane | ND(1) | ND(5.0) |
| 1,3-dichloropropylene | ND(1) | ND(4.2) |
| Ethylbenzene | ND(1) | ND(14) |
| Methyl bromide | ND(10) | ND(20) |
| Methyl chloride | ND(10) | ND(200) |
| Methylene chloride | ND(1) | ND(28) |
| 1,1,2,2-tetrachloroethane | ND(1) | ND(44) |
| Tetrachloroethylene | ND(1) | ND(32) |
| Toluene | ND(1) | ND(4.6) |
| 1,2-trans-dichloroethylene | ND(1) | ND(19) |
| 1,1,1-trichloroethane | ND(1) | ND(8.4) |
| 1,1,2-trichloroethane | ND(1) | ND(5.0) |
| Trichloroethylene | ND(1) | ND(9.2) |
| Trichlorofluoromethane | ND(10) | ND(58) |
| Vinylchloride | ND(10) | ND(15) |

[a]ND = not detected at level noted in parentheses.
[b](—) = not determined.

use only steam-cleaned, new PVC casing, the sections of which have been threaded so that they may be screwed together. The outside joint should be sealed with Teflon tape. Testing over the past four years has shown that such treatment will not result in any detectable compound leaching into the contiguous water.

### *Fiberglass Reinforced Epoxy Casing*

The typical three-week dwell-time study carried out in fiberglass reinforced epoxy casing prepared for field installation revealed no detectable quantities of the substances involved in manufacture. Two types of powder were also subjected to leachate analyses. One such study lasted for 72 h employing a 60-mesh powder. Subsequent analyses failed to reveal detectable quantities of any of the 129 priority pollutants, gross $\alpha$, gross $\beta$, or radium, or any of the elements sought

TABLE 7—*Base/neutral compound analyses (mg $L^{-1}$) determined in the three leachates from the fiberglass reinforced epoxy well casing.*

| | Powder Leaching Period | | |
|---|---|---|---|
| Substance | 72 h 50 g | 3 Weeks 150 g | Well Casing 3 Weeks |
| Acenaphthene | ND(10)[a] | ND(1.4) | ND(1) |
| Acenaphthylene | ND(10) | ND(0.8) | ND(1) |
| Anthracene | ND(10) | ND(0.9) | ND(1) |
| Benzidene | ND(10) | (—)[b] | ND(1) |
| Benzo(*a*)anthracene | ND(10) | (—) | ND(1) |
| Benzo(*a*)pyrene | ND(10) | ND(0.3) | ND(1) |
| 3,4-benzofluoranthene | ND(10) | (—) | (—) |
| Benzofluoranthene | ND(10) | ND(2.0) | ND(1) |
| Benzo(*ghi*)perylene | ND(10) | ND(0.8) | ND(1) |
| Benzo(*b*)fluoranthene | ND(10) | ND(0.9) | ND(1) |
| Benzo(*k*)fluoranthene | ND(10) | (—) | ND(1) |
| Bis(2-chloroethoxy)methane | ND(10) | ND(1.4) | ND(1) |
| Bis(2-chloroethyl)ether | ND(10) | ND(5.8) | ND(1) |
| Bis(2-chloroisopropyl)ether | ND(10) | ND(1.9) | ND(1) |
| Bis(2-ethylhexyl)phthalate | ND(10) | (—) | ND(1) |
| 4-bromophenylphenyl ether | ND(10) | ND(1.7) | ND(1) |
| Butyl benzyl phthalate | ND(10) | ND(0.7) | ND(1) |
| 2-chloronaphthalene | ND(10) | (—) | ND(1) |
| 4-chlorophenylphenyl ether | ND(10) | ND(0.9) | ND(1) |
| Chrysene | ND(10) | ND(2.2) | ND(1) |
| Dibenzo(*a,h*)anthracene | ND(10) | ND(1.3) | ND(1) |
| 1,2-dichlorobenzene | ND(10) | ND(1.1) | ND(1) |
| 1,3-dichlorobenzene | ND(10) | (—) | ND(1) |
| 1,4-dichlorobenzene | ND(10) | ND(0.5) | ND(1) |
| 3,3′-dichlorobenzidene | ND(10) | ND(0.4) | ND(1) |
| Diethyl phthalate | ND(10) | 4.8 | ND(1) |
| Di-*n*-butyl phthalate | ND(10) | (—) | ND(1) |

[a]ND = not detected at level noted in parentheses.
[b](—) = not determined.

in this leachate. Detectable substances encountered in a leachate of 200-mesh powder were either below detection limits set by EPA, were not regulated or clearly entered the leachate as a result of some type of contamination.

Presently, the use of composite epoxy-fiber reinforced casing has been approved for use for drinking water wells and piping to transport drinking water by the Colorado State Health Department, EPA Region VIII, City of Denver Environmental Division and Jefferson County Environmental Division.

The analytical results described in this paper support the safety of using reinforced fiberglass epoxy as a well casing for drinking water, as well as its use in Superfund sites that require long-term monitoring.

*Acknowledgments*

Dr. W. E. Walles of the Central Research Plastic Laboratory of Dow Chemical U.S.A. contributed the treated PVC pipe. Mr. Glen Boggs of Analytical and Environmental Chemistry of Health & Environmental Sciences of Dow Chemical U.S.A. provided the analytical results of the

TABLE 8—*Base/neutral determined (μg L⁻¹) in the three leachates from the fiberglass reinforced epoxy well casing and powder.*

| Substance | Powder Leaching Period 72 h 50 g | Powder Leaching Period 3 Weeks 150 g | Fiberglass Reinforced Epoxy Well Casing 3 Weeks |
|---|---|---|---|
| 2,4-dinitrotoluene | ND(10)[b] | ND(2.2) | ND(1) |
| 2,6-dinitrotoluene | ND(10) | ND(4.5) | ND(1) |
| Di-*n*-octyl phthalate | ND(10) | (—)[c] | ND(1) |
| 1,2-diphenylhydrazine[a] | ND(10) | (—) | ND(1) |
| Fluoranthene | ND(10) | ND(0.9) | ND(1) |
| Fluorene | ND(10) | ND(1.3) | ND(1) |
| Hexachlorobenzene | ND(10) | ND(0.9) | ND(1) |
| Hexachlorobutadiene | ND(10) | ND(0.9) | ND(1) |
| Hexachlorocyclopentadiene | ND(10) | ND(0.8) | ND(1) |
| Hexachloroethane | ND(10) | ND(5.0) | ND(1) |
| Indeno (1,2,3-*cd*) pyrene | ND(10) | ND(5.1) | ND(1) |
| Isophorone | ND(10) | ND(0.9) | ND(1) |
| Naphthalene | ND(10) | ND(0.5) | ND(1) |
| Nitrobenzene | ND(10) | ND(1.0) | ND(1) |
| *n*-nitrosodimethylamine | ND(20) | ND(100) | ND(1) |
| *n*-nitrosodi-*n*-propylamine | ND(10) | ND(300) | ND(1) |
| *n*-nitrosodiphenylamine | ND(10) | ND(300) | ND(1) |
| Phenanthrene | ND(10) | ND(0.7) | ND(1) |
| Pyrene | ND(10) | ND(0.7) | ND(1) |
| 1,2,4-trichlorobenzene | ND(10) | ND(1.3) | ND(1) |

[a]As Azobenzene.
[b]ND = not determined at level noted in parentheses.
[c](—) = not determined.

PVC leachates. Laboratory assistance from the personnel of Mammalian and Environmental Sciences of Dow Chemical U.S.A. is gratefully acknowledged. The Rocky Mountain Analytical Laboratory furnished the analytical results of the 72-hour leachate study of 60-mesh fiberglass reinforced epoxy casing particles. The Analytical Laboratory of Dow Chemical U.S.A. supplied the analyses of the three-week 200 mesh leachate study of the fiberglass epoxy casing material, and ERG of Ann Arbor, Michigan provided the analytical results from the three-week leachate study of the well casing itself. The two powders, as well as the steam-cleaned fiberglass reinforced epoxy casing, were contributed by Mr. Jeff Hunkin, Sr., of Groundwater Sampling Inc. of Englewood, Colorado. The author is grateful to all these people and the analytical laboratories that made this study possible.

## References

[*1*] Barcelona, M. J., Gibb, J. P., and Miller, R. N., EPA-600/2-84-024, Environmental Protection Agency, Washington, DC, Jan. 1984.

[*2*] "Guidelines Establishing Test Procedures for the Analysis of Pollutants under the Clean Water Act, Final Rule and Interim Final Rule and Proposed Rule. Corrections," *Federal Register,* Vol. 49, 1984, pp. 43234–43442; Vol. 50, pp. 690–697.

[*3*] Boettner, E. A., Gall, G. L., Hollingsworth, Z., and Aquino, R., EPA-600/1-81-062, Environmental Protection Agency, Washington, DC, Sept. 1981.

[*4*] Curran, C. M. and Tomson, M. B., *Groundwater Monitoring Review,* Vol. 3, Summer 1983, pp. 68–71.

TABLE 9—*Acid compounds and compounds involved in manufacture ($\mu g\ L^{-1}$) sought for in the three leachates from the fiberglass epoxy well casing and powder.*

| Substance | Powder Leaching Period | | Fiberglass Reinforced Epoxy Well Casing |
|---|---|---|---|
| | 72 h 50 g | 3 Weeks 150 g | 3 Weeks |
| 2-chlorophenol | ND(10)[a] | ND(1.3) | (—)[b] |
| 2,4-dichlorophenol | ND(10) | ND(2.2) | (—) |
| 2,4-dimethylphenol | ND(10) | ND(0.7) | (—) |
| 4,6-dinitro-o-cresol | ND(25) | ND(4.1) | (—) |
| 2,4-dinitrophenol | ND(25) | ND(6) | (—) |
| 2-nitrophenol | ND(10) | ND(2.4) | (—) |
| 4-nitrophenol | ND(10) | ND(2.1) | (—) |
| *p*-chloro-*m*-cresol | ND(10) | (—) | (—) |
| Pentachlorophenol | ND(10) | ND(2.5) | (—) |
| Phenol | ND(10) | 31 | ND(1) |
| 2,4,6-trichlorophenol | ND(10) | ND(1.9) | (—) |
| Cyclohexene oxide | (—) | 4 | ND(1) |
| Bisphenol A | (—) | 480 | ND(1) |
| Epichlorohydrin | (—) | ND(680) | (—) |
| Methyltetrahydrophthalic anhydride | (—) | ND(10) | (—) |

[a]ND = not determined at level noted in parentheses.
[b](—) = not determined.

[5] Junk, G. A., Svec, H. J., Vick, R. D., and Avery, M. J., *Environmental Science and Technology,* Vol. 8, Dec. 1974, pp. 1100–1106.
[6] Reich K. D., Trussell, A R., Lieu, F. Y., Leong, L. Y. C., and Trussell, R. R. in *Proceedings,* American Water Works Association Conference, Paper No. 31-4, June 1981, pp. 1249–1260.
[7] Sosebee, J. B., Jr., Geiszler, P. C., Winegardner, D. L., and Fisher, C. R. in *Hazardous and Industrial Solid Waste Testing: Second Symposium, ASTM STP 805,* American Society for Testing and Materials, Philadelphia, 1983, pp. 38–50.
[8] Wang, T. C. and Bricker, J. L., *Bulletin of Environmental Contamination and Toxicology,* Vol. 23, 1979, pp. 620–623.
[9] Miller, G. D. in *Proceedings,* Second National Symposium on Aquifer Restoration and Ground Water Monitoring, National Water Well Association, May 1982, pp. 236–245.
[10] Barcelona, M. J., Helfrich, J. A., and Garska, E. E., *Analytical Chemistry,* Vol. 57, 1985, pp. 460–464.
[11] Standard Specifications for IPS Rigid Poly (Vinyl Chloride) (PVC) Plastic Pipe, Schedules 40, 80 and 120, ASNI/ASTM D1785-83, American Society for Testing and Materials, Philadelphia, 1983, pp. 89–101.
[12] Peters, T. L., *Analytical Chemistry,* Vol. 52, 1980, pp. 211–213.
[13] Ostle, B., *Statistics in Research,* Iowa State College Press, Ames, IA, 1954.
[14] National Academy of Sciences, *Drinking Water and Health,* National Academy of Sciences Printing and Publishing Office, Washington, DC, 1977.

*Jerry N. Jones*[1] *and Gary D. Miller*[2]

# Adsorption of Selected Organic Contaminants onto Possible Well Casing Materials

**REFERENCE:** Jones, J. N. and Miller, G. D., **"Adsorption of Selected Organic Contaminants onto Possible Well Casing Materials,"** *Ground-Water Contamination: Field Methods, ASTM STP 963,* A. G. Collins and A. I. Johnson, Eds., American Society for Testing and Materials, Philadelphia, 1988, pp. 185–198.

**ABSTRACT:** In order to assess and monitor the impact of man's activities on ground-water quality, the nature, concentration, and behavior of pollutants in the subsurface must be determined. To do this, it is necessary to collect representative samples of water from the subsurface environment. Ideally, the material chosen as a well casing for monitoring wells should not add or remove constituents from the ground-water sample collected.

The objective of this study was to determine if specific trace organics adsorb and subsequently leach from various well casing materials upon exposure to those specific pollutants for specified periods of time. The types of well casings studied included thermoplastics, stainless steel and fluoroplastics. Of the pollutants studied, only 4-nitrophenol and 2,4,6-trichlorophenol showed appreciable adsorption for all well casing materials. Subsequent leaching of pollutants was not observed in any measurable quantities.

**KEY WORDS:** adsorption, well casing, ground water, monitoring wells, contamination, leaching

Variables which may cause biased interpretations in ground-water quality monitoring programs can be classified into two main groups, contaminant distribution and sample integrity [*1*].

Contaminant distribution is concerned with the distribution and extent of the sampling network. Each monitoring well represents a measurement of ground-water quality at a specific location. Therefore, the proper evaluation of ground-water chemistry data for an area is dependent upon the collective interpretation of single well data over the entire area of concern.

In maintaining sample integrity during ground-water sampling it is essential that water samples be collected in a manner that will clearly indicate the actual subsurface contaminant concentrations. This may seem to be a relatively simple task, but in fact it is very complicated. Some of the more familiar processes which may cause a loss in sample integrity are:

1. Contamination of the geologic material near the intake zone of the monitoring well as a result of well drilling and construction.
2. Contamination by materials used in installation of the monitoring well.
3. Contamination by sampling materials.
4. Sample degassing and volatilization.

[1] Chemist, U. S. Environmental Protection Agency, R. S. Kerr Environmental Research Laboratory, Ada, OK 74820.

[2] Assistant director and research program coordinator, Hazardous Waste Research and Information Center, State Water Survey Division of Illinois Department of Energy and Natural Resources, Champaign, IL 61820.

5. Contamination from the atmosphere.
6. Contamination by human factors.

Ground-water monitoring wells must be drilled and installed employing techniques that mitigate potential contamination of the aquifer and thus increase the probability of obtaining truly representative ground-water samples. In general, monitoring well casings should be made of materials resistant to suspected contaminants. It is the objectives of this research to investigate the uptake and release of pollutants, with a broad range of solubilities, in contact with potential well casing materials that are most likely to be used for monitoring ground-water contamination.

**Adsorption Mechanisms**

Adsorption is the adherence of a substance (adsorbate) onto the surface of another substance, called the adsorbent. Absorption, on the other hand, is the penetration of one substance (absorbate) into the inner structure of another substance, called the absorbent [*2*]. Since both of these processes frequently occur simultaneously, the term sorption is often used to encompass both adsorption and absorption. Throughout this report adsorption and sorption are used synonymously since the predominant mechanism of pollutant removal is considered to be surface removal or adsorption [*3*].

The degree of solubility is the most significant factor in determining the intensity of the lyophobic nature of the solute for the solvent. The more a substance likes the solvent system (hydrophylic), the less likely it is to be adsorbed. Conversely, a hydrophobic (water disliking) substance will more likely be adsorbed from aqueous solution. Therefore, one would expect an inverse relationship between solubility and adsorption. Solubility and adsorbability relationships will, however, vary between different classes of compounds. It has been shown that adsorption increases with increasing octanol-water partition coefficients (log of adsorption constant) and that the octanol-water partition coefficient may be a more significant quantity than solubility due to a broader application [*4*]. The significant point is that properties of chemicals such as pK, solubility, or partition coefficients are particularly significant in determining the extent and rate of adsorption [*5*].

A second primary driving force for adsorption results from an affinity of the solute for the solid. This affinity of solutes for the solid may be considered in three types of adsorption phenomena:

1. Electrical attraction of the solute to the adsorbent (exchange).
2. van der Waals' force (physical).
3. Adsorption of a chemical nature (chemisorption).

Adsorption due to electrical attraction of the solute to the adsorbent is often called exchange adsorption and falls within the realm of ion exchange. Exchange adsorption is the process by which ions of a substance concentrate at specific charged sites on the surface of the adsorbent.

Adsorption due to van der Waals' forces is termed physical adsorption, which represents cases where the adsorbed molecule is not fixed to a specific surface site but is free to move about on the adsorption interface.

Chemical adsorption occurs when the adsorbate and adsorbent chemically interact. Molecules which are chemically adsorbed are not free to move about the adsorption interface.

Physical adsorption is normally predominant at low temperatures and is characterized by low energies of adsorption. Conversely, chemical adsorption predominates at higher temperatures due to chemical reactions proceeding more rapidly at high temperatures.

Most adsorption processes are a result of a combination of all three forms of adsorption and normally it is not easy to distinguish between chemical and physical adsorption [*6*].

Many factors influence adsorption. These factors are discussed in relation to the following general categories:

1. Surface area.
2. Nature of the adsorbate.
3. pH.
4. Temperature.
5. Mixed solutes.
6. Nature of the adsorbent.

Because adsorption is a surface phenomenon, the extent of adsorption is a function of the available adsorptive surface area. Therefore, the more finely divided and the more porous the solid, the greater normally the amount of adsorption.

As previously stated, the solubility of the solute is, to a large extent, a controlling factor for adsorption processes. The effects noted relative to solubility-adsorption relationships can be seen by conceptualizing that some form of solute-solvent attraction must be overcome before adsorption can occur. The stronger the solute-solvent attraction, the greater the solubility; therefore, a smaller amount of adsorption occurs.

In general, the solubility of any series of organic compounds in water decreases with increasing chain length. The longer chain compounds decrease in solubility because, as more carbon atoms are added, the compounds become more hydrocarbon-like. This principle is stated in Traube's rule, which says that "adsorption from an aqueous solution increases as a homologous series is ascended" [*6*].

Polarity of adsorbate is another factor influencing adsorption. A general rule for predicting the effects of solute polarity on adsorption is that a polar solute will prefer the phase which is more polar. Simply stated, in adsorption "like" adsorbs "like," or polar surfaces prefer polar adsorbates and nonpolar surfaces prefer nonpolar adsorbates [*7*]. Since water is a polar molecule, polar solutes tend to be more soluble in water than nonpolar solutes.

The pH of a solution may significantly influence the extent of adsorption of organic pollutants from an aqueous solution. In general, adsorption of typical organic pollutants from water is increased with decreasing pH [*8*].

Temperature is another factor which can influence adsorption. Adsorption reactions are normally exothermic; therefore, the extent of adsorption generally increases as the temperature decreases [*6*].

Mixed solutes in a solution can also significantly affect the extent of adsorption of pollutants onto a surface. The pollutants may mutually enhance adsorption, may act relatively independently, or may interfere with one another. Similarly, since the adsorption of one substance tends to reduce the number of adsorption sites, the concentration of pollutants becomes important because it is the pollutant concentrations which provide the driving force to produce adsorption of the various substances. The presence of other solutes in an aqueous mixture can adversely affect the adsorption of the first compound [*9*].

The physicochemical nature of the adsorbent can have profound effects on both the rate and extent of adsorption. Every solid is a potential adsorbent but will react differently depending upon physical structure, chemical surface characteristics, and environmental conditions.

## Organic Adsorption and Leaching

The adsorption and leaching of organic compounds with various container and potential well casing materials have been studied to a much lesser extent than have inorganics (that is, trace metals). Organics are adsorbed by organic matter in porous media [*10*] and to a smaller degree on mineral surfaces [*11*]. Organics can also be adsorbed by a range of man-made surfaces to include well casing material. For example, the adsorption of dichlorodiphenyltrichloroethane (DDT) onto most surfaces, including stainless steel, has been documented [*12*]. Another example is the investigation of the interaction of *m*-cresol with two types of plastic tubing [*13*]. Industrial grade polyvinyl chloride (PVC) tubing and Bev-a-line very high temperature (VHT) tubing were exposed to an *m*-cresol solution for varying lengths of time up to two hours. The *m*-cresol was progressively lost to the PVC tubing in amounts proportional to the increasing contact time, while no measurable loss of *m*-cresol to the VHT tubing occurred.

Researchers have determined that organic contamination ranging from 1 to 5000 ppb by weight was detected in "organic free" water which had flowed through tubes of polyethylene, polypropylene, black latex, six formulations of PVC, and a plastic garden hose [*14*]. Among the contaminants found from PVC were *o*-cresol, naphthalene, butyloctylfumarate and butyl-chloroacetate. The amount of plasticizers in PVC tubing is approximately 40% by weight and represents an essentially inexhaustible supply of contamination. Should this supply of plasticizers eventually be depleted, the contamination will decrease but the tubing will have lost its flexibility and will therefore need to be replaced. However, in the case of relatively rigid polyethylene and polypropylene tubes, it was found that the tubes may eventually become noncontaminating. It should be noted that rigid PVC well casing contains a much smaller content of plasticizers and is thus much less likely to leach contaminants into contacting water.

In one recent laboratory experiment, "organic free" water was passed through five types of plastic tubing, including polypropylene, polyethylene, polytetrafluoroethylene (TFE), vinyl chloride and vinyl acetate copolymer, PVC (with glue), and PVC (non-glued) [*15*]. The results indicate low levels of leachates from all the tubing tested except TFE, which disclosed no leachates. Between one and five contaminants were found leached into the water from nonglued PVC, polyethylene, and polypropylene at less than 0.5 ppb. Numerous contaminants were found leached from glued PVC and Tygon. Two organic chemical spikes, naphthalene and para-dichlorobenzene, were not retained (sorbed) significantly by any of the tubing except Tygon.

A recent field and laboratory study included assessment of the potential for contamination of ground-water samples by six PVC primers and one PVC adhesive [*16*]. Tetrahydrofuran (THF), methylethylketone (MEK), methylisobutylketone (MIBK), and cyclohexanone were found to be leached into the water surrounding bonded joints. Even though a well was developed after installation and bailed prior to sampling, detectable levels of these compounds were found in well water several months after installation. After immersion of freshly cemented PVC pipe for five days, up to 110 ppm of MEK was found. After five washings and a 24-h equilibrium, the concentration of MEK was reduced to about 0.54 ppm. Similar trends were noted for the other three pollutants studied although the presence of each constituent and their relative concentrations varied significantly between solvent brands and manufacturers. All four compounds could be introduced into monitoring wells by one or more of the brands of adhesive. Concentrations of contaminants declined significantly with successive washes with deionized water, from a mean total concentration of 77800 ppb after five days of soaking to a mean total of 591 ppb after being washed four more times with deionized water and soaked for an additional 24 h.

Field sampling with a PVC bailer constructed several weeks before sampling resulted in samples with low levels of THF and MEK. With a PVC bailer constructed 30 min prior to sampling, 16000 ppb cyclohexanone, 3800 ppb MEK and 1400 ppb THF were found in a ground water sample. Contaminants in PVC solvents can mask the presence of other volatile priority pollutants that elute from the chromatograph at similar retention times. For example, MEK and 2,3-dichloroethane, cyclohexanone and bromoform, and THF and 1,1-dichloroethane all co-elute with the Environmental Protection Agency (EPA) analytical protocol. All these problems can most simply be circumvented by using well casings and bailers constructed with threaded joints [*16*].

While PVC is the most common type of synthetic well casing material in use, other synthetic materials have been used or could be used, including acrylonitrile butadiene styrene (ABS), styrene-rubber (SR), polyethylene, polypropylene, TFE, and polyvinylidene fluoride (PVDF). When sampling for organic contaminants, the generally recommended order of choice of well casing and sampling materials is glass, TFE, stainless steel, polypropylene, polyethylene, and other plastics, metals, and rubber [*12*]. However, the same researchers indicated that detailed experimental data on materials selection are simply not available.

The leaching of residual vinyl chloride monomer (RVCM) has been an area of concern and intense investigation since the discovery of a link between vinyl chloride monomer and a rare form of liver cancer (hemangiosarcoma) in humans. Through a massive research effort a limit of 10-ppm RVCM was established based on the conclusion that vinyl chloride migration from the wall of nominal 25.4-mm-diameter (1 in.) Schedule 40 PVC pipe (containing 10-ppm RVCM) into stagnant water retained in the pipe will be undetectable at a 2.0-ppb sensitivity in analysis. Today at least 98% of the PVC manufactured in North America contains less than 10-ppm RVCM and most contains less than 1 ppm RVCM [*17*].

The water leachability of components of PVC and CPVC pipes has been investigated in another recent study [*18*]. The extract water was analyzed for a number of volatile and extractable compounds and data indicate that the following organics were consistently found in significant concentrations:

| CEMENT SOLVENTS | VOLATILE ORGANICS |
|---|---|
| MEK | Dichloromethane |
| THF | Carbon tetrachloride |
| Cyclohexanone | Tetrachloroethene |
| DMF | Trichloroethene |
| | Toluene |

The cement solvents were found in the ppm range but concentrations decreased rapidly with rinsing. The volatile organics were typically found in the 1 to 10 ppb range except for toluene, which typically was detected under 1 ppb. These levels will not necessarily be found when cement solvents are avoided.

Investigations of water leached from PVC and CPVC pipes with and without cemented joints were performed in another study [*19*]. This testing was to determine whether leaching of chloroform and carbon tetrachloride was caused by the solvent cementing of joints. Chloroform was not found in the leachate from either PVC or CPVC. However, carbon tetrachloride was found at very low concentrations in the leachate from the CPVC pipe.

Schedule 40 PVC, polyethylene, and polypropylene well casing materials were examined by being exposed under batch conditions to two metal pollutants and six volatile organic pollutants [*20*]. Under the conditions of the experiment it was generally found that Schedule 40 PVC caused fewer monitoring interferences with volatile organics than did polyethylene or polypropylene. Also, no extractable pollutants were leached from the well casing materials under the conditions of the study.

Well casing materials can bias ground-water sampling efforts by adsorption of constituents, which results in lesser concentrations in the sample than in the ground water, by leaching of adsorbed constituents, leaching of casing constituents, and by permeation of constituents through the well casing from outside. The latter three mechanisms can all result in higher concentrations of chemicals in the sample than in the ground water. There are virtually no published field data comparing the adsorption of leaching that occurs with different well casing materials and there are only limited laboratory data. Most of the research is with tubing or piping that may have more plasticizers than well casing materials. Some research used laboratory water and high-purity chemical contaminants that are not indicative of field conditions, and no researchers have found a way to simulate field sampling conditions in the laboratory for comparative study of well casing influence. Generally recommended lists of well casing materials have been published with the acknowledgment that they are not based on detailed experimental data.

## Classes of Well Casing Materials

Well casings could be fabricated from almost any material. Under improper conditions, many of these materials could contribute to biased ground-water sampling either through adsorption or leaching of contaminants. The well casing materials selected for this study are representative of the general classes of materials that have a history of use in ground-water monitoring or are likely candidates for use nationwide in sampling ground water for organic contamination.

### *Plastics*

Plastics fall into two major categories, called thermosets and thermoplastics. Thermosets are those plastics which, once shaped, cannot be reshaped even if reheated. Thermoplastics can be softened and rehardened by heating and cooling. This allows thermoplastics to be easily molded or extruded into pipes and fittings [*21*].

The plastics most commonly used in ground-water monitoring fall into the thermoplastics category and include [*1*].

acrylonitrile butadiene styrene (ABS)
polyvinyl chloride (PVC)
styrene-rubber (SR)
polyethylene (PE)
polypropylene (PP)
polytetrafluoroethylene (Teflon—TFE)

Thermoplastics are resistant to electrical and electrochemical effects and are generally resistant to attack by naturally occurring chemical compounds. For this reason, thermoplastics have been extensively used in monitoring inorganic ground-water constituents [*1*]. Because thermoplastics are very susceptible to attack by organic compounds, however, extreme care must be exercised in monitoring situations where organic pollutants are present.

### *Metals*

The most commonly used metals in ground-water monitoring are iron, stainless steel, brass, aluminum, and copper [*1*]. Metal well casings (predominantly iron and stainless steel) are very strong and rigid; however, they can contaminate ground-water samples in a number of ways. Metal contaminants can be leached from metal well casings into the water supply. Metallic materials will also release in water the by-products of deterioration processes such as oxidation and corrosion. As previously discussed, metal well casing can also serve as an adsorbing surface for a wide variety of organic and inorganic ground water constituents.

### *Glass*

When considering inertness, glass has been shown to be the material of choice for use in sampling water for organic compounds [*12*]. In spite of its nature, glass can serve a useful purpose at several points in the sampling equipment. Glass is considerably less expensive than other materials of similar inertness such as TFE and is therefore most useful as sample container material. Due to its fragility, however, glass is not used to make well casings.

## Procedure

### *Well Casing Materials*

The particular well casing materials used in this study were chosen because of their availability and reported use in ground-water quality monitoring programs. They were:

1. PVC-40 (polyvinyl chloride—Schedule 40)
PVC-80 (polyvinyl chloride—Schedule 80)
3. ABS (acrylonitrile butadiene styrene)
4. Teflon (3 types)
   a. TFE—tetrafluoroethylene
   b. PFA—perfluoroalkoxy
   c. FEP—fluorinated ethylene propylene
5. Stainless steel (304)
6. Kynar (PVDF = polyvinylidene fluoride)

*Polyvinyl chloride (PVC)*—Polyvinyl chloride is produced by end-to-end polymerization of vinyl chloride ($CH_2{=}CHCl$). The resulting structure is —$CH_2$—CHCl—$CH_2$—CHCl—[*22*]. The suspension aids for PVC are normally cellulose ethers or polyvinyl alcohol, and the polymerization catalysts are typically organic peroxides. Both types of material may occur in small residual quantities in the finished polymer. The peroxides will degrade to other compounds [*23*].

The production process involves the combination of PVC resin with various types of stabilizers, lubricants, pigments, fillers, processing aids, and plasticizers (such as titanium dioxide, calcium carbonate, fatty acid-metal salt, oxidized polyethylene wax), all of which are potential sources of leached pollutants [*23*].

*Acrylonitrile Butadiene Styrene (ABS)*—ABS is the name given to a unique family of engineering polymers. The acronym is derived from the names of the three monomers used to produce the polymers: acrylonitrile ($CH_2$= C— N), butadiene ($CH_2$—CH—CH=$CH_2$), and styrene (vinylbenzene). The three-monomer system can be tailored to end-product needs by varying the ratios in which they are combined. Acrylonitrile contributes heat stability, chemical resistance, and aging resistance; butadiene imparts low-temperature property retention, toughness, and impact strength; and styrene adds luster (gloss), rigidity, and processing ease [*24*]. The additives for ABS are similar to those for other polymers: antioxidants, lubricants, pigments, and fillers. Typically, ABS uses carbon black as a pigment.

*Stainless Steel (304)*—Stainless steel is often chosen as a well casing material due to its strength and resistance to rusting and corrosion. Stainless steel has the potential, however, to leach heavy metals into the aqueous environment. These include chromium, nickel, molybdenum, iron, titanium, cobalt, and tungsten. There is also the possibility of an iron oxide coating developing over the surface of the steel. This coating can have unpredictable and changeable effects upon the adsorption capacity [*25*].

*Fluoroplastics*—Fluoroplastics are plastics based on polymers made from monomers containing one or more atoms of fluorine. All fluoroplastics are made by free radical initiated polymerization or copolymerization of the monomers. Fluorocarbon plastics are those made from perfluoro monomers. The fluorocarbon plastics chosen for this study were tetrafluoroethylene (TFE), fluorinated ethylene propylene copolymer (FEP), and perfluoroalkoxy resin (PFA). Another important fluoroplastic chosen to study was polyvinylidene fluoride (PVDF—not a fluorocarbon). The fluoroplastics were readily available and TFE and FEP have a history of use in ground-water monitoring programs [*26*].

Fluoroplastics, and especially fluorocarbon plastics, have unique properties which include high resistance to harsh chemical environments and high temperatures, useful electrical properties, good mechanical properties, low wettability, and low coefficients of friction.

Tetrafluoroethylene (TFE) is made in larger quantities than all other fluoroplastics combined. TFE has an exceptionally low coefficient of friction that produces the antistick properties of its surface. In fact it has the lowest coefficient of friction of any solid [*28*]. TFE and FEP have been used as well casing and screen materials for ground-water monitoring wells and are considered by some to be ideal materials for monitoring well construction [26].

Fluorinated ethylene propylene (FEP), in general, has properties similar to those of TFE. FEP is not as tough as TFE and its strength properties decrease more rapidly with increasing temperature. FEP retains most of the electrical properties of TFE, is insoluble in essentially all organic solvents, and is resistant to corrosive organic reagents.

Perfluoroalkoxy resin (PFA), has mechanical properties at elevated temperatures similar to TFE and much better than FEP. PFA is about equal to TFE in strength, hardness, and wear resistance. Its resistance to chemicals is similar to that of FEP.

Polyvinylidene fluoride (PVDF) is slightly less dense and has a lower melting point than TFE. In strength and resistance to wear, however, PVDF is generally similar to TFE [27].

*Pollutants Evaluated*

Pollutants used for this study were chosen from compounds that, according to the literature, have been and are most likely to be found in ground water. These compounds are all priority pollutants and are less volatile and more polar than those examined in previous studies [20,29]. Another consideration in choosing the pollutants is their wide range of solubilities in water. Table 1 lists the pollutants evaluated and their solubilities in water.

*Exposure Solutions*

Solutions used for exposing the well casings to the pollutants were made from laboratory water, and nonpolluted and polluted ground water. All three of these types of water were inoculated with the pollutants and the well casings were placed in the appropriate vials for the appropriate experimentation.

Laboratory water free of interferences (organic free) was prepared with a Barnstead Nanopure II Cartridge System followed by a Barnstead Organicpure Water Purification System.

The nonpolluted ground water was taken from Byrd's Mill Spring in Fittstown, Oklahoma. Byrd's Mill Spring is the source of drinking water for the town of Ada, Oklahoma. Byrd's Mill water has been extensively analyzed by personnel of the Robert S. Kerr Environmental Research Laboratory, U.S. Environmental Protection Agency, in Ada, Oklahoma. Table 2 gives data gathered on Byrd's Mill Spring water [30].

The polluted ground water was taken from the northeastern part of Oklahoma in the Tar Creek watershed area. Since November 1979 the Tar Creek watershed has received highly mineralized acid mine discharges from flooded underground lead-zinc mines of the Picher Field in Ottawa

TABLE 1—*Pollutants evaluated.*

| Pollutant | Solubility in Water, mg/L | |
|---|---|---|
| 2,4,6-trichlorophenol | 800 | (25°C) |
| 4-nitrophenol | 16000 | (25°C) |
| Diethyl phthalate | 896 | (25°C) |
| Acenaphthene | 3.42 | (25°C) |
| Naphthalene | 31 | (25°C) |
| 1,3-dichlorobenzene | 123 | (25°C) |
| 1,2,4-trichlorobenzene | 30 | (25°C) |
| Hexachlorobenzene | 0.020 | (25°C) |

TABLE 2—*Data for Byrd's Mill Spring—nonpolluted ground water (background levels).*

| Constituent | Concentration |
|---|---|
| pH | 7.2 SU |
| Sodium | 5 mg/L |
| Potassium | 2 mg/L |
| Calcium | 77 mg/L |
| Magnesium | 47 mg/L |
| Alkalinity | 325 mg/L |
| Hardness | 340 mg/L |
| Conductivity | 600 μmhos/cm |
| Total dissolved solids | 345 mg/L |
| Ammonia | <0.01 mg/L |
| Nitrate | 1.1 mg/L |
| Ortho-phosphate | 0.010 mg/L |
| Total phosphorus | 0.015 mg/L |
| Sulfate | 7 mg/L |
| Chloride | 4 mg/L |
| Chlorobenzene | 0.5 μg/L |
| bis-(2-chloroethyl)ether | 1.3 μg/L |
| Phenol | 0.6 μg/L |
| 2-chlorophenol | 0.7 μg/L |
| 1,2-dichlorobenzene | 0.7 μg/L |
| Nitrobenzene | 0.7 μg/L |
| 2,4-dichlorophenol | 1.3 μg/L |
| 1,2,4-trichlorobenzene | 0.9 μg/L |
| 2,4,6-trichlorophenol | 1.3 μg/L |
| Diethyl phthalate | 0.7 μg/L |

County, Oklahoma. The pH of the specific samples used for the study was 2.95. Table 3 gives average data on a number of polluted ground-water samples taken for this study [*31*].

The exposure solutions were made by first weighing the compounds (0.0200 to 0.0300 g) into a conical-bottom reaction vial. The solid compounds dissolved in the liquid compounds at room temperature. To make the working exposure solution, 3.5 μL (3.5 μg) of the stock solution was injected into a 2-litre borosilicate flask filled with either "organic free," Byrd's Mill, or Tar Creek water (final concentration is approximately 250 ppb). A ground-glass stopper was then inserted into the flask, eliminating headspace, to prevent loss of more volatile compounds. The solution was then stirred for 48 h in the absence of light and then used in the respective experimental tasks.

### *Analytical Methodologies*

*Atomic Absorption and Emission Spectroscopy*—The metals analyses for the Tar Creek samples were performed on a Spectraspan IIIB D.C. Argon Plasma Emission Spectrometer with Dynamic Background Compensator. The low-level (<0.100 ppm) metals analyses were completed on a Perkin-Elmer 5000 Atomic Absorption Spectrophotometer equipped with a Model 500 Heated Graphite Atomizer.

*High-Performance (Pressure) Liquid Chromatography (HPLC)*—The HPLC analyses were completed on a modular unit made by Beckman. The pumps were Beckman Model 110, and a Hitachi Model 100-40 variable-wavelength ultraviolet (UV) spectrophotometer was used as the detection system. The HPLC column was a Beckman Ultrasphere O.D.S. reverse-phase column

TABLE 3—*Data*[a] *for Tar Creek—polluted ground water (background levels).*

| Constituent | Number of Samples | Mean | Standard Deviation | Minimum | Maximum |
|---|---|---|---|---|---|
| Specific conductance (μmhos/cm) | 11 | 3659 | 441 | 3150 | 4530 |
| pH (SU) | 11 | 2.5 | 0.4 | 1.9 | 3.3 |
| Dissolved oxygen (mg/L) | 9 | 3.5 | 2.4 | 1.4 | 8.9 |
| Iron (mg/L) | 4 | 110 | 17.3 | 92 | 133 |
| Zinc (mg/L) | 4 | 88.8 | 30.6 | 56 | 130 |
| Cadmium (mg/L) | 4 | 0.362 | 0.074 | 0.260 | 0.430 |
| Lead (mg/L) | 3 | 0.199 | 0.081 | 0.121 | 0.282 |
| Chromium (mg/L) | 1 | 0.014 | . . . | 0.014 | 0.014 |
| Fluoride (mg/L) | 8 | 6.60 | 0.50 | 6.00 | 7.39 |

[a] Mine discharge from a large subsidence pit filled with mine water located in the southwest part of Commerce, OK.

with 5-μm particle size. The column dimensions were 4.6 mm inside diameter by 25 cm in length. The chromatography was performed using an isocratic 60% acetonitrile solution with a flow rate of 2.0 mL/min, and at a wavelength of 226 nm.

*Extraction and Elution Procedures*—An adaptation of a minicolumn procedure for concentration of organics from aqueous solution as reported in the literature was used to simplify the concentration procedure [*32,33*]. The basic procedure involves extracting the organics from a 30-mm sample by passing the sample through a minicolumn (TFE, 1.2 to 1.8 mm by 25 mm) containing a 50:50 mixture of XAD2:XAD8 resins. The sorbed organics are then eluted into 2 mL of a 15%:85% acetone:hexane solution for analysis.

## Experimental Work

### *Adsorption (Uptake) Experiment*

The objective of the adsorption experiment was to determine what percentage of the pollutants in the neutral pH ground water (unpolluted from Byrd's Mill Spring), exposed to the well casings, would be adsorbed onto the surface of the well casings.

Segments measuring 7.62 cm (3 in.) of each well casing were exposed to the pollutants in 40 mL borosilicate glass exposure chambers with TFE-lined caps. Batches of exposures were sampled initially, after one, three, and six weeks for a total of three batches of 36 vials plus eight initial samples. Duplicate exposure chambers were prepared for each well casing and the chambers were sacrificed upon sampling to eliminate head space as a variable.

The adsorption experiment results are given in Table 4. The results are indicative of adsorption after six weeks of exposure; however, the observed percentages of adsorption occurred within the first week of exposure. There were no appreciable changes in adsorption percentages from the first week to the sixth week except for 2,4,6-trichlorophenol (TCP). TCP adsorbed partially after one week and totally adsorbed after three weeks as indicated.

This experiment indicates that for the experimental conditions and compounds used the TFE well casings might be less likely to adsorb pollutants. At these concentration levels, however, the TFEs and PVC-40 and 80 exhibited very little difference in the amounts of adsorption.

TABLE 4—*Percentage of pollutant adsorbed per well casing after six weeks of exposure.*

| Casing Material | 4-Nitrophenol | 2,4,6-Trichloro-phenol[a] | Diethyl Phthalate | Naphthalene | Acenaphthene |
|---|---|---|---|---|---|
| PVC-40 | 100 | 0 | 0 | 48 | 59 |
| PVC-80 | 100 | 9 | 0 | 49 | 68 |
| ABS | 0 | 44 | 5 | 0 | 48 |
| Stainless steel | 100 | 31 | 0 | 19 | 58 |
| Teflon-PFA | 100 | 0 | 0 | 53 | 71 |
| Teflon-FEP | 100 | 18 | 9 | 79 | 83 |
| Teflon-TFE | 68 | 53 | 0 | 51 | 66 |
| Kynar-PVDF | 100 | 56 | 0 | 61 | 55 |

[a] Percentages adsorbed after 1 week—100% adsorption occurred after six weeks of exposure.

*Leaching (Release) Experiment*

The objective of the leaching (release) experiment was to determine what portion, if any, of the adsorbed pollutants, after a six-week exposure period, would leach or be released back into noncontaminated ground water.

Experimental design and analytical methods were the same as for the adsorption experiment, the only difference being that after the well casings were exposed and allowed to adsorb pollutants for six weeks the contaminated ground water was replaced with nonpolluted ground water. The water from the exposure chambers was then sampled initially, after 1 h, 4 h, 24 h, 4 days, and 14 days. These samples were analyzed to allow an assessment of the rate of release of pollutants back into the noncontaminated ground water.

As can be seen from Table 5, very little leaching of adsorbed pollutants actually occurred over the 14-day period. PVC-80 and PFA-TFE were the only well casings exhibiting any leaching and both leached naphthalene. For all practical purposes, only zero to trace amounts of the sorbed pollutants were released into noncontaminated ground water.

*Adsorption Site Experiment*

The objective of the adsorption site experiment was to determine if a proportionally greater amount of adsorption occurred at the rougher, cut ends of PVC-80 well casing than at the smooth,

TABLE 5—*Percentages of pollutants leached of percentages adsorbed.*[a]

| Casing Material | 4-Nitrophenol | 2,4,6-Trichloro-phenol[a] | Diethyl Phthalate | Naphthalene | Acenaphthene |
|---|---|---|---|---|---|
| PVC-40 | 0 | 0[b] | 0[b] | 0 | 0 |
| PVC-80 | 0 | 0 | 0[b] | 60 | 0 |
| ABS | 0[b] | 0 | 0 | 0[b] | 0 |
| Stainless steel | 0 | 0 | 0[b] | 0 | 0 |
| Teflon-PFA | 0 | 0[b] | 0[b] | 100 | 0 |
| Teflon-FEP | 0 | 0 | 0 | 0 | 0 |
| Teflon-TFE | 0 | 0 | 0[b] | 0 | 0 |
| Kynar-PVDF | 0 | 0 | 0[b] | 0 | 0 |

[a] Two-week leaching period after six-week adsorption.
[b] None adsorbed.

TABLE 6—*Percent adsorption of pollutants in low-pH (3.0) polluted ground water.*[a]

| Casing Material | 4-Nitrophenol | 2,4,6-Trichloro-phenol[a] | Diethyl Phthalate | Naphthalene | Acenaphthene |
|---|---|---|---|---|---|
| PVC-40 | 15 | 0 | 0 | 100 | 46 |
| PVC-80 | 100 | 0 | 0 | 23 | 0 |
| ABS | 60 | 0 | 76 | 72 | 40 |
| Stainless steel | 23 | 28 | 48 | 0 | 18 |
| Teflon-PFA | 0 | 0 | 29 | 90 | 0 |
| Teflon-FEP | 0 | 17 | 47 | 17 | 0 |
| Teflon-TFE | 17 | 14 | 35 | 0 | 64 |
| Kynar-PVDF | 31 | 28 | 38 | 0 | 21 |

[a] Six-week exposure.

finished surface of the well casing material. To make this determination, segments of PVC-80 were threaded at each end, filled with exposure solution (laboratory water), and capped.

The samples analyzed were exposed for one week in the threaded exposure chambers. After computing the surface areas available for both the adsorption and the adsorption site experiment it was found that the percentage adsorbed per square centimetre was independent (within 5 to 10%) of the surface area of the rough ends of the well casing material exposed in the first adsorption experiment.

The rough edges were only a small percentage (3 to 5%) of the total surface areas available for adsorption. Since the extent of adsorption is a function of the available adsorptive surface area, the observed adsorption percentages are not surprising [*6*]. It was necessary, however, to make this determination so that consistent adsorption, no matter what rate and extent, could be shown throughout the entire length of the well casing segments.

### *Low-pH Adsorption/Leaching Experiment*

The objective of the low-pH adsorption/leaching experiment was to determine the differences in rate and extent of adsorption using a low-pH (3.0) polluted ground water. This ground water, as stated previously, was taken from acid mine drainage in the Tar Creek area of northeastern Oklahoma.

Data from the low-pH experiment are given in Tables 6 and 7. The data indicate that, with the exception of ABS, all the other well casings showed less adsorption in the low-pH contaminated ground water than in the neutral-pH noncontaminated ground water (see Tables 4 and 6). A possible explanation could be stronger binding and more preferential complexing of the experimental pollutants with other pollutants in the contaminated ground water.

Another more probable explanation is the consideration of the pK values of the pollutants in

TABLE 7—*Percentage of pollutants leached of percentage adsorbed in low-pH (3.0) polluted ground water.*

| Casing Material | 4-Nitrophenol | 2,4,6-Trichloro-phenol[a] | Diethyl Phthalate | Naphthalene | Acenaphthene |
|---|---|---|---|---|---|
| PVC-40 | 0 | 0 | 0 | 100 | 0 |
| Stainless steel | 0 | 0 | 0 | 0 | 0 |
| Teflon-TFE | 0 | 0 | 0 | 0 | 0 |

[a] Two-week leaching period after six-week adsorption.

relation to the pH of the total solution. A definite relationship between the extent of adsorption, pH, pK has been shown, with a maximum adsorption occurring when the pH is approximately equal to the pK [5]. If the pH is decreased further, the hydrogen ion concentration increases and adsorption tends to decrease, suggesting a replacement of the adsorbed compound by the more preferentially adsorbed hydrogen ions.

## Conclusions

General conclusions from this experimentation and a review of the literature of previous studies are:

1. In the neutral- and low-pH ground waters, only 2,4,6-tri-chlorophenol and 4-nitrophenol showed appreciable adsorption in all well casings.
2. Leaching of adsorbed pollutants did not occur to a measurable extent in any of the well casing/pollutant systems.
3. The rate and extent of adsorption were not measurably different on the rough cut ends of the well casings than on the smooth finished surfaces.
4. Less adsorption of pollutants was exhibited for low-pH contaminated ground water than for the neutral-pH noncontaminated ground water.
5. There is no clear advantage to the use of one particular well casing material over the others for organics monitored in this study.
6. Probably of greater influence to sample integrity and representativeness than well casing material selection are the well purging procedures, sampling device selection and composition, and sample storage devices.
7. Solvents should be avoided for joining sections of casing materials in monitoring wells.
8. The amount of adsorption generally correlates with the solubility of the chemical independent of the well casing material.
9. Other influencing factors which have not been assessed may also bias sampling efforts. The most important of these may be biological growth on the casing surface, which can then uptake, transform, and release contaminants.

## References

[1] Gilham, R. W. et al., "Groundwater Monitoring and Sample Bias," API Publication 4367, American Petroleum Institute, Washington DC, June 1983.

[2] Hawley, G. G., *The Condensed Chemical Dictionary,* Van Nostrand Reinhold, New York, 1977.

[3] Reynolds, T. D., *Unit Operations and Processes in Environmental Engineering,* Brooks/Cole Engineering Division, Monterey, CA, 1982, pp. 187–188.

[4] Ward, T. M. and Holly, K., *Journal of Colloid Science,* Vol. 22, 1966, pp. 220–227.

[5] Tinsley, I. J., *Chemical Concepts in Pollutant Behavior,* Wiley-Interscience, New York, 1979, pp. 19–31.

[6] Weber, W. J., Jr. *Physicochemical Processes for Water Quality Control,* Wiley, New York, 1972, pp. 199–249.

[7] Cheremisinoff, P. N., *Pollution Engineering,* July 1976, pp. 24–32.

[8] Myers, A. L. and Zolandy, R. R., *Activated Carbon Adsorption of Organics from the Aqueous Phase—Volume I,* Ann Arbor Science, Ann Arbor, MI, 1980.

[9] Weber, W. J., Jr., in *Proceedings* Third International Conference on Water Pollution Research, Vol. I, Water Pollution Control Federation, 1966.

[10] Chiou, C. T., Peters, L. J., and Freed, V. H., "A Physical Concept of Soil-Water Equilibria for Nonionic Organic Compounds," *Science,* Vol. 206, No. 16, 1979, pp. 831–832.

[11] Rogers, R. D., McFarlane, J. C., and Cross, A. J., *Environmental Science and Technology,* Vol. 14, No. 4, 1980, pp. 457–460.

[12] Pettyjohn, W. A. et al., *Ground Water,* Vol. 19, No. 2, March-April 1981, pp. 180–189.

[13] "Publications on the Analysis of Spilled Hazardous and Toxic Chemicals and Petroleum Oils," U. S. Environmental Protection Agency, Industrial Environmental Research Laboratory, Cincinnati, 1980.

[14] Junk, G. A. et al., *Environmental Science and Technology,* Vol. 8, No. 13, Dec. 1974, pp. 1100–1106.
[15] Curran, C. M. and Tomson, M. B., *Groundwater Monitoring Review,* Vol. 3, No. 3, Summer 1983, pp. 68–71.
[16] Sosebee, J. B. et al., "Contamination of Ground Water Samples with PVC Adhesives and PVC Primer from Monitoring Wells," Environmental Science and Engineering, Inc., Gainesville, FL, Jan. 1982.
[17] "Vinyl Chloride; The Control of Residual Vinyl Chloride Monomer in PVC Water Pipe," Uni-Bell PVC Pike Associates, Dallas, TX, 1982.
[18] Montgomery, J. M. (as taken from), SRI International, HSH-4910, Menlo Park, CA, March 1983.
[19] B. F. Goodrich Laboratory Study (as taken from), SRI International, HSH-4910, Menlo Park, CA, March 1983.
[20] Miller, G. D. in *Proceedings,* 2nd National Symposium on Aquifer Restoration and Ground Water Monitoring, National Water Well Association, Worthington, OH, 1982, pp. 236–245.
[21] *Manual on the Selection of Thermoplastic Water Well Casing,* National Water Well Association and Plastic Pipe Institute, Worthington, OH, 1981, pp. 3–9.
[22] Jeziorski, R. J. and Wenzler, R. A. in *Modern Plastics Encyclopedia,* Vol. 59, No. 10A, 1982–1983, pp. 108–112.
[23] "Environmental Review of Proposed Expanded Uses of Plastic Plumbing Pipe," SRI International, HSH-4910, Menlo Park, CA, March 1983.
[24] Lantz, J. N. in *Modern Plastics Encyclopedia,* Vol. 59, No. 10A, 1982–1983, pp. 6–7.
[25] Harris, W. F., "Groundwater Quality Monitoring in the Tennessee Valley Region," TVA/ONR/WR-82/5, Tennessee Valley Authority, Norris, TN, Oct. 1981.
[26] Morrison, R. D., *Ground Water Monitoring Technology,* Timco MFG., Inc., Prairie Du Sac, WI, 1983, pp. 70–72.
[27] Sperati, C. A. in *Modern Plastics Encyclopedia,* Vol. 59, No. 10A, 1982–1983, pp. 35–50.
[28] *Nalgene Labware Products,* Nalgene Co., Sybron Corp., Rochester, NY, 1983, p. 80.
[29] *Federal Register,* Environmental Protection Agency, Part III, Monday, 3 Dec. 1979, pp. 69464–69575.
[30] Unpublished data given in personal communication with R. Cosby and B. Bledsoe of the U.S. Environmental Protection Agency, Robert S. Kerr Environmental Research Laboratory, Ada, OK, 1983.
[31] "Effects of Acid Mine Discharge on the Surface Water Resources in the Tar Creek Area, Ottawa County, Oklahoma," Oklahoma Water Resources Board EPA Grant No. CX810192-01-0, Oklahoma City, OK, March 1983.
[32] Tateda, A. and Fritz, J. A., *Journal of Chromatography,* Vol. 152, 1978, pp. 329–340.
[33] LeBel, G. L. et al., *Journal of the Association of Official Analytical Chemists,* Vol. 62, No. 2, March–April 1979, pp. 241–249.

*John F. Dablow, III,*[1] *Daniel Persico,*[2] *and Grayson R. Walker*[1]

# Design Considerations and Installation Techniques for Monitoring Wells Cased with Teflon PTFE

**REFERENCE:** Dablow, J. F., III, Persico, D., Walker, G. R., **"Design Considerations and Installation Techniques for Monitoring Wells Cased with Teflon PTFE,"** *Ground-Water Contamination: Field Methods, ASTM STP 963*, A. G. Collins and A. I. Johnson, Eds., American Society for Testing and Materials, Philadelphia, 1988, pp. 199–205.

**ABSTRACT:** A representative sample of ground water is essential for investigations where analytical testing results at the parts per billion level are needed to properly assess the constituents, concentration, and extent of contaminant plumes which impact ground-water resources. Monitoring wells constructed with casings of Teflon PTFE (polytetrafluoroethylene) can significantly improve the confidence level in a representative sample because of the inertness of the casing material to sorption or desorption of possible contaminating chemicals, particularly volatile organic compounds. Three strength-related properties—pullout resistance of the threaded couplings, compressive strength of the screen sections and flexibility of the casing string—must be considered during the design of monitoring wells cased with Teflon PTFE. Experimental data indicate that Teflon casing can be suspended to depths of 107 m with little risk of flush-threaded joint failure. Compressive strength testing shows that Teflon behaves predominantly elastically and is subject to only small screen opening deformation. This deformation can be overcome by specifying the manufactured screen opening to be larger than the opening required by the well design. Special installation techniques are essential when working with Teflon to assure a properly constructed monitoring well, including two techniques to eliminate "snaking" of the well casing: borehole centralizers, and the insertion method.

**KEY WORDS:** ground water monitoring wells, design considerations, compressive strength, flexibility, sorption, desorption, chemical resistance, well installation techniques

Recent improvements in the accuracy of analytical techniques in testing for the presence of contaminants in ground water have led to the need for improved field procedures during drilling, well installation, and sampling in order to assure that a representative sample of the ground water is obtained for analysis. One area to be considered during design of a monitoring system which can critically affect sample integrity is casing material selection. Well materials and sampling equipment constructed of Teflon[3] PTFE (polytetrafluoroethylene) provide a high degree of inertness to sorption and desorption of contaminants. These materials, when used in conjunction with strict drilling procedures and tight laboratory quality control/quality assurance standards, can assure that the analytical results reflect the actual ground-water conditions. This paper discusses design aspects and installation procedures developed during the construction of numerous wells cased with Teflon for application to the successful construction of PTFE ground-water monitoring wells.

[1] Vice president, chief geologist, and president and chief engineer, respectively, Hydro-Fluent, Inc.,

[2] Technical representative, E. I. duPont de Nemours & Co., Inc.,

[3] Dupont's registered trademark for its fluorocarbon resin.

## Monitoring Well Design Considerations—General

Many factors must be evaluated in order to design a successful ground-water monitoring system. These factors include:

1. Well use—ground-water sampling; piezometric measurements; or contaminant recovery.
2. Hydrogeologic environment.
3. Chemical characteristics and concentrations, if known, of contaminant.
4. Duration of well use—temporary or long term.

Once these factors have been fully evaluated, the specific well design parameters can be determined. The selection of materials and design of the well screen can have a significant impact on the integrity of ground-water samples. The following characteristics should be considered during the screen design [*1*].

1. Screens should be constructed from a material that is inert in the water being tested.
2. Screen openings should be maximized to facilitate rapid sample recovery.
3. Slot sizes should retain filter pack consistent with the capability to develop the well.
4. Slot openings should be nonplugging in design.
5. Slot openings, slot design, open area, and screen diameter should permit effective development.

Since many common screen and casing materials may react with contaminated ground waters, the screen and casing materials must be carefully selected to assure that a representative sample is collected. The following factors are critical to selection of the appropriate materials for the monitoring well system [*2*]:

1. Contaminants to be sampled.
2. Chemical reactiveness/inertness.
3. Strength of material.
4. Ease of installation.
5. Cost of material.

In cases where contaminant plumes consist of trace level organic compounds, the chemical reactiveness/inertness of the screen material becomes the determining factor for material selection. Monitoring wells constructed entirely of Teflon or hybrid wells with Teflon in the saturated zone and stainless steel as the riser pipe are not as susceptible to sorption and desorption of trace level organic compounds, in particular, as are wells cased entirely with polypropylene, polyvinylchloride (PVC), or galvinized steel [*3–5*]. Once the decision to use Teflon has been reached, two factors—strength of the material and ease of installation—must be thoroughly evaluated and addressed in the design specifications to assure a properly constructed monitoring well capable of producing representative ground-water samples.

## Teflon PTFE Strength Testing

### *Procedures*

Several strength-related properties of Teflon must be considered during the design process of the monitoring wells, including.

1. Pullout resistance of flush-threaded couplings.
2. Compressive strength of the screen section.
3. Flexibility of the casing string.

The following tests were conducted at the Technical Service Laboratory of E. I. duPont de Nemours & Co., Inc. of Wilmington, Delaware [*6*]. Compressive strength of screen of Teflon and thread pull strength of threaded casing of Teflon were evaluated using an Instron Corp. Model

1125 instrument. The data were collected by inputting to a variable speed chart recorder and a Zenith Data Systems microcomputer.

Compression testing was performed by loading screen samples of 0.025, (No. 10), 0.051 (No. 20), and 0.076 cm (No. 30) slot size to the desired load, with load adjustment as required to maintain the load during the course of testing. Ten slots from each sample were selected prior to testing, with the dimensions of these slots being measured, prior to loading, using a feeler gage. The dimensions of the selected slots were measured after the sample had been subject to the appropriate load. The average of these ten measurements was taken as a final dimension. Measurements were made at loadings of 0, 45, 91, 136, 182, and 227 kg with the load being maintained for 1 h. On selected samples the slot dimensions were measured immediately following load release as well as 1 h later.

The screened section consists of three rows of horizontal slots, the slotted sections comprising two thirds of the circumferential area of a 5.1-cm inside diameter, schedule 80 pipe. The vertical slot spacing was 6.5 mm. The samples which were tested were 25.4 cm in length.

Thread pull tests, or "tensile" pulls, were performed on threaded casing sections in order to ascertain the strength of the threaded sections (ability to support well string). The male and female sections were threaded together and the threaded assembly was connected to the Instron using special pipe inserts and clamps. The threaded assembly was subsequently put in tension, and pulled at a rate of 0.51 cm/min. The load, in kilograms, at which the threads first slipped was reported as the ultimate thread strength.

All tests were performed on commercially available 5.1-cm inside diameter, schedule 80 screen and casing made of Teflon. The threads were machined flush square and were 0.31 cm in width with 0.13-cm-deep flights. The threaded portions of the casing were 3.8 cm in length, with a total length of 5.2 cm each.

*Results*

The slot compression data were tabulated and a regression analysis was performed. The results of this analysis are shown in Table 1 and Fig. 1. The data support previous knowledge pertaining to the linear relationship, in Teflon, between compression and load. The regression analyses gave a correlation coefficient of 0.96.

The results as graphed are deceiving in that at or below a load of 91 kg the slot compression for all samples was in the range of 0.0025 cm. The linear character of the results over the majority of the load range would indicate a slot closure, at 91 kg, of approximately 0.008 cm, which is three times larger than that observed. This would suggest a nonlinear relationship between the load and deformation at low loadings.

Slot dimensions were measured immediately following and 60 min after load release. These evaluations showed that immediately following load release, Teflon screen 5.1-cm inside diameter, schedule 80 regains 65% of its dimensional loss, and within 1 h regains an additional 20% of this loss. As an example, a screen of the type tested with 0.076-cm slots will compress to slot widths of approximately 0.058 cm at a 227-kg load. Upon load release these slots will regain 0.011 cm to a 0.070-cm slot size. The same screen will have slot dimensions of 0.074 cm after 1 h. These data confirm the essentially elastic behavior of Teflon.

The thread pull (tensile) testing of the threaded Teflon casing showed the samples to resist an average of greater than 264-kg pullout force. This would indicate, given a safety factor of 2, that one could, with confidence, install a 107-m well string of 5.1-cm, schedule 80 Teflon, given a weight of approximately 0.05 kg/cm for casing of these dimensions.

## Monitoring Well Design Considerations—Teflon Casing

The experimental results on the load-bearing capacity of the threaded joints indicate that, including a safety factor of 2, the strength of these joints is sufficient to provide confidence in

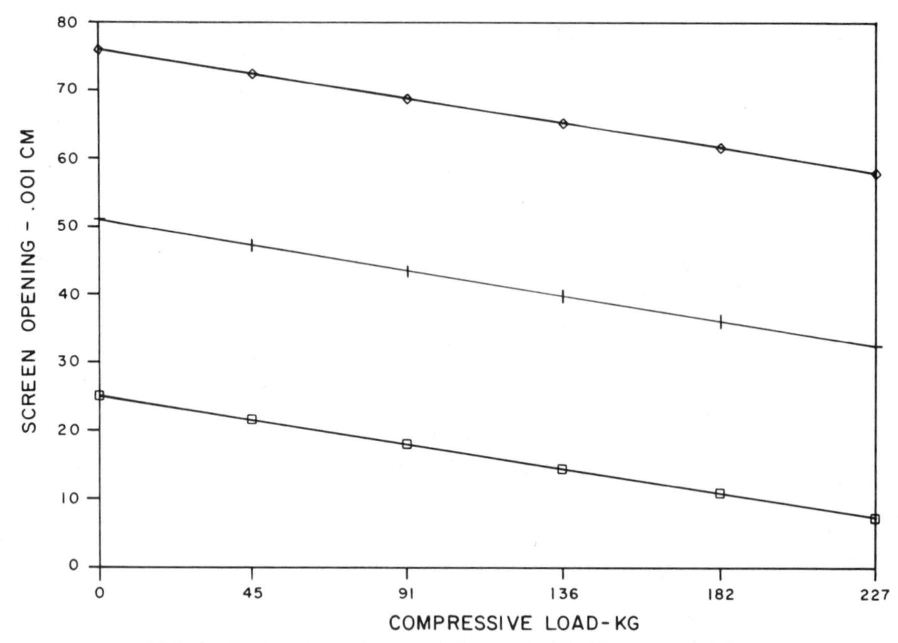


FIG. 1—*Static compression test: 5.1-cm, schedule 60 screen of Teflon.*

installing wells to 107 m with 5.1-cm inside diameter, schedule 80 pipe. A hybrid system consisting of screen and riser of Teflon and casing of PVC or stainless steel could be installed to considerably greater depths.

The experimental compression test results indicate that even under worst-case conditions when the full weight of the casing is supported by the screen resting on the bottom of the borehole, the screen will retain its integrity and approximately 75% of the original slot opening. If the casing is then suspended above the bottom of the borehole during installation, these tests would indicate that the screen will regain as much as 92% of the original slot opening. Detailed installation specifications requiring that the casing not rest on the bottom of the borehole would eliminate this substantial compressive load from the screen section.

If the designed screen opening must be strictly maintained after installation, a larger screen opening should be specified initially. Using the graphic results presented in Fig. 1, the anticipated screen opening deformation can be determined from the anticipated compressive load and added to the desired screen opening during the well screen design. The following equation can be used to calculate the compressive load on a 5.1-cm inside diameter, schedule 80 screen section for a monitoring well casing. The 5.1-cm inside diameter pipe is the standard commercially available pipe. If 10.2-cm inside diameter pipe is used, the constants should be multiplied by a factor of 3.2 to calculate the compressive load:

$$\text{Load (kg)} = 0.89 \text{ kg/m } L_{TS} + 1.63 \text{ kg/m } L_T + 1.04 \text{ kg/m } L_P$$

where

$L_{TS}$ = length of Teflon casing in saturated zone,
$L_T$ = length of Teflon casing in unsaturated zone, and
$L_P$ = length of PVC casing in unsaturated zone.

A second design method to minimize the screen section deformation, which can be used alternately or in conjunction with modification of the slot size, is to modify the specification for the spacing of the slots. In order to increase the unslotted pipe area and consequently the resistance to closing of the slots, as well as to maintain biaxial stiffness, 5.1-cm inside diameter Teflon screens should be specified with three 4.2-cm-long rows of evenly spaced slots separated by 2.1 cm. For 10.2-cm inside diameter pipe, the slots should be 5.4-cm-long with 3.7-cm blank areas. This provides a slightly larger blank area between the slots than typical PVC well screens.

The backfilling technique which is utilized to complete the well can significantly affect the amount of compressive load on the screens. If concrete is placed as the backfill above the screen section, significant sidewall friction will be developed within the borehole, thus effectively eliminating any compressive load on the screen section. Volclay[4] grout has been used recently as a replacement for concrete. This material has a relatively low strength and will not develop any significant long-term sidewall friction. Once the suspension of the casing is removed, the compressive load due to the weight of the casing will gradually be applied to the screens.

Another strength-related design consideration of Teflon is the propensity of Teflon casing to "snake" or become bowed when installed by conventional methods. The result of such "snaking" is the inability to lower sampling pumps and ground-water measurement devices past the bows in the casing. There are two installation techniques which can significantly improve casing straightness at the completion of construction. These techniques, which will be described in detail later, include:

1. Use of boring centralizers.
2. Insertion method—inserting a rigid PVC or steel pipe inside the casing during backfilling operations.

## Installation Procedures for Monitoring Wells Cased with Teflon

### *Preshipment Preparation*

The principal reason for specifying casing and sampling materials of Teflon is the high chemical stability and inertness to sorption and desorption of contaminants, particularly trace level organic compounds. In order to assure that uncontaminated materials are installed, specifications of materials should require a certification from the manufacturer, such as the du Pont quality seal, indicating that the following minimum standards are met:

1. The materials consist of 100% pure virgin PTFE resin and not reprocessed PTFE.
2. The components have been extruded and then sintered to drive off the volatile organic components.
3. The components have been rinsed with solvents to remove surface contaminants, and double rinsed with deionized water.
4. All materials should be packaged in such a manner as to prevent recontamination during shipment.

Once the Teflon materials have been delivered to the job site, the airtight packaging should be maintained intact until the pipe is lowered into the borehole. All well materials should be stored away from the drilling site to minimize the possibility of contamination by dust and chemicals present at the drill site.

### *Field Procedures*

For investigations that require monitoring wells of Teflon, the following procedures should be incorporated into the drilling program:

[4] Registered trademark of American Collard Co.

1. All drilling and sampling equipment that goes into the boring must be steam-cleaned prior to the start of each boring.
2. No drilling fluids or outside sources of contamination can be introduced into the borehole.
3. No lubricants may be used in the drilling operations which may contaminate the borehole vicinity.

Once the borehole is completed, one of the techniques—borehole centralizers or the insertion method—should be employed to improve well straightness.

The use of borehole centralizers is applicable when the borehole will stand open to the entire depth or when installing a well in a slant boring beneath a surface or subsurface structure. The centralizers are attached to the outside of the casing to keep the well string centered in the borehole. For 3.05-m lengths of Teflon casing, minimum spacing for the centralizers should be 1.5 m. Since most borings encounter caving conditions to some extent below the ground-water surface and the centralizers may in some instances impede the installation of the gravel pack, bentonite, and grout seals, this technique is not widely utilized.

The insertion method, when used in conjunction with suspension of the casing string above the bottom of the borehole, is the most reliable procedure for assuring that Teflon casing is straight in a monitoring well and that compressive stress on the screen is minimized. A thick-walled PVC pipe (schedule 80 or 120), with an outside diameter slightly less than the inside diameter of the Teflon casing, is inserted in the casing string after the string has been suspended slightly above the bottom of the boring. The PVC insert is left in place until the backfilling operation is completed. Backfilling of the well annular space can proceed using conventional techniques, once the straightness of the casing is verified. An alternative to PVC for 5.1-cm inside diameter pipe is standard AW drill rod which has been properly steam-cleaned. (A note of caution: The insert pipe or drill rod must not be supported directly on the bottom of the well casing as this may damage the bottom cap or plug.)

## Summary

The preliminary design of a monitoring well system is an essential part of the process of obtaining a representative sample of ground water. If the surface conditions, suspected contaminants, type of investigation and budgetary constraints justify the selection of Teflon for use as the screen and casing material, the following considerations should be included in the design and specification process.

1. Depth of installation less than 107 m for all Teflon casing.
2. Screen deformation of up to 25% will occur with increasing compressive load. Increasing the screen opening specification or specifying suspension of the casing during installation, or both, can minimize the amount of screen deformation.
3. Specifying virgin polytetrafluoroethylene resin for all components which will contact the ground water or sample.
4. Specifying the drilling and installation technique which will assure that the casing is installed straight (borehole centralizers, or insertion method).

Once the design phase of the investigation is complete, the field installation and sampling procedures should be carefully supervised by the project hydrogeologist. Field operations should be planned and implemented so that all possible sources of cross contamination are minimized or eliminated. The installation specifications should be strictly followed. All sampling, chain of custody, and laboratory analytical procedures should be closely monitored to assure that a representative sample is collected and that sample integrity is maintained.

## References

[1] Driscoll, F. G., "Ground Water and Wells," Johnson Division, St. Paul, MN, 1986, pp. 1108.
[2] Ramsey, R. H. and Montgomery, J. M., "Monitoring Ground Water in Spokane County, Washington" in *Proceedings*, Second National Symposium on Aquifer Restoration and Ground Water Monitoring, D. M. Nielsen, Ed., National Water Well Association, Worthington, OH, 1982.
[3] Barcelona, M. J., Helfrich, J. A., and Garske, E. E., *Analytical Chemistry*, Vol. 57, 1985, pp. 460–464.
[4] Schweitzer, P. A., *Handbook of Corrosion Resistant Piping*, Industrial Press Inc., New York, 1969.
[5] Sharom, M. S. and Solomon, K. R., *Canadian Journal of Fisheries and Aquatic Sciences*, Vol. 38, No. 2, 1981, pp. 199–204.
[6] "Teflon (R)—A Performance Guide for the Chemical Processing Industry," E. I. DuPont De Nemours & Company, Inc. Publication No. E-21623-2, Wilmington, DE, 1984.

*Franklin D. Patton*[1] *and H. Rodney Smith*[1]

# Design Considerations and the Quality of Data from Multiple-Level Ground-Water Monitoring Wells

**REFERENCE:** Patton, F. D. and Smith, H. R., "**Design Considerations and the Quality of Data from Multiple-Level Ground-Water Monitoring Wells,**" *Ground-Water Contamination: Field Methods, ASTM STP 963,* A. G. Collins and A. I. Johnson, Eds., American Society for Testing and Materials, Philadelphia, 1988, pp. 206–217.

**ABSTRACT:** The role of design in obtaining ground-water monitoring data is introduced. Several basic aspects of design are considered. These include the spatial distribution and length of monitoring zones and the percentage of the drill hole containing seals. Emphasis is placed upon multiple-level monitoring wells because, for most sites, they appear to be necessary to satisfy the requirements for establishing the three-dimensional distribution of natural piezometric and chemical data, as well as the influence of superimposed local variations. To facilitate the discussion of multiple-level monitoring wells, a proposed classification is presented.

Attention is given to mixing of fluids in the well bore annulus. Errors introduced by such mixing are shown to be related to the length of the monitoring zone with respect to the geometry and uniformity of the ground-water plumes. Analyses of samples affected by fluids mixing prior to and during sampling are shown to lead to significant errors in determining in-situ water quality.

The paper concludes that water samples and fluid pressures from short monitoring zones will generally be of a higher quality (that is, more representative of in-situ conditions) than data from longer monitoring zones.

**KEY WORDS:** ground water, monitoring wells, monitoring zones, packers, seals, quality data, multiple-level, fluid sampling, fluid pressures, piezometric levels

Quality assurance begins with design. Campbell and Mabey [*1*] have suggested four major elements necessary to establish quality ground-water data: (1) basis for measurement, (2) application of the measurement method, (3) statistical information, and (4) corroborative information.

The principal design activity is to establish the basis for measurement. However, design activities should also include consideration of the other elements noted above. Design activities, therefore, cover both theoretical and practical activities. For example, the selection of monitoring equipment, procedures to install the equipment, integrity testing, sampling, and analyses of the samples fall within "applications of the measurement method." These items should be carefully considered in the design, or the design objectives are unlikely to be achieved. Furthermore, the availability of reliable statistical information and corroborative information can be highly dependent on the design recommendations with regard to equipment and field procedures.

Figure 1 is a generalized flow chart suggesting the relationships between design and the other elements necessary to obtain quality data and to establish the data quality.

Design begins with a basic understanding of field conditions and monitoring needs. Among the more basic field design considerations are (1) the site and regional conditions; (2) the type, spatial

[1] President and geotechnical engineer, respectively, Westbay Instruments Ltd., 507 E. Third St., North Vancouver, BC, Canada V7L lG4.

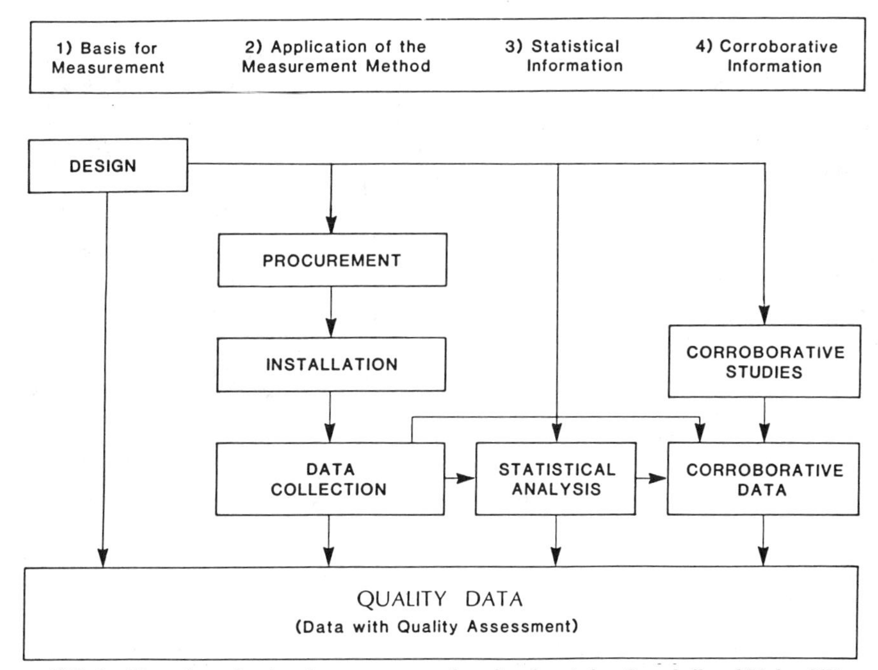


FIG. 1—*Flow chart showing the components of quality data (after Campbell and Mabey [1]).*

distribution, and geometry of monitoring zones; and (3) the selection of the equipment and procedures for installation and control required to establish the quality of the data.

Although all these items are important, the type, spatial distribution, and geometry of monitoring zones are the major topics to be covered in this paper. It seems apparent that the larger the number of monitoring zones and the greater the number of depth intervals monitored, the more complete the picture of the ground-water constituents and their spatial location. This requires either a large number of drill holes drilled to different depths with data obtained at a single level in each or fewer drill holes completed as multiple-level monitoring wells.

Multiple-level monitoring wells are receiving increased attention because they provide an opportunity for reducing costs while maintaining or significantly increasing the quantity of field data obtained. As this paper demonstrates, multiple-level monitoring wells can also address data quality requirements to an extent that is difficult for traditional single-level monitoring wells to achieve.

For most sites, multiple-level monitoring wells are necessary to satisfy the requirements for establishing the three-dimensional distribution of natural piezometric and chemical data, as well as the influence of superimposed local variations.

The present authors have found it convenient to classify multiple-level wells according to (1) whether the access tubes are open or closed; (2) whether one or several access tubes are used; and (3) whether the installations at a given location are clustered wells, a nested well, or a modular nested well. Figure 2 illustrates a classification using these attributes.

## Spatial Distribution of Monitoring Points

The geologic and, in particular, the hydrogeologic aspects of the site must be considered in the design of a ground-water monitoring system. Perhaps the most significant aspects of the hydrogeology

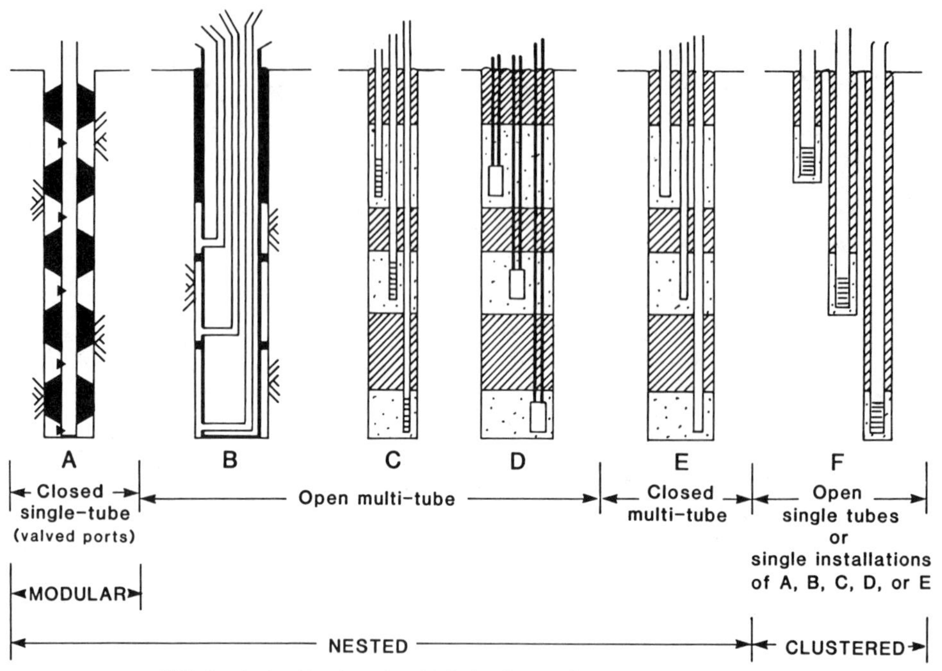


FIG. 2—*A classification of multiple-level ground-water monitoring wells.*

are the local and regional ground-water flow systems and their related ground-water chemistry. Imposed on the natural conditions are a set of artificial conditions resulting from a specific past, current, or future use of a site. The site conditions, both natural and imposed, are generally unknown or only partly known when monitoring wells are installed. Hence, it is necessary to have a sufficient spatial distribution of drill holes and monitoring points for the more important site conditions to be identified.

The term ''spatial distribution'' covers the plan coordinates of monitoring facilities as well as the maximum depth and distribution of monitoring points between the ground surface and the lowermost point. In considering the spatial distribution of monitoring points, the surrounding or regional conditions and the site conditions are important. For example, in Fig. 3*a* monitoring locations are distributed in and around a specific site of interest. This site has an area and a depth. Additional monitoring locations are distributed away from the site and below the site to establish the regional or background conditions as well as to confirm the presence or absence of hydrogeologic boundaries that may influence the data.

The depth and distribution of monitoring points should be sufficient to cover layering or other nonhomogeneous distributions of hydrogeologic units and any hydrochemical layering within individual units. With a simple source of contamination, simple hydrogeologic conditions, and a shallow site, perhaps as few as five locations within the site together with four to eight locations outside the site, each with five to ten monitoring points per location, may be enough to obtain an initial impression of the site conditions. However, for sites with complex contaminant distribution, complex hydrogeologic conditions, or for deep sites, appreciably more monitoring locations and monitoring points at each location are likely to be required. A progressive or staged approach to selecting the number and spacing of the monitoring locations, as was suggested in the August 1985 draft of the Resource Conservation and Recovery Act (RCRA) Ground-Water Monitoring

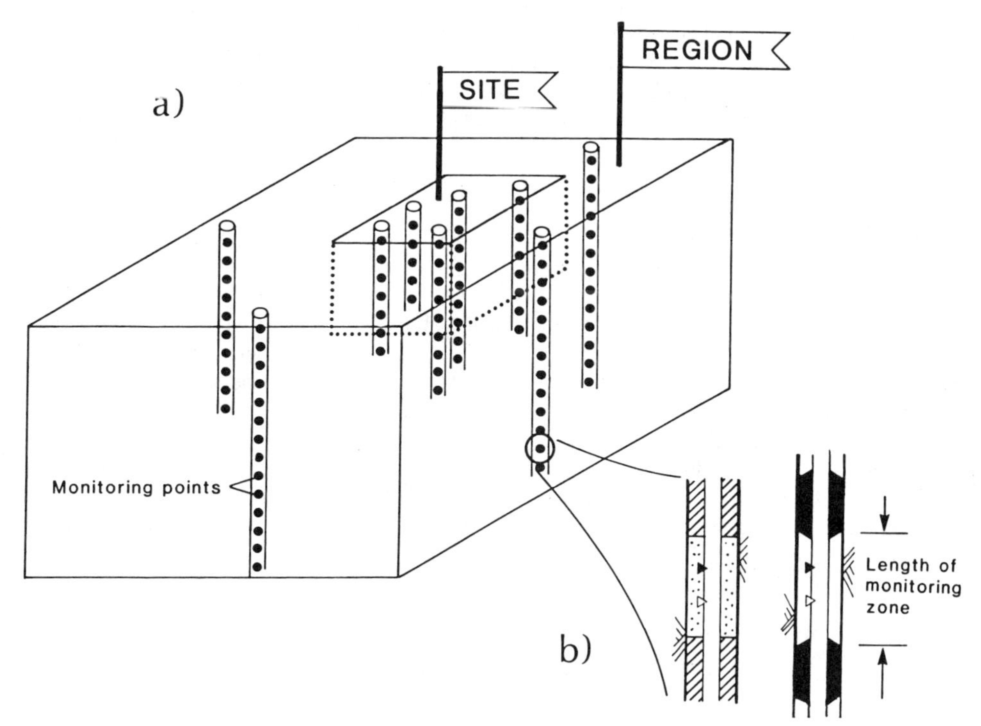


FIG. 3—*Spatial distribution of monitoring "points" or zones.*

Technical Enforcement Guidance Document [2], should provide a rational method of determining the number of monitoring locations for most sites.

The number of monitoring points installed and their depths at each location depend on many diverse considerations. These include (1) the number of hydrogeologic units and boundaries, (2) the distance from natural or imposed perturbations, and (3) the value of the resource under consideration.

Two simple empirical formulas were suggested by Patton [3] to determine the number of monitoring points required at one location in order to obtain representative distributions of fluid pressures in the saturated zone when several simple aquifers are present. The minimum number of piezometers required was suggested to be $2n + m$ where $n$ is the number of aquifers present and $m$ the number of aquitards present below the water table. For an "adequate" coverage the number of piezometers required was suggested to be $4n + 2m$. In most ground-water contaminant monitoring problems, however, both representative fluid pressures and fluid samples are required. Obtaining representative fluid samples of contaminant plumes can require appreciably more monitoring points than those indicated in the above relationships, which were developed for fluid pressure distributions only. The number of monitoring points should be adjusted to reflect distances to hydraulic or chemical perturbations and to the impact that small changes in the measured parameters could have on the significance of the data derived from the monitoring system. Consideration should also be given to the need for additional redundant data for corroborative and statistical purposes (see Fig. 1).

For assessment monitoring when individual contaminant plumes have been detected it is suggested in the RCRA document [2] that seven additional clusters of monitoring wells be installed on the downstream side of the contaminant source with monitoring points at a minimum of five different depths at each cluster location. As indicated in the RCRA document, such a monitoring array would be sufficient to monitor a plume only under very simple field conditions.

Whenever possible, the location of monitoring wells should be planned to facilitate the preparation of hydrogeologic sections and to meet modeling requirements.

## Geometry of Individual Monitoring Zones and Seals

Monitoring points exist only in an abstract sense and only at the conceptual design and modeling levels. In reality monitoring "points" are not points but cylindrical prisms or zones which have a length that is significant. The degree to which the shape of a monitoring zone departs from a point is generally inversely related to the quality of data to be derived from the associated monitoring system. Two examples of monitoring zones are shown in Fig. 3*b*. The monitoring zone is a cylinder whose outside diameter is roughly the diameter of the drill hole and whose length is determined by the distance between hydraulic seals completed on either end of the monitoring zone. The hydraulic seals can be provided by the bottom of the drill hole, by the beginning of a layer of low-permeability backfilling material (Fig. 3*b*, left-hand side), or by hydraulic packers (Fig. 3*b*, right-hand side).

In the following sections, the detrimental influence on samples and fluid pressures of the length of monitoring zones is discussed.

### *Sample Degradation Prior to Sampling Due to Fluid Mixing in the Well Bore Annulus*

Long monitoring (or completion) zones can result in mixing of fluids in the well bore annulus before they are sampled. This process can reduce the maximum level of contaminant that can be sampled and may raise the apparent contaminant level of uncontaminated layers. These processes occur when a nonuniform distribution of fluid chemistry is present within the monitoring zone.

Figure 4 illustrates conditions that would lead to a sampling problem. Figure 4*a* shows the hypothetical results of chemical analyses of samples from two monitoring wells. The results from one set of analyses (designated "b" in Fig. 4*a*) are from the monitoring well with a long monitoring zone shown in Fig. 4*b*. The results from another set of analyses (designated "c" in Fig. 4*a*) are from the well with several short monitoring zones shown in Fig. 4*c*. It would be expected that the results from the short monitoring zones would more closely reflect the in-situ values of contaminant concentrations shown on the right-hand side of Fig. 4 than would the results from the long zone. It is apparent that the shorter the monitoring zone, the more closely the results will reflect the true field conditions. The rectangle "b" in Figure 4*a*, within which analyses from the long monitoring zone might fall, indicates an apparent dilution of the maximum in-situ value by a factor of 10 to 100 times as well as an apparent increase in the minimum *in situ* value by a similar range. These differences, in both the magnitude and location of the actual values, can be significant in planning remedial treatment measures and in demonstrating compliance or noncompliance with established limits of contaminants in ground water.

### *Errors Introduced in Piezometric Levels*

The adverse influence of long completion zones on measurements of piezometric levels should also be considered. Long monitoring zones produce a subsurface cylindrical pocket with an artificially induced hydrostatic pressure distribution. The magnitude of the piezometric level recorded within a zone is a function of the pressures and transmissivities of all the layers intersected by the zone and can differ significantly from in-situ values. Thus, the errors in the piezometric levels determined from measurements taken from a long monitoring zone where the actual fluid pressure distribution is nonhydrostatic can be significant. The differences between the measured

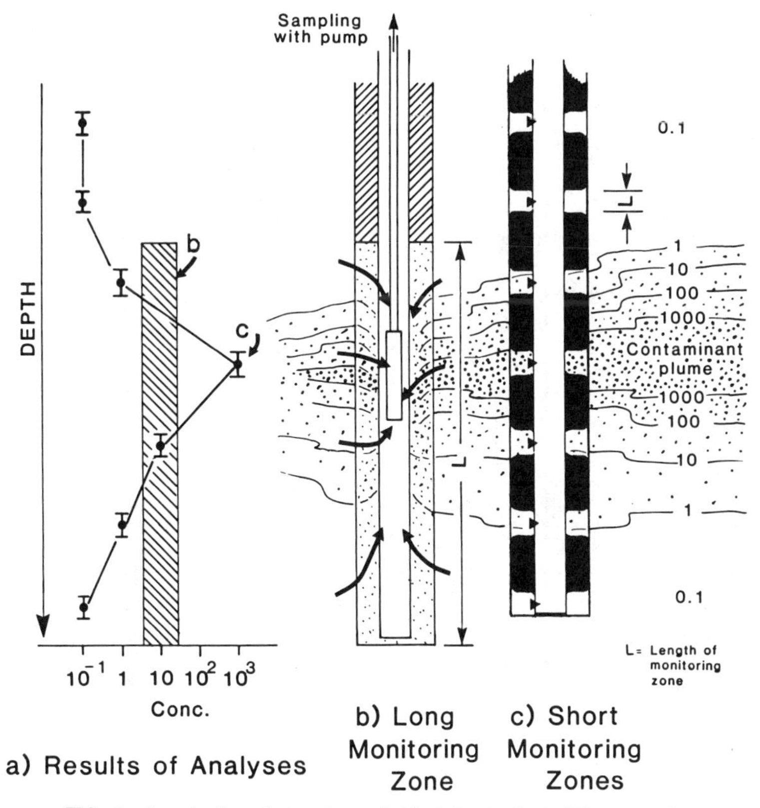


FIG. 4—*Sample degradation due to fluid mixing in the well bore annulus.*

piezometric levels and the actual levels becomes of greater interest when one realizes that a nonhydrostatic pressure distribution is the norm and hydrostatic pressure distributions are the exception. The importance of reliable measurements of piezometric levels in a contamination problem is that they are a fundamental parameter in defining the direction and rate of groundwater movement.

Figure 5 taken from Patton [*4*] shows an example of a common type of error in piezometric levels that can occur due to long monitoring zones. In Fig. 5*a* an open single-tube (standpipe) monitoring well with a long monitoring zone has been placed in an area with a downward hydraulic gradient. The water level in this monitoring zone is a function of the relative rate of the flow of the fluids moving into and out of the monitoring zone. In the example shown, the well would be said to be "dry" as the water level in the well falls below the bottom of the access tube placed in the monitoring zone, although the well obviously extends below the water table. The error would be greatly reduced if the monitoring zones are much shorter as in Figs. 5*b*, 5*c*, and 5*d*. Figure 5*b* shows a nest of standpipes (open multiple-access tubes). Figure 5*c* shows a "continuous piezometer" that can provide a continuous profile of fluid pressures but cannot provide fluid samples. Figure 5*d* shows a closed single-tube (modular) monitoring well installed with numerous short monitoring zones.

Any of the monitoring wells shown in Figs. 5*b*, 5*c*, and 5*d* would provide measurements of piezometric pressure superior to those available from the well in Fig. 5*a* simply because of their shorter monitoring zones. Thus, a large number of short monitoring zones may be necessary to

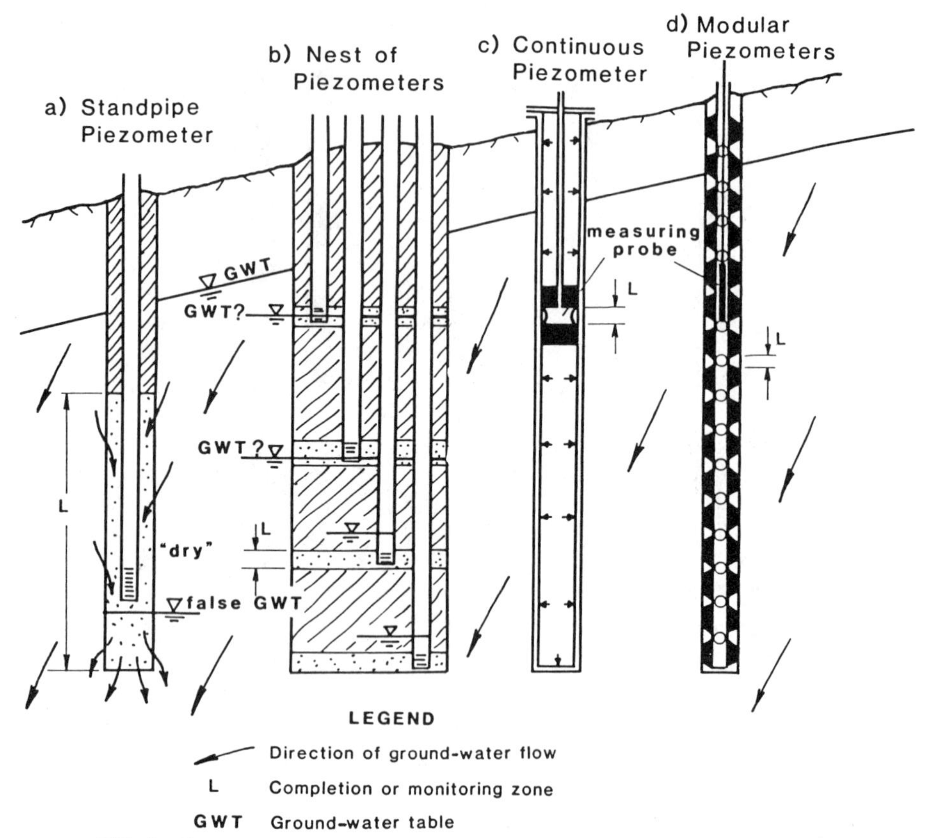


FIG. 5—*Errors in piezometric levels due to long monitoring zones (from Patton* [*4*]*).*

define the actual distribution of piezometric levels in the field. Figure 5 also indicates how something as simple as the water table may not be easy to establish in the field.

Patton [*4,5*] has reported errors exceeding 100 m (330 ft) in the piezometric levels recorded in standpipe piezometers in mountainous terrain. These errors were due to the presence of long monitoring zones. However, most contaminant monitoring sites have low topographic relief and in such places errors in piezometric levels are more likely to be on the order of 0.01 to 3 m (0.03 to 10 ft) in a single monitoring zone. These smaller errors can still be significant if they are large enough to produce substantial errors in interpreting the direction and rate of ground-water flow.

## Percentage of Drill Hole Containing Seals

A monitoring zone is no better than its confining seals. The seals should be able to withstand the differential hydraulic pressures that could develop between adjacent monitoring zones. Seals should also be sufficiently long that the monitoring zones (or cluster of monitoring zones) do not produce a vertical flow path which would significantly disrupt the natural ground-water flow regime.

From the viewpoint of the seal quality, it generally follows that the greater the sealed length of the drill hole between monitoring zones the better. Ideally, the drill hole would be sealed completely

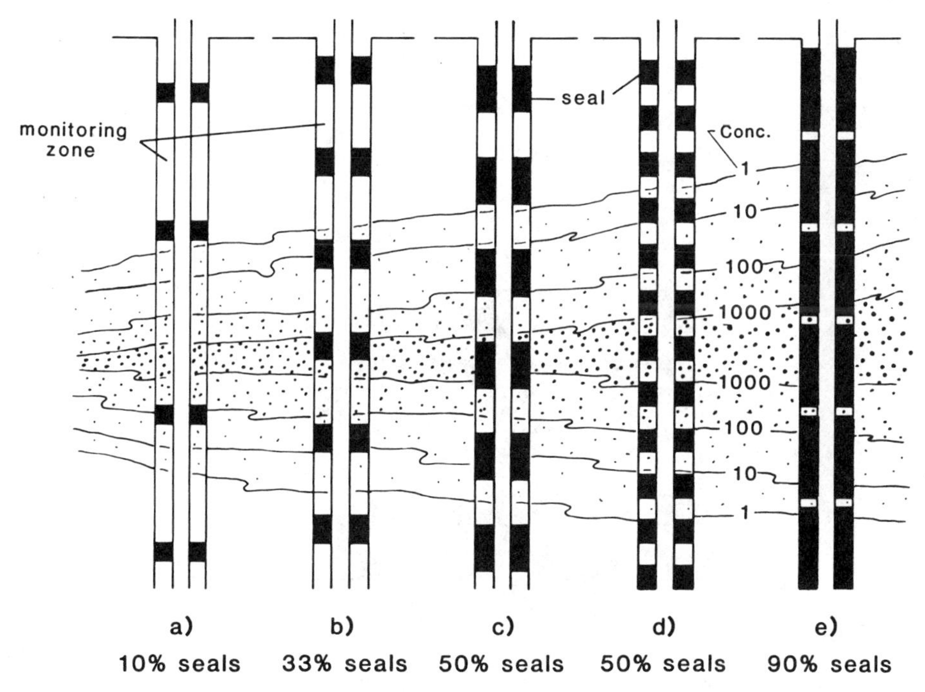


FIG. 6—*Variations in the percentage of drill hole containing seals.*

between monitoring zones. Yet if this is carried to the extreme, long seals could prevent sampling of key horizons. Also, if each monitoring well had a single monitoring zone with the rest of the drill hole sealed, so many monitoring wells could be required for an adequate density of monitoring points that the cost of the drilling, instrumentation, and installation would become prohibitive.

Figure 6 shows several cases with different percentages of the drill hole containing seals. The sealed length of a drill hole can vary from 10%, as in Fig. 6*a*, to 90%, as in Fig. 6*e*.

In the initial phase of an investigation many installations appear to be adequate on a qualitative basis when some 20 to 50% of the part of the drill hole that contains the monitoring zones is sealed (see Figs. 6*b*, 6*c*, and 6*d*). However, the percentage of the drill hole to be sealed should be compatible with the site conditions. A design objective would be to keep the percentage of seals as high as possible while minimizing the length of the monitoring zones. For example, the design of the installation in Fig. 6*d* would appear to be superior to the installation in Fig. 6*c*, although the percentage of seals is identical (50%) in both cases. The installation shown in Fig. 6*d* is better because of its shorter completion zones.

## Types of Seals and Integrity Checking

A great variety of sealing methods is available for monitoring wells. Each method has various capabilities and associated advantages and disadvantages. Several types of seals are shown in Fig. 7. Some seals have a direct influence on water quality, whereas others have only the possibility of influencing water quality. For example, cement grouts (Fig. 7*b* and 7*e*) tend to increase the pH of the surrounding fluids. Bentonites (Fig. 7*a*) tend to capture metallic cations due to their high cation-exchange capacity. Bentonites also tend to decompose in fluids with low pH, resulting in appreciably higher permeabilities than are normally associated with bentonites at pH values from

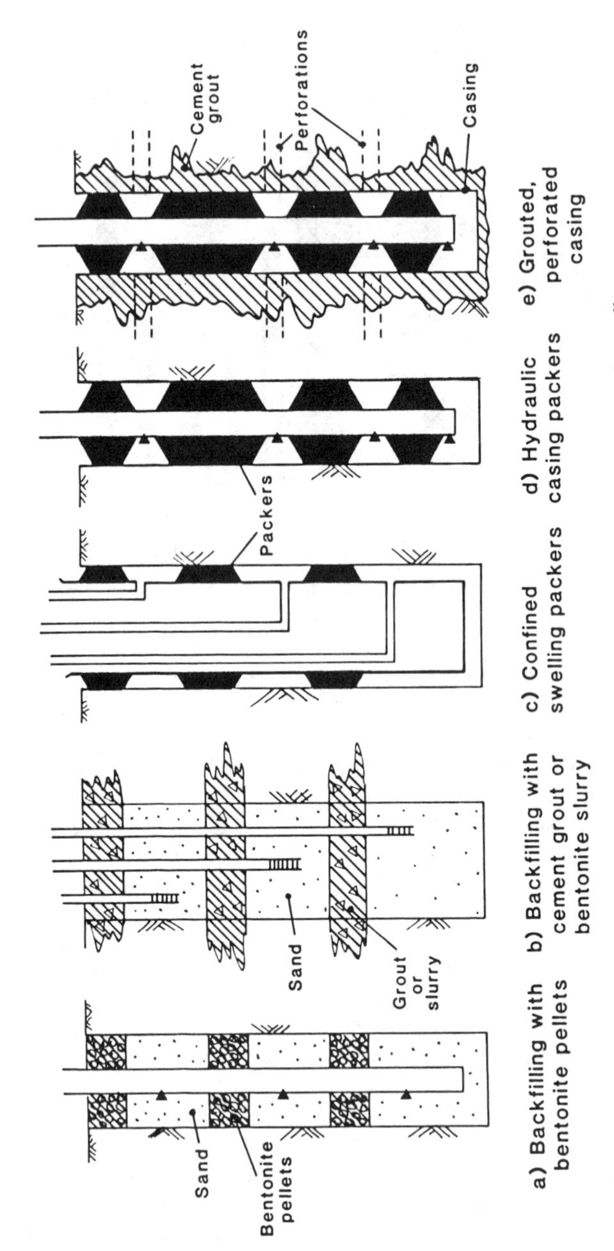


FIG. 7—*Some drill hole seals available for multiple-level monitoring wells.*

5 to 8. Polymer grouts can add undesirable organic compounds to the natural ground waters. Seals made out of plastic, rubber, and metal (Figs. 7*c*, 7*d*, and 7*e*) may be unsuitable where strong organic solvents or other fluids corrosive to metals are present. These disadvantages can be significant at some sites, but where the solubility of the elastomers used for the sealing glands is not objectionable, packers using these materials can have several advantages which are not available with grouts and bentonite seals.

A principal advantage of hydraulic inflatable casing packers (Figs. 7*d* and 7*e*) is that one can obtain a direct measurement of the sealing pressure during the inflation process as well as at intervals during the life of the monitoring well. This pressure is a quantitative measurement of one aspect of the integrity of a seal. Another benefit of packers is that the position of a packer in a drill hole is known with relative certainty, whereas it is common to have some uncertainty regarding the position of the top and bottom of the cement, bentonite, and grout seals.

Checking the capability of each seal to withstand differential fluid pressures can be an important part of a demonstration of the integrity of each monitoring zone. Whether such a capability can be proven depends upon the equipment installed and the hydrogeologic parameters of the natural formations surrounding the seal.

In some types of monitoring wells (modular, nested, multiple-level wells) it is possible to monitor fluid pressures above and below every seal regardless of the number of seals. Some types of monitoring wells also provide the opportunity to artificially induce differential piezometric levels on either side of each seal. Measurements and comparisons of the piezometric levels acting on either side of a seal with or without an artificial pressure surge applied to one side can be used to demonstrate the hydraulic integrity of the seal. If the hydraulic conductivity of the formation is sufficiently low opposite a seal or if natural differences in piezometric levels on either side of a seal can be detected, then it is possible to demonstrate the integrity of a seal for most purposes. However, if the natural formation has a high hydraulic conductivity, it may not be possible to demonstrate that a good seal has in fact been achieved.

The selection of the type of seal is likely to vary from one site to another to meet local conditions. This selection process is a major design activity that directly affects the quality of the ground-water data.

## Other Design Considerations

There are many other design considerations which affect the quality of the data obtained from a specific monitoring well. Each type of monitoring well has restrictions which limit its suitability in various situations. However, no single ground-water monitoring well design will be suitable for handling all sites. The greater the flexibility of the monitoring well design, the fewer the restrictions that will be present. For example, small-diameter access tubes may result in lower capital costs but may not facilitate reliable fluid pressure measurements in the monitoring zones, particularly in deep drill holes and when the piezometric levels are far below the ground surface. Also, the effects of temperature and other density variations are difficult to establish reliably in small tubes. A lack of knowledge of these factors can result in the design of a monitoring well that provides piezometric data which are not representative of the conditions.

What is considered "small diameter" will vary among hydrologists depending on their experience. However, access tubes below 2.5 cm (1 in.) in diameter generally do not permit fishing operations or other routine inspection, maintenance, and calibration activities.

Should fluid samples be kept at formation pressures? Is there a necessity for pumping prior to each sample being taken? What is the minimum volume of sample required? Can the integrity of each component of the monitoring system be checked? These and other questions should be addressed in the design phase of the project. Specific monitoring equipment and field procedures must also be selected at that time. These are among the many design decisions that will have a direct bearing on the character of the primary data obtained and upon the additional verification and corroborative data that will permit the quality of data from a ground-water monitoring installation to be assessed.

## Conclusions

The following conclusions can be drawn from the discussions presented in this paper:

1. The quality of the field data from a ground water monitoring system is closely related to the spatial distribution of monitoring points. This distribution should be adequate to characterize each significant hydrogeologic unit and boundary present.
2. Numerous short monitoring zones will generally provide more representative water samples and fluid pressure measurements than a smaller number of long monitoring zones.
3. Degradation of the sampling fluids can occur prior to and during sampling because of fluid mixing in the well bore annulus. It is possible that this mixing can dilute the contaminants so that the maximum values of contaminants determined from analyses of the water samples are on the order of 10 to 100 times less than those actually present in the portion of the ground water intercepted by the monitoring zone. In addition, such presampling degradation can account for an apparent concentration of contaminants in parts of monitoring zones where no such concentrations actually exist.
4. The presence of long completion zones can result in significant errors in the piezometric levels measured. This in turn can lead to errors in interpreting directions and rates of ground-water flow.
5. The percentage of a drill hole that is filled with seals is an important design consideration. Increased use of seals reduces the possibility of the drill hole acting as a conduit. For an initial assessment of a site, 20% to 50% is a reasonable percentage. However, this could change when more field data have been obtained. When a high percentage of the drill hole is sealed, important data may be lost unless many monitoring zones are installed.
6. The sealing methods and sealing materials have the potential to cause detrimental effects on the quality of water samples taken from adjacent monitoring zones. Because of the great variety of chemicals that can be present, no ideal sealing method exists that can handle all field situations. Thus, sealing methods should be chosen for each site individually. That is, they should be site-specific. The initial decisions should be reviewed as new data become available.
7. Hydraulic inflatable packers can provide a direct measurement of the sealing pressure, a feature not available with other sealing methods.
8. Integrity checks of the hydraulic sealing capability should be undertaken for all drill hole seals. These checks can be facilitated with some types of multiple-level monitoring wells. However, such checks may not be feasible when the formation surrounding the seal has a high hydraulic conductivity.
9. Multiple-level wells can be an effective method of monitoring ground-water conditions when the equipment meets basic design requirements of the site.

## References

[1] Campbell, J. A. and Mabey, W. R., "A Systematic Approach for Evaluating the Quality of Groundwater Monitoring Data," *Ground Water Monitoring Review*, National Water Well Association, Fall 1985, pp. 58–62.

[2] Resource Conservation and Recovery Act Ground-Water Monitoring Technical Enforcement Guidance Document (Draft), Section 1.2.1 Site Characterization Boring Program, Aug. 1985, pp. 1–9 to 1–15.

[3] Patton, F. D., "Groundwater Instrumentation for Mining Projects," Chapter 5 in *Proceedings*, 1st International Mine Drainage Symposium, Denver, CO, Miller Freeman Publications Inc., San Francisco, CA, May 1979, pp. 123–153.

[4] Patton, F. D., "The Role of Instrumentation in the Analysis of the Stability of Rock Slopes" in *Proceedings,* International Symposium on Field Measurements in Geomechanics, Zurich, Switzerland, 5–8 Sept. 1983, pp. 719–748.
[5] Patton, F. D., "Climate, Groundwater Pressures and the Stability Analyses of Landslides" in *Proceedings,* 4th International Symposium on Landslides, Vol. 3, Toronto, Canada, Sept. 1984, pp. 1–17.

The final version (2) was issued in September 1986 after this paper was presented. The final version of this reference does not contain some of the specific recommendations given in the draft edition which was referenced in this paper.

F. D. Patton/H. R. Smith
March 1988

# Water Sampling
# (Saturated and Unsaturated Zones)

*Michael J. Barcelona,*[1] *John A. Helfrich,*[1] *and Edward E. Garske*[1]

# Verification of Sampling Methods and Selection of Materials for Ground-Water Contamination Studies

**REFERENCE:** Barcelona, M. J., Helfrich, J. A., and Garske, E. E., "**Verification of Sampling Methods and Selection of Materials for Ground-Water Contamination Studies,**" *Ground-Water Contamination: Field Methods, ASTM STP 963,* A. G. Collins and A. I. Johnson, Eds., American Society for Testing and Materials, Philadelphia, 1988, pp. 221–231.

**ABSTRACT:** The development of proven sampling methods which incorporate the use of noncontaminating materials has progressed steadily in the past five years. Once the hydrogeologic conditions and potential contaminant source characteristics have been identified at a particular study site, the selection of drilling methods, suitable sampling mechanisms, and appropriate materials must be made for the chemical constituents of interest. The results of laboratory and field testing of sampling mechanisms, tubing and well casing materials, and sampling procedures have identified elements of sampling protocols which are effective in reducing sampling bias and imprecision. The proper selection of these elements bears on the accuracy and precision of all subsequent measurements, since field blanks, sample splits, replicate samples, and standards can only account for errors which occur *after* sample collection.

The limited laboratory and field studies of sampling error strongly suggest that poor borehole seals and improper well purging techniques are serious sources of bias for chemical constituent determinations. Sampling mechanisms and materials are also potential sources of bias, particularly for pH and gas-sensitive parameters and volatile organic compounds. Quality assurance must begin with the decisions made in establishing a sampling point (that is, well or dedicated installation) and collecting samples.

**KEY WORDS:** ground water, sampling procedures, hydrology, water quality, water chemistry

Monitoring wells are the most common sampling points for ground water collected for chemical analysis. It should be clear that the design requirements on monitoring wells for chemical work may differ substantially from those for piezometers, which may be used only for water level or hydraulic conductivity testing. Every monitoring investigation has specific chemical and hydrogeologic conditions which should be considered before wells are drilled and construction methods, sampling mechanisms, or other sample handling procedures are chosen. These decisions are important for the following reasons:

1. to avoid using methods/materials that are either excessive or inadequate for the situation;
2. to avoid unnecessary expense, either short-term, less expensive solutions which may suffice for a particular situation, or in the long term, because more durable or inert materials may be needed;
3. to tailor the network design to the situation so that the highest-quality data are obtained for the smallest investment of time or resources; and

[1] Head, Aquatic Chemistry Section, and chemists, respectively, Illinois State Water Survey, 2204 Griffith Drive, Champaign, IL 61820.

4. to collect valid data that will provide convincing evidence in court and minimize the need for additional expense in either defending the data or in obtaining better data.

In order to obtain the highest-quality data, it is also necessary to recognize potential sources of error or bias that are present in well installation and sampling procedures. Ideally, one would like to maintain the magnitude of sampling error below that of analytical error (for example, bias of ±20%). Sample handling and analytical errors which occur after the samples are brought to the surface can be corrected for by the use of field standards, blanks, and procedural standards. Only proper choices of well installation procedures, sampling materials, and sampling mechanisms will minimize the bias which may occur during well purging and sample collection. These sources of error can be quite large compared to analytical error and lead to grossly inaccurate results.

## Review of Known Sources of Error

### *Well Construction and Development Effects*

Well drilling techniques and construction materials can have long-lasting and deleterious effects on monitoring data [*1,2*]. The selection of a particular drilling technique should depend on the type of geology at the site, the expected depths of the wells, and both the availability and suitability of appropriate equipment for the contaminants of interest. Effective monitoring well design and construction require considerable care and an understanding of the hydrogeology and subsurface geochemistry of the site. Preliminary borings, well drilling experience and the details of the operational history of a site can be very helpful. The simplest, narrow diameter well completions which will permit development, accommodate the sampling gear and minimize the need to purge large volumes of contaminated water are preferred for routine monitoring activities. Helpful discussions of monitoring well design and construction are provided in several publications [*1,3–5*]. Regardless of the technique used, every effort should be made to minimize subsurface disturbance when constructing monitoring wells for chemical analysis. Innovative drilling techniques may be quite useful in some situations [*6*]. For critical applications, the drilling rig and tools should be steam-cleaned to minimize the potential for cross-contamination between formations or successive borings.

The use of drilling muds can cause problems for trace chemical constituent investigations because foreign organic matter will be introduced into the penetrated formations. Common "clay" muds contain some organic matter which is added to stabilize the clay suspension. The effects of drilling muds on ground-water chemistry have not been investigated in detail. Reports indicate, however, that the organic carbon introduced during drilling with "organic" muds can cause false water quality observations for periods of months or more [*1*].

This effect has also been observed at a sand and gravel aquifer field site in western Illinois [*8*]. Table 1 gives the effects of the use of a "biodegradable organic mud" drilling fluid on total organic carbon (TOC) determinations in ground water over a two-year period. In this case, the organic mud was used to hold the auger holes open during well construction. Background TOC levels were measured on water samples from driven sand points. Both the drilled wells and the sand points had been developed by swabbing and bailing. Total organic carbon values increased substantially in the drilled wells and remained more than threefold higher than measured background levels for two years. The volatile fraction of the TOC more than doubled, resulting in an apparent tenfold increase in volatile organic carbon during the first year, before returning to near previous levels. The high level of TOC is attributed to the presence of the "organic mud" itself while the volatile carbon is believed to result from stimulated microbial activity due to the introduction of the organic-rich substrate. The data in the table also show a 25% reduction in TOC in bladder-pumped samples from those previously obtained with a bailer. Organic particulates collected by the bailer artificially increased the TOC levels.

TABLE 1—*Effects of drilling fluids and sampling mechanism on TOC determinations.*

| | TOC[a] | Volatile Organic Compounds[a] (% of TOC) | Number |
|---|---|---|---|
| BACKGROUND WELLS | | | |
| 1st year | 4.38 ± 11%[b] | 0.7 (16%) | 12 |
| 2nd year | 4.23 ± 37% | 0.6 (14%) | 10 |
| "ORGANIC MUD" DRILLED[c] | | | |
| 1st year | 15.5 ± 27% | 6.2 (40%) | 12 |
| 2nd year (bailer samples) | 12.0 ± 31% | 0.7 (6%) | 10 |
| 2nd year (bladder pump) | 8.3 ± 60% | 0.7 (8%) | 6 |

[a] $mgC \cdot L^{-1}$.
[b] Relative standard deviation as percent of mean.
[c] Guar bean starch drilling aid.

The placement and grouting of monitoring wells are also potentially serious sources of error. The annular space between the casing and the borehole above the gravel pack should be backfilled with a sealant to prohibit the intrusion of foreign water into the screened interval. Care in selecting the grouting materials is important to ensure that the grouting does not interfere with the chemical constituents which are to be determined.

Borehole cuttings, bentonite, clay, and cement grout are commonly used sealants but each has limitations. Borehole cuttings are the least desirable because they are hard to compact during placement. Improper placement may cause bridging of the material which will later slump and destroy the integrity of the seal. Bentonite pellets are frequently used but their swelling properties are reduced in highly mineralized water. Additives are available that can reduce this problem but their effectiveness has not been well documented.

Cements or cement-bentonite mixtures that expand are preferred. However, contamination by cement grouts has been reported [*9,10*]. The usual symptoms are very high solution pH (that is, pH 10 to 12), alkalinity (> 90% hydroxyl), conductivity and calcium ion ($Ca^{++}$), which persist even after extensive well development and pumping. The problem arises when the cement grout is placed in the saturated zone and subsequently fails to set up. Components of the cement leach into the borehole, altering both solution chemistry and the distribution of chemical species. This problem usually can be avoided by removing standing water from the well borehole before placing the cement or by placing the cement with the use of a tremie pipe. It is also recommended that a sand pack be placed at least one foot above the screened interval and isolated from any cement grout by a bentonite seal.

## Purging Stagnant Water

Ground water which remains in a well between sampling periods can undergo substantial chemical changes and is no longer representative of *in situ* aquifer conditions. Stagnant water must be removed from the well or isolated from the sampling mechanism to ensure sample integrity [*11*]. Various purging strategies have been reported in the literature [*1,12,13*]. The best approach is to calculate the purge volume requirement from the hydraulic performance of the well [*3,14*]. Then one can verify the calculated purge volume monitoring pH, redox potential (Eh), specific conductance, and temperature while purging the well. An in-line, flow-through electrode cell used in conjunction with a positive-displacement bladder pump can be very useful in this regard [*15,16*]. Sample collection may begin when these parameters stabilize to less than about a ±10% change over two successive well volumes pumped. In this way, one can minimize the amount of water that needs to be handled and a consistent basis for comparing the chemical composition of successive samples can be established. It should be clear that rules of thumb calling for the removal of a specific number of well volumes prior to sampling ignore hydrogeologic conditions and the hydraulic performance of the well.

### Well Casing Materials

The question of the best well casing material for a specific monitoring situation must be addressed by considering subsurface geochemistry as well as the nature and concentration of the contaminants of interest. Therefore, strength, durability and inertness should be balanced with cost considerations in the choice of well casing materials. Common well casing materials include polytetrafluoroethylene resin (PTFE) (that is, Teflon[2]), polyvinyl chloride (PVC), stainless steel, and other ferrous materials. The strength, durability, and potential for sorptive or leaching interferences on chemical constituent determinations have been reviewed in detail for these materials [*2,3*]. Unfortunately, there is very little documentation of the actual severity or magnitude of well casing interferences from field investigations.

Polymeric materials have the potential to sorb (that is, either adsorb or absorb) dissolved chemical constituents and to leach components of polymer formulations or previously sorbed substances. Similarly, ferrous materials may adsorb dissolved chemical constituents and leach metal ions or corrosion products. The potential for problems in both cases is real, yet incompletely understood.

PTFE is the well casing material least likely to cause significant errors for either organic or inorganic chemical constituents. It is sufficiently durable for monitoring installations less than 150 m deep. For deeper installations, it can be linked to another material (for example, stainless steel) above the highest seasonal, stagnant water level.

Stainless steel well casing can be expected to be—under noncorrosive conditions—the second least likely material to cause significant error for organic chemical constituents. The release of iron, chromium, or nickel may occur under corrosive conditions. Sorption effects may also be significant sources of error for chemical constituents after corrosion processes have altered the virgin surface.

Rigid PVC well casing with National Sanitation Foundation approval may be used in monitoring well applications when noncemented or threaded joints are used and organic chemical constituents are not expected to be of present or future interest [*2*]. Significant losses of strength, durability and inertness (that is, sorption or leaching) may be expected under conditions where organic contaminants are present in high concentrations. Recent field results indicate that PVC casing can exert significant but unpredictable effects on trace level organic constituents compared with stainless steel and PTFE materials [*10*]. PVC should perform adequately for inorganic chemical constituent studies when organic constituents are not present in high concentrations.

These recommendations are preliminary since they are based largely on laboratory studies and only limited field experience. However, it may be several years before sufficient data can be collected to better estimate material performance. Field verification of the suitability of a well casing material for a specific situation may entail construction of "referee" wells of a more inert material in close proximity to both upgradient and downgradient sampling points. The added cost involved in this option relative to using an initially cheaper material is minimal compared with the cost of defending questionable monitoring data in the legal arena.

[2] Registered trademark of E. I. duPont de Nemours and Co.

## Sampling Mechanisms and Tubing

Five basic types of sampling devices are available for monitoring well situations: grab samplers (for example, bailers and thief samplers), gas-drive devices, positive-displacement pumps, suction, and gas-lift devices [*17,18*]. Field and laboratory studies [*11,14,19*] have shown that bailers and positive-displacement bladder pumps can yield representative samples for certain inorganic species. Gas-lift devices, on the other hand, will strip dissolved gases from the water and alter pH and redox conditions and thereby cause a change in species composition. Other studies have shown that bladder pumps provide high recoveries and consistent results for volatile organic samples [*20*]. Significant bias and imprecision have been observed with volatile organic samples obtained using gas-drive displacement pumps and suction pumps [*20,21*]. Based on considerations of sample sensitivity to mechanism effects, recommendations for sampling mechanisms are provided in Table 2.

The tubing that transports the water sample from the pumping device to the surface is also a potential source of error or bias. An evaluation was conducted of five tubing materials—PTFE, polyethylene (PE), polypropylene (PP), PVC, and silicone (SIL)—in a static sorption test to determine the sorptive bias for several volatile halocarbons (chloroform, trichloroethylene, trichloroethane, and tetrachloroethylene) frequently found in contaminated ground-water samples [*22*]. These experiments were conducted in closed loops of tubing which had been cleaned and allowed to air dry prior to being filled with spiked solutions of the halocarbons. Initial solution concentrations ranged from 100 to 400 μg/L total halocarbons. The study revealed that all of these materials sorbed some of the test compounds during short exposure periods. The relative sorptive affinity of the five materials followed the order TFE $<$ PP $<$ PE $<$ PVC $<$SIL as shown in Fig. 1. Rates of sorption and desorption were not significantly affected by background levels of organic carbon (simulated by polyethylene glycol) or a range of ionic strengths ($10^{-3}$ to $10^{-1}$ M $NaHCO_3$).

The impact of sorptive processes occurring in sampling tubing under actual field conditions is difficult to predict, but based on the results of this study some preliminary estimates can be made. Hypothetical percent sorptive losses of halocarbons predicted from the experimental rates of sorption are given in Table 3. The calculations assume that 15 m of tubing is used to convey ground water containing a 400-ppb mixture of halocarbons, at a flow rate of 100 mL/min, with an initial sorption rate of approximately 10 $\mu g/m^2/min$. For 6-mm inside diameter tubing, the percent losses range from 11 to 36%. The predicted losses are more dependent on the materials than on tubing diameter. At constant flow rates, however, the predicted losses increase with tubing of larger diameter. These losses for PTFE and PP tubing are in the range of controlled analytical bias for trace organic analytical methodologies. However, the errors involved with using PE, PVC or, SIL tubing are in excess of analytical error and may contribute to gross errors in trace organic determinations.

Most of the studies evaluating sampling mechanisms and materials were performed under controlled conditions which minimize the impact of mechanism or material errors. Nonetheless, it is clear that the potential for serious problems exists. The rates and extent of sorptive and volatilized losses of halocarbons suggest that well purging and preexposure of the tubing material will not completely alleviate the impact of these biases. A thorough evaluation of sampling mechanisms and tubing that would anticipate these kinds of problems should be part of the planning of all ground-water sampling protocols.

## Discussion

### *Controlling Sources of Error: Verification*

The preceding review indicated that there are many potential sources of error involved in ground-water sampling protocols. Well construction and development techniques can disrupt the subsurface

| Type of constituent | Example of constituent | | Positive-displacement bladder pumps | Thief, *in situ* or dual check valve bailers | Mechanical positive-displacement pumps | Gas-drive devices | Suction mechanisms |
|---|---|---|---|---|---|---|---|
| | | ↑ | ←------------------------- INCREASING RELIABILITY OF SAMPLING MECHANISMS | | | | |
| Volatile Organic Compounds<br><br>Organometallics | Chloroform<br>TOX<br>$CH_3Hg$ | INCREAS | Superior performance for most applications | May be adequate if well purging is assured | May be adequate if design and operation are controlled | Not recommended | Not recommended |
| Dissolved Gases<br><br>Well-Purging Parameters | $O_2$, $CO_2$<br><br>pH, $\Omega^{-1}$<br>Eh | ING SAM | Superior performance for most applications | May be adequate if well purging is assured | May be adequate if design and operation are controlled | Not recommended | Not recommended |
| Trace Inorganic Metal Species<br><br>Reduced Species | Fe, Cu<br><br>$NO_2^-$, $S^=$ | PLE SENS | Superior performance for most applications | May be adequate if well purging is assured | Adequate | May be adequate | May be adequate if materials are appropriate |
| Major Cations & Anions | $Na^+$, $K^+$, $Ca^{++}$<br>$Mg^{++}$<br><br>$Cl^-$, $SO_4^=$ | ITIVITY | Superior performance for most applications | Adequate<br><br>May be adequate if well purging is assured | Adequate | Adequate | Adequate |

TABLE 2—*Matrix of sensitive chemical constituents and various sampling mechanisms (from Ref 3).*

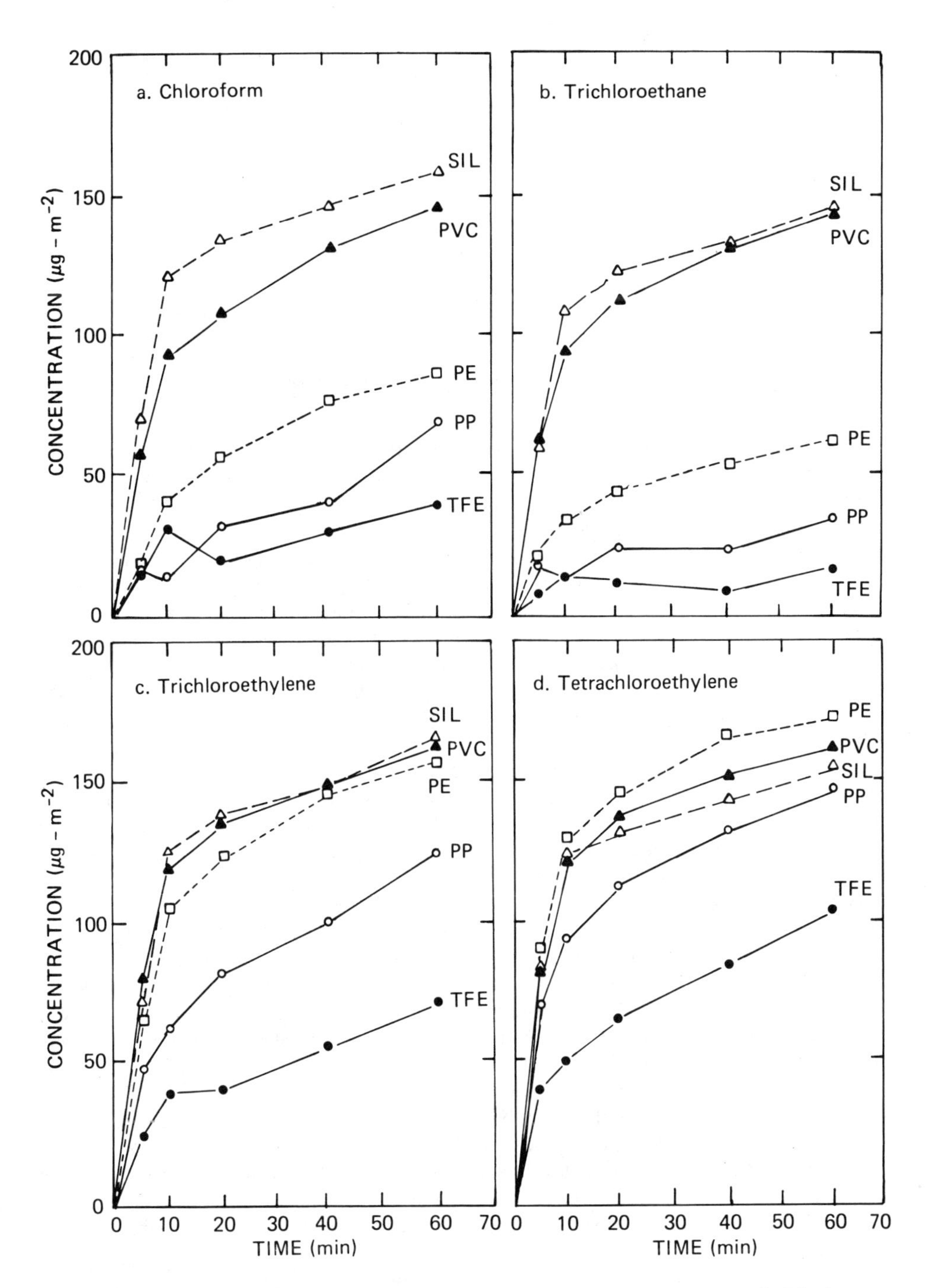


FIG. 1—*Concentration of sorbed chlorinated organics. The sorbed concentration* ($\mu g \cdot m^{-2}$ *of the four test compounds from distilled water solutions is shown as a function of time in exposure to tubing materials: (a) chloroform, (b) trichloroethane, (c) trichloroethylene, (d) tetrachloroethylene. Dissolved concentrations were initially between 90 and 120 ppb of each compound (from Ref* 22).

TABLE 3—*Predicted percent sorptive loss of chlorinated hydrocarbons due to tubing exposures*[a].

| Tubing Internal Diameter | % Loss | | | | |
|---|---|---|---|---|---|
| | PTFE | PP | PE | PVC | SIL |
| 6 mm (¼ in.) | 11 | 14 | 24 | 33 | 36 |
| | (9) | (14) | (17) | (40) | (36) |
| 9.5 mm (⅜ in.) | 16 | 22 | 36 | 50 | 56 |
| | (14) | (22) | (25) | (61) | (56) |
| 12.7 mm (½ in.) | 21 | 29 | 48 | 67 | 74 |
| | (18) | (29) | (33) | (81) | (74) |

[a] 400-ppb mixture of chloroform, trichloroethane, trichloroethylene, and tetrachloroethylene calculated on the basis of initial sorption rates on passage through 15 m of tubing at 100 mL·min$^{-1}$. Percent loss values are tabulated for the original solution (with) and without a 5-ppm organic carbon background.

environment and cause long-lasting effects on chemical results. A rigorous development procedure can sometimes minimize problems associated with poor hydraulic performance, turbidity, or minor contamination. However, one can rarely be sure when the problem is corrected because antecedent conditions in the subsurface environment frequently are unknown.

The issue of well casing material effects is a topic of continuing controversy. The lack of field data to support the "conservative approach" leaves room for many to argue the least-cost position of using less-expensive materials. Sampling mechanisms and associated tubing are other sources of potential bias. Loss or contamination of constituents from physical (volatilization, sorption/desorption) or chemical (precipitation, change in speciation) processes may approach the level of analytical error in uncontrolled sampling situations. All of these potential problems are real, but there are ways to minimize their impact.

The most effective way to minimize error in a ground-water monitoring situation is to document installation, development, purging, and sampling procedures and reevaluate the results as they become available. If the proper choices are made at the outset, the sources of serious error can be controlled. A summary of sources of error and their relative contribution to bias in selected chemical constituent determinations is given in Table 4. For example, drilling muds have been shown to have the potential to alter TOC results in water samples by a factor of 3 or more. The entry in the table suggests a gross positive error (+) which could be up to 300% higher than the true value. Other errors could yield a grossly high or low (±) value from the mean value. The table is not yet complete because much work remains to be done in evaluating the actual magnitude of the errors. However, improperly placed grouts or seals and inadequate purging of monitoring wells can contribute significantly to gross errors in chemical constituent determinations. Well casing effects are more subtle except where corrosive conditions attack ferrous materials and elevate dissolved iron levels. The negative bias caused by degassing or gas exchange due to the sampling mechanisms can effect both organic and inorganic chemical determinations. Finally, the potential sorptive loss effects of flexible sampling tubing exposures have been demonstrated to be more serious than analytical inaccuracies, which generally range from ±20% bias. As our understanding of the relative magnitudes of error sources improves, it may be possible to standardize sampling and analytical techniques for ground-water monitoring applications. For the time being, one should take a conservative approach and both document and take steps to minimize error whenever possible.

## Research and Routine Evaluation Needs

1. Installation, monitoring and sampling systems (for example, *in situ* monitors) which minimize disruption of the subsurface must be developed and validated.

TABLE 4—*Potential contributions of sampling methods and materials to error*[a] *in ground-water chemical results.*

| Parameter | Concentration, units | Drilling Muds | Grouts, Seals | Well Purging | Well Casing | Sampling Mechanism | Sampling Tubing | References |
|---|---|---|---|---|---|---|---|---|
| pH | 5–9 pH units | . . . | +, 4 to 5 units cement | ±, 0.1 to 4 units | . . . | gas lift +, 0.1 to 3 units | . . . | *10, 11, 14* |
| TOC | 0.5–25 $mg \cdot C \cdot L^{-1}$ | +, 300% | . . . | ±, 500% | ±, 200% | bailer +, 150% | . . . | Table 1, *10* |
| Fe(II) | 0.01–10 $mg \cdot L^{-1}$ | . . . | –,[b] 500% cement | –,[b] 1000% | +, 1000% iron, galvanized steel | gas lift –,[b] 500% | . . . | *2, 10, 11, 14* |
| Volatile | 0.5–15 $\mu g \cdot L^{-1}$ | . . . | . . . | ±, 10 to 100% | ±, 200% | . . . | . . . | *10* |
| Organic compounds | 80–8000 $\mu g \cdot L^{-1}$ | . . . | . . . | . . . | . . . | suction –,[b] 1 to 15% | –10 to 75% | *20, 21* |

[a] Bias values exceeding > ±100% denoted as gross errors (+ or –); other values expressed as percent of reported mean.
[b] No data available on the type and extent of error for this parameter.

2. Field evaluations of the seriousness of well purging and the effects of well casing materials on chemical constituent determinations need to be conducted.

3. Methods are needed to routinely verify and validate the field installation and operation of ground-water sampling systems.

4. Improved methods of sealing boreholes and verifying the integrity of the seals should be developed.

## Acknowledgment

The authors would like to thank their colleagues at the Illinois State Water Survey, Pamela Beavers in particular, for help in preparing the manuscript. Also, the support of the U.S. Environmental Protection Agency-R. S. Kerr Environmental Research Laboratory/Ada, Oklahoma and Environmental Monitoring Systems Laboratory/Las Vegas, Nevada is gratefully appreciated. The comments of the reviewers were very helpful in preparing the final manuscript.

## References

[1] Scalf, M. R., McNabb, J. F., Dunlap, W. J., Cosby, R. L., and Fryberger, J., *Manual of Ground-Water Quality Sampling Procedures,* National Water Well Association, Worthington, OH, 1981.

[2] Barcelona, M. J., Gibb, J. P., and Miller, R. A., "A Guide to the Selection of Materials for Monitoring Well Construction and Ground-Water Sampling," Illinois State Water Survey Contract Report No. 327, USEPA-RSKERL, EPA-600/52-84-024, U.S. Environmental Protection Agency, 1983.

[3] Barcelona, M. J., Gibb, J. P., Helfrich, J. A., and Garske, E. E., "Practical Guide for Ground-Water Sampling," Illinois State Water Survey Contract Report No. 374, USEPA-RSKERL under cooperative agreement CR-809966-01, U. S. Environmental Protection Agency, Ada, OK, Aug. 1985.

[4] *Ground Water and Wells,* E. E. Johnson, Inc., St. Paul, MN, 1966, p. 440.

[5] Wehrmann, H. A., *Ground Water Age,* April 1983, pp. 35–38.

[6] Yare, B. S., *Ground Water,* Vol. 13, No. 2, 1975, pp. 151–154.

[7] Brobst, R. B., "Effects of Two Selected Drilling Fluids on Ground Water Sample Chemistry," Monitoring Wells, Their Place in the Water Well Industry Educational Session, National Water Well Association, National Meeting and Exposition, Las Vegas, NV, Sept. 1984.

[8] Barcelona, M. J., *Ground Water,* Vol. 22, No. 1, 1984, pp. 18–24.

[9] Dunbar, D., Tuchfeld, H., Siegel, R., and Sterbenz, R., *Ground Water Monitoring Review,* Vol. 5, No. 2, 1985, pp. 70–74.

[10] Barcelona, M. J. and Helfrich, J. A., "Effects of Well Construction Materials on Ground Water Samples," *Environmental Science and Technology,* Vol. 20, No. 11, 1986, pp. 1179–1184.

[11] Schuller, R. M., Gibb, J. P., and Griffin, R. A., *Ground Water Monitoring Review,* Vol. 1, No. 1, 1981, p. 42.

[12] Fenn, D., Cocozza, E., Isbister, J., Braids, O., Yare, B., and Roux, P. "Procedures Manual for Ground Water Monitoring at Solid Waste Disposal Facilities," EPA/530/SW611, U.S. Environmental Protection Agency, Cincinnati, OH, 1977.

[13] Nacht, S. J., *Ground Water Monitoring Review,* Summer 1983, pp. 23–29.

[14] Gibb, J. P., Schuller, R. M., Griffin, R. A., "Procedures for the Collection of Representative Water Quality Data from Monitoring Wells," Cooperative Groundwater Report 7, Illinois State Water Survey and Illinois State Geological Survey, Champaign, IL, 1981.

[15] Grisak, G. E., Jackson, R. E., and Pickens, J. F., "Monitoring Groundwater Quality: The Technical Difficulties," *Water Resources Bulletin,* San Francisco, CA, June 12–14, 1978, pp. 210–232.

[16] Garske, E. E. and Schock, M. R., "An Inexpensive Flow-Through Cell and Measurement System for Monitoring Selected Chemical Parameters in Ground Water," *Ground Water Monitoring Review,* Vol. 6, No. 3, 1986, pp. 79–84.

[17] Nielsen, D. M. and Yeates, G. L., *Ground Water Monitoring Review,* Vol. 5, No. 2, 1985, pp. 83–99.

[18] *Ground Water Monitoring Review,* Vol. 5, No. 3, 1985, pp. 33–45.

[19] Stolzenberg, T. R. and Nichols, D. G., "Preliminary Results on Chemical Changes in Ground Water Samples due to Sampling Devices," Report to Electric Power Research Institute, Palo Alto, CA, EA-4118, by Residuals Management Technology, Inc., Madison, WI, June 1985.

[20] Barcelona, M. J., Helfrich, J. A., Garske, E. E., and Gibb, J. P., *Ground Water Monitoring Review,* Vol. 4, No. 2, 1984, pp. 32–41.
[21] Ho, J. S-Y, *Journal of the American Water Works Association,* Dec. 1983, pp. 583–586.
[22] Barcelona, M. J., Helfrich, J. A., Garske, E. E., *Analytical Chemistry,* Vol. 57, No. 2, 1985, p. 460 (correction: *Analytical Chemistry,* Vol. 57, No. 13, 1985, p. 2752).

*Andrew W. Panko*[1] *and Peter Barth*[2]

# Chemical Stability Prior to Ground-Water Sampling: A Review of Current Well Purging Methods

**REFERENCE:** Panko, A. W. and Barth, P., **"Chemical Stability Prior to Ground-Water Sampling: A Review of Current Well Purging Methods,"** *Ground-Water Contamination: Field Methods, ASTM STP 963,* A. G. Collins and A. I. Johnson, Eds., American Society for Testing and Materials, Philadelphia, 1988, pp. 232–239.

**ABSTRACT:** Sample preservation and analysis of ground waters follow rigid protocols governed by well-established quality assurance/quality control procedures. Well purging requirements are less rigorously defined, making chemical sampling critical in obtaining representative data.

Purging a well for chemical sampling causes local changes in water chemistry in terms of temperature, pH, solubility of gases, and redox potential. Subtle changes over a more widespread area may also occur. Before sampling a new or existing well, hydraulic and chemical stability should be established between the aquifer water and the water within the well. This becomes more important as the concentration of contaminants being evaluated becomes lower or nears the detection limit of the test. The actual physical process of retrieving representative water samples, therefore, is complicated by the chemical implications of short-term aquifer dewatering.

In developing new wells, stability must be verified so that all traces of drilling fluid (where it has been used) and drilling debris are removed from the well and only aquifer water is sampled. This is typically performed by making volume-by-volume analysis of pH, specific conductance, and temperature until successive samples show consistent values. Properly developed existing wells usually reach chemical stability relatively rapidly.

Each monitoring well or set of wells has its own unique hydraulic and chemical characteristics. All wells, after first being drilled, must be purged to determine when equilibrium is reached. It cannot be assumed that removal of a fixed number of casing volumes of water will guarantee stability. Once this stability has been achieved, further tests are still required. These tests will determine the volume of water which must be removed to obtain a representative formation water sample, for each individual well or group of wells, in the same aquifer or overburden situation. It is essential that the evaluation of the hydraulic character of a well be made in conjunction with water chemistry. This evaluation should be made in light of the level of analytical accuracy required, and prior to establishing a standard number of volumes to be removed for purging prior to sampling for any well.

**KEY WORDS:** ground water, sampling, water quality, purging, chemical analysis, observation wells, well completion, water removal

There is no specific standard method for preparing a new or existing monitoring well for water quality sampling. Present guidelines are largely derived from well development procedures used in the water well industry and laboratory quality assurance procedures used in environmental sampling. Over the past 15 years, advances in instrumentation have permitted the identification of minute changes in ground-water quality and improved the understanding of the impact of well development and water quality sampling on the collection of a "representative" sample.

[1] Project scientist, Acres International Ltd., P.O. Box 1001, Niagara Falls, ON L2E 6W1, Canada.

[2] Project manager, Acres International Corp., 140 John James Audubon Parkway, Amherst, NY 14228–1180; currently, senior projects manager, ICF-SRW Associates, Inc., Robinson Plaza II, Suite 200, Pittsburgh, PA 15205.

Site investigations and environmental monitoring associated with regulatory requirements emphasize "representative" aspects of water quality samples. In ground-water sampling, the achievement of chemical stability is critical to obtaining a "representative" sample. This requires a thorough knowledge of the hydraulic characteristics of a well and general quality of the water as identified by indicator parameters. For new wells, hydraulic and chemical data on an individual well are initiated with well development. Water quality sampling of existing wells requires the establishment and routine collection of a sufficiently large database so that variations in well characteristics and conditions affecting water quality are identified.

**Well Development**

Historically, well development was essentially defined as the hydraulic development of a well to produce particulate-free water. Johnson [*1*] defines well development as "those steps in completing a water well that aim to remove the finer material from the aquifer, thereby cleaning out, opening up, or enlarging passages so that water can enter the well more freely."

A benchmark for determining the adequacy of development, according to the U.S. Environmental Protection Agency (EPA) [*2*], is "largely a matter of experience and judgment, and as a general rule, if interrupted overpumping or rawhiding is used as a final method of development, the degree of development can be estimated from sand samples caught in an Imhoff cone."

These definitions were adequate for defining the development of potable water wells in high-yielding aquifers, but are not entirely satisfactory in the case of new monitoring wells. This is particularly true for wells in low-yielding cohesive soils. In this paper, well development is defined as the process by which a monitoring well is stimulated to

1. Enhance hydraulic communication with the geologic strata of concern.
2. Remove particulate matter and fluids (when used) remaining from well drilling and construction.
3. Stabilize chemical changes which may have occurred during well drilling and construction; in the case of low-permeability material, the sampling zone should be vertically isolated.

Well development can be quantified by the collection of both hydraulic and water quality data. For example, the construction of a monitoring well in rock requires the collection of data on the quantity and quality of water lost and recirculated during drilling. A measure of the quantity of water lost during drilling provides an indication of the volume of drilling fluid to be removed during well development. Where the chemical quality of drilling fluids differs from that of the formation water, chemical data can be used to identify when chemical stability is being approached. Recently, the EPA [*3*] has issued directives which discourage the use of drilling methods that introduce fluids and particulates into the formation under study.

Even augering of soils in the zone of saturation can increase the temperature and conductivity of the local ground water. Figure 1 shows plots of chemical changes in water quality for new wells during development and for existing wells during purging. The point at which temperature, conductivity, or pH become asymptotic with the *x*-axis indicates, in terms of well volume purges, the point at which the well was approaching chemical stability.

Figure 1*a* shows data for a new monitoring well installed in a silty-clay till to a depth of 5 m. Temperature equilibrates first, after six volumes, followed by conductivity at the seventh volume. Note that there is still some fluctuation of pH even after nine and ten volumes of water have been removed from the well.

Figure 1*b* shows the changes of pH and conductivity (temperature was not measured) during the development of a large-diameter (25-cm) domestic well drilled through 12 m of clay till and finished in 3 m of fractured dolomite. As shown, equilibrium of pH and specific conductance have been established by the time ten volumes of water are removed. Little or no change occurs after this, even after the removal of 90 volumes of water.

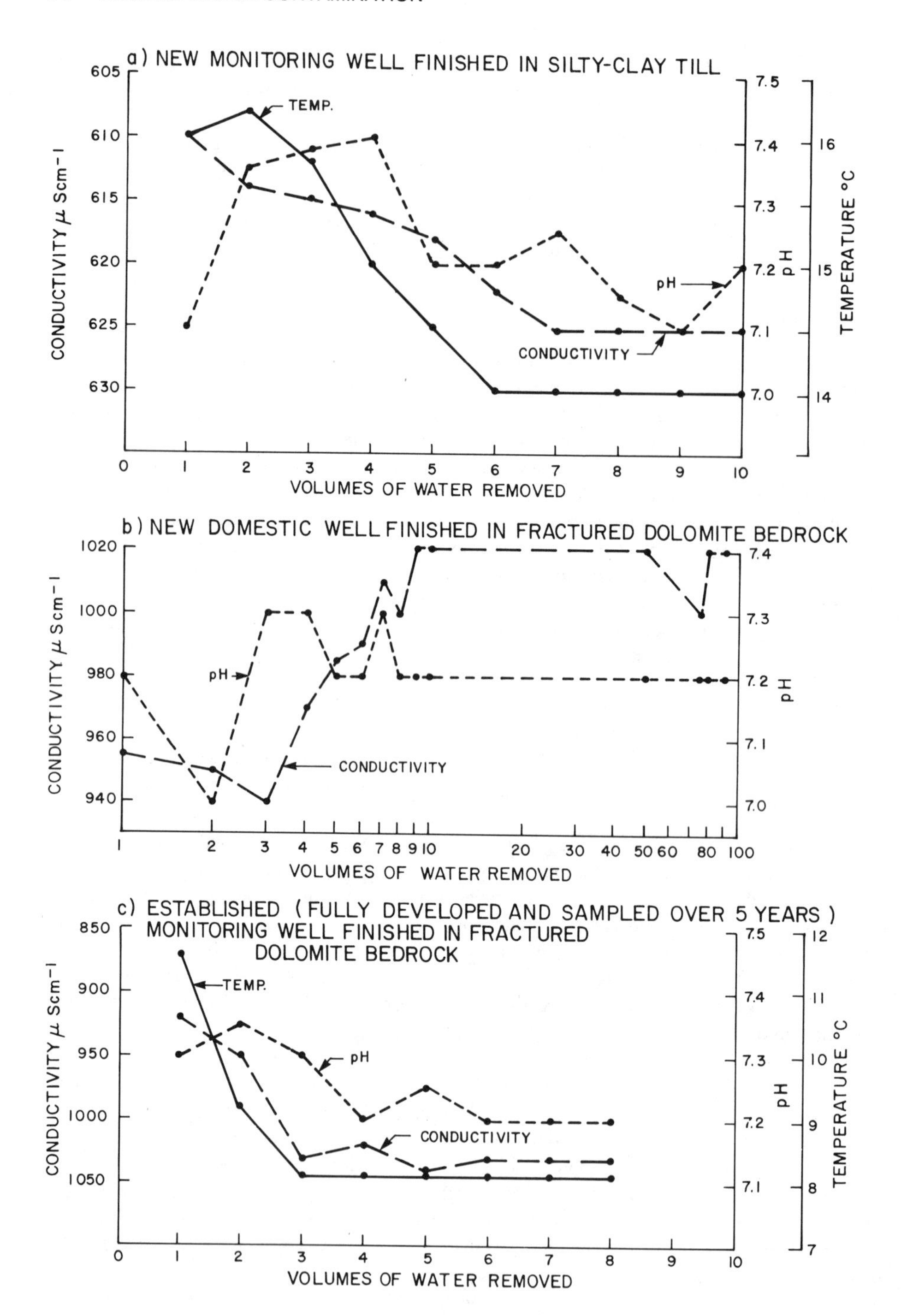


FIG. 1—*Volumes of water removed versus indicator parameters.*

Figure 1*c* shows that established wells can reach equilibrium faster than development wells. The well described here reached chemical equilibrium in terms of temperature and conductivity after just four volumes and in terms of pH after six volumes. This well was drilled through several metres of lacustrine sediments and till, and finished in about 2 m of fractured dolomite.

## Purging of Existing Wells

Existing monitoring wells must be purged to remove chemically stagnant water from the well casing and sand pack prior to collecting water quality samples. The quantity of water purged will vary depending on geologic characteristics and well construction.

Over the past several years, general guidelines were proposed regarding the volume of water which should be purged prior to the collection of samples. The EPA *Procedures Manual for Ground Water Monitoring at Solid Waste Disposal Facilities* [*4*] suggests that at least one casing volume be removed, but recommends that three to five are more preferred. Gibb et al. [*5*] states that "the general rule of thumb of pumping 4 to 6 well volumes will in most cases produce samples representative of aquifer water." The EPA [*6*] states that purging a minimum of four to ten bore volumes is a common procedure.

The use of these guidelines in the absence of adequate hydraulic and chemical data on an individual well may result in erratic data that do not accurately represent the actual conditions. To develop a representative data base, the characteristics of an individual well must be established prior to collection of the first sample and must be monitored throughout its useful life to detect hydraulic and chemical changes.

## Chemical Stability Within Wells

The chemical stability within a well is initially dependent on the characteristics of fluids within the strata of concern, and procedures used to install the well. These initial conditions are further influenced by water quality sampling activities. Chemical changes which can alter the chemical stability of casing water include:

(*a*) pH alteration resulting from carbon dioxide stripping, poor well cementing, and sample degassing due to pressure changes;

(*b*) oxidation of geologic media via dewatering and the use of airlift pumping systems;

(*c*) thermal changes resulting from conductive contact of ground water within the well casing and air; and

(*d*) sorption of trace metals and organics on particulate matter in poorly developed or poorly constructed wells.

A more detailed discussion of these factors is given by the EPA [*4*], Gibb et al. [*5*], Scalf [*6*], Hem [*7*], Wood [*8*], Freeze [*9*], and Back [*10*].

## Current Technology

The evaluation of chemical stability in ground water prior to sampling is a two-step process involving the analysis of hydraulic and chemical data on an individual well.

### *Hydrologic Tests*

Prior to water quality sampling, a short-duration pumping test should be performed in any new or existing monitoring well that has not been previously studied. The duration of this test will depend on the recharge capacity of the well and available pumping equipment. Gibb et al. [*5*] suggest that a brief two-to-three-hour test be performed and analyzed using a method outlined by Papadopolis and Cooper [*11*].

Hydrologic testing of a well which has been previously tested and whose hydraulic characteristics are known should include an evaluation of well recovery to 90% of the static water level. This may not be practical in wells that are pumped dry and when recovery may take several days. However, the percent recovery to the static water level should be recorded. Data on changes in well recovery over time are useful in evaluating changes in the chemical composition of the water and determining when a well should be abandoned or replaced. Supplemental pumping tests are suggested for reevaluating wells having non-seasonally-related changes in recovery. Well purging strategies are discussed in detail in a recent EPA report [*3*]. Examples are given on well purging strategies and pumping rates.

### *Chemical Tests*

Chemical analysis can be divided into tests for general water quality parameters, or "indicator" parameters, such as temperature, pH, specific conductance, dissolved oxygen (DO), oxidation-reduction potential (Eh), turbidity, total organic carbon (TOC), hydrogen sulfide ($H_2S$), and specific water quality parameters. Specific parameters are defined as those chemical parameters directly related to a target element source or other particular concerns.

Although it is important to choose the correct parameters to assure that chemical stability has been achieved prior to sampling, it is not always practical economically or logistically. Ideally, it is necessary to analyze for the specific parameter of concern. For example, a study monitoring a ground-water plume of sodium chloride emanating from a road salt storage facility would likely encompass the use of the easily measured specific parameters of chloride or specific conductance or both. Both of these parameters are readily measured in the field by portable instrumentation. On the other hand, in a study monitoring polynuclear aromatics (PNAs) or other organics in ground water, it is impractical to measure PNAs during well purging. This is due to the inherent difficulty of analyzing organics outside the laboratory. Therefore, in the case of organics or inorganics not easily analyzed in the field, the more easily measured indicator parameters are substituted.

### *Indicator Parameters*

The most common indicator parameters used to evaluate the gross characteristics of water quality during well development and purging include the following:

*pH*—This key parameter measures the solution's hydrogen ion concentration and is reported as the negative log of the concentration. The test is rapid, reliable, and reproducible and has no shortcomings as long as the stagnant water in the well casing has a pH sufficiently different from the surrounding formation.

*Conductivity*—also known as specific conductance at 25°C, is a measure of a water's ability to conduct a charge and typically reflects the concentration of ionic species present. Like pH, conductivity is easily measured and gives reliable and reproducible results.

*Temperature*—This is the easiest parameter to measure accurately and can be the most useful, especially when used in conjunction with pH and conductivity. Temperature can be measured using a thermocouple or other downhole or flowcell-coupled probe, or by simple thermometer if the well is evacuated on a volume-by-volume basis.

The above three tests provide an initial indication of the quality of the water being evacuated from a well during development or purging. Depending on site conditions, these key indicator parameters can be supplemented by the following:

*Eh*—The test for Eh is a measure of the oxidation-reduction potential of a water and is best performed downhole or in a flow-through cell to avoid contact with the atmosphere. Because of the difficulty in measurement and low sensitivity of the test, fairly large differences in Eh between formation water and drilling fluid or casing water must be present.

*DO*—The test for dissolved oxygen can serve as a backup for Eh for identifying an oxidizing environment. DO measurement is generally more reliable and reproducible than Eh.

$H_2S$—This parameter can serve as a backup for Eh measurement for identifying a reducing environment.

*Turbidity*—Because of variable pump height and agitation during well development and purging, the use of turbidity as an indicator of well development is limited. For specific applications, however, it can provide an indication of the quantity of particulate matter in a sample and may indicate qualitative changes in development.

### *Specific Parameters*

Specific parameters are those directly related to contamination or supply problems and include a wide range of inorganic and organic chemicals. Specific parameters are usually more difficult to measure in the field using EPA-approved laboratory procedures. For this reason, it is easier to establish that a sample is representative of the formation being tested by using the easily measured indicator parameters.

It is difficult to couple sophisticated instrumentation to pump discharge lines. A volume-by-volume subsample approach can be used for atomic absorption, auto-analyzer, or ion chromatography where little or no sample preparation is necessary. Organics processed by gas or liquid chromotagraphy introduce additional problems associated with complex sample preparation.

Certain colorimetric tests can also be readily performed on a volume-by-volume subsample basis. Although many are not approved by the government agencies for official reporting, they can provide useful data on the completeness of well development and purging. Normally, field tests would be followed by sampling and analysis in a laboratory.

### *Equipment and Procedures*

Of the indicator parameters, temperature, pH, conductivity, Eh, and DO are easily measured on a volume-by-volume or flow-through basis.

Temperature may equilibrate, becoming warmer or cooler, depending on weather, well depth, and casing materials. Steel casing, compared with polyvinyl chloride (PVC), conducts heat relatively well. Thus on a hot day, heat can be conducted to casing water and surrounding shallow ground water along the casing, and less so by the air heating up within a sealed well casing. The opposite effect is true on very cold days, to the extent that water may freeze within the upper portions of the casing.

An in-line thermometer or thermocouple in a flow-through cell is the ideal means of relating temperature to volume of water removed. Alternatively, if the well is pumped one volume at a time, the temperature of a subsample of each volume purged must be measured immediately. Generally, three successive volumes showing the same temperature provides a reasonable degree of assuredness that formation water is being sampled.

pH and specific conductance can both be measured similarly to temperature, in a flow-through cell with calibrated electrodes, or on a volume-by-volume basis. It is desirable that all three parameters be measured.

Eh and DO require somewhat more careful sampling procedures. It is not practical to use a bailer or vacuum pump to sample since bailing introduces air into the sample and vacuum pumping may promote degassing of the water. If it is not possible to measure Eh and DO in situ down the hole, a submersible or peristaltic pump should be used to sample the well during pumpings. In this case, the Eh electrode and DO probe must be inserted in a flow-through cell.

Ideally, a pump connected to a wellhead flow-through cell containing the necessary probes, in conjunction with continuous monitoring instrumentation and a flowmeter, can be used for fast recharging wells. Less exotic but more practical when testing slowly recharging shallow wells involves the withdrawal of water from the well on a volume-by-volume basis, retaining for analysis the first portion of each volume.

The yield of the well will best define how the well should be purged. Where possible, a well should be evacuated completely. A sample of the first volume evacuated is retained, the well allowed to recover to some reasonable percentage of its static water level (ideally 90-plus %), and the evacuation and sampling process repeated. When the value of the parameter of interest stabilizes (within the limits of the test's accuracy) and the values obtained on at least three successive tests on three successive casing volumes are the same, the well is likely sufficiently purged and sampling can commence.

For high-yielding wells that cannot be evacuated to dryness, the EPA [*4*] does not recommend bailing without prepumping. Instead, two procedures are suggested, both of which are readily adaptable to chemical monitoring during the purging procedure. The hardware of choice here is a flow-through cell placed in line at the wellhead in conjunction with a flowmeter. The inlet line must be placed just below the surface of the water and the water pumped out at a rate equal to the well recharge rate. This can be accomplished by a water level indicator and variable-speed pump. As the water in the casing is removed, formation water will enter the screen and move up to the surface where it is being removed from the system. When no more change is noted in the values of the test parameters, it is likely that only true formation of aquifer water is flowing through the monitoring cell. At this time a sample is taken directly from the cell discharge, or, if required, a bailer or other sampling tool is used to obtain a sample.

As an alternative method, the EPA [*4*] suggests placing the inlet line of the sampling pump near the bottom of the screened portion. The well is then pumped at its own recovery rate, with the water entering the screen and flowing directly to and through the inlet tube. One disadvantage of this method is that small fluctuations in recovery rate or pumping may allow stagnant water standing in the casing above the screen—especially in wells having relatively short screened sections—to be drawn to the inlet tube. This instantly contaminates the formation water. Further, this method does not adequately address problems associated with contaminants of density different from that of water which could float on top of the water in the casing or sink to the bottom of the well.

Once the volume of water to be purged has been established, testing for indicator parameters should be repeated for that well for each subsequent sampling.

When the volume of water required to purge the well has been established, it is important not to exceed this amount. Overpumping a well can result in the introduction of water from other sources that could dilute, concentrate, or otherwise provide erroneous data. Overpumping will not produce a water sample representative of the formation that contains the screened interval of the well.

## Conclusions

The achievement of chemical stability in monitoring wells prior to water quality sampling requires a thorough knowledge of the hydraulic and chemical characteristics of any wells designated for sampling. Data required to understand the peculiarities of individual wells should be obtained during well development activities, or the first sampling period for existing wells.

The suggested procedure involves performing a two-to-three-hour pumping test on the well while simultaneously analyzing for indicator parameters and specific parameters. As a minimum, data on temperature, pH, and specific conductance should be obtained. Other indicator parameters, such as Eh, DO, $H_2S$, and turbidity, may enhance the database for wells in which fluids are introduced during installation, or which exhibit erratic behavior. Analysis of specific parameters provides information on the concentration variability during sampling.

A wide variety of portable analytical equipment is available consisting of direct-reading instrumentation for temperature, pH, specific conductance, Eh, and DO; spectrophotometers for the analysis of metals and other inorganic parameters; and ion, gas, and liquid chromatographs for analyzing ionic and organic parameters. This type of equipment allows site personnel to collect quantitative data at the site prior to initiating a more formalized water quality monitoring program. In some instances, specific parameters analyzed on site may be used as part of a quality assurance program.

The objective of any water quality sampling program is to quantify chemical changes that occur during well development, purging, and sampling, and to avoid dependence on qualitative guidelines and rules of thumb.

*Acknowledgment*

This paper was made possible through the funding and support of Acres International Ltd. The authors thank the following for their contributions in the field, laboratory, and office: Bob Adams, Dan Barton, Tony DiFruscio, Pam Penner, Leslie Smythe, and Tony Tawil. We also thank Mr. and Mrs. John Warner for providing access to obtain samples for the domestic well portion of our experiments.

## References

[1] *Ground Water and Wells,* Johnson Division, UOP Inc., St. Paul, MN, 1972.

[2] "Manual of Water Well Construction Practices," U.S. Environmental Protection Agency, Office of Water Supply, EPA 570/9-75-001, Washington, DC, 1975.

[3] "Practical Guide for Ground Water Sampling," U.S. Environmental Protection Agency, EPA/600/2-85/104, Washington, DC, 1985.

[4] "Procedures Manual for Ground Water Monitoring at Solid Waste Disposal Facilities," U.S. Environmental Protection Agency, SW-611, Washington, DC, 1980.

[5] Gibb, J. D., Schuller, R. M., and Griffon, R. A., "Procedures for the Collection of Representative Water Quality Data from Monitoring Wells," Illinois State Water Survey and Geological Survey Cooperative Ground Water Report No. 7, 1981.

[6] Scalf, M. R., McNabb, J. F., Donlap, W. I., Cosby, R. L., and Fryberger, J., "Manual Ground Water Quality Sampling Procedures," U.S. Environmental Protection Agency, Robert S. Kerr Environmental Research Laboratory, Ada, OK, 1981.

[7] Hem, J. D., "Study and Interpretation of the Chemical Characteristics of Natural Water," U.S. Geological Survey, Water Supply Paper 2254, U.S. Government Printing Office, Washington, DC, 1985.

[8] Wood, W. W., "Guidelines for Collection and Field Analysis of Ground Water Samples for Selected Unstable Constituents," *Techniques of Water Resources Investigations by the United States Geological Survey,* Book 1, Chapter 2, U.S. Government Printing Press, Washington, DC, 1976.

[9] Freeze, R. A. and Cherry, J. A., *Ground Water,* Prentice-Hall, Englewood Cliffs, NJ, 1979.

[10] Back, W. and Freeze, R. A., *Chemical Hydrogeology.* Vol. 73, Hutchinson Ross Publishing Co., distributed through Van Nostrand Rheinhold, New York, 1983.

[11] Papadopolis, I. S. and Cooper, H. H., "Drawdown in a Well of Large Diameter," *Water Resources Research,* Vol. 3, No. 1, 1967, pp. 241–244.

*Jay Unwin*[1] *and Van Maltby*

# Investigations of Techniques for Purging Ground-Water Monitoring Wells and Sampling Ground Water for Volatile Organic Compounds

**REFERENCE:** Unwin, J. and Maltby, V., "**Investigations of Techniques for Purging Ground-Water Monitoring Wells and Sampling Ground Water for Volatile Organic Compounds,**" *Ground-Water Contamination: Field Methods, ASTM STP 963,* A. G. Collins and A. I. Johnson, Eds., American Society for Testing and Materials, Philadelphia, 1988, pp. 240–252.

**ABSTRACT:** An experiment designed to detect and quantify contamination of simulated ground-water samples with stagnant water in a monitoring well found that an average of about 2 to 4% of the water pumped from locations above the screen and an average of about 1% of the water pumped from within the screen of the monitoring well came from the stagnant water above the pump inlet even when effects of drawdown were precluded. All runs showed peaks of contamination up to 8% or higher. The amount of contamination is highly variable over a short time frame. An experiment to identify the effects of the sampling device and the tendency of a compound to evaporate from water solution on the loss of the compound during the ground-water sampling process revealed that the potential for losses rises as the tendency to evaporate from solution increases, and that up to 30% loss can occur under certain circumstances. Less than 10% loss is more typical with the techniques normally used for such sampling. Little indication of differences between commonly used sampling devices was found.

**KEY WORDS:** ground water, monitoring well, sampling, purging, tracer study, volatile organic compounds (VOC), pump, bailer

Greatly expanded ground-water monitoring activity in the past few years, due to increased regulatory and research interest in protecting the ground-water resource, has elevated the importance of a number of unresolved technical issues. Questions about monitoring system design, materials of construction, sampling devices and practices, and data analysis remain to be answered.

## Monitoring Well Purging

One question which any field investigator faces almost immediately upon deciding to sample ground water is what to do about the stagnant water within a monitoring well above the screened section of the well. This water has been isolated from the aquifer at least since the last time the well was sampled. During that time, the chemical quality of the water may have been altered by direct introduction of foreign material into the well or by interactions with the well casing or at the interface with the atmosphere. Even without such alterations, the stagnant water would not reflect any changes in the ground-water quality that may have occurred since the last time the well

[1] Research engineer and research assistant, respectfully, National Council of the Paper Industry for Air and Stream Improvement, Inc. (NCASI), Western Michigan University, Kalamazoo, MI 49008.

was sampled. Because the investigator cannot be certain which, if any, of these influences have occurred or whether inclusion of some of the stagnant water in a sample from the well would significantly change the conclusions that might be drawn from the data, the safe thing to do is to prevent or minimize such inclusion.

One common technique used to deal with this issue is to pump sufficient water from the well to ensure adequate purging of the stagnant water from the well prior to sampling. The volume of water to be removed is usually based on the volume of the well. Anywhere from 1 to 20 times this volume of water is usually removed in order to purge the well. Questions about this technique relating to placement of the pump inlet and mixing of the water column due to movement of a bailer through the water have not been thoroughly addressed in the literature. Another common approach to the problem recommended by the U.S. Geological Survey [*1*] is to pump a well until parameters such as pH, specific conductance, and temperature stabilize. This approach ignores the possibility that a near constant contribution of stagnant water may be established. It also fails to address contributions of stagnant water that are too small to noticeably affect the measured parameters, but which may significantly alter the outcome of a trace organics analysis. Gibb [*2*] suggests using knowledge of the time-drawdown characteristics of a well to determine the fraction of water being pumped at any time that is attributable to drawdown. This approach fails to account for contributions of stagnant water from any source other than drawdown.

Previous work in this area by Unwin [*3,4*] has shown the general behavior of a tracer within the well bore while purging the well in various ways. It shows that, even in the absence of drawdown, potentially significant quantities of stagnant water can enter the pump inlet from above. This fact has implications particularly for the purging technique described by Gibb which does not consider any other source for stagnant water except from drawdown. The experiment described in this paper was designed to take a more detailed and quantitative look at the hydraulic behavior of a monitoring well during purging by examining truly trace concentrations while better defining the fraction of stagnant water that is drawn into the pump inlet as a function of bore volumes pumped (or time) and inlet placement.

## Volatile Organic Compound Sampling

One of the more commonly raised questions regarding ground-water sample integrity concerns the loss of volatile organic compounds (VOC) during sampling. Many sampling devices submit the water to reduced pressure. Others expose the water to a gas/water interface. Still others submit the water to excessive turbulence. Finally, all sampling methods cause the water to come into contact with the materials of construction of the sampling device. These are all factors with the potential to cause stripping of volatile organic compounds from a sample.

The most comprehensive investigations of this question published to date have been by Ho [*5*], Barcelona [*6*], and Unwin [*7,8*]. A summary of the work conducted by the author is presented in this paper.

## Well Purge Experimental Procedure

### *General*

This experiment consisted of a series of runs in which the stagnant water (above the screen) in a physical model of a typical ground-water monitoring well was spiked with a tracer compound. The well was then sampled with the pump inlet in various positions while the tracer concentration in the effluent was constantly monitored to determine what fraction of the water collected was from the stagnant zone above the pump inlet. The general experimental setup is illustrated in Fig. 1.

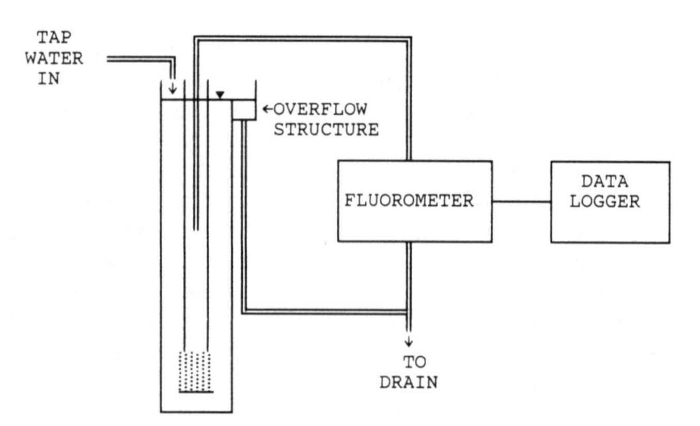


FIG. 1—*Experimental setup for the well purging experiment.*

The three pump inlet positions used were: midway between the water surface and the top of the screen (mid-depth), three quarters of the distance between the water surface and the top of the screen (¾ depth), and 5 cm (2 in.) below the top of the screen (midscreen). The two positions above the screen were chosen in order to investigate the effect of varying the position of the pump inlet within the stagnant zone. The position within the screened section was used in order to investigate whether sampling from within the screened section is a practical way to avoid contamination of samples with stagnant water.

Most researchers agree that drawdown is a major factor in causing stagnant water to enter a pump inlet. The objective of this research was to examine factors other than drawdown that could cause stagnant water to enter the pump inlet. Therefore, any effects of drawdown were eliminated in this experiment by conducting all runs at constant head. Thus, the behavior observed in this experiment is what might occur in the absence of significant drawdown, or after drawdown was essentially complete for a particular well/pumping rate. As one example of the applicability of this approach, consider the previously mentioned purging method of Gibb [*2*]. This method involves using knowledge of the drawdown characteristics of a well in combination with certain equations to calculate at what point in the purging process the contribution of stagnant water due to drawdown has reached an acceptably low level (for example, 5%). This point will usually occur when drawdown is essentially complete. It is therefore important to understand any other processes which may operate at that point to cause stagnant water to be collected since the equations consider no other factors except drawdown.

### *Equipment and Materials*

Tap water from the Kalamazoo, Michigan ground-water-based municipal system was used in all runs. The tracer used was a fluorescent dye, Rhodamine WT, at an initial concentration in the stagnant zone of approximately 5 ppb.

The dye concentration was measured with a Turner Model 111 fluorometer equipped with No. 546 primary and No. 590 secondary filters and a flow-through door (Turner Cat. No. 110-880) for continuous measurement of a flowing stream. The system was calibrated and found linear over a range of 0.1 to 5.0 ppb. Calibration and all measurements were done at a slit width of $\times 30$.

The 0 to 10-mV output signal from the fluorometer was acquired by a data logging system consisting of a Compaq portable personal computer equipped with a Data Translation Model 2801 data acquisition and control adapter board and appropriate software.

The model well is illustrated in Fig. 2. It was built from 5-cm (2-in.) inside diameter drawn

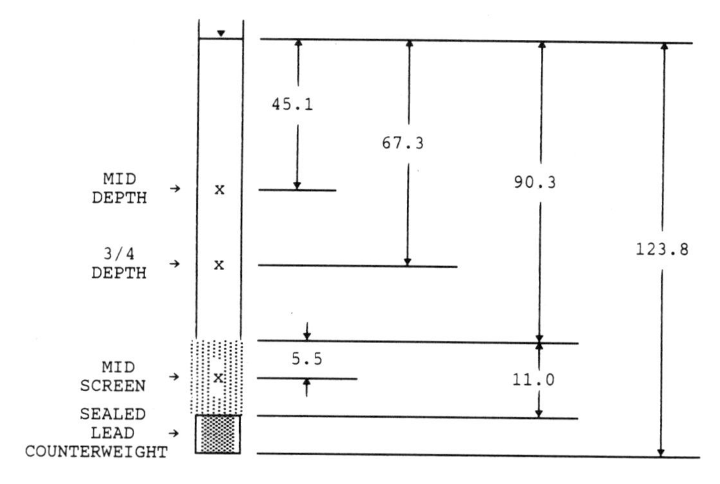


FIG. 2—*Schematic of the model well (dimensions in centimetres, not to scale).*

acrylic tubing. The screened section was perforated by one hundred and twenty-six 6-mm (¼ in.) and one hundred and twelve 3-mm (⅛ in.) evenly spaced holes to give about 25% open area. The constant-head tank in which the well was immersed was constructed from 16.5-cm (6½-in.) inside diameter cast acrylic tubing, and it incorporated an overflow structure at the top to facilitate maintenance of a constant water level.

Water was pumped from the well to the fluorometer through a 2.7-m (9 ft) length of 6-mm (¼-in.) inside diameter polyethylene tubing by a peristaltic pump using a 30-cm (12 in.) length of medical-grade silicon rubber tubing in the pump head.

## *Detailed Procedure*

A run was initiated by installing an inflatable rubber packer in the well just above the top of the screen and removing all the water above the packer. This water was then replaced with the tracer solution to a level slightly below the static water level so that the volume displaced by the packer would be replaced by fresh water moving into the screened section rather than by spiked water moving down into the screened section when the packer was removed. The packer was then slowly deflated and gently removed. The pump inlet was then carefully placed at the chosen depth for the run. Pumping at a rate of 1.0 L/min (0.26 gpm) and data acquisition at a sampling rate of 15 per minute commenced simultaneously.

At the end of a run, the inlet line was raised to pump from slightly below the water surface until the measured dye concentration in the effluent had returned to zero. At this point another run could be started.

## *Calculations*

For the runs where it is applicable, the number of bore volumes pumped is based on the volume below the pump inlet (that is, the end of the sample line) and above the top of the well screen. Because the measured dye concentration in fresh tap water was zero for all runs, the fraction of the pumped volume that came from the stagnant zone above the pump inlet is simply the ratio of the measured concentration in the pumped water to the initial concentration in the stagnant zone. For the two runs in which the inlet was located above the screen, the initial concentration was

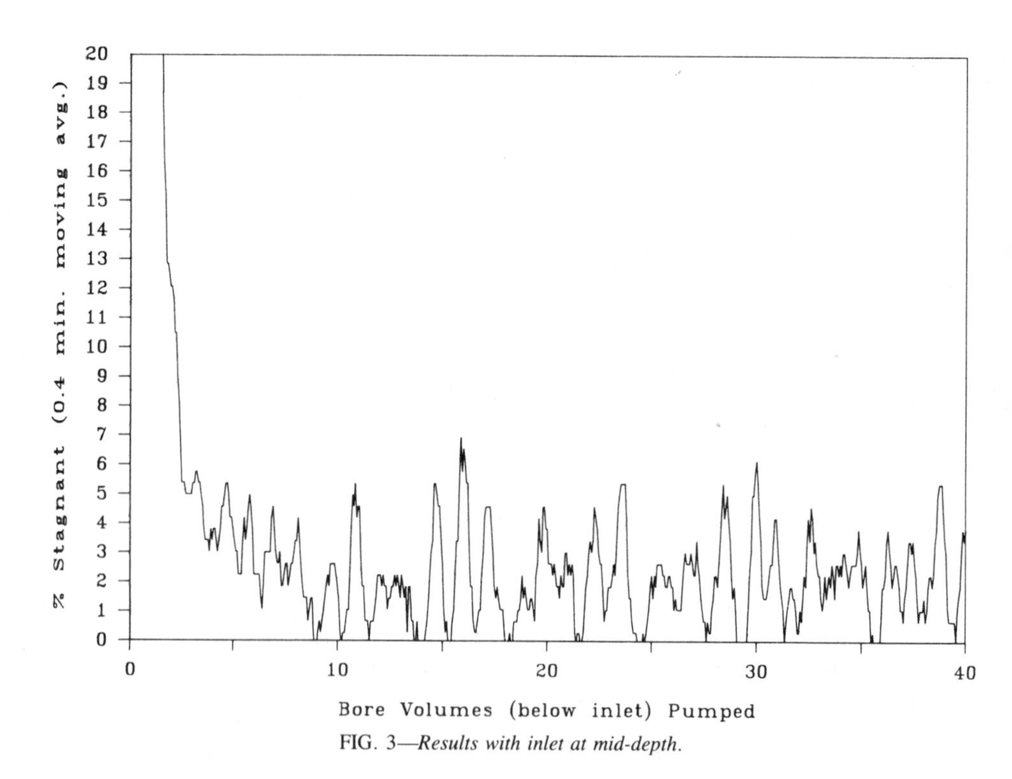


FIG. 3—*Results with inlet at mid-depth.*

taken as the peak value recorded at the start of the run when pure tracer solution would have been pumped. For the run in which the inlet was located within the screen, the initial concentration was taken as 5.0 ppb.

*Well Purge Results*

Figures 3 and 4 present, graphically, the results of the runs which involved pumping from mid-depth and ¾ depth, respectively, within the stagnant zone. The two figures are qualitatively similar in that they both show a drop in the initial peak at 3 to 5 bore volumes followed by periods of relatively high variability in the input from the stagnant zone that lasts for the duration of the run.

There are two quantitative differences between the results from the two runs that bear further scrutiny. First, the frequency of the oscillation in the percent stagnant water when pumping from ¾ depth seems to be less than that when pumping from mid-depth. This effect disappears if the data are plotted over a time domain rather than a bore volume domain. Figure 3 is already essentially plotted over the time domain since for that inlet position the pump rate was about one bore volume per/minute. In Fig. 5, the data for the run at ¾ depth are plotted over the time domain at the same scale as in Fig. 3 in order to facilitate visual comparison of the plots. The fact that the frequencies of the oscillations become more comparable when plotted over the time domain indicates that the phenomenon which accounts for the oscillations is independent of the inlet placement. Earlier experiments at a higher, more visible, dye concentration provided evidence that the phenomenon involved is related to turbulence and subsequent mixing around the pump inlet.

The other notable quantitative difference between the two runs is that the steady-state input for the run at ¾ depth seems to be higher than that for the run at mid-depth. The average values for

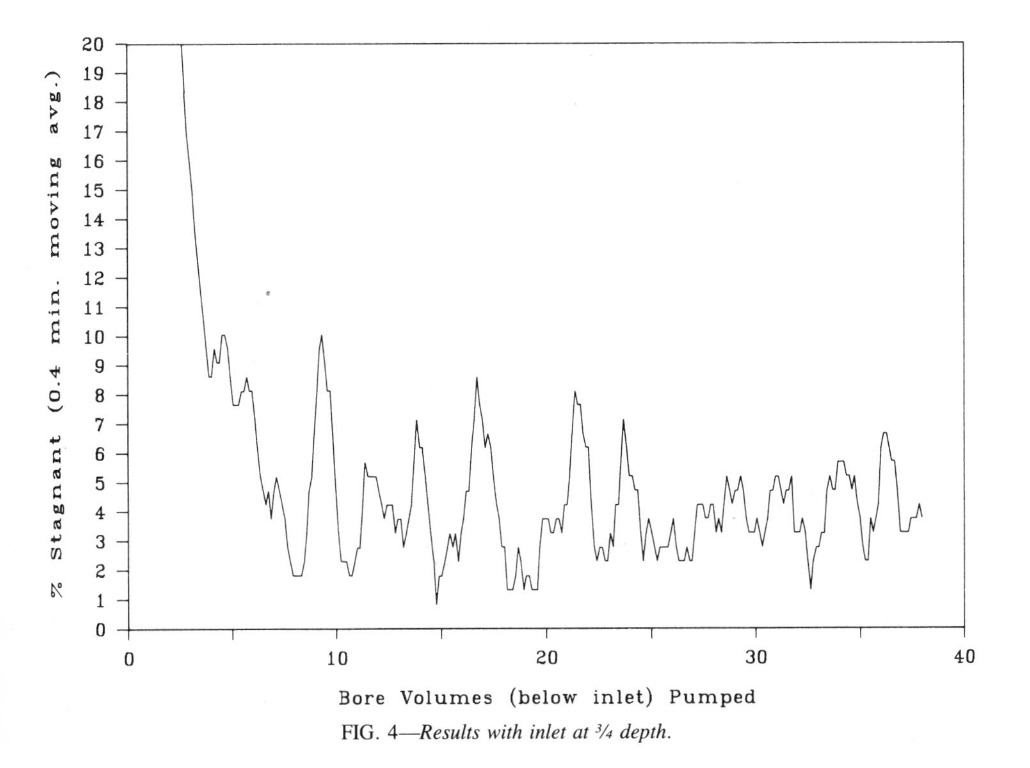


FIG. 4—*Results with inlet at ¾ depth.*

percent stagnant water over the range of 10 to 30 bore volumes are 2.0 for the run at mid-depth and 3.8 for the run at ¾ depth.

The results from the run involving pumping from within the screened section are presented in Fig. 6. This plot shows that the stagnant water was not excluded by placing the pump inlet within the screened section. Indeed, the peak contributions of stagnant water experienced while pumping from within the screen are comparable to those observed while pumping from the positions within the stagnant zone. The mean contribution was, however, lower. Over the time from 0 to 46 min the average percent stagnant water while pumping from within the screen was 0.95.

## VOC Sampling Experimental Procedure

### *General*

This experiment simulated sampling ground water for five different volatile organic compounds using four different sampling devices. The experiment was designed to provide information about the effects of two variables, the type of sampling device and the tendency of the compound to evaporate from water solution, on the ability to obtain a representative VOC sample from a ground-water monitoring well.

### *Equipment and Materials*

The volatile organic compounds used in this experiment are summarized in Table 1. These compounds were chosen primarily because of the wide range they cover in their tendency to evaporate from a solution with water. One measure of this tendency is the Henry's Law constant,

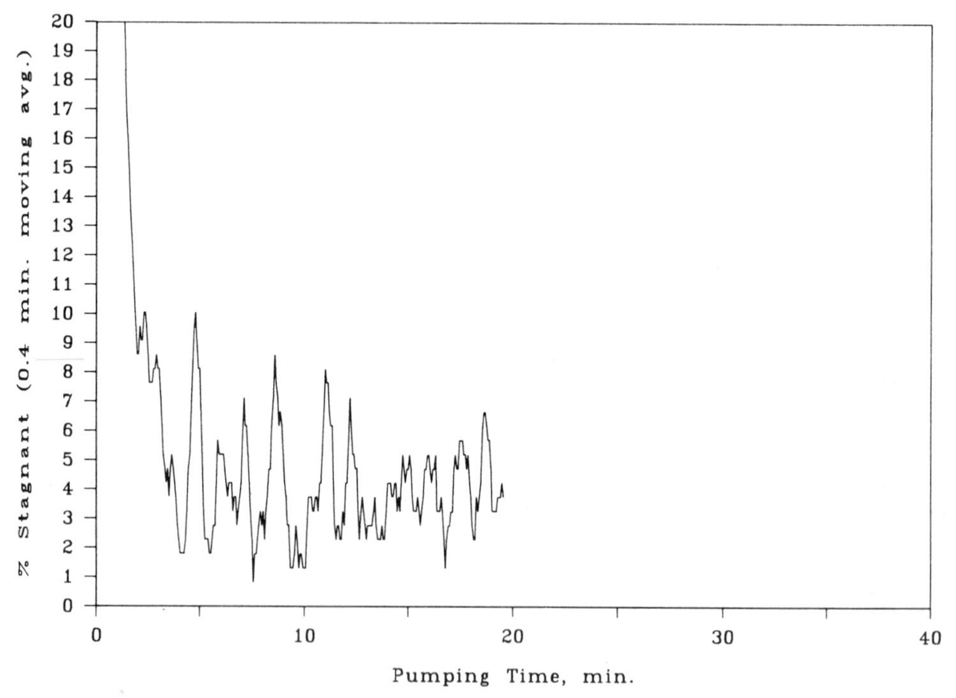


FIG. 5—*Results with inlet at 3/4 depth plotted against time.*

$H$, shown for each compound in Table 1. This is an appropriate measure for these purposes because it is essentially the ratio of the water solubility and the vapor pressure of the compound, so the two major factors that could affect the compound's tendency to either leave or stay in solution are included in the constant. The values for $H$ in the table were derived from data published by the Environmental Protection Agency [*9*]. For the units assigned to $H$ in the table, a higher value of $H$ indicates a greater tendency for the compound to evaporate from the solution. Thus, diethylether has the least tendency to evaporate from water solution and tetrachloroethylene has the greatest tendency to do so.

Table 2 summarizes the sampling devices used. The evacuated flask sampler was constructed from a 2000-mL (0.5 gal) glass sidearm Erlenmeyer flask with a single-hole rubber stopper through which a glass tube extended to within 0.5 cm (0.2 in.) of the bottom of the flask. Vacuum was applied to the sidearm (about 640-mm (25 in.) Hg in this experiment) to cause the sample to rise within the line attached to the glass tube and thus be deposited in the flask. Because it subjects the sample to moderate turbulence, atmospheric contact, and reduced pressure, this sampling technique was included to represent a "worst case" method. The bailer was homemade entirely from PTFE, a fluorocarbon polymer. It was 61 cm (24 in.) in length with a 2.5-cm (1.0 in.) inner diameter. The gas squeeze pump, which was also homemade, was designed around the principle of pumping water by squeezing a bag or diaphragm with gas pressure to displace the water it contains upward through the sample line. A number of commercially available ground-water samplers use this principle. Design and performance details for the particular squeeze pump used in this experiment have been published elsewhere [*4*]. The Johnson-Keck SP81 pump was used as purchased after cleaning according to the manufacturer's suggestion. This pump was the model that was available in early 1981. The VOC grab sampler was a homemade, all-aluminum device

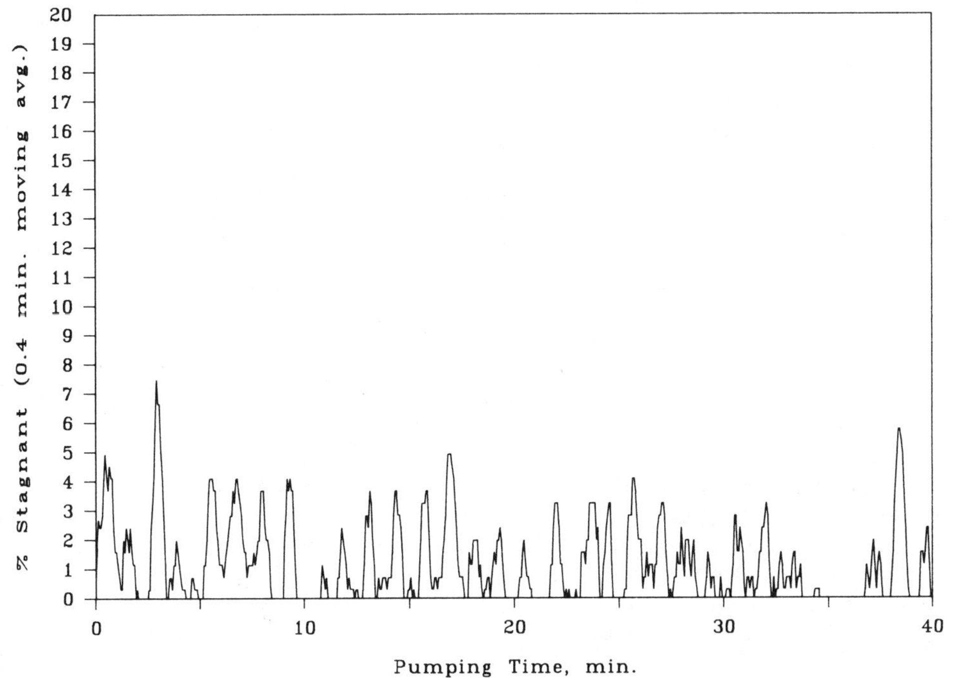


FIG. 6—*Results with inlet at midscreen.*

designed to allow a sample vial to be uncapped and filled while submerged. This method was used to take reference samples against which samples from the other devices could be compared.

Figure 7 depicts the simulated monitoring well setup from which the samples were withdrawn. The reservoir was a thoroughly detergent and steam cleaned painted steel 210-L (55 gal) drum. The well casings (two were used in the actual setup), which extended to the floating cover but not into the water, were constructed from 5-cm (2-in.) inside diameter polyvinyl chloride ID (PVC) pipe. These casings shielded sample lines from the sun, and the bailer from the wind. The floating cover was a 15-cm (5.9 in.) thick layer of washed and rinsed plastic foam packing material. The three sampling devices that required sample line used equal lengths of cleaned and rinsed FEP (a fluorocarbon polymer) lined 6-mm (¼-in.) inside diameter polyethylene tubing. The vertical rise from the water surface to the sample collection point was about 7 m (23 ft).

TABLE 1—*Compounds used in the VOC sampling experiment.*

| Compound | Formula | $H$ (@ 25°C), atm m³/mol |
|---|---|---|
| Diethylether | $C_2H_5OC_2H_5$ | 0.000906 |
| Chloroform | $CHCl_3$ | 0.00339 |
| Toluene | $C_6H_5CH_3$ | 0.00593 |
| Trichloroethylene | $CCl_2CHCl$ | 0.0117 |
| Tetrachloroethylene | $CCl_2CCl_2$ | 0.0287 |

TABLE 2—*Sampling devices used in the VOC sampling experiment.*

| Device | Mode of Operation | Contact Materials |
|---|---|---|
| Evacuated flask | suction lift | glass, air |
| Bailer | grab, bottom loading | PTFE, air |
| Gas squeeze pump | gas compression of bag, submersible | PTFE, stainless steel |
| Johnson-Keck SP81 | electric, progressing cavity, submersible | synthetic rubber, stainless steel |
| VOC grab sampler | submerged removal of vial cap | glass, PTFE, air |

*Detailed Procedure*

The reservoir was spiked with about 1 ppm of each of the volatile compounds. The spiking was done in such a way as to ensure that the compounds present in the reservoir would be entirely dissolved.

The samples were collected with each device three separate times to provide replicates for statistical analysis. The order in which the different devices were used was randomized to eliminate any bias that systematic ordering might have introduced. All samples were taken from a location about 0.6 m (2 ft) below the water surface. Each time a sample was collected with one of the test devices, a reference grab sample was taken concurrently from as near the test device inlet as possible. The evacuated flask and bailer were filled once, emptied, then filled again before samples

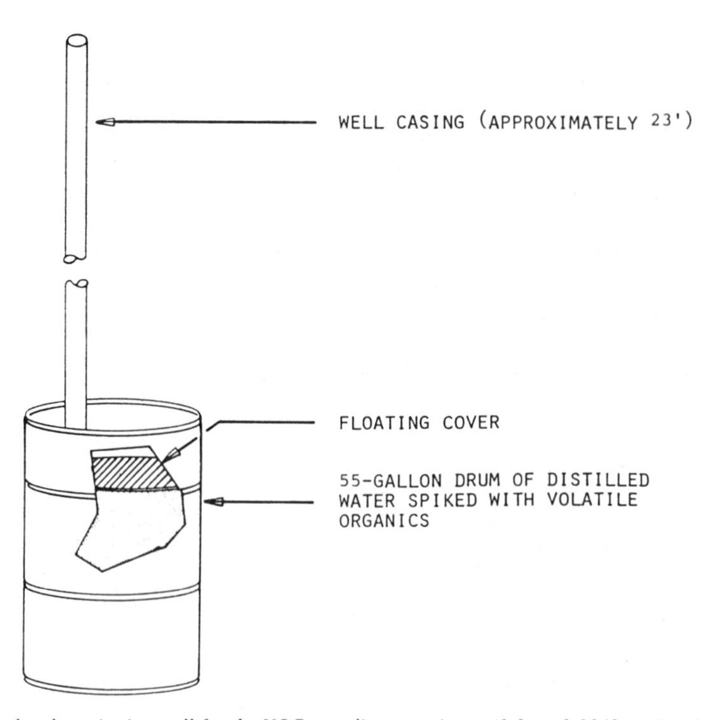


FIG. 7—*Simulated monitoring well for the VOC sampling experiment (1 ft = 0.3048 m; 1 gal = 3.7854 L).*

were taken from them. A portion of the bailer contents was discarded prior to sample collection. The submersible pumps were run long enough to displace the pump volutes and lines at least three times prior to sample collection. Samples from the flask and bailer were slowly poured directly into the sample vials. Samples from the submersible pumps were taken directly from the sample line, the end of which was inserted into the vial and slowly raised to follow the water surface as the vial filled. The flow rate from the pumps was adjusted to about 60 mL/min (0.02 gal/min) either by adjusting the gas driving pressure in the gas squeeze pump or by restricting the discharge line on the Johnson-Keck pump.

All samples were collected in standard 40-mL (0.010 gal), PTFE-lined, septum-capped borosilicate glass vials. No visible bubbles were present in the full sample vials. Samples were held in refrigerated storage prior to analysis. Aliquots of the samples were diluted and analyzed by purge and trap followed by separation in a gas chromatograph and quantitation by mass spectrometer. Laboratory quality control procedures and a 33-day storage stability study ensured proper instrument performance and stability of samples awaiting analysis.

*Calculations*

Each replicate for this experiment consisted of a reference sample and a test sample. The percent loss of volatile compound was the observed variable in the experiment. This was calculated as

$$\text{percent loss} = \frac{\text{reference concentration} - \text{test concentration}}{\text{reference concentration}} \times 100$$

The data were submitted to a two-way analysis of variance to determine the statistical significance of the effects of the type of sampler used and of the compound involved as well as any interactions between these two factors. Where the effect due to the sampler was statistically significant and the interaction was not, the relative performance of the individual samplers was examined using Student's $t$-test with the Bonferroni correction for multiple comparisons [*10*]. Actual levels of significance (probabilities of observed outcomes assuming no true difference between means) were calculated for each comparison. Statistically significant effects due to the compounds were not examined by individual comparisons, but rather by graphical presentation.

*VOC Sampling Results*

Table 3 and Fig. 8 present results of the experiment. Two-way analysis of variance on these data indicated highly statistically significant ($p < 0.001$) effects due to both the device used and also the compound sampled. Interaction between the two factors was also statistically significant ($p = 0.011$). Figure 8 indicates that losses during sampling increase as the tendency of the compound to evaporate from water increases (that is, as $H$ increases). This effect is clearest for the evacuated flask, but it seems to be present for the other devices too.

Though the presence of interaction between factors precluded individual comparisons between devices, it is clear from Fig. 8 that one major effect due to the devices is that the evacuated flask caused greater losses of volatile organics than any of the other three methods. This result, which was not unexpected, demonstrates that the experiment was at least sensitive enough to distinguish that much difference between devices.

In an attempt to distinguish between the other three devices, the evacuated flask data were removed and another two-way analysis of variance was performed. Again, the device and compound factors were highly statistically significant ($p = 0.003$ and $0.004$, respectively) but there was no statistically significant interaction between the two factors ($p = 0.150$) so individual comparisons between the three devices could proceed. The results of these comparisons are summarized in Table 4. The performance of the bailer is neither statistically different from that of the Johnson-

TABLE 3—*Results from the VOC sampling experiment.*[a]

| | Ether | Chloroform | Toluene | Trichloro-ethylene | Tetrachloro-ethylene |
|---|---|---|---|---|---|
| | | | FLASK | | |
| $\bar{x}$ | 14.07 | 20.03 | 23.07 | 26.67 | 30.27 |
| $s$ | 1.29 | 1.91 | 3.36 | 1.60 | 2.05 |
| | | | BAILER | | |
| $\bar{x}$ | 0.61 | 1.26 | 2.89 | 3.49 | 5.76 |
| $s$ | 2.12 | 2.56 | 2.84 | 1.66 | 3.52 |
| | | | SQUEEZE | | |
| $\bar{x}$ | −0.11 | −0.42 | 0.65 | −0.14 | 0.64 |
| $s$ | 4.28 | 0.73 | 5.70 | 6.77 | 3.09 |
| | | | JOHNSON-KECK | | |
| $\bar{x}$ | −2.11 | 0.85 | 7.28 | 8.37 | 8.44 |
| $s$ | 1.10 | 0.74 | 2.95 | 1.00 | 1.91 |

[a] Based on three replicates.

Keck pump nor from that of the squeeze pump. The difference in performance between the squeeze pump and the Johnson-Keck pump is statistically significant, but not highly so.

## Discussion and Conclusions

The results of the two experiments presented here show that the method of purging a well and the method used to sample a well can have an effect on the quality of ground-water samples

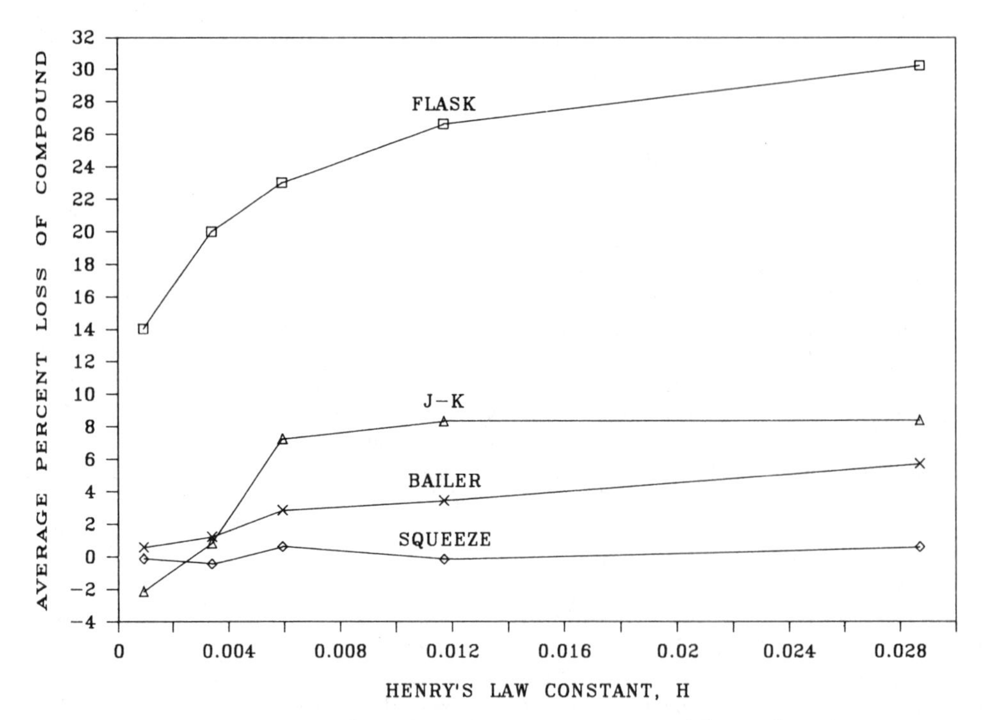


FIG. 8—*Results from the VOC sampling experiment (means of three replicates).*

TABLE 4—*Levels of significance[a] for comparisons of devices in the VOC sampling experiment.*

| | Johnson-Keck | Squeeze |
|---|---|---|
| Bailer | 0.690 | 0.129 |
| Squeeze | 0.027 | . . . |

[a] Individual two-tailed *t*-tests with Bonferroni correction for multiple comparisons.

obtained. In particular, the well purging experiment demonstrates that a purging technique which involves pumping from below the water surface for some period of time followed by sample collection from the same location may lead to contamination of the sample with stagnant water even if the well has reached equilibrium with regard to drawdown. The degree of contamination is highly variable (0 to 8% by volume) over a time frame measured in minutes and it appears to be caused by turbulence around the pump inlet. If the stagnant water in a well contains even trace concentrations of compounds for which the chemical analysis is extremely sensitive, then failure to avoid this phenomenon might produce misleading results. Not only might unrepresentative values for constituents be obtained, but since the introduction of stagnant water is so variable over such a short time, multiple samples from the same well could potentially show more variability than they would were the stagnant water excluded. This has implications for ground-water quality data analysis.

With regard to sampling for volatile organic compounds, it is clear that the evacuated flask or any other sampling device that operates by a similar principle is unsuitable for such sampling. The results of the experiment also indicate that as the tendency of the compound to evaporate from water solution increases, so does its potential for loss during sampling. This experiment produced no evidence for a difference in VOC sampling performance between the bailer and the squeeze pump. There is some statistical evidence that the performance differs between the squeeze pump and the Johnson-Keck pump, but this evidence is not strong and should be corroborated before it is taken to reflect the true situation.

*Acknowledgments*

We thank Mr. Steven Norton of the National Council of the Paper Industry for Air and Stream Improvement, Inc. (NCASI) for his assistance in carrying out both experiments. We also thank Dr. Gerald Sievers, professor of mathematics, Western Michigan University, for his assistance in designing and carrying out the statistical aspects of the VOC sampling experiment.

## References

[*1*] "Ground Water" in *National Handbook of Recommended Methods for Water-Data Acquisition,* Chapter 2, U.S. Geological Survey, Reston, VA, 1980.

[*2*] Gibb, J., Schuller, R., and Griffin, R., "Procedures for the Collection of Representative Water Quality Data from Monitoring Wells," Illinois State Water Survey and Illinois State Geological Survey Cooperative Ground-Water Report No. 7, Champaign, IL, 1981.

[*3*] Unwin, J. and Huis, D. in *Proceedings,* Third National Symposium on Aquifer Restoration and Ground-Water Monitoring," National Water Well Association, Worthington, OH, 1983, pp. 257–262.

[*4*] Unwin, J., "A Guide to Groundwater Sampling," Technical Bulletin No. 362, National Council of the Paper Industry for Air and Stream Improvements, Inc., New York, Jan. 1982.

[5] Ho, J., *Journal of the American Water Works Association,* Vol. 75, No. 11, Nov. 1983, pp. 583–586.
[6] Barcelona, M., Helfrich, J., Garske, E., and Gibb, J., *Ground Water Monitoring Review,* Vol. 4, No. 2, Spring 1984, pp. 32–41.
[7] Unwin, J., "A Laboratory Study of Four Methods of Sampling Groundwater for Volatile Organic Compounds," Technical Bulletin No. 441, National Council of the Paper Industry for Air and Stream Improvements, Inc., New York, Aug. 1984.
[8] Unwin, J. in *Proceedings,* Fourth National Symposium and Exposition on Aquifer Restoration and Ground-Water Monitoring, National Water Well Association, Worthington, OH, 1984, pp. 214–220.
[9] "Treatability Manual Volume 1: Treatability Data," Report No. EPA 600/2-82-001a, U.S. Environmental Protection Agency Office of Research and Development, Sept. 1981.
[10] Snedecor, G. and Cochran, W., Statistical Methods, The Iowa State University Press, Ames, IA, 1980, p. 116.

*James H. Ficken*[1]

# Recent Development of Downhole Water Samplers for Trace Organics

**REFERENCE:** Ficken, J. H., "**Recent Development of Downhole Water Samplers for Trace Organics,**" *Ground-Water Contamination: Field Methods, ASTM STP 963*, A. G. Collins and A. I. Johnson, Eds., American Society for Testing and Materials, Philadelphia, 1988, pp. 253–257.

**ABSTRACT:** Studies concerning the movement and fate of organic chemicals, including hazardous substances, in ground water have underscored the need to use boreholes to obtain samples of ground water that are neither contaminated with foreign materials nor diluted with water from other sources. Additionally, the water samples need to be taken in a manner that prevents the escape of volatile substances. The samples need to be kept close to the original state while in shipment to the water quality laboratory and at the laboratory where the water must be removed from the sample holders without the escape of volatiles or reaction with the air. Within the past year the U.S. Geological Survey has identified its needs for data concerning organics in ground water, and established design criteria for the development of prototype samplers. Certain essential sampler design criteria are: The sampler must go down a 5.1-cm (2 in.) diameter well; the practical sampling depth is at least 61 m (200 ft); the sample holder must be flushed before a sample is taken; and the sample must be kept at the same hydrostatic pressure as when collected.

The Survey is developing prototype samplers using the designated design criteria—a piston sampler operating from a well-logging head, a pumping sampler using a commercial downhole gear pump, a hand-pump sampler, and two electric motor-driven samplers using logging heads. These samplers flush a sample cartridge with the water to be sampled and finally seal the sample prior to being brought to the surface. Later, a design criterion was added that all flush water be removed from the well. New design schemes were drafted that accomplished that requirement. One of the design schemes was a manual sampler which uses a downhole reservoir to collect the flush water. Other design schemes have been envisioned but were found not to be practical. A special sample cartridge that will fit these samplers has been developed to allow removal of the water under pressure. This design helps prevent the escape of volatile substances from the water sample. Fabrication materials include stainless steels, fluoroelastomers, and other fluorinated plastics.

**KEY WORDS:** samplers, sampling, ground water, water wells, water pollution, organic wastes, hazardous materials, organic compounds, volatility

The U.S. Geological Survey is involved in studies concerned with the contamination of ground water by hazardous and toxic substances. Some of these substances are organic compounds that may exist in the ground water in concentrations as low as nanograms per litre. Even in trace amounts some of these compounds are considered dangerous to human health. It is important therefore that the movement in the ground water and ultimate fate of these substances be better understood. Water samples collected from boreholes drilled into the aquifer of interest offer a way to study the distribution and movement of trace organic substances provided that water samples containing concentrations of these substances are representative of aquifer conditions. The water sample must not have suffered a change in the concentration of the constituents of interest from

[1] Hydrologist, U.S. Geological Survey, Water Resources Division, Hydrologic Instrumentation Facility, National Space Technology Laboratories, 39529.

the time of sample collection to the time of analysis. Because the concentrations are low, extreme care must be taken to prevent contamination or dilution of the water sample.

Some of the organic compounds are volatile and may escape from a water sample when collected at depth and then depressurized as the sample is raised to the surface. Some constituents may react with air if the sample is exposed. A water sample therefore, when collected at depth, needs to be sealed to prevent the escape of volatile substances and to prevent dilution and contamination. In addition, some way needs to be provided to allow the water sample to be shipped to a laboratory and analyzed with no significant change in the concentrations of the constituents of interest.

This paper describes recent design concepts for samplers capable of collecting sealed downhole water samples and also describes prototype samplers recently developed by the U.S. Geological Survey from some of these design concepts.

## Sampling Needs and Design Criteria

Design criteria or specifications evolve directly from the needs or performance desired of the sampler. Some of the needs are as follows:

1. The sampler must go down a borehole and collect a water sample.
2. The sampler material must not contaminate the water sample.
3. The sampler must be capable of being lowered to some maximum depth.
4. The sample holder (cartridge or cylinder) must be flushed before the final sample is taken.
5. The sample holder must be sealed at depth to prevent the escape of volatiles or contact with air.
6. The pressure must not be lowered significantly on the water sample and cause outgassing bubbles.
7. The water sample must be capable of shipment to the laboratory for analysis.

The design criteria evolving from the sampling needs depend on more specific uses or needs. For example, the design criteria that evolve from the needs of hazardous waste studies are as follows:

1. Borehole or well minimum diameter, 5.1 cm (2 in.).
2. Sampler material of stainless steel and fluorinated plastics.
3. Maximum water depth 61 m (200 ft).
4. Sample holder flush equal to three volumes of sampler.
5. Sample holder must be sealed at depth.
6. The formation of bubbles must be minimized in the water sample.
7. Sample holder must be shippable.

Some of the design criteria evolving from other study needs would necessarily be somewhat different than the above. Studies involving deep wells might require water depths in excess of 610 m (2000 ft). High temperatures might require the use of special materials in the construction of the sampler.

In addition to the specific design criteria, other general criteria associated with the design of field equipment also apply. These are:

1. The equipment should be lightweight and portable, hand carried if possible.
2. The equipment should be simple and easy to operate.
3. The need for power should be minimized.
4. In order to eliminate gas cylinders and gas pumps, no designs involving the use of gas-operated equipment are considered.

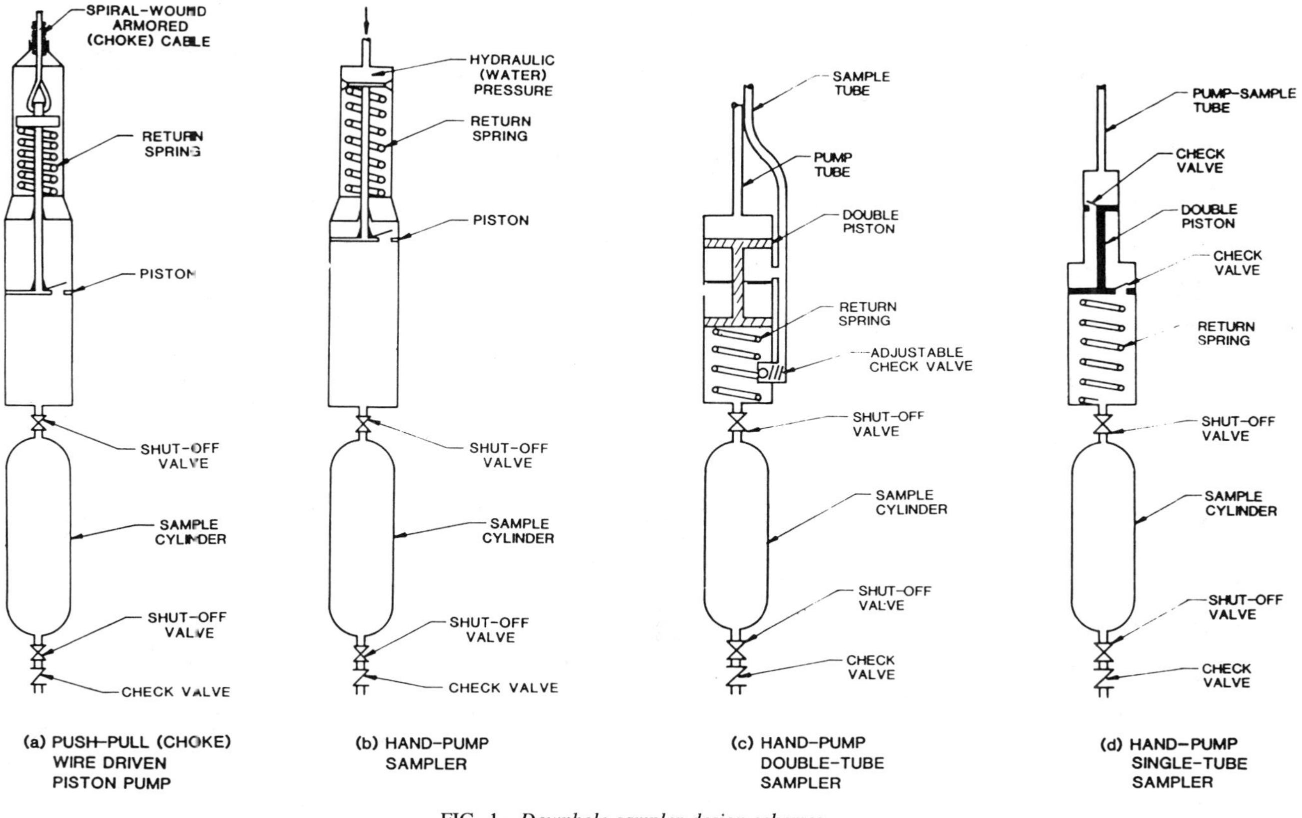


FIG. 1—*Downhole sampler design schemes.*

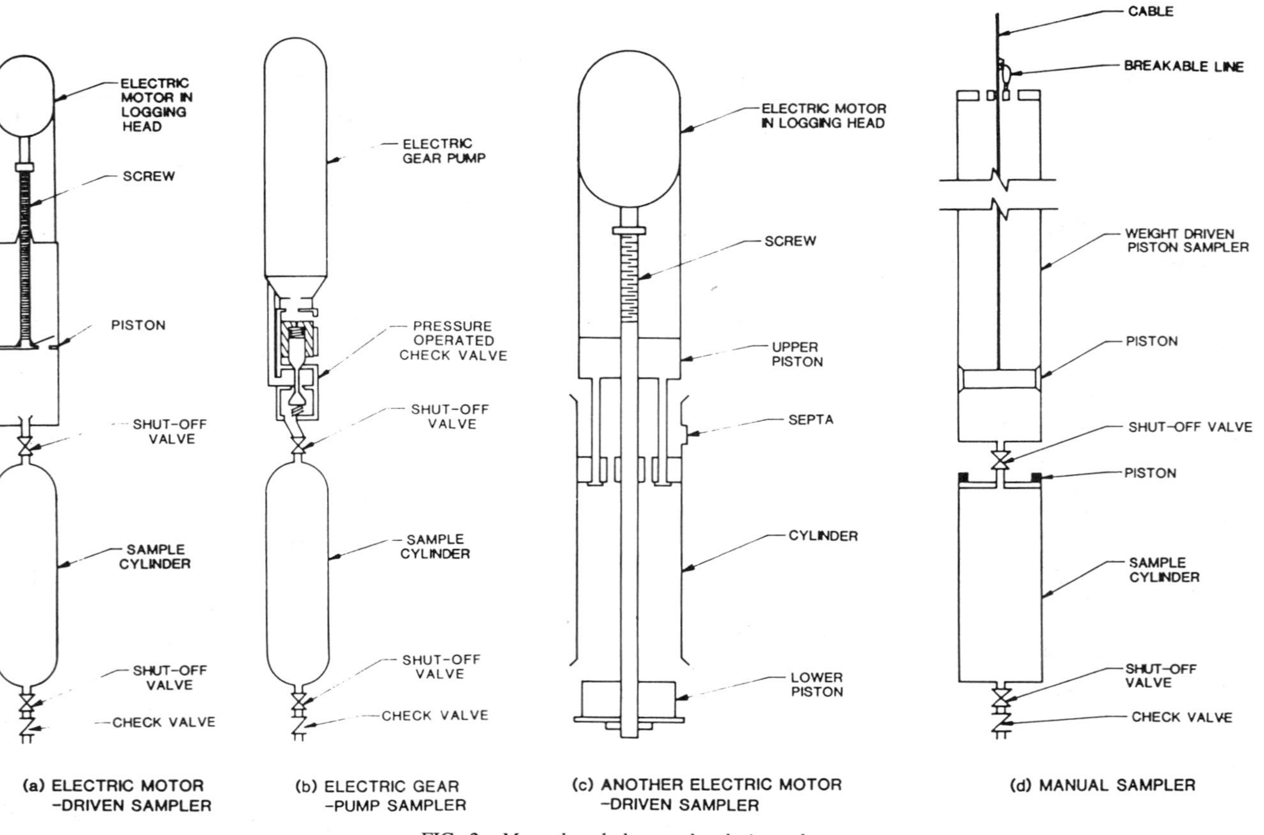


FIG. 2—*More downhole sampler design schemes.*

Many design schemes can be generated once the criteria have been established. Figure 1 shows four downhole sampler design schemes. All use a piston to pump water through a detachable sample cylinder. No choke wire driven samplers (Fig. 1*a*) have been built because of the difficulty in obtaining low friction and easily cleaned choke wire for a 61-m (200 ft) depth. A prototype has been built of the hand-pump driven sampler (Fig. 1*b*). However, the return spring had insufficient force to overcome the pressure generated by 61 m (200 ft) of water. Both the choke wire and the hand-pump designs circulated water from the borehole through the sample cylinder and back into the well. The design of the double-tube sampler (Fig. 1*c*) offers a way of balancing the pressure of 61 m (200 ft) of water that the spring must overcome. The hand-pump single-tube sampler (Fig. 1*d*), moves more water up than is moved down to compress the return spring. The double-tube sampler and the single-tube sampler both pull water up through the same cylinder and expel it at the surface. One of the difficulties is obtaining relatively flexible tubing of noncontaminating, easily cleaned material that will not expand significantly at the working pressures. In addition, the tubing must have a safe working pressure above the maximum working pressure. No prototype, double-tube or single-tube samplers are being built.

Figure 2 shows four more sampler design schemes. The electric motor-driven sampler (Fig. 2*a*) originally developed by K. E. Stevens, Geological Survey, New Mexico, and A. J. Boettcher, Geological Survey, Colorado, operates from a cable-suspended logging head originally designed for a different type of sampler (written communication, A. J. Boettcher, 1985). An electric motor drives a threaded shaft, causing a piston to move up and down. A valve closure is provided at each end of the piston travel. Sampling is achieved by cycling the motor one direction and then the other, causing water to be pumped through the sample cylinder. Two samplers have been built—one for wells of about 10.2 cm (4 in.) in diameter or larger and the other for wells having diameters as small as 3.8 cm (1.5 in.). The sampler is capable of collecting sealed samples at depths greater than 610 m (2000 ft) of water. The electric gear-pump driven sampler shown in Fig. 2*b* uses a commercially available electric gear pump 3.8 cm (1.5 in.) in diameter. Water pressure developed by the pump is used to open a slide valve, causing water to be drawn through the sample cylinder. When the pump is stopped, a spring returns the slide valve to the closed position which seals the sample cylinder. Prototypes built utilizing the gear pump have not operated satisfactorily because of excessive friction in the slide valve and rapid wear of the fluoroplastic gears in the pump. Shown in Fig. 2*c* is another sampler driven by the type of logging head used for the aforementioned sampler (Figure 2*a*). This sampler is of small diameter and was originally used for sampling of ground water for inorganic constituents (written communication, W. T. Sonntag, 1985). When the electric motor in the logging head is activated, a threaded shaft is driven upward, causing two plugs or pistons to seal an open cylinder. This sample cylinder is not removable but is provided with a septum to allow water sample removal by needle and syringe. A prototype has been built for sampling trace organics in ground water and is being evaluated by the Survey.

Figure 2*d* shows the design of a manual sampler which uses a piston or plunger to cause water to pass through the sample cylinder as the assembly moves downward under its own weight when a line or tether is broken. This is achieved by lowering the assembly to the desired depth, then breaking the tether or line by a sharp tug on the cable. Also shown in Fig. 2*d* is a new type of sample cylinder fabricated with a piston in one end. A screw press is used to force water from the sample cylinder directly into analytical instrumentation.

Prototype samplers for collecting samples of ground water containing trace organics are being evaluated by U.S. Geological Survey personnel. They continue to investigate designs capable of meeting the sampling needs for trace organics in ground water.

*Thomas E. Imbrigiotta,[1] Jacob Gibs,[1] Thomas V. Fusillo,[1] George R. Kish,[1] and Joseph J. Hochreiter[1]*

# Field Evaluation of Seven Sampling Devices for Purgeable Organic Compounds in Ground Water

**REFERENCE:** Imbrigiotta, T. E., Gibs, J., Fusillo, T. V., Kish, G. R., and Hochreiter, J. J., **"Field Evaluation of Seven Sampling Devices for Purgeable Organic Compounds in Ground Water,"** *Ground-Water Contamination: Field Methods, ASTM STP 963,* A. G. Collins and A. J. Johnson, Eds., American Society for Testing and Materials, Philadelphia, 1988, pp. 258–273.

**ABSTRACT:** Seven different sampling devices—(1) a bladder pump, (2) a helical-rotor submersible pump, (3) a gear submersible pump, (4) an open bailer, (5) a point-source bailer, (6) a syringe sampler, and (7) a peristaltic pump—were evaluated to determine their ability to recover purgeable organic compounds (POCs) from ground water under field conditions. Significant differences among the mean concentrations obtained with each sampler were found for selected POCs at each of three sites. The overall order of the sampling devices from highest to lowest recovery using data from all three sites yielded the following: (1) gear submersible pump, (2) point-source bailer, (3) open bailer, (4) helical-rotor submersible pump, (5) bladder pump, (6) syringe sampler, and (7) peristaltic pump. The overall standardized mean concentrations of Samplers 1 through 5 were closely grouped, indicating these devices were very similar in their ability to recover POCs. A similar overall order of the coefficients of variation indicated no clear difference between the sampling devices in the precision with which they recovered POCs. However, the three samplers with the lowest coefficients of variation (highest precision) were all positive-displacement pumping devices whereas the three samplers with the highest coefficients of variation (lowest precision) were all grab-sampling devices.

**KEY WORDS:** ground-water sampling, purgeable organic compounds, volatile organic compounds, sampling techniques, water quality, water chemistry

Increased emphasis on the need to sample wells for analysis of organic compounds has triggered the development and production of a wide variety of ground-water sampling devices. Environmental investigators have questioned the ability of all these devices to obtain equally representative samples for the analysis of purgeable organic compounds (POCs). Such compounds present special sampling and analytical difficulties because of their low molecular weights, low solubilities in water, low boiling points, and natural tendency to degas in open systems.

Most of the previous evaluations of sampler recovery efficiency for POCs have been conducted under controlled laboratory conditions. Ho [*1*] tested a peristaltic pump in the laboratory and found significant losses, particularly for highly volatile POCs (Henry's law constants >0.01 atm-m$^3$/mol), with sample lifts more than 4.8 m (16 ft) and pumping rates greater than 2.6 L/min (0.7 gpm). Unwin [*2*] found that a peristaltic pump had significantly higher losses of five POCs from a spiked solution than those of a bailer, a bladder pump, and a helical-rotor submersible pump.

[1] Hydrologist, environmental engineer, hydrologist, hydrologist, and hydrologist, respectively, U.S. Geological Survey, 810 Bear Tavern Road, Suite 206, West Trenton, NJ 08628.

Barcelona and others [3] carried out a comprehensive laboratory comparison of 14 commercially available sampling devices for their performance in collecting dissolved gases and POCs. They found significant bias and poorer precision in sampling with gas-displacement, mechanical positive-displacement, and suction-lift pumping mechanisms. Bladder pumps were rated best for most applications, whereas bailer performance was rated satisfactory but heavily dependent on the expertise of the sampling personnel.

Houghton and Berger [4] conducted a field comparison of several types of sampling devices but did not evaluate POC recovery. They did observe increased losses of volatile inorganic species, such as dissolved carbon dioxide, ammonia, mercury, molybdenum, and selenium, using air-lift, submersible-centrifugal, and peristaltic pumps. They rated the bladder pump best in avoiding these problems.

Recently, Pearsall and Eckhardt [5] completed a field comparison of seven sampling devices for recovery of trichloroethylene and 1,2-dichloroethylene. They found no significant difference in paired comparisons between the concentrations of these two compounds recovered by two types of helical-rotor submersible pump, an impeller submersible pump, and a centrifugal pump. However, reduced trichloroethylene concentrations were obtained with a peristaltic pump. Bailer recoveries of trichloroethylene and 1,2-dichloroethylene concentrations in the 76 to 96 μg/L range did not differ significantly from those of a helical-rotor submersible pump. However, significantly reduced bailer recoveries were observed for these compounds in the 23 to 29 μg/L concentration range.

The objective of this study was to conduct a comprehensive statistical comparison of commercially available ground-water sampling devices for the recovery of purgeable organic compounds in a field environment. Seven different samplers were compared at each of three field sites for a wider diversity of purgeable organic compounds than in any of the previously mentioned investigations.

## Methods

### *Sampling Devices Evaluated*

The sampling devices evaluated in this study included (1) a bladder pump, (2) a helical-rotor submersible pump, (3) a gear submersible pump, (4) an open bailer, (5) a point-source bailer, (6) a syringe sampler, and (7) a peristaltic pump. These samplers represent three general classes of commercially available and commonly used sampling devices in ground-water monitoring: positive-displacement pumps (1–3), grab samplers (4–6), and suction-lift pumps (7).

The bladder pump tested is a noncontact gas-driven pump constructed completely of polytetra fluoroethylene (Teflon[2]) and Teflon-coated materials. It uses compressed gas to alternately expand and contract a flexible bladder to force successive pump volumes past a check valve and up through a discharge line [6]. This pump produces a noncontinuous flow at a maximum rate of approximately 3.8 L/min (1.0 gpm).

The helical-rotor submersible pump employs an electric motor to turn a helical stainless-steel rotor against a semi-flexible Viton stator to create a progressing-cavity pumping head [7]. This pump is capable of delivering a continuous stream of water yielding 2.3 L/min (0.6 gpm) at 16 m (52 ft) of static lift head. The samples contact only stainless-steel, Viton, and Teflon surfaces in this pump.

The gear submersible pump has a set of meshing Teflon gears that form the basis of the pumping system [8]. The gears, driven by an electric motor, act as paddle wheels which push the water along the internal pump walls from the intake point to the pump discharge. The close tolerance of the Teflon gears prevents water from passing back between them. The sample contacts only the

[2] Use of commercial or brand names is for identification purposes only and does not constitute endorsement by the U.S. Geological Survey.

stainless steel pump body and the Teflon gears and discharge tubing. A pump rate of approximately 1.9 L/min (0.5 gpm) is attained at 16 m (50 ft) of static lift head.

The open bailer used in this study is constructed entirely of Teflon. This device consists of a cylinder with a one-way valve on the bottom that allows water to pass through while it travels downward [*6*]. After the bailer is lowered to the desired sampling depth, the direction of travel is reversed and a slug of water is retained by the one-way valve. The bailer volume is approximately 0.7 L (0.2 gal). The retrieved sample is transferred from the bailer to the sample container by means of a Teflon bottom-emptying device which opens the one-way valve.

The point-source bailer differs from the open bailer only in that a second one-way valve is positioned at the top of the bailer barrel [*6*]. The top one-way valve effectively seals off the sample in the bailer as it is retrieved. This prevents exchange with water higher in the well, reduces the potential for POC degassing during upward travel, and prevents contamination of the sample with particulates scraped from the well casing. The volume of the point-source bailer is approximately 0.7 L (0.2 gal). The top one-way valve is dislodged with a small rod to allow displacement of the sample with air during sample transfer using a bottom-emptying device.

The syringe sampler tested in this study consists of a sliding Teflon piston in a stainless-steel barrel [*9*]. The piston is machined to very close tolerances that create a seal with the barrel. The syringe sampler is lowered to the desired sampling depth with the piston positioned at the bottom of the stainless-steel barrel. The space above the piston is evacuated through the attached pressure tubing. As the piston is drawn toward the top of the barrel, the sample is drawn through a screen, past a one-way valve, and into the bottom of the barrel. After retrieving the syringe, a positive pressure is applied to drive the piston downward and force the sample out a different one-way valve into a sample container. The volume of the sample retrieved is approximately 0.8 L (0.2 gal).

The peristaltic pump creates a vacuum to draw water up from a well to the collection point at the surface [*8*]. The system evaluated in this study consists of Teflon intake tubing, a glass Erlenmeyer receiving flask, and a peristaltic pump with silicone tubing in the pump head. The peristaltic pump evacuates the system and draws water up the intake line into the receiving flask. When filled, the flask is opened and the sample is transferred to a sample container. This system was used so that the samples would contact only Teflon and glass surfaces and not the silicone tubing in the pump head, which has been found by Ho [*1*] and Barcelona and others [*10*] to sorb POCs from solution. The peristaltic pump has a relatively low pumping rate of about 0.6 L/min (0.2 gpm) and is limited by atmospheric pressure and pump losses to a lift of about 9 m (27 ft).

A final pump used in this study, though not evaluated for purgeable organic compound recovery, was an impeller submersible pump. This pump was used at one of the test sites to flush the casing prior to the sampler evaluation because of an interest in determining the effect of pumping rate on the recovery of POCs. The impeller pump used was a 10-cm-diameter (4 in.) multiple-stage submersible pump of the type commonly found in domestic wells. It pumped at an approximate rate of 38 L/min (10 gpm) at a static head lift of 2 m (6 ft). Water-contact surfaces in this pumping system included stainless steel, various rigid plastics, and flexible polyvinyl chloride.

*Experimental Design*

Three wells with water known to contain POCs were chosen for intensive field sampling tests—one in Cape Cod, Massachusetts, one in northern New Jersey, and one in southern New Jersey. All three wells are screened in shallow, unconfined aquifers consisting mainly of unconsolidated sands and gravels, and had hydrostatic pumping heads of approximately 2 m (6 ft). The sample recovery depths were 15 m (50 ft) in the Cape Cod well, 4 m (12 ft) in the northern New Jersey well, and 6 m (19 ft) in the southern New Jersey well.

At each site, 15 to 28 replicate samples were collected with each sampler. The replicates were prevent bias. For example, in the first round at Cape Cod, four replicates were taken with the open bailer, four replicates with the helical-rotor submersible pump, four replicates with the gear submersible pump, and so on until all seven samplers were used. The second round was conducted similarly, this time with seven replicates obtained successively with the seven samplers in a different randomly picked order. Table 1 summarizes the experimental design and also lists the POCs detected at each of the three test sites.

### *Field and Laboratory Methods*

At all sites a submersible pump was used to flush standing water from the well casing. In accordance with the procedures outlined in Wood [*11*] and Claassen [*12*], temperature, pH, specific conductance, and dissolved oxygen concentrations were monitored at 5 or 10-min intervals to determine the inorganic chemical stability of the water being pumped from the well. Chloride concentration was measured, using a specific ion electrode, as an additional indicator of the inorganic water quality. Ultraviolet absorbance at 254 nm (nanometres) was also measured in the field using a Hitachi Model 100–20 single-beam UV-VIS (ultraviolet-visible) spectrophotometer. Absorbance at this wavelength is characteristic of unsaturated aliphatic and aromatic organic compounds. Therefore, this measurement was used as an indicator of organic water quality.

Chemical stability was deemed to have been achieved when all measurements varied within 5% for three successive readings. The time needed to attain chemical stability always far exceeded the time necessary to flush the recommended minimum of three casing volumes (D. A. Rickert, U.S. Geological Survey, written communication, 1985).

TABLE 1—*Experimental design and purgeable organic compounds detected.*

| Site | Round | Day | Replicates | Flushing Pump[a] | Sampling Order[a] | Purgeable Organic Compounds Detected |
|---|---|---|---|---|---|---|
| Cape Cod, | 1 | 1 | 4 | B | E B G A F D C | 1,1-dichloroethylene |
| Massachusetts | 2 | 2 | 7 | B | C B D G E F A | trans-1, 2-dichloroethylene |
| | 3 | 2 | 4 | . . . | D A B C E G F | trichloroethylene |
| | | | | | | tetrachloroethylene |
| Northern | 1 | 1 | 8 | B | B A F C D E G | 1,1-dichloroethane |
| New Jersey | 2 | 2 | 8 | H | E D F A C B G | 1,1-dichloroethylene |
| | | | | | | trans-1, 2-dichloroethylene |
| | | | | | | 1,1,1-trichloroethane |
| | | | | | | trichloroethylene |
| | | | | | | tetrachloroethylene |
| Southern | 1 | 1 | 14 | B | B G A C D F E | vinyl chloride |
| New Jersey | 2 | 2 | 14 | B | B A D C E G F | 1,1-dichloroethane |
| | | | | | | trans-1,2-dichloroethylene |
| | | | | | | chloroform |
| | | | | | | trichloroethylene |
| | | | | | | benzene |
| | | | | | | toluene |
| | | | | | | ethylbenzene |
| | | | | | | chlorobenzene |
| | | | | | | *o*-1,2-dichlorobenzene |
| | | | | | | *m*-1,3-dichlorobenzene |

[a] A = peristaltic; B = helical-rotor submersible; C = bladder; D = syringe; E = open bailer; F = point-source bailer; G = gear submersible; H = impeller submersible.

Samplers were cleaned prior to sampling and between rounds with distilled deionized water that was sparged in the field with high-purity helium to remove any remaining purgeable compounds. Samples of the wash water were collected before each sampler use to check for POC cross-contamination.

All samples were concentrated by purge and trap and analyzed for purgeable priority pollutant concentrations by gas chromatography with the following detectors: a Hall electrolytic conductivity detector (GC/HALL) in series with a photoionization detector (GC/PID), and a mass spectrometer (GC/MS). This corresponds to U.S. Environmental Protection Agency Methods 601, 602, and 624, respectively [*13*].

*Statistical Methods*

For each purgeable organic compound detected at a site, a mean concentration for each sampler was determined. Unforseen nonrandom effects other than those due to sampler were observed in these results. The effect of round was particularly notable. This effect, which presumably was due to the use of differing well flushing rates prior to each round, will be discussed in more detail later. Inasmuch as round was not one of the effects being tested in the experiment, it was eliminated from the statistical analyses by standardizing the concentration data using the following formula

$$X(\mathrm{std}) = X - \bar{X}(\mathrm{round})/s(\mathrm{round})$$

where

$X(\mathrm{std})$ = standardized concentration,
$X$ = observed concentration,
$\bar{X}(\mathrm{round})$ = mean concentration for a compound in a round, and
$s(\mathrm{round})$ = standard deviation of the concentration for a compound in a round [*14*].

Using this formula, the standardized data for each round has a mean of zero and a standard deviation of one.

The standardized data were compared using a simple one-way analysis of variance to determine if the sampling devices exhibited significantly different standardized mean concentrations at the 95% confidence level. If so, this was attributed to differences in their ability to recover each purgeable organic compound. If the analysis of variance showed there was a significant difference between samplers for a compound, the Tukey's multiple comparison test was employed to determine which standardized means differed. The Tukey's test was used because it is one of the most effective multiple comparison tests at preserving the chosen significance level and minimizing Type I errors [*14*]. The Tukey's test resulted in groupings of samplers whose standardized mean concentrations did not differ at the 95% confidence level for a particular compound. All statistical tests were conducted using a software package called SAS [*15,16*].

The use of standardized data facilitated the combination of results from compounds with widely different concentrations. To summarize the results for a sampling site, the standardized data from all compounds that exhibited a significant difference between sampling devices were included in a one-way analysis of variance. For example, at Cape Cod, 10 to 13 replicates for each of the 7 samplers for each of the 4 compounds were included. The associated Tukey's test resulted in comparison of overall standardized means for each sampler and groupings of samplers whose overall performances were similar at the site. The standardized mean concentrations were converted to a 1 to 7 scale for ease of understanding and presentation in the tables.

**Experimental Results and Discussion**

Because this was a field study, the true concentrations of purgeable organic compounds in water at each well were not known and percent recoveries could not be calculated as in a laboratory study. It was therefore assumed the highest mean POC concentration obtained with a sampler was the most accurate. This assumption was not unreasonable because (1) losses of POCs would be more likely than increases simply because of their physical and chemical properties, and (2) all samplers tested were constructed of generally inert materials (Teflon, stainless steel, Viton and glass), so contamination from the pumps was minimized.

Averages of field water-quality measurements and POC concentrations for all samplers by round at the three test sites are given in Table 2. Several general characteristics of the ground-water quality at each of the sites should be noted.

First, water from the well in Cape Cod, which was situated in a plume of contamination caused by sewage effluent [*17*], represents the simplest POC matrix sampled; only four chlorinated solvent purgeables were identified by GC/MS.

Water from the northern New Jersey well, which was contaminated by wastewaters from a metal plating/degreasing operation (E. F. Vowinkel, U.S. Geological Survey, written communication, 1986), is a slightly more complex POC matrix; six purgeable organic compounds were

TABLE 2—*Steady-state field properties or constituents and mean POC concentrations by round at the test sites. [Concentrations are expressed in micrograms per litre unless otherwise noted.]*

| | Date, Round No., and Site | | | | | | |
|---|---|---|---|---|---|---|---|
| | 7-17-84 1 | 7-19-84 2 | 7-19-84 3 | 8-14-84 1 | 8-15-84 2 | 12-5-84 1 | 12-6-84 2 |
| | Cape Cod, MA | | | Northern NJ | | Southern NJ | |
| Steady-state field properties or constituents | | | | | | | |
| pH (units) | 7.0 | 7.0 | 7.0 | 6.5 | 6.4 | 6.1 | 6.1 |
| water temperature (°C) | 12.5 | 12.0 | 12.0 | 18.4 | 18.9 | 15.5 | 16.0 |
| specific conductance (μS/cm @ 25 °C) | 227 | 228 | 223 | 805 | 865 | 1780 | 1780 |
| Dissolved oxygen (mg/L) | 0.2 | 0.6 | 0.3 | 0.5 | 0.3 | 0.2 | 0.2 |
| chloride (mg/L) | NM[a] | NM | NM | 140 | 145 | 560 | 640 |
| ultraviolet absorbance (units) | NM | NM | NM | 0.014 | 0.035 | 0.404 | 0.559 |
| Purgeable organic compounds | | | | | | | |
| vinyl chloride | <1[b] | <1 | <1 | <1 | <1 | 64 | 70 |
| *1,1*-dichloroethane | <1 | <1 | <1 | 51 | 79 | 11 | 11 |
| trans-*1,2*-dichloroethylene | 180 | 170 | 170 | 6.5 | 9.8 | 76 | 100 |
| *1,1*-dichloroethylene | 5.8 | 5.8 | 5.4 | 37 | 53 | <1 | <1 |
| chloroform | <1 | <1 | <1 | <1 | <1 | 10 | 15 |
| *1,1,1*-trichloroethane | <1 | <1 | <1 | 700 | 940 | <1 | <1 |
| benzene | <1 | <1 | <1 | <1 | <1 | 66 | 66 |
| trichloroethylene | 89 | 84 | 82 | 41 | 60 | 27 | 28 |
| tetrachlorethylene | 79 | 75 | 72 | 180 | 270 | <1 | <1 |
| toluene | <1 | <1 | <1 | <1 | <1 | 20 | 21 |
| ethylbenzene | <1 | <1 | <1 | <1 | <1 | 25 | 27 |
| chlorobenzene | <1 | <1 | <1 | <1 | <1 | 220 | 200 |
| *1,2*-dichlorobenzene | <1 | <1 | <1 | <1 | <1 | 15 | 14 |
| *1,3*-dichlorobenzene | <1 | <1 | <1 | <1 | <1 | 16 | 21 |
| Number of measurements | 25 | 31 | 28 | 55 | 40 | 50 | 82 |

[a] NM = not measured.
[b] <1 = lower limit of quantitation.

identified by GC/MS. This water had the highest total concentration of purgeable organics with levels exceeding 1 mg/L. The data presented show significant differences between the mean POC concentrations obtained for each of the rounds at this site. These differences were apparently caused by the different pumping rates used during well flushing. The overall effect of the differing pumping rates will be discussed in more detail later.

Water from the southern New Jersey well, which was contaminated by a chemical waste recycling facility (J. J. Hochreiter, U.S. Geological Survey, written communication, 1986), was by far the most complex POC matrix of the three sampled. Eleven POCs were identified by GC/MS at this site. This water contained a diverse mixture of purgeable organic compounds ranging from the very volatile vinyl chloride (Henry's law constant = 0.08 atm-m$^3$/mol) to the less volatile dichlorobenzenes (Henry's law constants = 0.002–0.004 atm-m$^3$/mol). Ground water at the southern New Jersey site was the only one of the three sampled which contained detectable concentrations of aromatic POCs (benzene, toluene, ethylbenzene, chlorobenzene, 1,2-dichlorobenzene, and 1,3-dichlorobenzene). The presence of aromatic compounds is reflected also in the higher ultraviolet absorbance measurements given in Table 2.

### *Cape Cod, Massachusetts Site*

Statistical analysis of the Cape Cod results was done on 10 to 13 replicate sample concentrations for each of the 7 samplers for each of 4 POCs. The difference in numbers of replicates was caused by the loss of a few data values due to analytical difficulties.

Analysis of variance of the Cape Cod data indicated that the standardized mean concentrations of the samplers differed significantly (at $p \leq 0.05$) for all four compounds identified at this site. The Tukey's test results are given in Table 3. The samplers are shown in order from highest to lowest standardized mean concentrations for each purgeable organic compound. Samplers within the same grouping bracket are not significantly different at the 95% confidence level in their ability to recover the particular POC from the samples. The range of the actual mean concentrations are given for each compound to indicate the levels at which these groupings occurred.

The bladder pump, the open bailer, and the helical-rotor submersible pump consistently recover the highest concentrations of all four purgeable organic compounds at the Cape Cod test site. The

TABLE 3—*Tukey's test results: individual purgeable organic compound and overall orders and groupings of samplers for the Cape Cod site.*

| | Individual Compounds[a,b] | | | | Overall | |
|---|---|---|---|---|---|---|
| Order | 1,1-dichloro-ethylene | Trans-1,2-dichloro-ethylene | Tri-chloro-ethylene | Tetra-chloro-ethylene | Order | Sampler |
| 1 | BLAD | BLAD | HR | HR | 1.52 | BLAD |
| 2 | OB | OB | OB | BLAD | 1.87 | HR |
| 3 | HR | HR | BLAD | OB | 2.02 | OB |
| 4 | PSB | GEAR | GEAR | GEAR | 3.77 | GEAR |
| 5 | GEAR | SYR | PSB | PSB | 5.05 | PSB |
| 6 | SYR | PSB | SYR | SYR | 5.81 | SYR |
| 7 | PER | PER | PER | PER | 7.00 | PER |
| Range (µg/L) | 6.4–4.7 | 190–160 | 95–68 | 91–58 | | |

[a] BLAD = bladder pump; OB = open bailer; HR = helical-rotor submersible pump; PSB = point-source bailer; GEAR = gear submersible pump; SYR = syringe sampler; PER = peristaltic pump.

[b] Sampling devices within a bracket do not differ significantly at the 95% confidence level in their standardized mean POC concentration.

peristaltic suction-lift pump recovers the lowest concentrations of all compounds and is the only pump not grouped with the two top samplers in any of the comparisons. However, for all four compounds there is considerable overlap in the groupings within each sampler comparison. For example, for 1,1-dichloroethylene and trichloroethylene, all samplers except the peristaltic pump are grouped together. The groupings are useful in comparing the relative performance of pairs of samplers.

To summarize the data for the Cape Cod site, a one-way analysis of variance was done on the standardized data as described earlier in the "Statistical Methods" subsection. The overall site order and grouping of the samplers given by the Tukey's test are given in the last two columns of Table 3. There is more differentiation between samplers when all four POCs are considered together rather than separately. This overall site order and grouping corroborates the previous general observations that the bladder pump, the helical-rotor submersible pump, and the open bailer are the most effective samplers and that the peristaltic pump is the least effective sampler at recovering POCs.

The point-source bailer ranking is probably low because of an operational problem encountered in emptying this sampling device at this first test site. Initially, the bailer was not equipped with a deflecting rod to displace the top ball check valve and allow the sample to drain out. Consequently, this bailer could not be emptied from the bottom so samples were poured from the top into the 40-mL sample vials. Aeration of the samples with associated degassing of the purgeable organics probably took place during this sample transfer. Prior to any further field testing, the bailer was returned to the manufacturer and the deflecting rod was installed. At the final two test sites, the point-source bailer was properly drained using the bottom-emptying device, and the standardized means were equal to or higher than those of the open bailer.

### *Northern New Jersey Site*

Statistical analysis of the northern New Jersey results was done on 7 to 16 replicate concentrations for each of the 7 samplers for each of 6 POCs. The difference in numbers of replicates was caused by analytical difficulties and the effects of sampler order and pumping rate which resulted in the elimination of the data from one round for each of two samplers. As mentioned previously, the well flushing rate affected the results at this site. A detailed discussion of these results will be deferred to the "Effect of Pumping Rate" subsection.

Analysis of variance of the remaining northern New Jersey data showed that the standardized mean concentrations of the samplers differed significantly (at $p \leq 0.05$) for five of the six purgeable organic compounds identified at this site. The sixth POC was not significantly affected by sampling device primarily because its concentrations were near the detection limit of the gas chromatographic analysis procedure. At low concentrations, analytical variation makes up a much greater portion of the total variation in the results, causing greater overlap in the mean concentrations of all the samplers. At this site, the average coefficient of variation for the five significantly affected compounds was 20%, compared to 51% for the low concentration compound.

The Tukey's test results for the five POCs significantly affected by sampling device are given in Table 4. The point-source bailer and the gear submersible pump are the two most effective samplers for all compounds at the northern New Jersey site. The bladder and the peristaltic pumps consistently recovered the lowest standardized mean concentrations for all five compounds. As with the first test site, there is considerable overlap in all the groupings for all five compounds.

The overall site order and grouping of the samplers are shown in the final two columns of Table 4. These groupings do not overlap. They confirm the observations made on the individual compound comparisons, that the point-source bailer and the gear submersible pump are the most effective samplers and the bladder and peristaltic pumps are the least effective samplers for POCs at this site.

TABLE 4—*Tukey's test results: individual purgeable organic compound and overall orders and groupings of samplers for the northern New Jersey site.*

| | Individual Compounds[a,b] | | | | | Overall | |
|---|---|---|---|---|---|---|---|
| Order | 1,1-dichloroethylene | 1,1-dichloroethylene | 1,1,1-trichloroethane | Trichloroethylene | Tetrachloroethylene | Order | Sampler |
| 1 | PSB | PSB | PSB | PSB | GEAR | 1.03 | PSB |
| 2 | GEAR | GEAR | GEAR | GEAR | PSB | 1.55 | GEAR |
| 3 | OB | HR | HR | OB | OB | 3.34 | OB |
| 4 | SYR | SYR | SYR | SYR | HR | 3.42 | HR |
| 5 | HR | OB | OB | HR | SYR | 3.51 | SYR |
| 6 | BLAD | BLAD | BLAD | BLAD | BLAD | 5.75 | BLAD |
| 7 | PER | PER | PER | PER | PER | 7.00 | PER |
| Range (μg/L) | 70–48 | 49–32 | 920–630 | 56–42 | 260–170 | | |

[a] BLAD = bladder pump; OB = open bailer; HR = helical-rotor submersible pump; PSB = point-source bailer; GEAR = gear submersible pump; SYR = syringe sampler; PER = peristaltic pump.

[b] Sampling devices within a bracket do not differ significantly at the 95% confidence level in their standardized mean POC concentrations.

The poor performance of the bladder pump was surprising because this pump was one of the most effective samplers at Cape Cod and it was highly rated in previous studies [*2–4*]. The northern New Jersey well had a static water level of approximately 2 m (6 ft) below land surface and was sampled at a depth of 4 m (12 ft). The bladder pump was 2 m (6 ft) long so that the top of the sampler was at or near the top of the water column in the well. Any drawdown during sample collection would have caused the pumping level to decline below the top of the bladder pump. A loose connection between the bladder pump and the discharge line may have allowed air to leak into the system, or the pump body may have been unable to fill completely with water, resulting in a gas head space for POCs to partition into. Either possibility could lead to pumpage of a mixture of air and water, thereby promoting aeration and stripping of the volatiles from the sampled water. It is believed that the poor bladder pump performance at this site was caused simply by the operational problem that the height of the water column in the well was insufficient to keep a pump of this size submerged.

### *Southern New Jersey Site*

Statistical analysis of the southern New Jersey results was done on 12 to 23 replicate concentrations for each of the 7 samplers for each of 11 compounds. The difference in numbers of replicates was caused by analytical difficulties and by an effect of sampling order/pumping rate similar to that experienced at the northern New Jersey site that resulted in the loss of a round of data for two samplers. This problem will be discussed in the "Effect of Pumping Rate" subsection.

Analysis of variance of the southern New Jersey data indicated that the standardized mean concentrations of the samplers differ significantly (at $p \leq 0.05$) for only 6 of the 11 purgeable organic compounds identified at this site. The Tukey's test results for the six affected compounds are given in Table 5. The groupings show much overlap between samplers at this site. No sampler or group of samplers was consistently more or less effective than the others. Indeed, the peristaltic pump, which had previously been the least effective sampler at the first two sites, was the most effective pump for three of the six compounds at this site.

The overall order and grouping of samplers at this site (last 2 columns in Table 5) show that

TABLE 5—*Tukey's test results: individual purgeable organic compounds and overall orders and groupings of samplers for the southern New Jersey site.*

| | Individual Compounds[a,b] | | | | | | Overall | |
|---|---|---|---|---|---|---|---|---|
| Order | Vinyl chloride | 1,1-dichloro-ethane | Chloro-form | Tri-chloro-ethylene | Benzene | Ethyl-benzene | Order | Sampler |
| 1 | PSB | PSB | PER | PSB | PER | PER | 2.95 | PER |
| 2 | GEAR | OB | HR | OB | BLAD | HR | 3.11 | BLAD |
| 3 | BLAD | GEAR | BLAD | BLAD | HR | BLAD | 3.49 | PSB |
| 4 | OB | BLAD | OB | GEAR | GEAR | PSB | 3.98 | OB |
| 5 | PER | SYR | GEAR | PER | PSB | GEAR | 4.05 | GEAR |
| 6 | HR | PER | SYR | SYR | OB | OB | 5.14 | HR |
| 7 | SYR | HR | PSB | HR | SYR | SYR | 6.45 | SYR |
| Range (μg/L) | 83–69 | 17–11 | 17–12 | 39–28 | 69–58 | 29–22 | | |

[a] BLAD = bladder pump; OB = open bailer; HR = helical-rotor submersible pump; PSB = point-source bailer; GEAR = gear submersible pump; SYR = syringe sampler; PER = peristaltic pump.

[b] Sampling devices within a bracket do not differ significantly at the 95% confidence level in their standardized mean POC concentrations.

all samplers except the syringe are included in the top group. Even the syringe overlaps with most other samplers.

The performance of the peristaltic pump at this site is very different from its previous performances at the other sites. The effect of temperature on the operation of the peristaltic pump may offer one possible explanation. Both the northern and southern New Jersey wells are shallow and unconfined and sampled at similar depths. The only difference is that the northern New Jersey well was sampled in August, when the ambient air temperature was 29°C (85°F), whereas the southern New Jersey well was sampled in December when the ambient air temperature was close to 0°C (32°F). Solar heating of the discharge line and the receiving flask at the surface probably combined with the high air temperature to produce favorable conditions for degassing of purgeables from the water samples during the northern New Jersey sampling. At the southern New Jersey sampling site the weather was cloudy and the air temperature was lower than that of the sampled ground water. Therefore, conditions were optimal for the POCs to stay in solution and be recovered in the samples. This may partially explain why all the samplers overlap so much in the Tukey's test results at the southern New Jersey site.

### *Overall Order and Grouping*

To summarize the results for all three sites, an overall order was calculated using the standardized mean concentrations for each of the 7 samplers for each of the 15 compounds that exhibited significant differences between sampling devices (4 compounds at the Cape Cod site, 5 compounds at the northern New Jersey site, and 6 compounds at the southern New Jersey site). The standardized means of each sampler were averaged over all 15 compounds and converted to a 1 to 7 scale for ease of understanding and presentation. An analysis of variance and Tukey's test were not used in this overall comparison because the known differences in results between sites would result in a significant sampler by site interaction term in any analysis of variance model. The results are given in the left half of Table 6.

The gear submersible pump, point-source bailer, open bailer, helical-rotor submersible pump, and bladder pump are all closely grouped; their scores differing by only 0.64 on a scale from 1

TABLE 6—*Overall orders of sampling devices based on standardized mean concentration data and coefficient-of-variation data from all three well sites.*

| Standardized Mean Concentration Data | | Coefficient-of-Variation Data | |
|---|---|---|---|
| Overall Scores | Sampler | Overall Percentages | Sampler |
| 3.01 | gear submersible pump | 17.6 | gear submersible pump |
| 3.02 | point-source bailer | 17.7 | bladder pump |
| 3.15 | open bailer | 20.4 | helical-rotor submersible pump |
| 3.51 | helical-rotor submersible pump | 20.6 | peristaltic pump |
| 3.65 | bladder pump | 20.8 | syringe sampler |
| 5.12 | syringe sampler | 22.3 | open bailer |
| 5.15 | peristaltic pump | 22.3 | point-source bailer |

to 7. This implies that overall there are only minor differences in the ability of the five sampling devices to recover the purgeable organic compounds sampled at the three test sites. In addition, because of previously discussed operational problems, the overall scores of the point-source bailer and the bladder pump are lower than they could have been. Thus, the actual difference between the highest and lowest scores for these five samplers may be even less.

The peristaltic pump and the syringe sampler are the only two samplers that consistently recover lower POC concentrations than the others. Previous studies [*1–3*] have also found reduced recoveries of purgeable organic compounds with the peristaltic pump in laboratory tests. Presumably the losses are due to the POCs degassing into the vacuum created by the suction-lift mechanism by which the pump operates. The reduced recoveries of the syringe sampler were apparently caused by failure of the seal between the Teflon piston and the syringe barrel. Suspended particulate matter in the well may have caused this seal to wear and then leak, exposing the sample to reduced pressures when the sampler was being filled or to increased pressures when the sampler was being emptied. Degassing or stripping of the purgeables from the water samples apparently occurred during these sampling operations.

The coefficients of variation of the mean concentration data are a measure of the precision with which the sampling devices are able to recover purgeable organic compounds. To summarize the coefficient of variation data from all three sites, an overall order was calculated identically to that of the mean concentration data using the actual coefficient of variation percentages obtained for each sampler for each compound. These results are presented in the right half of Table 6. No clear groupings are obvious based on these values, with less than 5% separating the most and least precise samplers. There is no indication that a significant difference exists in POC recovery precision between the sampling devices. However, the three samplers with the lowest coefficients of variation (highest precision) in Table 6 are all positive-displacement pumps, whereas the three with the highest coefficients of variation (lowest precision) are all grab sampling devices. Thus, depending on the general class of sampler used, there may be a slight effect on the precision of POC sampling. Barcelona and others [*3*] noted that grab samplers were subject to greater variability in POC recoveries due to their heavy dependence on the experience of the sampling personnel and the care with which they are used.

### *Effect of Pumping Rate*

Two different effects caused by pumping rate were observed in this study. First, the effect of different well flushing rates prior to sampling was purposely tested at the northern New Jersey

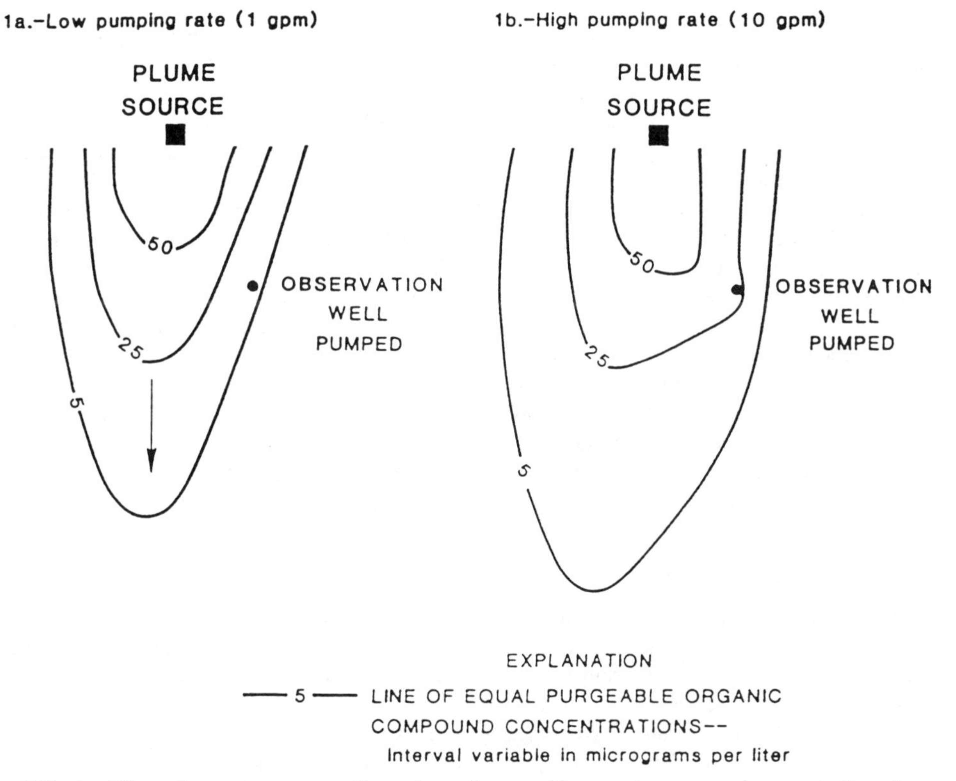


FIG. 1—*Effect of pumping rate on lines of equal purgeable organic compound concentrations in a hypothetical plume.*

site. Secondly, the effect of a greatly reduced flushing rate was observed unintentionally in sampling for POCs at the southern New Jersey site.

At the northern New Jersey site, the well was flushed at pumping rates of 3.8 L/min (1 gpm) the first day and 38 L/min (10 gpm) on the second day. When field water-quality measurements stabilized each day, the sampling was carried out using the seven sampling devices in a randomly determined order. The mean concentrations of the five POCs determined to be significantly affected by sampling device were 21 to 41% higher on the second day than on the first day for the sampler with the highest overall recoveries. The difference between the mean concentrations for the most effective and least effective samplers ranged from 28 to 40% for these five POCs. Thus, the pumping rate at which a well is flushed prior to sampling for POCs can be of the same importance as the type of sampling device used.

The probable explanation of the difference in mean POC concentrations at the two pumping rates is illustrated by the hypothetical plumes shown in Fig. 1. In Fig. 1*a*, a low pump rate [3.8 L/min (1 gpm)] is used to flush a well and has a certain minimal effect on the shape of the contaminant concentration contours. In Fig. 1*b*, a higher pump rate [38 L/min (10 gpm)] is used to flush the same well and, due to its larger cone of depression and increased hydraulic stress on the system, distorts the contaminant concentration contours and draws in a different (higher) concentration than the lower pump rate. It is probable that such a scenario occurred at the northern New Jersey well. However, depending on the well location with respect to the plume and the site hydrogeology, it is also possible for POC concentrations to decrease with increased pumpage. In such a case, the higher pumping rate would increase the radius of the cone of depression and draw in less-contaminated ground water from a larger area of the aquifer.

At the southern New Jersey site, the well was flushed the first day at a uniform low rate [approximately 3.8 L/min (1 gpm)] until stable field measurements indicated water of unchanging water quality was being produced. Sampling then took place with the seven sampling devices in a randomly determined order. As an example, the mean concentrations of trichloroethylene obtained for each of the samplers is plotted in Fig. 2 in the order in which they were used. In this random order, the first four samplers (B, G, A, and C) were pumping devices and the last three (D, F, and E) were grab samplers. The mean concentration of trichloroethylene remained near the stabilized level for all four pumping devices and for the first grab sampler. However, the mean concentrations for the final two grab samplers decreased by an average of 39%. This change in water quality was confirmed by corresponding decreases in the field water-quality measurements (specific conductance, chloride, and UV absorbance). Thus, within one hour of the cessation of pumping, the trichloroethylene concentration contours apparently changed from a situation similar to that illustrated in Fig. 1*b* to that in Fig. 1*a*. The rapid decrease in POC concentrations may have been aided by the fact that the plume is not areally extensive.

## Summary and Conclusions

Significant differences among the mean concentrations recovered with seven samplers were found for selected POCs at each of the three test sites. The relative order of sampler effectiveness varied between sites and among compounds detected at any single site.

The gear submersible pump, point-source bailer, open bailer, helical-rotor submersible pump, and bladder pump were all closely grouped, suggesting a lack of any real differences among them in their ability to recover the POCs sampled under the conditions present at the three sites. The peristaltic pump and the syringe sampler were the only samplers that consistently recovered lower POC concentrations than the others. Presumably the peristaltic pump losses were caused by degassing of the POCs into the vacuum created by the suction-lift mechanism by which the pump operates. Reduced POC concentrations obtained with the syringe sampler were apparently caused by leakage of the gastight seal between the Teflon piston and the syringe barrel, which exposed the samples to both very low and very high pressures during the sampling process.

The overall order of the samplers based on their coefficients of variation indicated no clear difference between the sampling devices in the precision with which they recovered POCs. However, the three samplers with the lowest coefficients of variation (highest precision) were all positive-displacement pumping devices, whereas the three samplers with the highest coefficients of variation (lowest precision) were all grab-sampling devices.

Operating conditions, such as mechanical problems, ambient air temperature, and position of the sampler in the water column of the well, can have significant impacts on the ability of any sampling device to recover purgeable organic compounds. In addition, scale effects, such as the size of the plume being sampled, the distance from the source to the sampling well, and the pumping rate with which the well is flushed, can significantly influence the recovery of POCs at a well site. In this study, these factors were as important as the type of sampling device used to take the samples.

The results of these field experiments should have a great deal of transfer value to many sites where sampling for purgeable organic compounds is of interest. Each of the sites was hydrologically similar in that all were unconfined sand and gravel aquifers of relatively shallow depths [4 to 15 m (12 to 50 ft)]. It is not uncommon to find ground-water contamination problems involving purgeable organic compounds in aquifers such as these due to their high permeability. The POCs detected in this study had a wide range of volatilities and included the compounds most commonly

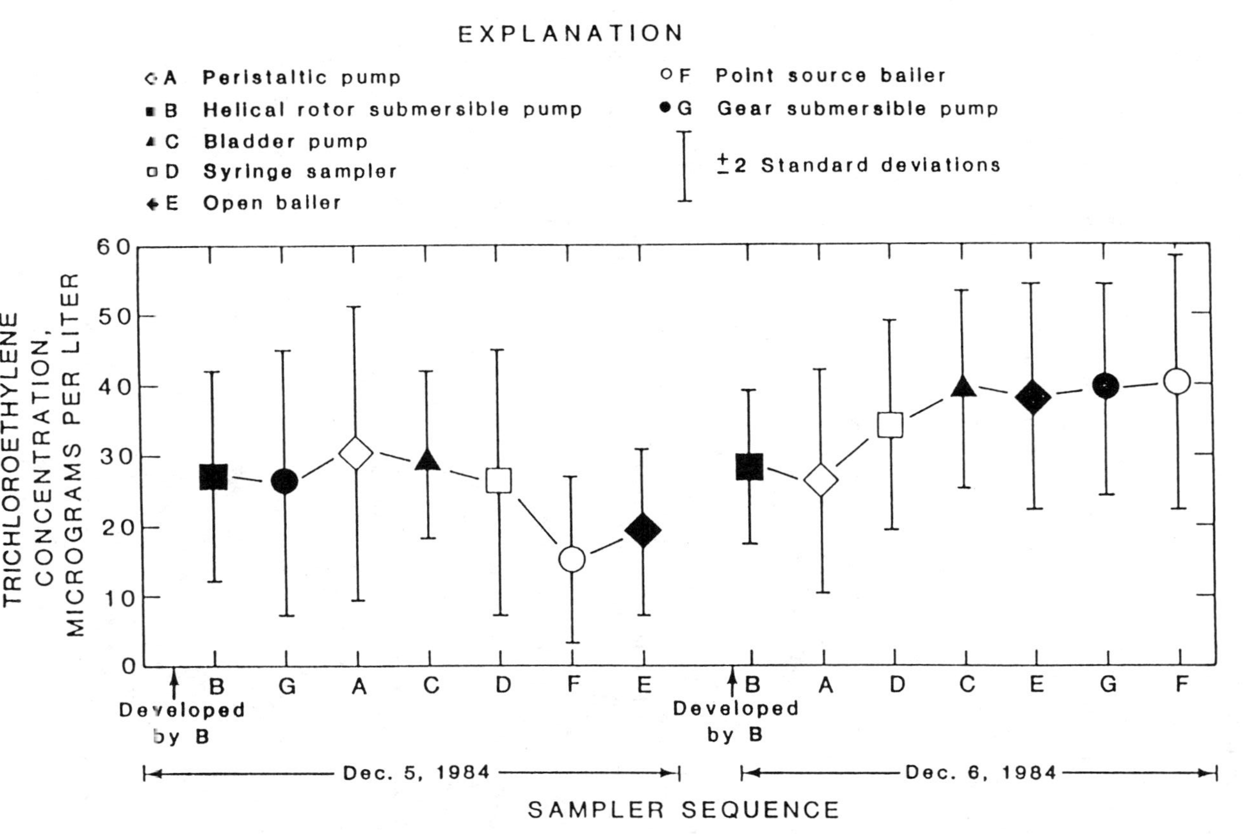


FIG. 2—*Vcriation of mean trichloroethylene concentrations by sampler at the southern New Jersey test site, 5 and 6 Dec. 1984.*

found in contaminated ground water such as dichloroethylene, trichloroethylene, and tetrachloroethylene.

The ideal sampler for purgeable organic compounds would: (1) subject the sample to a minimum of turbulence, (2) not expose the sample to negative pressure or vacuum, (3) not heat the sample, and (4) be constructed of nonreactive materials such as stainless steel, Teflon, Viton, or glass. The top five sampling devices—the gear submersible pump, point-source bailer, open bailer, helical-rotor submersible pump, and bladder pump—all meet these criteria and may be used with proper care to collect POC samples in suitable situations. If a well is 5 cm (2 in.) in diameter with a limited amount of water to flush from the casing, then one of the positive-displacement pumps can be used to both flush and sample the well for POCs. However, if the well has a larger diameter and a larger volume of water to flush, a nonideal pump that has a higher pumping rate may be used to flush the casing, after which a bailer or a positive-displacement pump can be used to collect the sample for POCs.

*Acknowledgments*

This research was funded by the Toxic Waste—Ground-Water Contamination Program of the U.S. Geological Survey. The authors would like to thank James A. Kammer for his help with the laboratory analyses and the data reduction.

## References

[*1*] Ho, J. S.-Y., "Effect of Sampling Variables on Recovery of Volatile Organics in Water," *Journal of the American Water Works Association,* Nov. 1983, pp. 583–586.

[*2*] Unwin, J., "A Laboratory Study of Four Methods of Sampling Ground Water for Volatile Organic Compounds," NCASI Technical Bulletin No. 441, National Council of the Paper Industry for Air and Stream Improvement, New York, Aug. 1984.

[*3*] Barcelona, M. J., Helfrich, J. A., Garske, E. E., and Gibb, J. P., "A Laboratory Evaluation of Ground Water Sampling Mechanisms," *Ground Water Monitoring Review,* Vol. 4, No. 2, Spring 1984, pp. 32–41.

[*4*] Houghton, R. L. and Berger, M. E., "Effects of Well-Casing Composition and Sampling Method on Apparent Quality of Ground Water" in *Proceedings,* Fourth National Symposium and Exposition on Aquifer Restoration and Ground Water Monitoring, National Water Well Association, Columbus, OH, 23–25 May 1984.

[*5*] Pearsall, K. A. and Eckhardt, D. A., "Effects of Selected Sampling Equipment and Procedures on Trichloroethylene Concentrations in Ground-Water Samples," *Ground Water Monitoring Review* Vol. 7, No. 2, Spring 1987, pp. 64–73.

[*6*] Morrison, R. D., *Ground Water Monitoring Technology, Procedures, Equipment, and Applications,* Timco Manufacturing, Inc., Prairie du Sac, WI, 1983.

[*7*] *Instruction Manual for the SP-Series Submersible Sampling Pump,* Keck Geophysical Instruments, Inc., Okemos, MI, 1985.

[*8*] *Pump Applications Guide 1984–1985,* Cole-Parmer Instrument Company, Chicago, 1984.

[*9*] "Procedures and Equipment for Ground Water Monitoring in Small Diameter Wells," Industrial and Environmental Analysts, Inc., Research Triangle Park, NC, 1984, pp. 9–11.

[*10*] Barcelona, M. J., Helfrich, J. A., and Garske, E. E., "Sampling Tubing Effects on Ground Water Samples," *Analytical Chemistry,* Vol. 57, Feb. 1985, pp. 460–464.

[*11*] Wood, W. W., "Guidelines for Collection and Field Analysis of Ground-Water Samples for Selected Unstable Constituents," U.S. Geological Survey Techniques of Water Resources Investigations, Book 1, Chapter D2, 1976.

[*12*] Claassen, H. C., "Guidelines and Techniques for Obtaining Water Samples that Accurately Represent the Water Chemistry of an Aquifer," U.S. Geological Survey Open-File Report 82-1024, Lakewood, CO, 1982.

[*13*] *Methods for Organic Chemical* Analysis of Municipal and Industrial Wastewater, EPA-600/4-82-057, J. E. Longbottom and J. J. Lichtenberg, Eds., U.S. Environmental Protection Agency, Cincinnati, OH, July 1982.

[*14*] Zar, J. H., *Biostatistical Analysis,* Prentice-Hall, Englewood Cliffs, NJ, 1974.

[15] *SAS User's Guide: Basics 1982 Edition,* SAS Institute Inc., Cary, NC, 1982.
[16] *SAS User's Guide: Statistics 1982 Edition,* SAS Institute Inc., Cary, NC, 1982.
[17] *Movement and Fate of Solutes in a Plume of Sewage-Contaminated Ground Water, Cape Cod, Massachusetts: U.S. Geological Survey Toxic Waste–Ground-Water Contamination Program,* U.S. Geological Survey Open-File Report 84-475, D. R. LeBlanc, Ed., Boston, MA, 1984.

*Bengt-Arne Torstensson*[1] *and Andrew M. Petsonk*[2]

# A Hermetically Isolated Sampling Method for Ground-Water Investigations

**REFERENCE:** Torstensson, B-A. and Petsonk, A. M., "**A Hermetically Isolated Sampling Method for Ground-Water Investigations,**" *Ground-Water Contamination: Field Methods, ASTM STP 963,* A. G. Collins and A. I. Johnson, Eds., American Society for Testing and Materials, Philadelphia, 1988, pp. 274–289.

**ABSTRACT:** Ground-water studies are often associated with the use of actions or materials which can lead to introduction of errors or contamination. For instance, transfer of liquids from a bailer to a laboratory vial can permit volatile chemicals sampled from water in an open well to escape, or can give rise to external contamination or human contact with the fluids.

A set of devices, the BAT® Groundwater Monitoring System, has been developed which specifically eliminates these and other difficulties. All individual elements of the system are hermetically sealed, permitting *in situ* conditions (for example pressures and dissolved gases) to be maintained. Hydraulic interconnections and fluid transfers are achieved through use of hypodermic needles, flexible septa, and internally generated pressure gradients. The system permits pressurized sampling of both liquids and gases from virtually any depth in groundwaters, open or sectioned-off wells, and piping systems, as well as directly from surface waters. Groundwater may be extracted *in situ* from either the saturated or vadose zones of soil or rock formations.

Samples are taken in the field with the same presterilized and pre-evacuated vials that will be sent to the laboratory; no further transfers, pumps, or piping mechanisms are required. Since the samples are sealed and pressurized, the total gas content will remain the same as that at the sampling site, and the samples can truly be said to be representative of the *in situ* conditions.

The *in situ* system also contains additional above-ground components for use in ground-water investigations. These measure ground-water pressures or soil suctions, perform tests of hydraulic conductivity, and allow tracer tests to be run. All of these devices use needles and septa to access the only below-ground piece of equipment. This—a permanently installed and sealed filter tip—contains no electronics, and only a single moving part: a flexible septum.

**KEY WORDS:** ground water, surface water, sampling, monitoring, representative samples

Standardization of ground-water studies requires attainment of a high degree of control over as many phases of the studies as possible. The development of a large body of protocols for specific chemical analyses performed in the laboratory (for example Environmental Protection Agency (EPA) Methods 601, 602 [*1*] etc.) has provided a much needed boost in this direction. Similar protocols are sorely lacking in other areas of ground-water investigation, although some progress has been made in preparing guidelines for ground-water monitoring point location, construction, and sampling procedures.

This paper presents a family of devices, the BAT® Groundwater Monitoring System [*2*], which was developed specifically to address many of the above concerns. In particular, this system allows one to

1. use simple, reliable, in-ground ports for performing ground-water studies;

[1] President, BAT Envitech AB, S-182 36 Danderyd, Sweden.
[2] President, BAT Envitech Inc., Long Beach, CA 90807.

2. install test ports using controlled procedures;
3. use the ports for a variety of different tests, using a range of temporarily connected adapters;
4. perform all tests in a hermetically isolated environment, allowing exacting control over most test variables; and
5. transfer samples or other test data directly to the laboratory, using techniques analogous to those used to do the test in the field.

## Description of the Method

Tests which may be performed using the hermetically isolated method include sampling of pore liquid and gas, measurement of pore pressures and tensions, *in situ* measurement of hydraulic conductivity, and tracer testing.

The manner in which the system is applied is illustrated in Fig. 1. It may be seen that all links in the chain connecting ground water via a test adapter to the laboratory are isolated from the external physical, chemical, and human environment.

Note that the technique shown in Fig. 1 allows use of several kinds of access ports, test adapters, laboratory procedures; and transfer devices. However, all of these are interconnected using the same types of mechanisms. The means by which this is accomplished is shown in Fig. 2. It may be seen that the equipment facilitates transfer of liquids, gases, and pressures via a system of hypodermic needles and septa located in the various devices.

As mentioned above, a variety of test adapters has been developed for use with this system, and these adapters are discussed in the sections which follow. In addition, two different means of providing an access port to ground water have been utilized, and these are also described below.

## Equipment

Components of the hermetically isolated system have been designed to use one of two types of ground-water access ports. One port is a sealed filter tip placed *in situ* in soils at the end of an

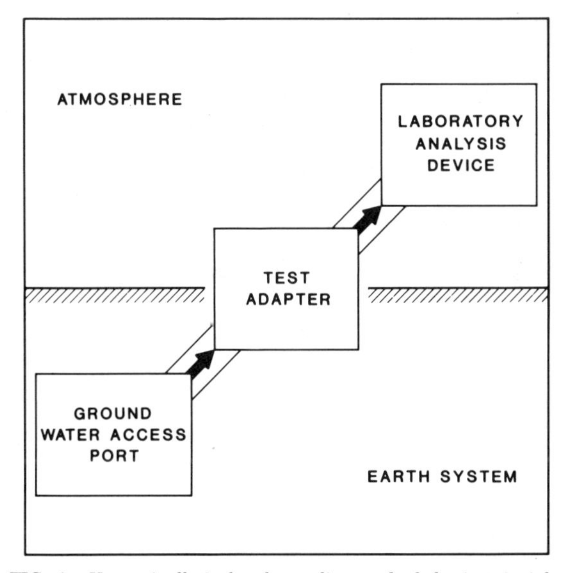


FIG. 1—*Hermetically isolated sampling method: basic principle.*

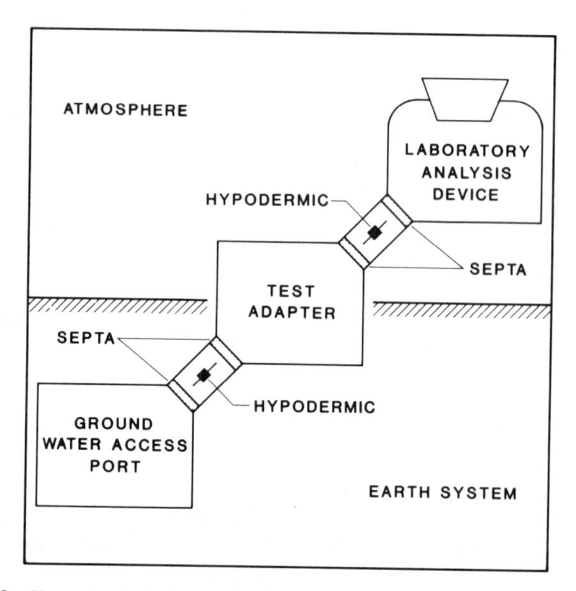


FIG. 2—*Hermetically isolated sampling method: sample transfer mechanisms.*

extension pipe; the other is a conventional (monitoring) well or other borehole. Certain adapters have been developed for use with each of these access methods.

## In Situ *Sampler*

The *in situ* ground-water sampling mechanism fits into the system concept as shown in Fig. 3. It consists of three basic components; see Fig. 4:

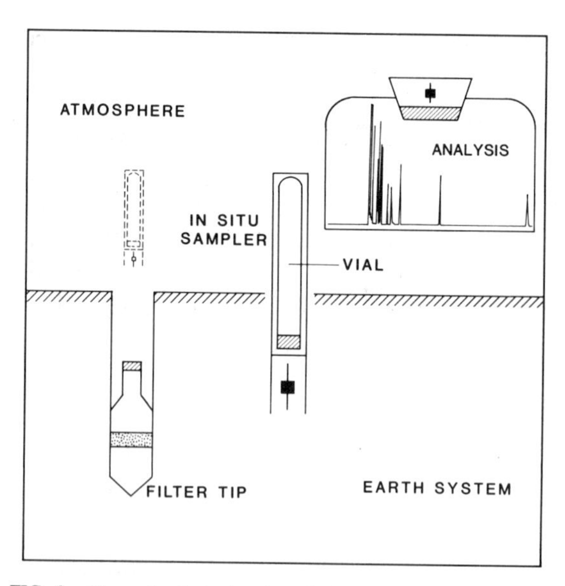


FIG. 3—*Hermetically isolated method:* in situ *sampling system.*

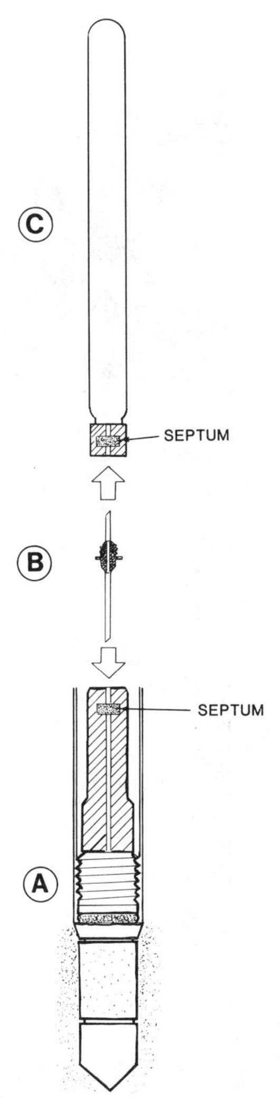


FIG. 4—*Components of* in situ *sampling system: (*A*) filter tip attached to extension pipe; (*B*) double-ended hypodermic needle; (*C*) sample vial.*

(a) a permanently installed, sealed filter tip which is attached to an extension pipe,
(b) a disposable, double-ended hypodermic needle, and
(c) a presterilized, evacuated sample vial of glass.

Both the filter tip and the sample vial are sealed with a specially developed septum. The vial is mounted in a portable sampling probe together with the needle. When lowered down the extension pipe (Fig. 5), the probe connects to the cap of the filter tip, causing the ends of the needle to penetrate the septa in the tip and the sample vial, respectively.

The hypodermic needle provides a temporary, leakproof hydraulic connection between the filter tip and the sample vial. Due to the vacuum in the vial, ground water or gas or both are therefore drawn from the formation via the filter tip into the vial.

Some fluid must be purged from the system prior to actual sampling. However, since the tip is

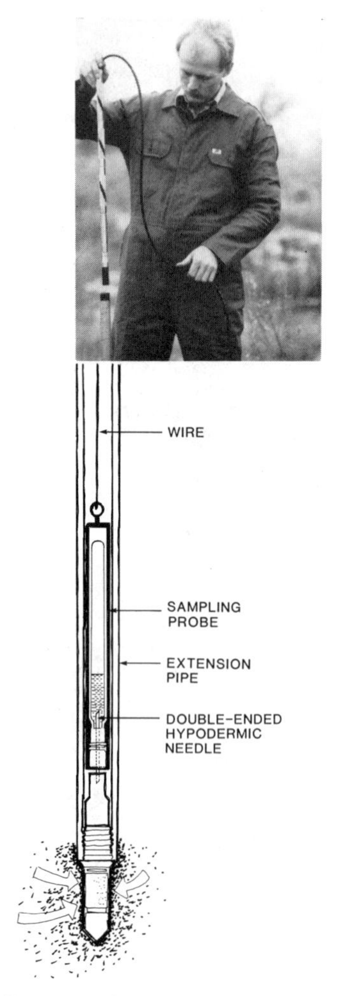


FIG. 5—*Connecting the* in situ *sampler to the filter tip.*

in direct contact with formation fluids, and contains only a very small volume itself, the amount required to be purged is also minimized.

When the probe is disconnected from the filter tip, the septa in both the filter tip and the sample vial automatically reseal. A special mechanism causes the needle to first be disconnected from the sample vial, ensuring no loss of sampled fluid or pressure. Tests have shown that the septum in the filter tip can be pierced hundreds of times without loss of its automatic self-sealing capability.

The basic *in situ* system is designed for 2.54 cm (1 in.) nominal inside diameter extension pipes, and permits use of sample vials having a volume of 35 mL (1.183 fl oz). By use of larger-diameter pipes [for example, 3.8 or 5.08 cm (1½ or 2 in.)] the sample volume can be increased to between 150 and 500 mL (5.07 and 16.90 fl oz).

There are also a number of special add-on adapters which facilitate use of the *in situ* sampler. For instance, sample vials may be cascaded using double-ended vials. Both two- and three-vial chains have been tested. (This technique is described below for the well sampler.)

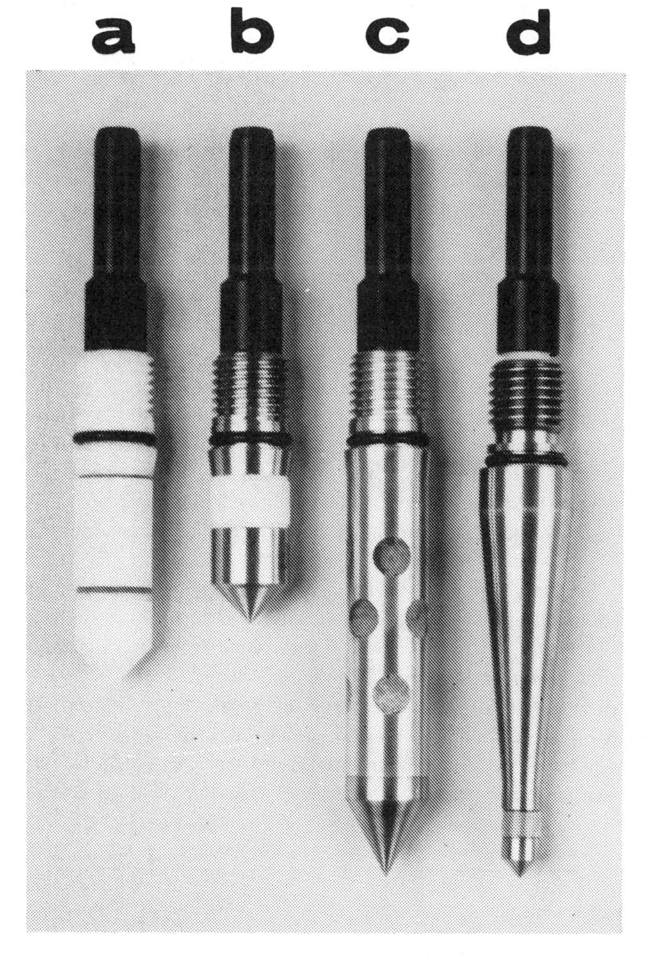


FIG. 6—*Filter tips:* (a) *standard tip with thermoplastic body;* (b) *stainless steel tip;* (c) *heavy duty tip;* (d) *probe tip.*

To aid metals analysis, an adapter carrying a 0.2 or 0.45-μm pore-size millipore® filter may be inserted between the transfer device and the sample vial. Injection of a minute amount of acid into the evacuated vial just prior to sampling can also aid sample preservation for such analyses.

*Filter Tips*—BAT filter tips have been developed in a variety of configurations and materials (Fig. 6);

1. The standard filter tip has been designed with special consideration given to requirements for both high chemical inertness and adequate mechanical strength. It has a body of thermoplastic and a filter of high-density polyethylene, sintered ceramic, or porous polytetrafluoroethylene (PTFE). The ceramic filter has an air entry value which permits sampling and pressure measurement in the vadose zone.
2. The stainless steel (ASTM 316 H) tip provides additional installation strength.

3. The heavy-duty tip has its filter molded inside the body of stainless steel or brass. This tip is designed for use in coarse soils such as gravels and tills.

4. The probe tip is used for special purposes, for example, sampling and pressure measurement in thin soil layers.

As was mentioned above, a filter tip is attached to an extension pipe, and is normally installed by pushing it under a static load down to the desired depth. The heavy-duty tip can be driven into the soil using a light hammer. Installations can, of course, also be made in predrilled holes.

*Well Sampler*

The hermetically isolated method also encompasses a sampler for wells as shown in Fig. 7. A detail of the sampling apparatus itself appears in Fig. 8. In use, an evacuated, presterilized sample vial is placed into the sampler housing. This housing also contains a septum at its lower end, and a double-ended hypodermic needle. A piston at the upper end of the housing is connected by a length of tubing to a hand pump at the surface. When a small air pressure is applied through the tubing, the piston expands, forcing the ends of the needle to penetrate the septa in the vial and at the base of the sampler housing. Flow is initiated immediately, and continues until the vial is filled and pressurized. By opening the tubing at the surface, the piston contracts, allowing the vial to reseal.

Using a vial that has been sufficiently evacuated, it will be visibly filled by the sampled liquid. If, however, total elimination of head space is desired, a configuration similar to that mentioned for the *in situ* sampler may be used (Fig. 9). Two sample vials are cascaded together using an additional hypodermic needle and septum. Upon completion of sampling (Fig. 10), the lower vial (A) will be entirely full, with all head space restricted to the upper vial (B). This technique may also be utilized when it is desirable to obtain simultaneous, duplicate samples.

Two versions of the well sampler are currently in use, one for ''shallow'' and one for ''deep'' applications. Where appropriate, these samplers may be used in either open or sectioned-off wells,

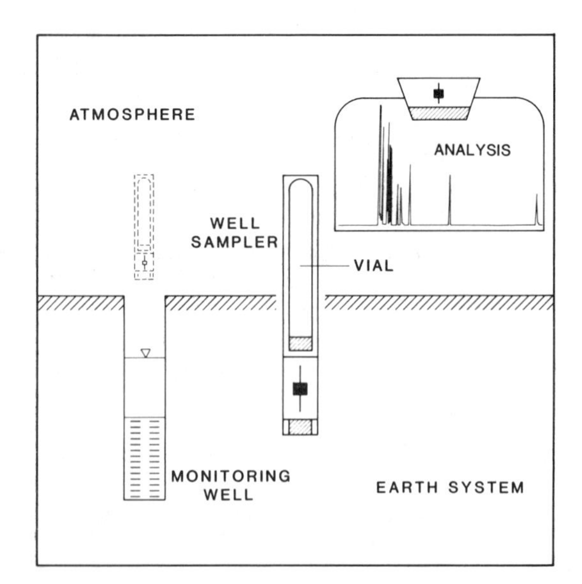


FIG. 7—*Hermetically isolated method: well sampling system.*

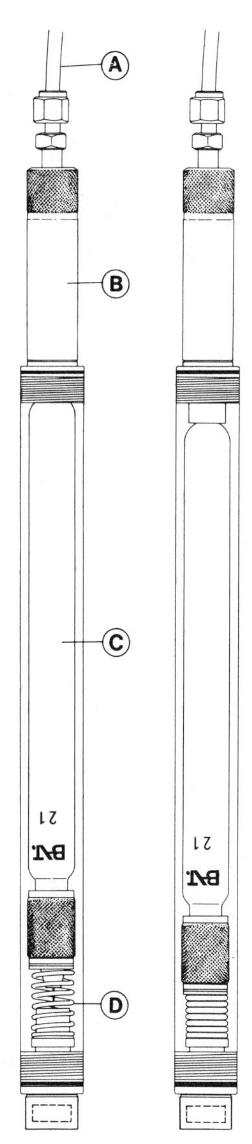


FIG. 8—*Components of well sampler:* (A) *activation tubing;* (B) *piston;* (C) *sample vial;* (D) *double-ended hypodermic needle.*

or in surface waters. Due to their small capacity, these devices are not intended for evacuation or purging of wells and boreholes.

*Shallow Well Sampler*—The sampler for shallow wells may be used at depths of up to 200 m (656 ft) below the water table or piezometric level. This system utilizes a glass or stainless steel sample vial, and a housing constructed of stainless steel. The activation tubing to the surface is normally made of a tetrafluoroethylene-based material. A pressure of only approximately 800 kPa (8 bars) is required to activate the sampler.

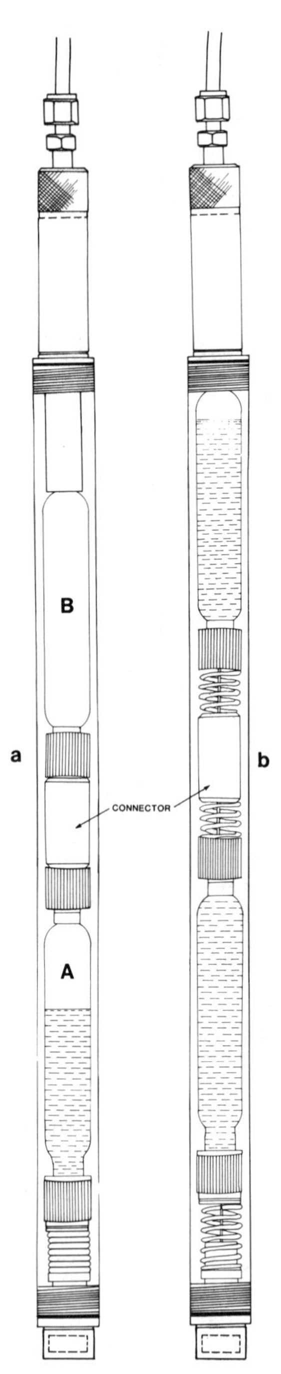


FIG. 9—*Well sampler with cascaded sample vials:* (a) *during sampling;* (b) *after completion of sampling.*

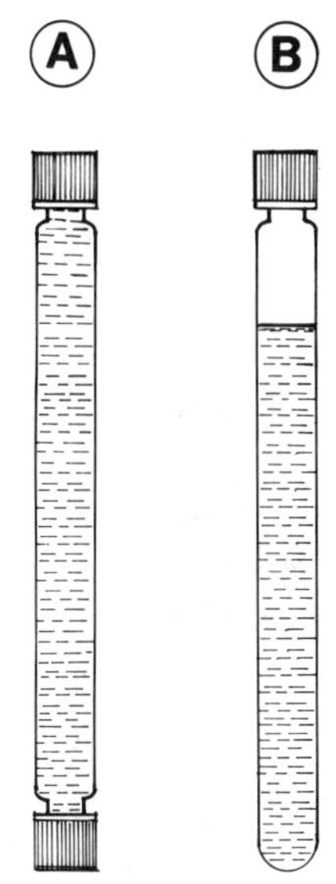


FIG. 10—*Samples ready for transfer to laboratory:* (A) *without head space;* (B) *with head space.*

*Deep Well Sampler*—The deep well sampler is intended for wells in which the water pressure head can reach values between 20 and 30 GPa (approximately 2000 to 3000 m of water column[3]). It has the same basic design as the shallow sampler, the differences being primarily in the materials used in construction of the device.

*Other Applications*

The *in situ* BAT system allows connection to the filter tip of a host of devices for performing a variety of tests on soil and soil pore contents other than just sampling of pore fluids. Adapters have been developed for measurement of pore pressures and tensions, and *in situ* measurement of hydraulic conductivity [*2,5,7*]. The latter device may also be used for tracer and hydraulic fracture testing.

The pore pressure adapter is seen in Fig. 11. It uses a silicon strain gage-type pressure transducer, connected to a single-ended hypodermic needle via a liquid-filled chamber. By lowering this adapter down the extension pipe to the filter tips, pressures or tensions or both may be sensed almost instantaneously, with a resolution of 0.01 m $H_2O$ (98 kPa).

The permeameter (Fig. 12), performs rapid, discrete measurements of both saturated and unsaturated hydraulic conductivities. It uses a transducer to measure the rate of change in gas pressure as liquid flows into or out of the otherwise closed test chamber. This rate reflects the flow rate in accordance with Boyle's law. A detailed discussion of the theory and application of this device is found in Ref. *7*.

Data required for computing the hydraulic conductivity may be collected either manually or automatically. The results are usually initially analyzed in the field simultaneously with data generation (Fig. 13).

[3] 10.33 m $H_2O$ = 1 atm.

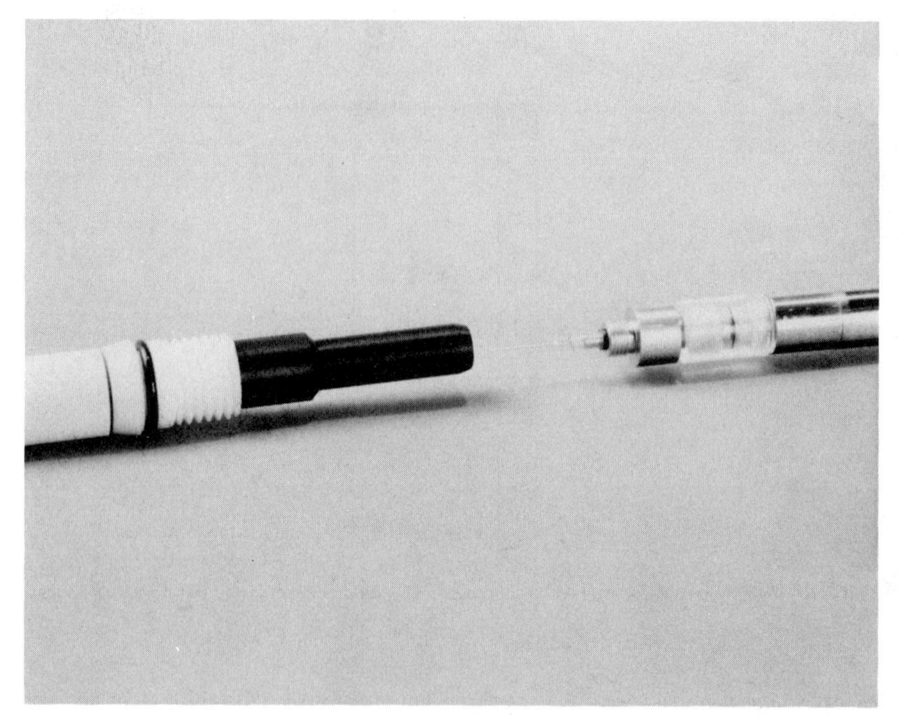

FIG. 11—*The pore pressure probe consists of a silicon strain gage-type pressure transducer equipped with a hypodermic needle.*

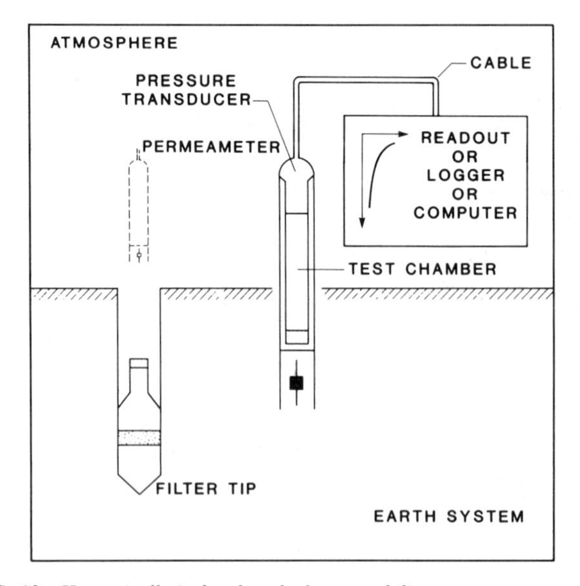


FIG. 12—*Hermetically isolated method; permeability measurement system.*

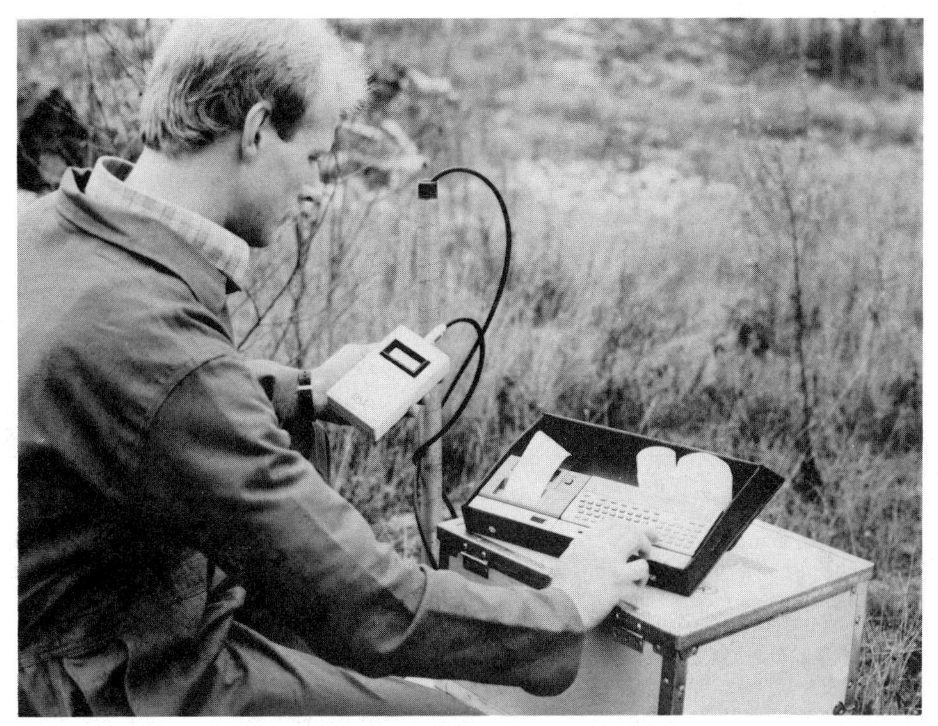

FIG. 13—*Permeability field data analysis.*

## Experimental Results

### *Field Testing of Sampling Devices*

The sampling mechanisms described above have been field tested at a number of sites.

*Deep Well Sampler*—Tests of the deep well sampler are being carried out within the framework of the Swedish program for disposal of high-level nuclear waste in crystalline rock. These tests are part of a project to characterize the chemistry of deep ground waters. The ability to analyze for dissolved gases at great pressures was a primary factor motivating selection of this instrument [*3*].

In use, two deep samplers are employed within a packer section, which is part of a package of downhole test equipment. The packers seal off a test section which is then pumped. After pressure recovery, the samplers are activated, and pressurized water samples are recovered. The two samplers can be activated simultaneously, or separately at different test sections.

Data from this project unfortunately have not yet been released for publication.

*Shallow Well Sampler*—Tests with the shallow well sampler were conducted under the auspices of The Earth Technology Corporation at two facilities in California being investigated for limited subsurface contamination by various organic compounds. Parallel, duplicated samples were obtained from purged monitoring wells using both a conventional tetrafluoroethylene (TFE) bailer and an acrylic version of the BAT sampler. These samples were analyzed using EPA Methods 601 and 602 for volatile and aromatic organics. Results for the detected compounds of interest are listed in Tables 1 and 2.

TABLE 1—*Comparative organic sampling analysis results for Site P (concentrations in μg/L).*

| Parameter | Monitoring Well: Bailed W-18 | Monitoring Well: BAT sampler W-18 | BAT Filter Tips, *In Situ* Sampler: P 3 | BAT Filter Tips, *In Situ* Sampler: P 6 |
|---|---|---|---|---|
| EPA Method 601 | | | | |
| tetrachloroethene | 6.2 | 11 | N/D[a] | . . . |
| trichloroethylene | 43 | 56 | N/D | . . . |
| other EPA Method 601 | N/D | N/D | N/D | . . . |
| EPA Method 602 | | | | |
| benzene | 25 | 30 | 3.7 | 1.4 |
| chlorobenzene | 150 | 190 | 93 | N/D |
| ethylbenzene | 2.3 | 3.1 | 1.6 | N/D |
| toluene | 1.2 | 1.1 | 1.6 | N/D |
| other EPA Method 602 | N/D | N/D | N/D | N/D |

[a] N/D = not detected (<1 μg/L).

It may be seen that in all cases, samples taken using the 35-mL (1.183 fl oz) size BAT well sampler led to analysis results comparable to those for bailed samples [40-mL (1.352 fl oz) size]. Most interesting, however, are the results from Site P (Table 1). Here the BAT samples showed that levels of most detected volatile compounds were 20% to 35% higher than would have been assumed from the bailed samples. For one parameter, the increase was 77%.

These limited data therefore seem to verify the ability of the hermetically isolated method to reduce loss of volatile organics due to atmospheric contact. Actual levels of volatiles are probably even higher than those detected here, since the sample vials contained some head space (a result of incomplete evacuation), and no head space analysis was performed. It will, however, be necessary for other comparative field tests to be performed before definite conclusions can be drawn in this regard.

In Situ *Sampler*—The *in situ* sampler was also used at both of these sites. Prior to installation at Site P, determinations of stratigraphy were done using a static cone penetration test. The selected positions of the filter tips relative to the wells and generalized lithologies are shown in Fig. 14. Only some of the tips shown were used in this study. Samples were analyzed using EPA Methods 601 and 602, and these results are also indicated in Tables 1 and 2.

There are obvious differences between the samples gathered *in situ* and those extracted from well water. While in general many of the same parameters were detected, the levels of material

TABLE 2—*Comparative organic sampling analysis results for Site S (concentrations in μg/L).*

| Parameter | Monitoring Wells: Bailed W-3 | Monitoring Wells: Bailed B-6 | Monitoring Wells: BAT Sampler W-3 | Monitoring Wells: BAT Sampler B-6 | BAT Filter Tip, *In Situ* Sampler S-1 |
|---|---|---|---|---|---|
| EPA Method 601 | | | | | |
| *1,1,1*-trichloroethane | N/D[a] | N/D | N/D | 7.7 | 7.9 |
| other EPA Method 601 | N/D | N/D | N/D | N/D | N/D |
| EPA Method 602 (all) | N/D | N/D | N/D | N/D | . . . |

[a] N/D = not detected (<1 μg/L).

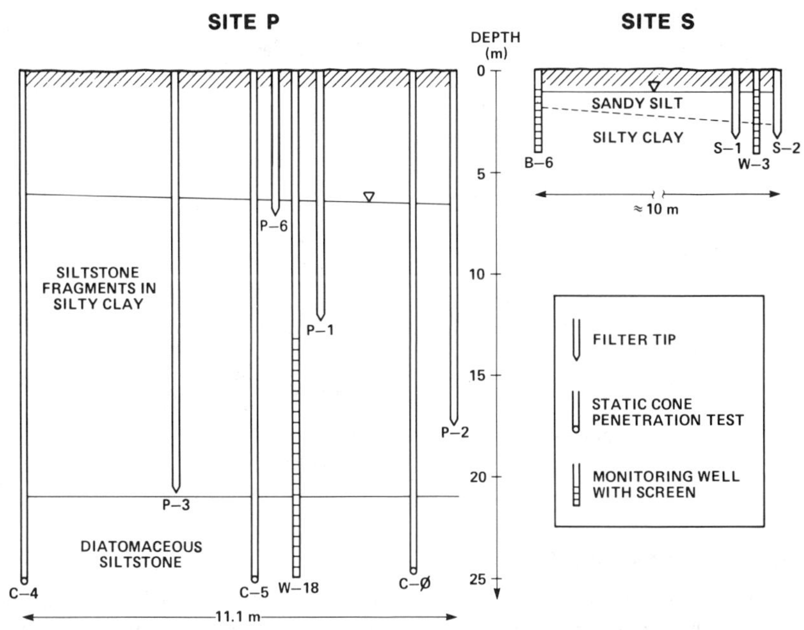


FIG. 14—*Comparative sampling test sites; generalized vertical profiles.*

identified are not in good agreement. This discrepancy is due primarily to the discrete nature of the *in situ* filter tips compared to the integrated effect of a screened well. In the heterogeneous type of soil at Site P shown in Fig. 14, additional filter tips or "chemical sounding" would be required to accurately locate the source of each of the contaminant constituents. This effect can become even more pronounced in highly stratified soils.

It should be noted that stratigraphic information can be determined *in situ* using pore pressure sounding or, as was done at Site P, cone penetration testing. The soil type can be further verified by performing a permeability test simultaneously or in conjunction with sampling.

It is also interesting to note that the slight contamination of the soil at Site S was discovered using only the hermetically isolated sampling methodology, and that similar levels were noted at two different locations (see Fig. 14). Again, this is most probably due to soil heterogeneities. The ability of the *in situ* sampler to extract water from the clay soil at this site underscores its usefulness in locating what was probably a remnant of an older, single-occurrence spill.

A number of other testing programs using the *in situ* sampler have published results. Torstensson [*2*] used the device to analyze flow conditions in a glacial clay sequence at Verka, Sweden. These data were used to verify assumptions about the age of pore water in the clays, and confirmed that very little flow had occurred since deposition of the sediment some 6000 years before. Chloride levels in samples from various depths correlated well with other data for rates of deposition and times of transgression/regression of proto-Baltic Sea waters; see Fig. 15.

The Swedish State Power Board has used the *in situ* system to investigate the content of methane in the pore water in several peat bogs in Sweden. This study [*4*] concluded that the *in situ* system provided reliable and highly reproducible results.

Haldorsen et al. [*5*] have used the *in situ* device to obtain samples from unsaturated agricultural soils in Norway. This work, together with pore tension and unsaturated hydraulic conductivity measurements, is intended to, among other things, provide data for a model of nitrogen migration.

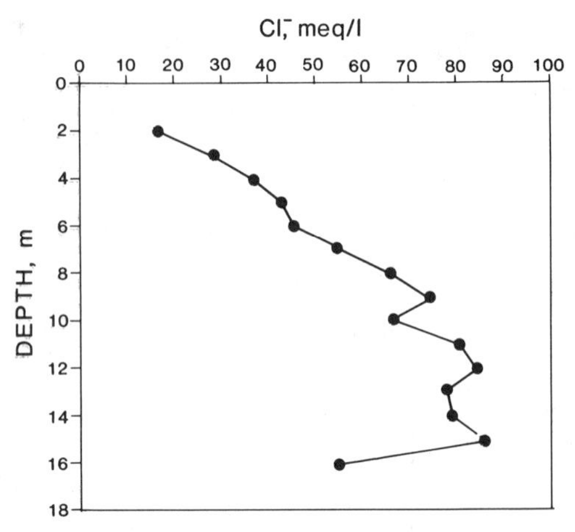


FIG. 15—*$Cl^-$ profile at the Verka test site (from [2]).*

*Laboratory Testing of Sampling Devices*

The *in situ* filter tips have undergone a large body of laboratory tests to determine the degree, if any, of their influence on sampled waters. Both the raw materials composing the tip have been tested, as well as the assembled system. Tests related to inorganic constituents were carried out at the Royal Institute of Technology in Stockholm, Sweden [*6*]. These tests have demonstrated that for properly prepared filter tips

1. most simulated ground-water solutions do not leach material from the tip,
2. components of the solutions are not attenuated by the filter or tip in significant amounts, and
3. backflushing the filter tip with distilled water, in the lab or in the field, removes all traces of previously sampled material.

Stevens Institute of Technology in Hoboken, New Jersey is currently investigating organic materials in this regard.[4]

Extensive tests of different septa have also been carried out in order to determine a material composition which will have optimum characteristics with respect to both chemical inertness and self-sealing capability.

## Conclusions

The above description has illustrated the underlying principles and benefits associated with use of a hermetically isolated sampling methodology. Such a technique allows great control over factors which could influence sampling results, and inhibits the introduction of additional factors.

In the comparative case study presented above, the result of use of this technique is dramatically seen. The isolated sampling methodology clearly identified the more extensive degree of contamination present and showed that, in order to minimize costs associated with possible cleanup operations, the location of the strata which contributed to this situation needed to be more clearly defined.

[4] G. Korfiatis, personal communication.

*Acknowledgments*

The authors wish to acknowledge the efforts of the numerous individuals and organizations who have been and continue to be involved with evaluation of the hermetically isolated sampling method. In particular, we would like to thank The Earth Technology Corporation for contributing field data to this report.

## References

[*1*] ''Methods for Chemical Analysis of Water and Wastes,'' EPA-600/4-79-020, U.S. Environmental Protection Agency, March 1979.

[*2*] Torstensson, B-A., ''A New System for Ground-Water Monitoring,'' *Ground-Water Monitoring Review*, Vol. 4 No. 4, 1984, pp. 131–138.

[*3*] Wickberg, P., ''Characterization of Deep Groundwater Chemistry; Equipment for Gas Sampling,'' presented to the Annual Meeting of the Geologic Association of Canada, Fredricton, NB, Canada, 1985.

[*4*] Sandstedt, H. and Brolin, L., ''Vyrmetan Projektet. Sammanst(llning av Analysresultat fr)n Gasprovtagningar,'' Internal Report (in Swedish), The Swedish State Power Board, Stockholm, 1985.

[*5*] Haldorsen, S., Petsonk, A.M., and Torstensson, B-A., ''An Instrument for *In Situ* Monitoring of Water Quality and Movement in the Vadose Zone,'' *Proceedings*, National Water Well Association Conference on Characterization and Monitoring of the Vadose Zone, Denver, 1985.

[*6*] Seman, P-O., ''Landfill Leachate Attenuation in Soil and Ground Water; I: Soil Water Sampling by Suction—Evaluation of Current Methods,'' Research Report Trita-Kut 1039, Department of Land Improvement and Drainage, Royal Institute of Technology, Stockholm, 1985.

[*7*] Petsonk, A.M., ''The BAT Method for *In Situ* Measurement of Hydraulic Conductivity in Saturated Soils,'' thesis in Hydrogeology, University of Uppsala, Sweden, 1984.

*Craig R. Oural,[1] Sam B. Upchurch,[1] and H. Ralph Brooker[1]*

# Sampling Interaquifer Connector Wells for Polonium-210: Implications for Gross-Alpha Analysis

**REFERENCE:** Oural, C. R., Upchurch, S. B., and Brooker, H. R., "**Sampling Interaquifer Connector Wells for Polonium-210: Implications for Gross-Alpha Analysis,**" *Ground-Water Contamination: Field Methods, ASTM STP 963,* A. G. Collins and A. I. Johnson, Eds., American Society for Testing and Materials, Philadelphia, 1988, pp. 290–303.

**ABSTRACT:** Ground water in peninsular Florida commonly has gross-alpha radioactivity in excess of 15 pCi/L. This activity, which is unsupported by radium, results from $^{210}Po$. Comparison of polonium with gross-alpha activity in samples collected and preserved by present methods indicates that gross-alpha analyses are inconsistent and frequently underestimate total alpha activity.

In interaquifer recharge wells, polonium decreases with depth owing to sorption and mixing with host water. Therefore, sampling depth affects polonium and gross-alpha activity determination. $^{210}Po$ analysis is sensitive to sample collection, preservation, and preparation techniques. With few exceptions, sampler selection has little impact on polonium recovery. Bailers and air-lift samplers give unreliable polonium recovery. Sample preparation methods are critical to polonium recovery. Filtration and sample fixation must be in the field.

**KEY WORDS:** uranium daughters, radium, polonium, recharge wells, ground-water samplers, ground-water chemistry, Florida, phosphate mining

As a means of mitigating water withdrawals from the Floridan aquifer, the central Florida phosphate industry utilizes passive, interaquifer recharge wells, which are screened in the surficial sand aquifer and drain into the Floridan carbonate aquifer. By removing water from the surficial aquifer and transferring it into the Floridan, reduction in the impact of withdrawals on the Floridan aquifer and partial dewatering of land to be mined are achieved. Recharge-well design includes isolation of the phosphatic ore, which is known to contain uranium and its daughters [*1*]. Regulatory agencies require use of recharge wells to mitigate withdrawals and monitoring of the quality of water recharging the Floridan. Water being transported to the Floridan is expected to conform to public drinking water standards.

A chronic problem detected by the monitoring program is the failure of water samples to meet gross-alpha radiation standards of 15 pCi/L. Gross-alpha activities as high as 2300 pCi/L have been reported [*2*]. In many cases these samples are found to be within standards for radium and, thus, there is alpha radiation that is not supported by decay of radium. The excess gross-alpha radiation is due to $^{210}Po$.

Gross-alpha radiation analyses show considerable variation from well to well [*2*] and within a single well through time. In the absence of standard sampling and sample preparation methods, the apparent spatial and temporal variability has been attributed to both sampling error and real

[1] Research associate, Department of Physics, professor of geology, and associate professor of physics, University of South Florida, Tampa, FL 33620.

variation. Most of this variability is due to incomplete recovery of $^{210}$Po. The purpose of this paper is to

1. demonstrate the effects of sampling methodology and sample treatment on $^{210}$Po analyses, and
2. propose new standards for sample acquisition and treatment for $^{210}$Po analysis.

## Previous Work

$^{238}$U, the progenitor of $^{226}$Ra, $^{222}$Rn, and $^{210}$Po, constitutes 99.3% (by mass) of the uranium present in the District [*3,4*]. The remaining 0.7% is $^{235}$U. $^{232}$Th, the ancestor of $^{228}$Ra, has an activity that is nearly two orders of magnitude less than that of uranium in the host rock. Thus, one can assume that $^{238}$U and its daughters constitute the primary sources of radiation in the District.

Upchurch et al. [*5,6*] studied the effects of recharge wells on ground-water chemistry, including water from recharge wells. They concluded that water in the recharge wells chemically reflects the surficial aquifer. Kimrey and Fayard [*7*] studied twelve recharge wells in the District. They found that 50% of the wells exceeded gross-alpha radiation standards.

The Polk County Health Department [*2*] conducted a study of the radiation in 71 recharge wells. They found that 35% of the wells exceeded the 5 pCi/L limit for radium and 51% exceeded the 15 pCi/L limit for gross-alpha activity. There seemed to be no correlation of gross-alpha or $^{226}$Ra activities with sample depth. And, inconsistencies in use of sampling vessels and procedures weakened comparisons between samples. This study indicated the need for a study in which sampling and analytical techniques are standardized.

## Methods

### *Plan of Study*

A recharge well was selected that is known to exhibit unsupported gross-alpha activity. In order to optimize the sample-handling procedure, the effects of samplers, the variation of radiation with depth, and the effects of filtering and sample preservation techniques were determined.

Seven phosphate-mining companies participated in a controlled sampling exercise using their routine procedures. We sampled at a fixed depth and with a standard sampler before and at the end of each day's sampling. The companies sampled the well according to their own practices. Their samples were split. Half was sent to the commercial laboratory that normally does their radiation work. The other half was analyzed by us using procedures described below.

### *Laboratory Equipment*

The hardware used by us includes complete alpha- and gamma-detection systems with special software and spectroscopic procedures developed for environmental analyses [*8*]. The alpha-detection system consists of Tennelec TC-256 alpha spectrometers, each with low-background, cleanable, 400-mm$^2$ surface-barrier detectors.

### *Chemical Procedures*

All procedures are standard methods used in research laboratories [*8*]. Commercial laboratories utilized Environmental Protection Agency (EPA)-approved methods for gross-alpha and radium analyses.

Uranium is concentrated by coprecipitation with $Fe(OH)_3$. The uranium is separated from thorium and radium by anion exchange and then scavenged by theonyltrifluoroacetone (TTA) in the presence of benzene. The benzene solution is then evaporated on a stainless-steel planchet for counting [8,9]. $^{232}U$ is used as an internal standard.

Polonium is also coprecipitated with $Fe(OH)_3$. After resolubilization, it is allowed to plate out on a polished silver disk for about 3 h at 85°C [10]. The silver disk is then counted by alpha spectroscopy. $^{209}Po$ is used as an internal standard.

Radium and thorium isotopes are quantitatively scavenged by passing the water through manganese-impregnated acrylic fibers [11]. The fiber is then ashed and packaged for analysis by gamma spectroscopy.

### *Field Methods*

To sample the cascading water at the top of the well, an open bailer, constructed of 10-cm polyvinyl chloride (PVC), was used. To sample standing water, a 10-cm PVC sampler that could be closed to receive water from a specific depth was constructed [8].

The apparatus used for filtration consists of a Geofilter, 2.4-L polymethylmetacrylate cylinder with a 0.45 -μm Micropore filter [8].

## Study Well

The recharge well selected was well KR-98Bx, which is located in the NW$\frac{1}{4}$, s. 4, T31S, R23W on the Kingsford Mine of the International Minerals and Chemical Corp. (IMCC). Well KR-98B is chemically and radiologically typical of recharge wells in the District [2]. The well consists of 6.7 m of 25-cm PVC casing, 6.1 m of slotted casing in the surficial aquifer, 24.7 m of PVC casing to the top of the Floridan, and 130 m of uncased hole (Fig. 1).

Two measurements of flow during 1978 indicated 14 and 8 L/s at a depth of 26 m. In 1984 the well was found to have silted up to a depth of 62.5 m and flow was as little as 1 L/s.

## Causes of High Gross-Alpha Activities

Within the $^{238}U$ decay series, $^{210}Po$ contributes most of the alpha activity in the test well. An initial examination of the well revealed that there were 18.12 pCi/L $^{210}Po$, compared with 1.9 to 4.2 pCi/L $^{226}Ra$, 0.27 to 0.44 pCi/L $^{228}Ra$, 0.13 to 0.59 pCi/L $^{238}U$, and 0.18 to 0.49 pCi/L $^{234}U$. $^{210}Po$ is clearly out of equilibrium with its ancestors, and equals or exceeds the gross-alpha activity reported at the well on several occasions.

## Sample Preparation

### *Effects of Sample Preparation*

It was assumed that polonium can sorb onto container walls and particulates, decrease the available metal, and bias the analytical result. Since $^{210}Po$ analysis uses a $^{209}Po$ spike, adsorption can be tracked to some degree. Addition of nitric acid ($HNO_3$) was used to retard sorption on container walls. This is a common practice in the phosphate district and is required in metals treatment by EPA [12].

During routine monitoring by the companies the sample may or may not be fixed with acid or filtered at the time of collection, or both. Furthermore, the sample may set in the laboratory or in transit for an unspecified time. All of these procedures were thought to affect gross-alpha and polonium analysis results.

To test the effect of timing of preparation, a set of samples was filtered and acidified at differing

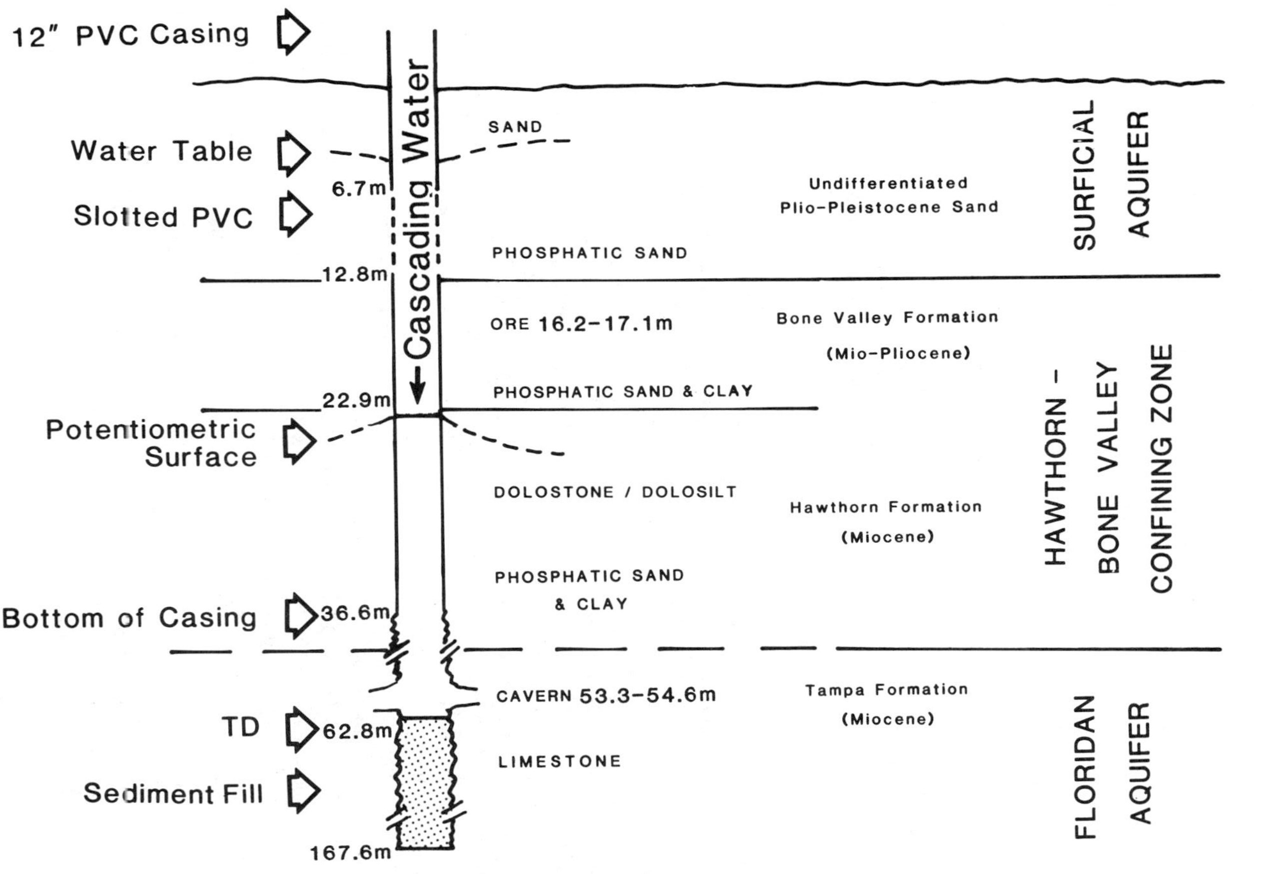


FIG. 1—*Diagram showing the design of Recharge Well KR-98B and the hydrostratigraphy of a typical recharge well in the central Florida phosphate district.*

TABLE 1—*Sorption of polonium-210 as a function of filtration and acidification time. All activities are in pCi/L at the 95% confidence level. Confidence limits are due to counting error alone.*

| Sample No. | Treatment[a] | Polonium-210, pCi/L |
|---|---|---|
| R98B-9 | filtered upon collection | 14.86 ± 0.30 |
| R98B-10 | filtered after 3 h | 4.90 ± 0.18 |
| R98B-13 | filtered after 24 h | 2.59 ± 0.14 |
| R98B-14 | filtered after 2 weeks | 8.78 ± 0.34 |
| R98B-11 | refrigerated upon collection, filtered after 24 h | 3.33 ± 0.14 |
| R98B-12 | refrigerated upon collection, filtered after 2 weeks | 1.43 ± 0.08 |

[a]Samples were collected at 46 m with a closed thief sampler, mixed thoroughly, then split. Acidification was with $HNO_3$ after filtration.

times. A sample was collected from a depth of 46 m with the closed thief sampler, thoroughly mixed and homogenized, then split into different samples. Table 1 gives the conditions of each sample treatment and the resulting Po activity.

Time of filtration is most important. Nearly 70% of the $^{210}Po$ activity was lost to suspended particulates in a 3-h period and 83% within 24 h. The 3-h interval included time to return the samples for analysis, during which they were agitated, heated, and possibly oxidized. Sorption is, therefore, very rapid and the final activity is highly dependent on turbidity of the sample and length of time before filtration. The test results clearly indicate the need for immediate filtration.

Use of different filter pore sizes was also tested. A sample was collected at a depth of 21 m. One split was filtered through Whatman No. 4 filter paper and then acidified. It had a $^{210}Po$ activity of 25.25 ± 0.64 pCi/L. The other was filtered through 0.45-μm Millipore paper. Acidification followed filtration. The $^{210}Po$ activity was 19.20 ± 0.64 pCi/L. Therefore, some of the $^{210}Po$ is fixed on micrometre-sized, filterable solids. Micropore (0.45 μm) filtration reduced the reported polonium activity by about 25%.

Sample R98B-14 (Table 1) is anomalously high, presumably because of ferric hydroxide ($Fe(OH)_3$) formed during the two-week interval. $Fe(OH)_3$ and polonium hydroxide ($Po(OH)_4$) precipitate from water samples as the water warms and equilibrates with the atmosphere. The $Fe(OH)_3$-$Po(OH)_4$ complex retains polonium in a form that can be recovered as the hydroxides are later dissolved.

Refrigeration seems much less important as a sample-preservation technique than rapid filtration, but additional studies of the effects of refrigeration are indicated. Refrigeration of the sample stored for two weeks inhibited precipitation of $Fe(OH)_3$ and the resulting polonium activity is the lowest of the sample set. Therefore, it appears that refrigeration may have a negative impact on polonium analysis and, until such time as further tests can be made, refrigeration is not recommended.

### *Acidification Studies*

Two tests of the impact of time of acidification (Tables 2 and 3) gave repeatable results. It was discovered that, while the amount of $^{209}Po$ spike recovered remained constant, the amount of $^{210}Po$ detected increased with the time between collection and acidification.

TABLE 2—*Effects of time of acidification on polonium recovery. Activities are at the 95% confidence level. Confidence limits are due to counting error alone. R98B-47 is a single water sample filtered on a 0.45-μm filter in the field and split six ways.*

| Sample No. | Treatment | Polonium-210, pCi/L |
|---|---|---|
| R98B-47-A1 | 1. acidify sample bottle<br>2. add water to bottle | 13.30 ± 0.34 |
| R98B-47-A2 | 1. acidify sample bottle<br>2. add water to bottle | 13.34 ± 0.34 |
| R98B-47-B1 | 1. add water to bottle<br>2. acidify in lab afer 3 h | 15.70 ± 0.38 |
| R98B-47-B2 | 1. add water to bottle<br>2. acidify in lab after 3 h | 16.02 ± 0.38 |
| R98B-47-B3 | 1. add water to bottle<br>2. acidify in lab after 3 days | 18.00 ± 0.50 |
| R98B-47-C | 1. add ammonium hydroxide to sample bottle<br>2. add water to bottle<br>3. acidify in lab after 3 h | 21.17 ± 0.48 |

In light of the experience with Sample R98B-14 (Table 1), it was decided to try using a base to preserve polonium through coprecipitation with $Fe(OH)_3$ (Tables 2 and 3). The sample was preserved with ammonium hydroxide ($NH_4OH$) because, first, it is used in the laboratory procedure to coprecipitate $Fe(OH)_3$-$Po(OH)_4$ and, second, ammonia ($NH_4^+$) has the strongest exchange capacity of the common cations and is used to wash exchangeable cations from clays. Thus, the $NH_4^+$ may serve to saturate container walls and clays and prevent sorption of polonium. This may not be the optimal method for polonium recovery, but the preliminary tests indicate that fixation with $NH_4OH$ increases the recovery of polonium in water samples when the samples are analyzed by spontaneous deposition of polonium on silver (Tables 2 and 3).

The causes of the differences in recovery of the $^{210}Po$ with time of acidification or the nature of the fixative or both are poorly known and need additional investigation. Polonium occurs in 4 valence states, of which +4 and +2 are most stable [*13*]. Bernabee and Sill [*14*] found that valence state is critical to polonium recovery. The best recovery by plating on silver is accomplished when the polonium is kept in the +4 state. They found that $Po^{+2}$ is highly volatile and should not be allowed to form in the sample. Organics in the sample were found to reduce the polonium and enhance losses by volatization. $HNO_3$ is used to oxidize organics and prevent reduction of the polonium. They also found that polonium dioxide ($PoO_2$) is refractory and is not recovered on the silver disks used in alpha spectroscopy. Divalent halides are volatile, so fixation with hydrochloric acid (HCl) includes the risk of formation of the volatile $PoCl_2$. In the presence of $NH_4OH$, polonium forms $Po(OH)_4$, which is stable and soluble in acid.

If the sample is stored without fixation in the presence of organics, which are abundant in the surficial aquifer waters and recharge wells, recovery of polonium may not be representative because of loss through volatilization. Also, warming of the sample causes $Fe(OH)_3$ to form, which should fix polonium in an available state. If, however, $PoO_2$ forms, the polonium analysis yields an activity less than the true activity.

In our experiments (Table 2), it appears that $PoO_2$ formation is inhibited or the oxide is destroyed in the interval prior to acidification. The spike, which is in the form of polonium nitrate

TABLE 3—*Effects of fixation with acid and base on recovery of polonium. Samples were homogenized and split. All samples were filtered on 0.45-μm paper. Activities are at the 95% confidence level, based on counting error alone.*

| Sample No. | Treatment | Polonium-210, pCi/L |
|---|---|---|
| R98B-48A | 1. acidify bottle<br>2. add water sample<br>3. spike in field | 7.44 ± 0.24 |
| R98B-48B | 1. acidify bottle<br>2. add water sample<br>3. spike in lab after 3 h | 7.88 ± 0.22 |
| R98B-48C | 1. place water in bottle<br>2. add spike in field<br>3. acidify in lab after 3 h | 11.00 ± 0.32 |
| R98B-48D | 1. add ammonium hydroxide in field<br>2. add water sample<br>3. add spike in field<br>4. acidify in lab after 3 h | 18.40 ± 0.36 |
| R98B-48E | 1. add ammonium hydroxide in field<br>2. add water sample<br>3. add spike in field<br>4. acidify in lab after 10 days | 31.13 ± 1.30 |
| R98B-48F | 1. add ammonium hydroxide in field<br>2. add water sample<br>3. add spike in lab after 3 h<br>4. acidify in lab after 10 days | 26.42 ± 0.68 |

($^{209}Po(NO_3)_4$), is not affected and recovery (or analytical efficiency) remains constant, while the sample $^{210}Po$ in a recoverable form increases.

In order for the $^{209}Po(NO_3)_4$ spike to accurately represent the $^{210}Po$, including refractory $PoO_2$, equilibration of the spike and $^{210}Po$ must occur. Our data suggest that equilibration is not necessarily complete, which results in failure to account for some of the refractory or volatile $^{210}Po$.

Fixation with $HNO_3$, as it is now used, involves risk of polonium loss. It either inhibits formation of the hydroxides, so that $PoO_2$ can form and reduce recovery, or it does not adequately oxidize the organics and volatization after reduction results. Additional $HNO_3$ fixative might resolve the latter problem.

*Impact on Gross-Alpha Analysis*

The acidification and volatization problems with polonium cast doubt on the gross-alpha radiation analytical method. This method [*15*] involves several steps that render polonium recovery questionable, including:

(a) Plastic or glass containers are allowed; Bernabee and Sill [*14*] found polonium sorption on glass to be problem.

(b) Filtration through a 0.45-μm filter is required, but no time of filtration is specified. Immediate filtration is necessary.

(c) Preservation is with 0.5 N $HNO_3$, but no amount of preservative is specified. $NH_4OH$ appears to be a better fixative for dissolved polonium.

(d) A shelf life of up to 1 year is allowed. Formation of $PoO_2$, volatization, and sorption problems necessitate shorter holding times.

(e) The liquid sample is evaporated on a planchet for counting. Evaporation, especially by heat application, can enhance polonium volatization.

Therefore, the gross-alpha radiation analytical procedure probably does not reflect the actual polonium activity in the water.

### *Sample Preservation Procedure*

Because of the high rate of polonium sorption on particulates, the suggested sample-preparation procedure for gross-alpha radiation or polonium analysis or both includes immediate filtration through a 0.45-μm filter at the time of collection.

Filtered samples for the analysis of polonium are immediately preserved with 1 to 2%, by volume, concentrated $HNO_3$. Choice of immediate preservation with $HNO_3$ will standardize the results and conform with the recommended polonium method (which suggests use of HCl) [*16,17*]. Samples for gross-alpha radiation analysis should be treated with $HNO_3$ in the same fashion. Standard gross-alpha analysis should have no problem in detecting $PoO_2$, so volatization and sorption are the major concerns.

The sample should be analyzed as soon as possible. If gross-alpha radiation analysis is to be undertaken, care should be exercised to prevent chemical reduction or volatization of polonium through heating of the sample.

Fixation with $NH_4OH$ or other procedures may be recommended upon further examination of the problem. Gross-alpha radiation results should be considered as minima, because of the unknown loss of polonium.

## Sampling Methods

### *Introduction*

Many different methods and devices are used to collect samples from recharge wells. Experiments were developed to identify variations in polonium analyses due to sampling depth and preparation methods. The first was to sample the well at different depths to determine if there is any stratification of polonium. The second was to invite interested mining companies to sample the well using their own procedures. By comparison of their results with controls, the different methods can be evaluated.

### *Polonium Stratigraphy*

To identify stratification of polonium in the well, five samples were collected. An open bailer was used to sample the cascading water at 15 m, and a closed thief was used for samples taken from depths of 23 m (top of standing water), 30, 46, and 61 m. These samples were filtered on 0.45-μm filters, and immediately acidified with 2% $HNO_3$ by volume.

There is a decline in $^{210}Po$ activity with depth (Fig. 2). The polonium activity is 77% lower at the bottom of the well than it is in cascading water. This reduction may be caused by any of three processes. First, major elements in recharge wells vary with depth [*6*], which suggests mixing that can dilute the polonium. Second, clays and organics can sorb polonium and there is an increase in turbidity with depth. Finally, Eh and pH change with depth, so it is possible that polonium valence and recovery efficiency vary with depth.

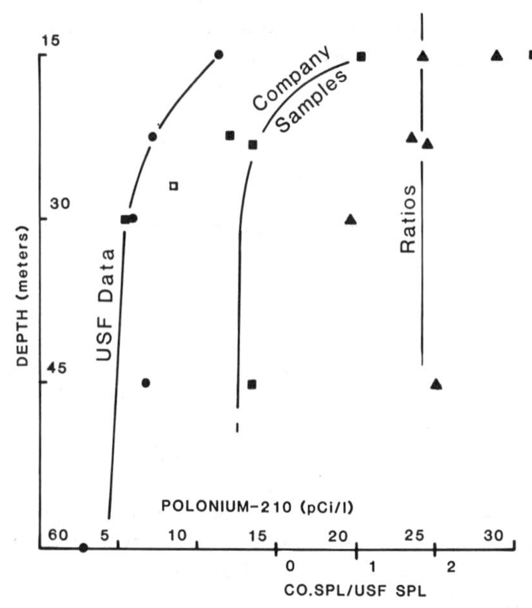


FIG. 2—$^{210}$*Po activities in Well KR-98B as a function of depth. Circles are USF samples filtered with 0.45 − µm paper. Squares are company samples analyzed by USF. The open square represents a laboratory-filtered sample; all others were field-filtered.* CO. SPL/USF SPL *represents the ratio of polonium activity measured in the company sample to that measured in the USF-collected sample.*

*Industry Sampling Methods*

The individual phosphate companies use different sampling methods, different sample-preparation techniques, and different analytical laboratories. To determine how these factors affect reported variations in gross-alpha activities, each of the seven participating companies sampled the well according to its usual sampling method and sample preparation technique. For control, the well was also sampled at a standard depth (23 m) with the open bailer before and after the company sampling; thus any changes in polonium with time or due to the sampling could be detected. Each company sample was split and both halves were treated according to the company's method. One half was analyzed for $^{210}$Po by the authors. The other half was sent to the company's usual laboratory for $^{226}$Ra and gross-alpha analyses.

By this technique, water quality and differences in technique can be controlled. Water samples taken before and after the company sampling provide control on changes in actual polonium activity. The samples taken by the company method and analyzed by the authors allow identification of differences in sampling and sample preparation technique, while the samples sent to the commercial laboratories allow evaluation of the effects of their analytical procedures.

*Temporal Variations*

The samples taken by us at the beginning and end of each company sampling day indicate that there was little variation in polonium activity with time (Fig. 3*a*). These data also show the consistency in the results of our analyses, provided that sample preparation was held constant.

Three samples were filtered approximately 3 h after collection, upon return to the laboratory. Some of the $^{210}$Po was lost through sorption on suspended solids or formation of $PoO_2$. All other

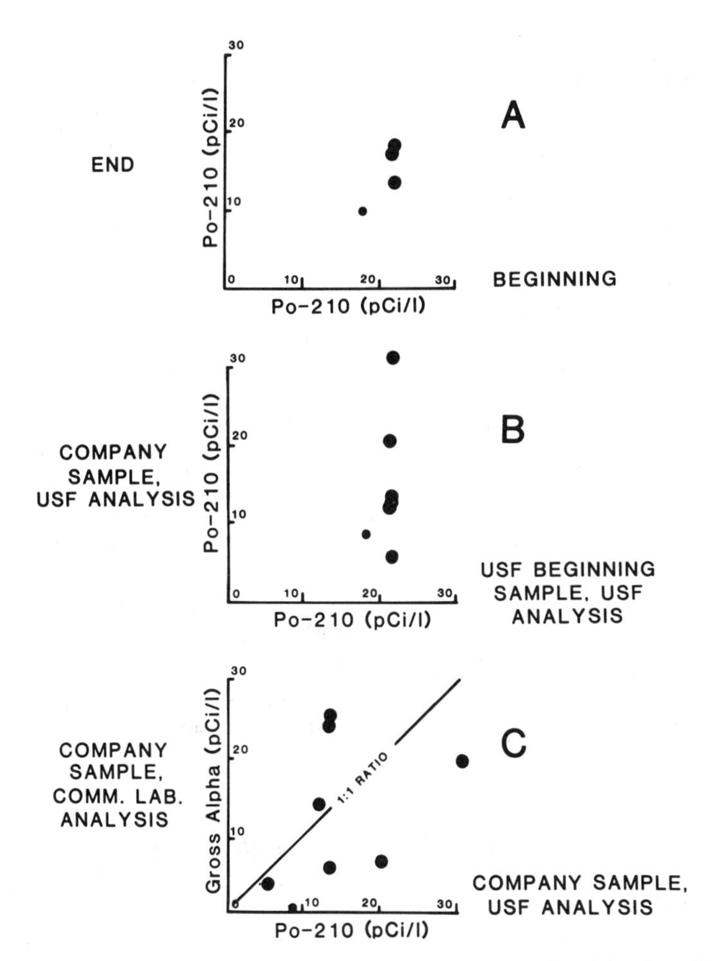


FIG. 3—(A) *Comparison of* $^{210}$*Po activities at beginning and end of each day of sampling.* (B) *Comparison of polonium activities in company-collected samples and USF sample collected at the beginning of each sampling day.* (C) *Comparison of polonium activity and gross-alpha radiation in company-collected samples. The small dot represents the sample filtered in the laboratory.*

samples were filtered in the field. The difference in activities from beginning to the end of the sampling day reflects the difference in time of acidification, which was in the laboratory 3 h after collection.

### *Comparison of Sampling Techniques*

The data from the companies varies significantly, but much of the variability can be explained. The companies used a number of different techniques for sampling (Table 4). Since specific designs are not relevant, the details of most designs are not discussed herein.

There is a great deal of variation in the $^{210}$Po analyses in the sample splits. These data, plotted against the University of South Florida (USF) beginning sample for the day of collection (Fig. 3*b*), show that the variability is not due to changes in water chemistry, but is in part due to sampling method and in part due to depth. Figure 2 shows these data plotted as a function of

TABLE 4—*Sampling devices and depths used by each participating company.*

| Company | Depth of Sample, m | Sampling Device | Sample Preparation and Analytical Laboratory |
|---|---|---|---|
| 1 | 15 (cascading water) | metal bailer | micropore filtration; Lab No. 1 |
| 2 | 15 (cascading water) | metal bailer | no preparation; Lab No. 2 |
| 3 | 23 (standing water) | LAB-LINE bailer | refrigerate in field; Lab No. 2 |
| 4 | 24 (standing water) | flow through, check ball thief | micropore filtration on 1 of 2 samples; Lab No. 3 |
| 5 | 27 | ISCO bladder pump | refrigerate in field; Lab No. 4 |
| 6 | 30 | air-lift sampler[a] | refrigerate in field; Lab No. 4 |
| 7 | 46 | stainless steel geophysical logger | no preparation; Lab No. 5 |

[a]This sampler consists of a length of 2.5-cm PVC pipe that extends to sample depth. An air hose is used to blow compressed air and water up the pipe. The water is highly aerated as it leaves the pipe.

depth. Three samples fall off of the curve. The other samples fall on a curve that parallels the curve for the USF data. The difference in position of these two curves is due to the differences in filtration method and problems in acidification mentioned above. The ratio of the data points at a given depth is also plotted (Fig. 2) and indicates that, with the exception of the three samples, the ratio is constant. Therefore, one can conclude that most of the variation in the samples is due to sampling at differing depths.

The three samples that do not conform to the depth curve indicate some of the problems with sampling. One sample (the 31.06-pCi/L sample from 15 m) was taken with a metal bailer. Bailers have two obvious problems. First, unless the water pouring into the well is homogeneous, there is a chance of collecting a nonrepresentative sample. Second, if the sampler scrapes the casing wall, microbes, clays, and scale can be incorporated into the sample. Rapid filtration should account for these particulates, but chemical changes, such as reduction or flocculation of hydroxides or both, could prejudice the results. The 5.45-pCi/L sample from 30 m is also anomalous. This sample was collected by airlift, which oxidized the iron and polonium. By the time it was filtered in the laboratory, much of the polonium had been removed with the $Fe(OH)_3$. The third sample, the 8.40-pCi/L sample taken from 27 m (the open box in Fig. 2), was filtered in the laboratory with Whatman paper, rather than in the field. Removal of the coprecipitated $Po(OH)_4$ during filtration may have reduced the polonium available for recovery and analysis.

*Commercial Laboratory Analyses*

The companies submitted their sample splits to their laboratories for $^{226}Ra$ and gross-alpha analysis. Five laboratories were used. The lag times between sample receipt and analysis are not

TABLE 5—*Results of the commercial laboratory analyses of radium-226 and gross-alpha radiation compared with polonium-210. See Table 4 for details of sample preparation.*

| Company No. | Laboratory No. | Depth, m | Gross-Alpha Radiation, pCi/L | Radium-226 pCi/L | Polonium-210 pCi/L |
|---|---|---|---|---|---|
| 1 | 1 | 15 | 19.3 ± 2.5[a] | 1.58 ± 0.25[a] | 31.06 ± 0.58 |
| 2 | 2 | 15 | 6.5 ± 1.9 | 2.3 ± 0.2 | 20.34 ± 0.48 |
| 3 | 2 | 23 | 13.9 ± 2.3 | 0.7 ± 0.1 | 12.00 ± 0.38 |
| | | | 25.0 ± 1.6[a] | 0.5 ± 0.8[a] | 13.53 ± 0.40 |
| 4 | 3 | 24 | 23.6 ± 1.6[b] | 1.3 ± 0.1[b] | |
| 5 | 4 | 27 | less than 1 | less than 1 | 8.40 ± 0.24 |
| 6 | 4 | 30 | 3.7 ± 1.1 | less than 1 | 5.45 ± 0.18 |
| 7 | 5 | 46 | 5.6 ± ??? | 1.22 ± ??? | 13.45 ± 0.34 |

[a]Reported as filtered upon collection. See Table 4.
[b]Reported as nonfiltered.

indicated on the laboratory reports. All participating laboratories are certified by the Florida Department of Health and Rehabilitative Services.

Comparison of these data (Table 5) with the polonium data taken on the USF sample split indicates that the depth relationship is not clearly represented. Laboratory 4, in particular, seems to report low results.

A test of analytical validity can be made by comparing the reported gross-alpha radiation with polonium activity (Fig. 3*c*, Table 5), which is the major alpha emitter. Note that the 1:1 ratio line indicates equal activities of the two variables. If recovery of all alpha-emitting radionuclides is efficient, gross-alpha activity must be greater than the polonium activity. Therefore, only those samples that fall above the 1:1 line are realistic. Those samples that fall below the line have low gross-alpha activities compared to polonium.

One cannot say that these inconsistencies are a result of the analytical laboratories alone. Samples from four different laboratories fall below the line. One laboratory, which reported analyses for two different companies, had a sample on each side of the 1:1 ratio line. One laboratory, which did work for two different companies, had both samples below the line and these samples were also inconsistent with the depth relationship in polonium (Fig. 2), so sampling and sample preparation are questionable. In this latter case, however, the sampling method produced small reductions in recoverable polonium activity compared to the gross-alpha activities reported.

Therefore, analytical procedures contribute to the uncertainties in determining radiation in recharge wells. With the exception of the results of one laboratory (Laboratory 4), it appears that these uncertainties are less than those introduced by sampling.

## Conclusion and Recommendations

Unsupported gross-alpha activity in the well studied is primarily due to $^{210}Po$, which is present in activities greater than predicted by $^{226}Ra$.

$^{210}Po$ and gross-alpha radioactivity measurements are extremely sensitive to sampling procedures, sample preparation techniques, and analytical procedures. Polonium decreases with depth in recharge wells, so sample depth influences the gross-alpha and polonium activities reported.

Clays [including silicates and $Fe(OH)_3$], particulate organics, and dissolved organics can interfere with polonium and gross-alpha radioactivity measurements. Clays and particulate organics can sorb polonium and reduce the reported gross-alpha activity. Oxidation of organics can reduce the Eh of the water sample and produce volatile, $Po^{+2}$ compounds, which also reduces the gross-alpha activity. Time of filtration and length of storage of the sample before preparation for analysis affect polonium recovery. While gross-alpha analysis may detect refractory polonium, risks of loss exist.

The methods of polonium radiation analysis and monitoring presently used introduce variability and reduce ability to detect, compare, and manage water quality. A standard method is needed to remedy this problem.

The following is recommended as a means of standardizing monitoring procedures. Note that the recommendation for fixation of the sample with $HNO_3$, which follows standard methods now in effect, may be an interim recommendation. We plan to continue evaluation of the effects of fixatives on polonium and gross-alpha radiation analyses and, if justified, make additional recommendations for changes in standard procedures to the American Society of Testing and Materials.

The recommended procedures for polonium and gross-alpha sampling are:

1. Samples should be collected with a closed, thief-type sampler. Airlift- and bailer-type samples are not acceptable substitutes.
2. Samples should be collected from the same horizon in all wells. It is recommended that this horizon be just below standing water in the well.
3. Samples should be filtered immediately upon collection with a 0.45-μm filter.
4. Samples should be stored in plastic containers at all times.
5. Immediate fixation of the sample to prevent loss of polonium is necessary. If gross-alpha analysis is to be undertaken, sufficient $HNO_3$ (2% concentrated, by volume) should be used to oxidize all soluble organics and minimize volatization of polonium. If polonium is to be analyzed, we recommend an interim procedure of fixation with $HNO_3$ as above to agree with standard methods now in use. Immediate fixation with $NH_4OH$ increases recovery of polonium and should be investigated further as a potentially superior method of sample fixation.
6. The sample should be analyzed promptly. Blind standards and replicate samples should be included frequently to insure quality control in the laboratory. Frequent $^{210}Po$ analysis is recommended to substantiate the source of high gross-alpha activities and to validate polonium in the gross-alpha procedure.

*Acknowledgments*

This material is based upon work supported by the Florida Institute of Phosphate Research under Award No. 82-05-014. Any opinions, findings, and conclusions or recommendations expressed in this paper are those of the authors and do not necessarily reflect views of the Florida Institute of Phosphate Research.

The following were most helpful during this study and we acknowledge them with thanks: Ken Weber and the Southwest Florida Water Management District; the International Minerals and Chemical Corp., especially Jay Allen and Jerry Tanner; Dr. Willard Moore and Dr. Jacqueline Michel of the University of South Carolina; Dr. Claude Sill of EG&G Idaho; Dr. James Cowart of Florida State University; the Florida Department of Environmental Regulation; and the participating companies (International Minerals and Chemical, Gardinier, Mobil Chemical, Estech, Brewster Phosphate, W. R. Grace, and AMAX).

## References

[*1*] Altschuler, Z. S., Clarke, R. S., and Young, E. J., "Geochemistry of Uranium in Apatite and Phosphorite," Professional Paper 314-D, U.S. Geological Survey. Reston, VA, 1958, pp. 45–90.

[2] "Florida Phosphate Recharge Well Analysis," Unpublished Report, Polk County Health Department, Office of Radiation Control, Florida Department of Environmental Regulation, Tallahassee, FL, 1983.
[3] Guimond, R. J., "The Radiological Aspects of Fertilizer Utilization," U.S. Environmental Protection Agency, Office of Radiation Programs, Washington, DC, 1977.
[4] Guimond, R. J. and Windham, S. T., "Radioactivity Distributions in Phosphate Products, By-Products, Effluents, and Wastes," U.S. Environmental Protection Agency, Office of Radiation Programs, Washington, DC, 1975.
[5] Upchurch, S. B., Linton, J. R., Spurgin, D. D., and Kaufmann, R. S., "Supplement to 'Impact of Phosphate Mining on Ground-Water Quality at the Kingsford Mine, Polk and Hillsborough Counties, Florida'—Radium in Ground Water," International Minerals and Chemical Corp., Bartow, FL, 1979.
[6] Upchurch, S. B., and Kaufmann, R. S., "Impact of Phosphate Mining on Ground-Water Quality at the Kingsford Mine, Polk and Hillsborough Counties, Florida," International Minerals and Chemical Corp., Bartow, FL, 1979.
[7] Kimrey, J. O. and Fayard, L. D., "Geohydrologic Reconnaissance of Drainage Wells in Florida," Water-Resources Investigations Report 84-4021, U.S. Geological Survey, Reston, VA, 1984.
[8] Oural, C. R., Brooker, H. R., and Upchurch, S. B, "Source of Gross-Alpha Radioactivity Anomalies in Recharge Wells, Central Florida Phosphate District," Florida Institute of Phosphate Research, Bartow, FL, 1986.
[9] Cowart, J. B. and Osmond, J. K., "Uranium Isotopes in Ground Water as a Prospecting Technique," U.S. Department of Energy, Grand Junction Office, CO, 1980.
[10] Flynn, W. W., *Analytica Chimica Acta*, Vol. 43, 1968, pp. 221–227.
[11] Moore, W. S., *Deep-Sea Research.*, Vol. 26, 1975, pp. 647–651.
[12] *Methods for Chemical Analysis of Water and Wastes,* Environmental Monitoring and Support Laboratory, U.S. Environmental Protection Agency, Cincinnati, OH, 1979, pp. xv–xvi.
[13] Bagnall, K. W., *Chemistry of the Rare Radioelements*, Academic Press, New York, 1957, pp. 1–94.
[14] Bernabee, R. P. and Sill, C. W., "Radiochemical Determination of Polonium-210 in Ores and Environmental Samples," Report ID0-12092, U.S. Department of Energy, Idaho Operations Office, Idaho National Engineering Laboratory, undated.
[15] *Annual Book of ASTM Standards, Part 31, Water,* American Society of Testing and Materials, Philadelphia, 1981.
[16] Rushing, D. E., *Analytical Chemistry*, Vol. 37, 1966, pp. 900–905.
[17] "Work Group 5 on Chemical and Physical Quality of Water," in *National Handbook of Recommended Methods for Water-Data Acquisition*, Office of Water Data Coordination, U.S. Geological Survey, Reston, VA, 1982, Chapter 5.

*Lorne G. Everett,*[1] *Leslie G. McMillion,*[2] *and Lawrence A. Eccles*[2]

# Suction Lysimeter Operation at Hazardous Waste Sites

**REFERENCE:** Everett, L. G., McMillion, L. G., and Eccles, L. A., **"Suction Lysimeter Operation at Hazardous Waste Sites,"** *Ground-Water Contamination: Field Methods, ASTM STP 963,* A. G. Collins and A. I. Johnson, Eds., American Society for Testing and Materials, Philadelphia, 1988, pp. 304–327.

**ABSTRACT:** Four questions relative to suction lysimeter performance have been raised by operators of land treatment systems. These questions deal with plugging of the porous segments, soil suction operational ranges, loss of volatile organics under negative pressure, and adsorption and screening of the various lysimeter parts. This paper describes the physical operation of a lysimeter, the characteristics of the lysimeters tested [high- and low-flow lysimeters with porous ceramic cups and lysimeters with porous polytetrafluoroethylene (PTFE) cups], and procedures followed in preparing the lysimeters for testing. It then presents the results of experiments dealing with the first three questions—plugging, operational ranges, and volatile organics loss. It was found that suction lysimeters placed in the vadose zone of most types of soils will initially drop off rapidly in intake rate, but will stabilize after about 15 L of moisture has been drawn through the porous cups. Packing a crystalline silica flour slurry around the cups of PTFE lysimeters negates most of the plugging associated with finer particles in soils. The effective operating range of ceramic lysimeters is between 0 and 60 centibars of suction, independent of the use of silica flour. The operating range of PTFE lysimeters without silica flour is extremely narrow, but with the use of silica flour is extended to about 7 centibars of suction. Volatile organics are lost from suction lysimeters, but the amount of the loss is difficult to estimate. The testing program indicates that before field installation all lysimeters should be checked for leaks using pressure techniques, and that lysimeters have "dead" spaces, that is, reservoirs of moisture, ranging from 34 to 80 mL, that cannot be extracted from the cups, and that must be taken into account when determining moisture collection rates. Information gathered during the test program is planned for inclusion in an Environmental Protection Agency (EPA) guideline document entitled "Unsaturated Zone Monitoring at Hazardous Waste Land Treatment Units."

**KEY WORDS:** vadose zone monitoring, suction lysimeters, unsaturated zone, ground-water monitoring, hazardous waste monitoring, volatile organics

The regulations issued by the Environmental Protection Agency (EPA) under the Resource Conservation and Recovery Act (RCRA) on December 18, 1978 (*Federal Register 1978*) require hazardous waste landfarm sites to implement soil pore-liquid monitoring programs. On May 19, 1980, EPA issued interim standards applicable to owners and operators of hazardous waste treatment, storage, and disposal facilities (*Federal Register, 1980*). A major emphasis in the interim standards was placed on monitoring procedures for evaluating the performance of land treatment facilities. The May 1980 rules and regulations introduced the term "land treatment" for use in place of "landfarming" and suggested that landfarms be referred to as "land treatment facilities" to imply that the waste is being treated and not disposed of at the facilities. The

[1] Manager, Natural Resources Program, Kaman Tempo, 816 State St., Santa Barbara, CA 93102.

[2] Branch chief and hydrologist, respectively, U.S. Environmental Protection Agency, Environmental Monitoring Systems Laboratory, P.O. Box 15027, Las Vegas, NV 89114.

regulations require that both unsaturated-zone and saturated-zone monitoring be used to assess the potential of waste constituent migration to ground water. Unsaturated-zone monitoring, as defined by the regulations, includes both soil pore-water and soil-core monitoring. EPA has developed a draft guideline document covering soil and pore-liquid monitoring at hazardous waste land treatment units [*1*].

Questions regarding the applicability and validity of soil pore-water monitoring results have been raised by operators of land treatment systems. The questions may be summarized as follows:

1. What kinds of soil texture will clog the porous segment of suction lysimeters?
2. What is the operational range of suction lysimeters?
3. What is the degree of loss of volatile organics under various negative pressures?
4. What is the effect of lysimeter components on sorption and micropore screening of hazardous constituents?

## Literature Review

Reporting on lysimetry applications has increased significantly during the past ten years. Porous materials and methods examined include porous disks or plates [2–4], hollow fibers [*5,6*], porous polytetrafluoroethylene (PTFE) cups [*7,8*], fritted glass filters and cups [*9–12*], "Swinnex" filter holders [*13*], and chemically inert, highly adsorbent materials such as ceramic points [*14*], sponges [*15*], polyethylene [*16*], and ceramic cups [*17–22*]; also, ceramic candles contained in sheet-metal troughs [*23–25*], and ceramic cups of very small size such as microsoil solution samplers [*26,27*]. A thorough historical review, including 197 references, concerning sampling of soil solutions by these and other methods has been done [*28*]. Sixty different vadose zone monitoring methods have been fully described [*22*] and interpreted as part of an overall ground-water monitoring program [*29*].

Difficulties such as sample pore water representation [*30*] the disruption of normal drainage patterns caused by suction [*32–34*], clogging [*26*], and operational limitations and constraints [*35*] have been identified.

## Objectives

The primary objective of this paper is to report and discuss the results of laboratory tests for clogging, operational range, and volatile organics loss in some common suction lysimeters. The second objective is to provide EPA with the scientific evidence to substantiate variances, given that these devices will or will not fail under certain field conditions.

## Purpose

The RCRA represents the first time that EPA has mandated the use of pore-water sampling. Specific monitoring equipment for this application, therefore, has not been fully developed; field and laboratory test data are highly lacking. Few individuals have experience upon which EPA can base its decisions on the use of this technique at hazardous waste sites. Consequently, the purpose of this research is to test, in the laboratory, specific operational conditions that may affect the validity of data obtained from suction lysimeters. Certain physical and chemical conditions may exist that preclude the use of suction lysimeter sampling. Quantitative understanding of the operation of the monitoring devices under these conditions must be developed so EPA can recommend, enforce, or provide variances to the existing RCRA requirements.

**Porous Segments in Lysimeters**

The vadose (unsaturated) zone consists of a mixture of soil particles, water that is held on the surface of the particles and in small capillary spaces between the particles, and interconnecting air passages that may be open to the atmosphere at the soil surface. Removing moisture for chemical analysis from the vadose zone requires the use of special porous materials. Simply exerting a suction on an open tube inserted into the vadose zone will not remove moisture because the interconnecting air passages in the soil will eventually direct the flow of air into the evacuated tube. However, by using a porous cup sealed to the end of the tube, samples can be removed by suction, provided the diameter of the individual pores in the porous cup does not exceed a critical value.

If the porous cup is fabricated from a hydrophilic material, such as ceramic, water will fill the pores of the cup completely. The water bonds to the porous ceramic and cannot be removed from the pores unless the air pressure differential across the wall of the cup reaches a critical value which is related to the pore size. If the porous cup is fabricated from a hydrophobic material such as PTFE, water will fill the pores of the cup, the bonding of the water to the hydrophobic material will be less, and slightly less vacuum will be required to remove the water.

A soil-water sampling device that can be used successfully in the vadose zone to withdraw moisture from the soil must incorporate a porous cup which has pores so small that the air in the soil, under atmospheric pressure, cannot enter even though a full vacuum is created within the sampler. Under these conditions, water from the capillary spaces in the soil will flow through the pores in the porous cup and into the sampler, but air will not enter.

The air pressure required to force air through a porous cup which has been thoroughly wetted with water is called the "bubbling pressure" or "air entry value." The smaller the pores in the cup the higher this pressure will be. The relation of the pore size to the bubbling pressure or air entry value is defined by the equation $D = 30Y/P$, where $D$ is the pore diameter measured in micrometres. $P$ is the bubbling pressure or air entry value measured in millimetres of mercury and $Y$ is the surface tension of water measured in dynes/centimetre.

The maximum size of the pores that will permit this action is as follows. At 20°C, the surface tension of water is 72 dynes/cm. The maximum air pressure is 1 atm (14.7 psi) or 760-mm mercury. In accordance with the equation $D = 30Y/P$, the maximum pore size in the porous cup would be $D = (30)(72)/(760) = 2.8$ μm. The pore size of ceramic cups is between 2 and 3 μm. If the pores of the wetted sampler cup do not exceed 2.8 μm in diameter, a full vacuum can be maintained within the sampler and the water films in the pores of the porous cup will not break down. If the pore size of the cup is twice this amount, namely 5.7 μm, the maximum vacuum that can be pulled within the sampler is 380-mm mercury (0.5 atm). Likewise, if the pore size is twice again as large, namely, 11.4 μm in diameter, the maximum vacuum that can be pulled without the cup leaking air is 190 mm mercury (0.25 atm). Since most pores used in PTFE suction lysimeters are between 70 and 300 μm in diameter, the bubbling pressure is only 0.04 atm. The bubbling pressure of low-flow ceramic cups is between 2.38 and 3.06 atm (35 and 45 psi), which translates to a suction of 233 centibars. The bubbling pressure of PTFE cups is between 0.051 and 0.068 atm (0.75 and 1 psi), which translates to a suction of 7 centibars.

Where porous materials are used in air-water systems such as suction lysimeters, the most direct method of evaluating the pore size of the material is through the use of air pressure. By thoroughly wetting the porous material and then exposing one side of it to increasingly higher air-pressure values while the other side is under water, we can readily observe when the air pressure becomes high enough to enter the pores and cause bubbling on the opposite side. The specific air pressure at which this bubbling occurs is a direct measurement of the pore size as defined by $D = 3Y/P$, and indicates directly the effectiveness of the porous materials to withstand air pressure differentials when in use. Evaluating pore size distribution by the mercury intrusion method or other means does not give direct information as to how the porous material will perform in the air-water system in which it is used.

PTFE pores are generally round and symmetrical, while ceramic pores are of various ragged shapes. The strength of the water menisci in the individual pores is a function of pore shape as well as overall size; consequently, an accurate measurement of the pressure at which the meniscus will break down and allow air to enter can be made only by direct measurement of the bubbling pressure or air entry value of the wetted porous material.

As shown in Fig. 1*a,* the pores within a suction lysimeter will not hold a vacuum in a dry condition. Air can move freely from the soil or silica flour surrounding the lysimeter through the pores into the instrument's interior. Thus, suction lysimeters should be installed in a wetted condition, and silica flour should be added as a slurry [0.45 kg (1 lb) to 150 mL (5 fl oz) of water]. Figure 1 is highly diagrammatic; each pore is representative of a tortuous route through the cup wall. In reality, millions of these tortuous pore routes are located throughout the cups. As shown in Figure 1*b,* the pores become completely filled when the cup has been placed in a wet environment and the pressure on both sides of the cup wall is 1 atm (14.7 psi). As shown in Fig. 1*c,* the surface tension of the wetted pore begins to change as a suction is developed within the cup. As noted in Fig. 1*d,* the radius of curvature of the surface tension decreases as the suction within the cup increases. The ability of water molecules to withstand these pressure gradients is the reason that air will not enter the cup even though the interior of the cup has been evacuated. Water, on the other hand, will freely move through the wetted pores under the gradient induced by the negative pressure within the cup; the surface tension is greatly increased by the reduction in pore size. As demonstrated in Fig. 1*e,* the surface tension in the pore can be broken by increasing the gradient across the cup wall to a pressure greater than the bubbling pressure of the cup.

Once a sample has been obtained in the suction lysimeter, as evidenced by a reduction in the suction gage, pressure must be applied to the lysimeter interior to push the sample to the surface. When a pressure is exerted on the porous segment, the meniscus will extend away from the center of the suction lysimeter. If too much pressure is applied within the suction lysimeter, the sample may be expelled through the pores. The meniscus behavior, therefore, is a function of the vacuum, pressure, pore size, porous material, soil moisture, and soil texture.

## Suction Lysimeters Tested

Each ceramic lysimeter tested comprised a ceramic bottom cup, a polyvinylchloride (PVC) body part, and a Neoprene pressure-fitting plug at the top. The ceramic lysimeters numbered S1 through S6 were low-flow sampling units (standard 2-bar cups). Ceramic lysimeters numbered S11 through S16 were high-flow sampling units (1-bar cups). All of the ceramic samplers were delivered new, had plastic sheaths surrounding the porous segment, and had a protective rubber band.

The PTFE units tested, numbered T1 through T6, had PTFE bottom cups and PVC body parts. The PTFE units numbered T9 through T11 had PTFE body parts as well as bottom cups. Unit T12 was an all-PTFE middle-entry suction lysimeter. All of the PTFE lysimeters were received in totally sealed plastic sheaths. Prior to starting the plugging experiments, all of the lysimeters were tested for flow rates, leaks, bubbling pressure, dead space, etc., to make sure that they were functioning properly.

## Bubbling Pressure Test Results

Each of the suction lysimeters used in the bubbling pressure tests was "wetted up." A minimum of 500 mL (16.9 fl oz) sample was pulled at 80-centibars vacuum into each unit. Each lysimeter was then submerged in 20 cm (7.87 in.) water at 73°C (22.7°F) for 2 h with the water inside the

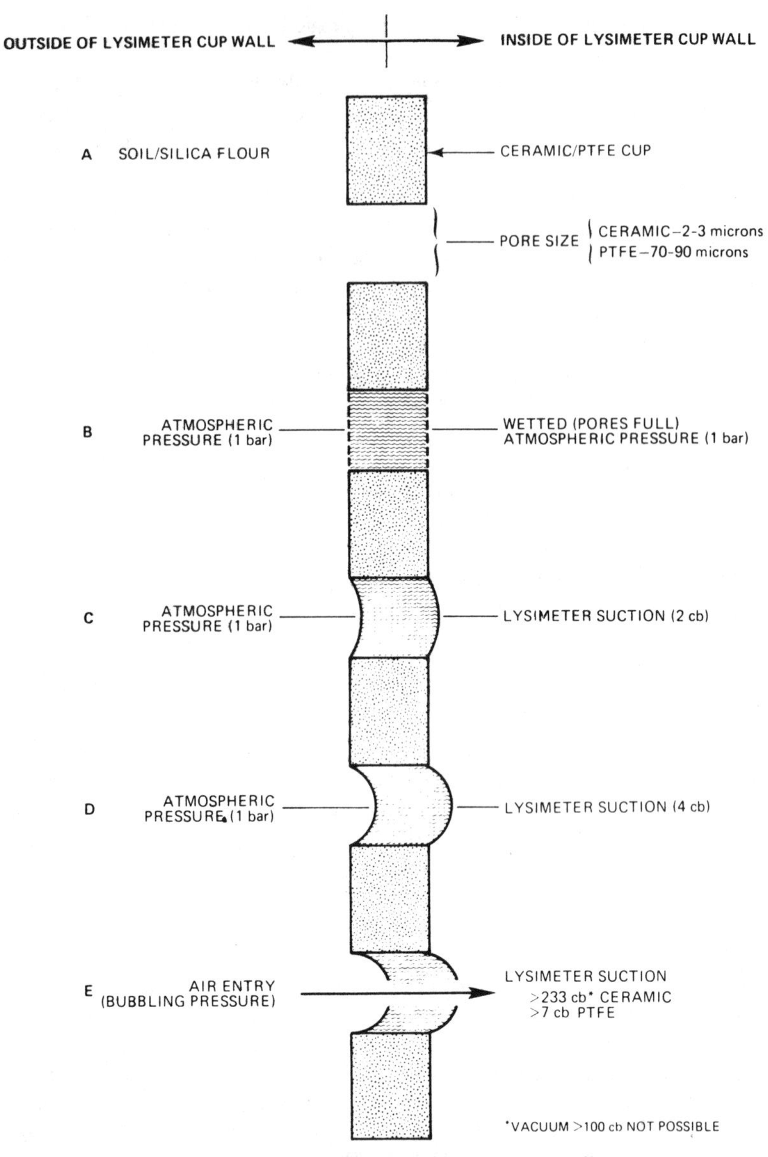


FIG. 1—*Diagrammatic view of lysimeter cup wall.*

unit. At the end of the 2-h period, the water inside each unit was removed and the units were ready for testing.

All pressure tests were conducted in a 45-L (1.52 fl oz) glass aquarium, measuring 80 by 25 by 40 cm (31.5 by 9.8 by 15.7 in.). The two pressure test gages used [0 to 1.02 atm and 0 to 4.08 atm (0 to 15 psi and 0 to 60 psi)] were developed by U.S. Gauge, Inc., and were accurate to 0.25%. The bubbling pressures for all the suction lysimeters investigated are shown in Fig. 2. As noted in Fig. 2, the PTFE/PVC units (T1 through T6) had a bubbling pressure of less than 0.068 atm (1 psi). The all-PTFE units (T9 through T11) had a bubbling pressure of less than

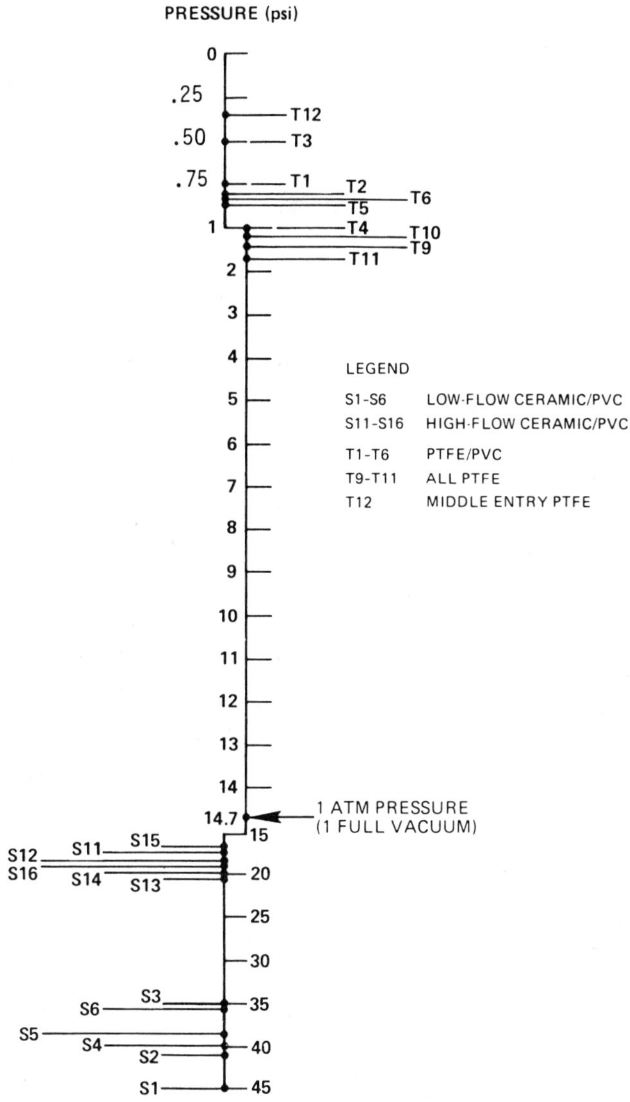


FIG. 2—*Bubbling pressures for various suction lysimeters.*

0.136 atm (2 psi). The middle-entry all-PTFE unit had the lowest tested bubbling pressure [between 0.017 and 0.034 atm (0.25 and 0.5 psi)].

The bubbling pressures of the low-flow ceramic and PVC units (S1 through S6) were between 2.38 and 3.06 atm (35 and 45 psi). In addition, the bubbling pressures of the high-flow ceramic and PVC units (S11 through S16) were in the range of 1.224 to 1.428 atm (18 to 21 psi). Since it is physically impossible to draw a vacuum greater than 1 atm (14.7 psi), the bubbling pressure of both the low-flow and the high-flow ceramic units appeared to be high enough not to be affected by air entry problems.

During the bubbling pressure tests, we noted that some of the PTFE units had air leaks at the

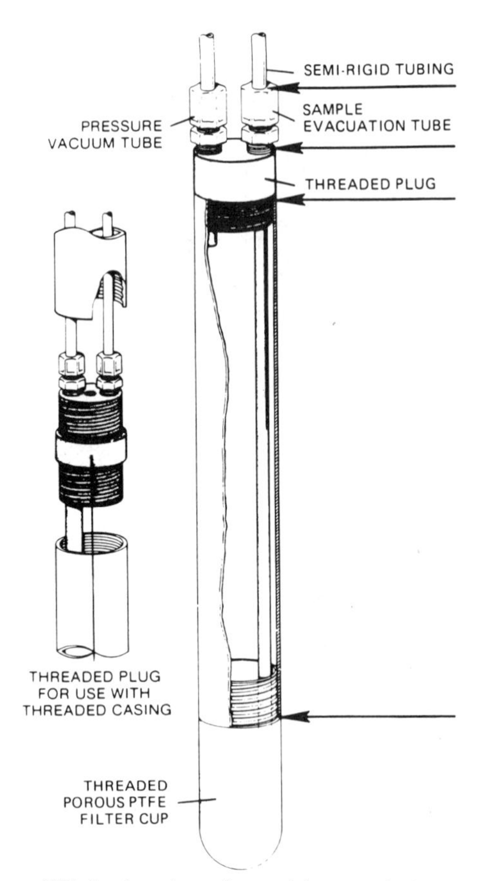


FIG. 3—*Locations of potential vacuum leaks.*

threaded areas of the lysimeter (see Fig. 3). After tightening each unit by hand and using PTFE tape, the leaks were eliminated. No leaks were found in any of the ceramic lysimeters. It is recommended, however, that prior to field installation pressure techniques be used to test all lysimeters for leaks.

## Vacuum Loss for Ceramic and PTFE Lysimeters (Without Silica Flour)

Ceramic low-flow lysimeters (S1 through S6) and ceramic high-flow lysimeters (S11 through S16) were used for this experiment. In addition, PTFE/PVC lysimeters (T1 through T6), the all-PTFE suction lysimeters (T9 through T11), and the PTFE middle-entry lysimeter (T12) were used. Each of the lysimeters used were "wetted up" by first pulling a sample of at least 500 mL (16.9 fl oz) through the porous segment at 80-centibars vacuum, and subsequently soaking the lysimeters in 20 cm (7.87 in.) of water for 2 h. Each lysimeter was fitted with a Bourdon vacuum gage and a vacuum line. Each porous segment was then placed in a plastic bag containing a wet paper towel to prevent dry-out in the laboratory. The suction lysimeters were evacuated to over 80 centibars of suction and were sealed.

As noted in Fig. 4, the PTFE/PVC units (T1 through T5) had an initial vacuum of 90 centibars. After less than 5 min, each of the PTFE units had less than 20 centibars of vacuum. Unit T5, for example, had less than 5 centibars of vacuum in 2 min. PTFE unit T12 (the middle-entry unit)

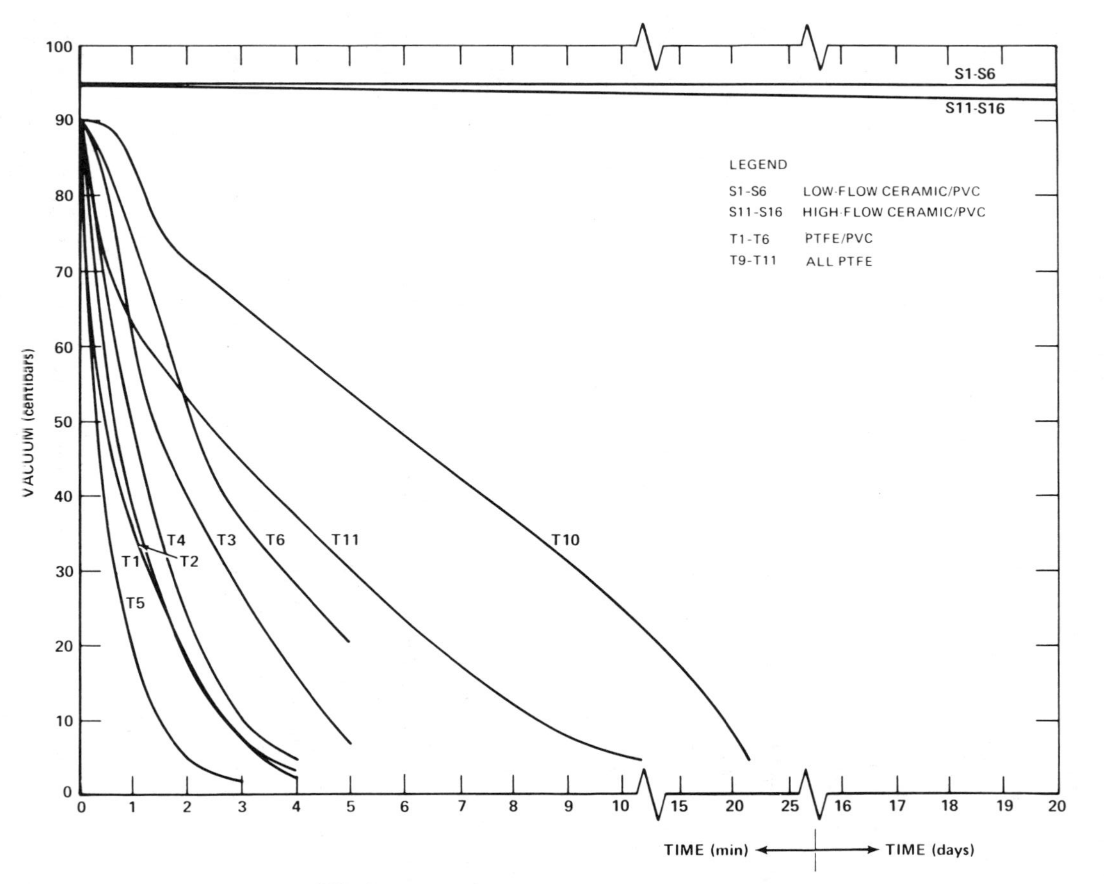


FIG. 4—*Vacuum loss experiments (without silica flour).*

could not hold a vacuum at all. The all-PTFE units (T9 through T11) had a vacuum reading of less than 5 centibars after 20 min.

The ceramic low-flow suction lysimeters (S1 through S6) had an initial vacuum of 95 centibars. As shown by Fig. 4, after 20 days of readings, the vacuum reading for all six units remained at 95 centibars. The initial vacuum reading for the ceramic/PVC high-flow suction lysimeters (S11 through S16) was 94 centibars. At the end of 20 days, the reading for each of the six high-flow suction lysimeters was 92 to 93 centibars. It was apparent that the ceramic lysimeters were able to sustain a continuous, high-level vacuum for at least 20 days. This is considerably longer than would be required in the field. After the experiment was over, one of the ceramic lysimeters was noted to have held 95 centibars vacuum for over 6 months.

The above tests show that PTFE units will not hold 10 centibars of vacuum without silica flour packed around the cups, but that the ceramic units will hold 95 centibars vacuum effectively without the silica flour.

## Lysimeter Pore Plugging Experiment

The suction lysimeter plugging experiments were conducted in 121-L (32 gal) Rubbermaid drum test chambers. White test chambers were used to allow additional visual inspection of water levels. Each of the 121-L (32 gal) test chambers was filled with 114 L (30 gal) of test soil. Steel drums were not used because of their possible interaction with the test soils and test solution. Interaction of the selected test chambers with either the soil materials or the test solution was considered negligible.

### *Test Chamber Soil Textures*

Five separate test chambers were used to contain the test soils. The soils tested range through each of the four categories used for percolation tests. Soils of known texture were identified in Santa Barbara, California. All of the soil samples were originally screened twice through 0.6 cm (0.236 in.) wire mesh to remove large pebbles, leaves, roots, and other organic matter. Samples of known texture were mixed by volume, and 114 L (30 gal) (dry weight) of soil combinations were placed in each of the five test chambers. To obtain totally mixed samples for each test chamber, the samples were mixed for 15 min in a portable cement mixer.

Standard textural analysis procedures were followed in determining the particle size analysis. The exact location of each of the soil textures is located on the soil classification chart (see Fig. 5).

### *Installation of Lysimeters*

Fifteen centimetres (5.9 in.) of soil was placed in the bottom of each test chamber. A thin-walled PVC casing 15 cm (5.9 in.) in diameter was placed on top of the soil layer. The 15-cm (5.9 in.) casing was selected to allow 5 cm (1.96 in.) of annular space between the inside diameter of the casing and the outside diameter of the lysimeter. The casing was then surrounded with test soil to the capacity of the container. A silica slurry, consisting of 200-mesh crystalline silica flour, was mixed with distilled water in the ratio of 0.454 kg (1 lb) silica to 150 mL (5 fl oz) water as recommended by the manufacturers (see Fig. 6). The silica packing material recommended for test procedures is 99.88% pure. The silica slurry was carefully mixed to ensure that no lumps existed in the mixture. Slurry was then poured into the casing to a depth of 8 to 10 cm (3.15 to 3.93 in.), the lysimeter was placed in the center of the casing, and slurry was added until the porous segment was covered to at least 5 cm (1.96 in.) above its top. Care was taken to ensure that the lysimeter remained centered, while excess slurry was added to allow for settling when the

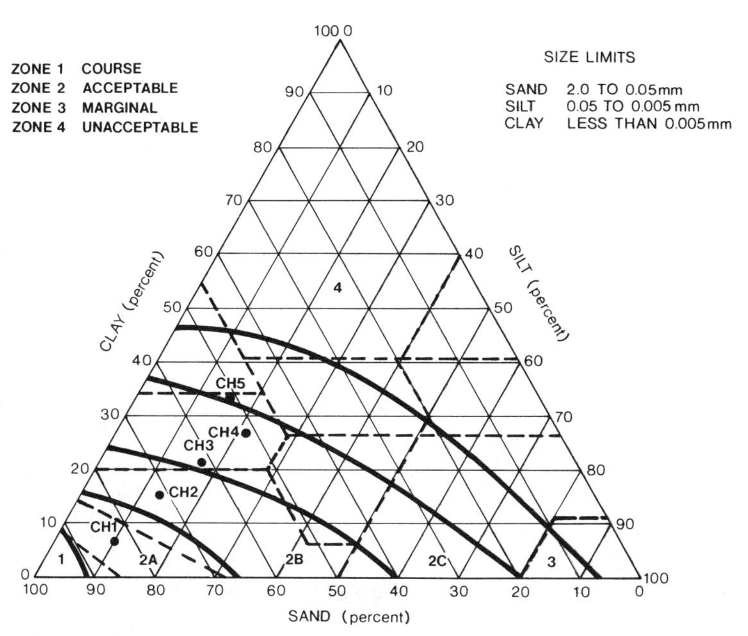


FIG. 5—*Soil percolation suitability chart showing location of soils in test chambers (CH).*

casing was removed and slumping occurred. The casing was removed by carefully twisting and pulling it slowly enough so that the silica flour did not adhere to the casing walls. Each lysimeter in each test chamber was installed in an identical manner. When all three lysimeters were installed in a test chamber, the remaining soil was added to top off the chamber. Each test chamber was then immediately filled with water. Over the next day, each of the test chambers was totally saturated. A standard tensiometer was then inserted in the middle of each test chamber so that the porous segment of the tensiometer was at the same depth as the porous segment of the suction lysimeters. The tensiometers indicated that each test chamber was 100% saturated. A vacuum was then immediately pulled on each of the suction lysimeters and the first sample discarded as packing moisture.

## *Plugging Experiment Results*

Discussions with numerous chemical laboratories predicate that about 500 mL (16.9 fl oz) of sample would be required to conduct a water quality analysis from a hazardous waste land treatment suction lysimeter. Since RCRA requires quarterly sampling, more than 2000 mL (4 × 500) (67.6 fl oz) will be required each year for analysis. Since most permits cover a period of 30 years of waste land treatment operation, a total of 60 L (2 fl oz) sample would be collected (2000 mL × 30) (67.6 fl oz). As a result, we determined that if 60 L (2 fl oz) sample could be pulled through the suction lysimeters, it would be indicative of the lysimeter performance over a 30-year period of activity. It is reasonable to conclude, however, that over a 30-year period other effects could contribute to the plugging of the porous segment of suction lysimeters.

Daily flow rates in each of the 15 suction lysimeters in the five test chambers were determined over a period of four months. The flow rate on the ceramic high-flow lysimeters (S11 through S16) ranged between 176.6 mL/h (5.97 fl oz) to a high of 208.3 mL/h (7.04 fl oz) over the entire experiment. The highest and lowest flows were recorded for the sand and clay test chambers

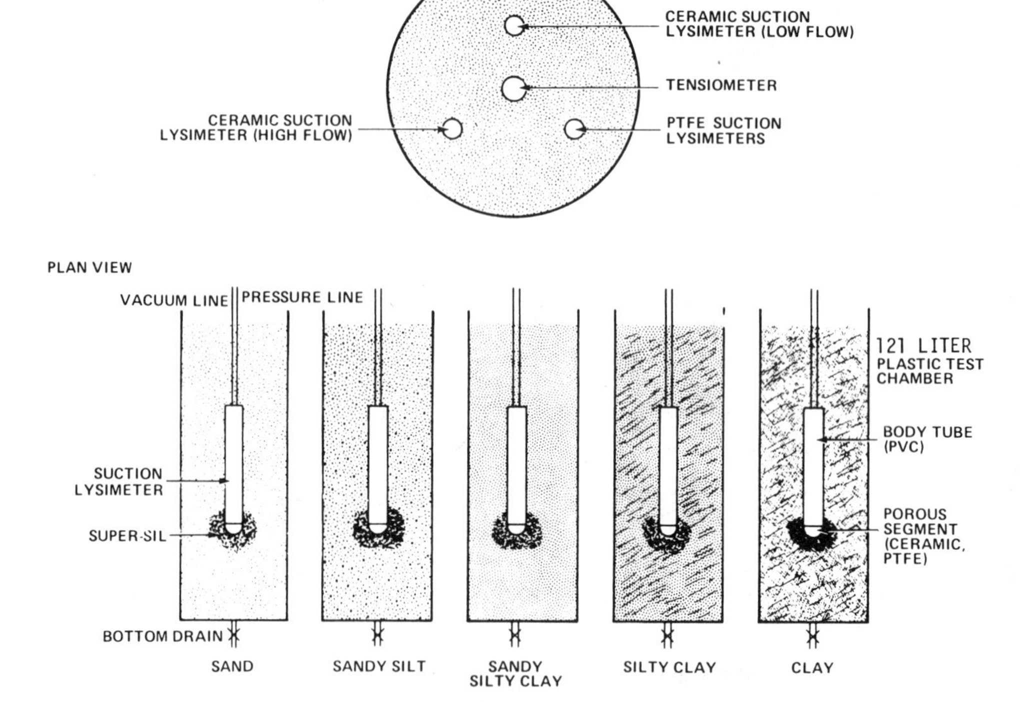


FIG. 6—*Test chambers (CH1–5) for plugging experiments.*

respectively. Statistical evaluation of the high-flow results indicated that the flow rate did not change over the 4-month period of the investigation. The sampling interval for the high-flow lysimeters ranged between 3 and 4 h. Since the maximum volume was collected each time, the sampling time precluded any potential variation in flow rates with time or soil texture. Based upon the continuous very high flow rates experienced over the 4-month test period, it was determined that ceramic high-flow suction lysimeters were not affected by plugging regardless of soil texture.

Using the sampling interval of 3 to 4 h, the flow rates for the five PTFE lysimeters ranged from 4.9 mL/h (0.165 fl oz/h) up to 6.2 mL/h (0.20 fl oz/h). Plots of the change in flow rate with accumulated volume passing through the PTFE lysimeters indicated a significant drop in flow rate between 0 and 15 L (0 and 3.96 gal) sample. After 15 L (3.96 gal) sample had been extracted, however, the flow rate in each of the PTFE lysimeters tended to stabilize. The flow rates for the PTFE units tended to stabilize between 2 and 10 mL/h (0.067 and 0.338 fl oz/h). Although these flow rates are small, they do indicate that a sample could be obtained in each of the soil textures over a 2- to 3-day period.

The flow rate in the ceramic low-flow units, numbered 1 through 5, showed a mean flow that ranged from 34.8 through 82.9 mL/h (1.176 through 2.803 fl oz/h). The lowest flow rates were recorded in test chamber No. 5. The flow rate for units 1 through 6 are plotted in Fig. 7. As noted with the PTFE units, the flow rate in the ceramic units dropped very rapidly between 0 to 15 L (0 and 3.96 gal) accumulated sample. After extraction of 15 L (3.96 gal), however, the flow rates stabilized. As can be seen from Fig. 7, a sufficient sampling could be obtained from these lysimeters over a 24-h period.

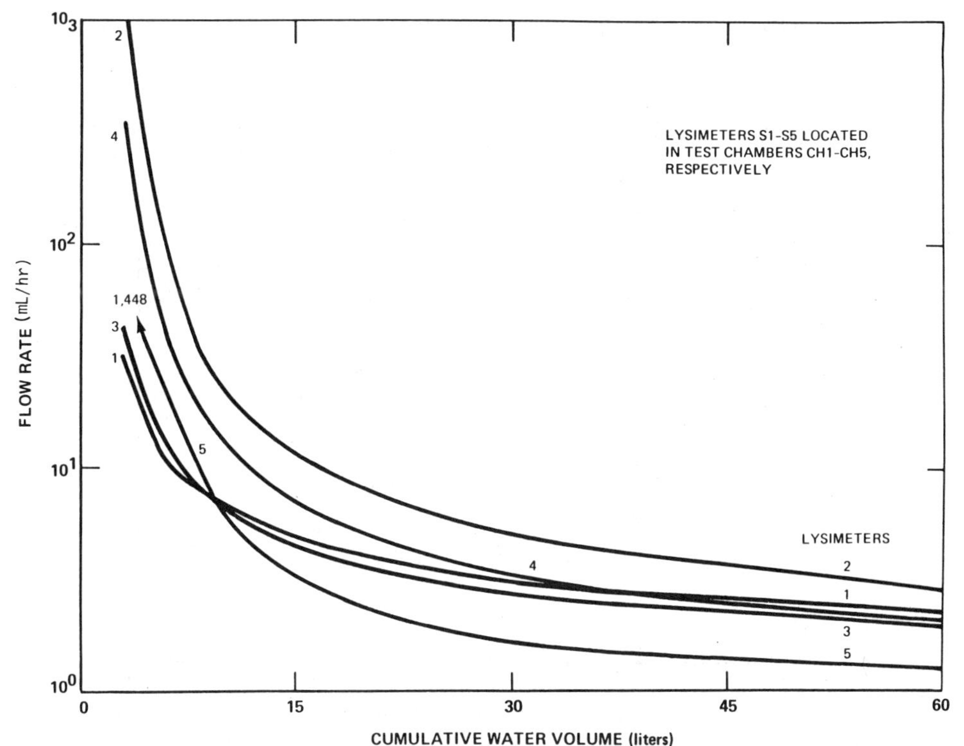


FIG. 7—*Reduction in flow rate as a function of soil texture.*

The results of testing units 1 through 6 were subjected to a linear regression represented by the equation $y = ax^b$, where $y$ = flow rate in ml/h, $x$ = cumulative volume in l, $b$ = regression slope, and $a = \log^{-1}$ regression intercept. The linear regression plot for each of the five low-flow lysimeters is given in Fig. 8. As evidenced by the high $r^2$ value, most of the variation is explained by the available data. The only test chamber that resulted in a significant reduction in flow was test chamber number 5. Although the flow rate in test chamber number 5 is significantly lower than in the other four test chambers, the suction lysimeters were still able to provide sufficient samples for analysis after 60 L had been extracted.

## Suction Lysimeter Operating Ranges

A highly instrumented test chamber (as shown in Fig. 9) was used to test the operational ranges of the lysimeters. The soil selected for the test chamber was an intermediate soil made up of sandy, silty clay. About 5 cm of soil was placed on the bottom of the test chamber. A 30 cm porous ceramic extractor plate was placed on top of the soil. The porous extractor plate had a pore size of less than 3 μm. A discharge line from the bottom of the plate extended to the top of the test chamber. Fifteen cm of soil were placed on top of the extractor plate. A PTFE, a ceramic low-flow, and a ceramic high-flow suction lysimeter were installed in the test chamber, as shown in Fig. 6. At least 5 cm of silica flour were placed around each of the porous segments on the suction lysimeters. The suction lysimeters were located within the test chamber to minimize their influence on each other. Seven porous (2-μm) extractor tubes were inserted in the test chamber in

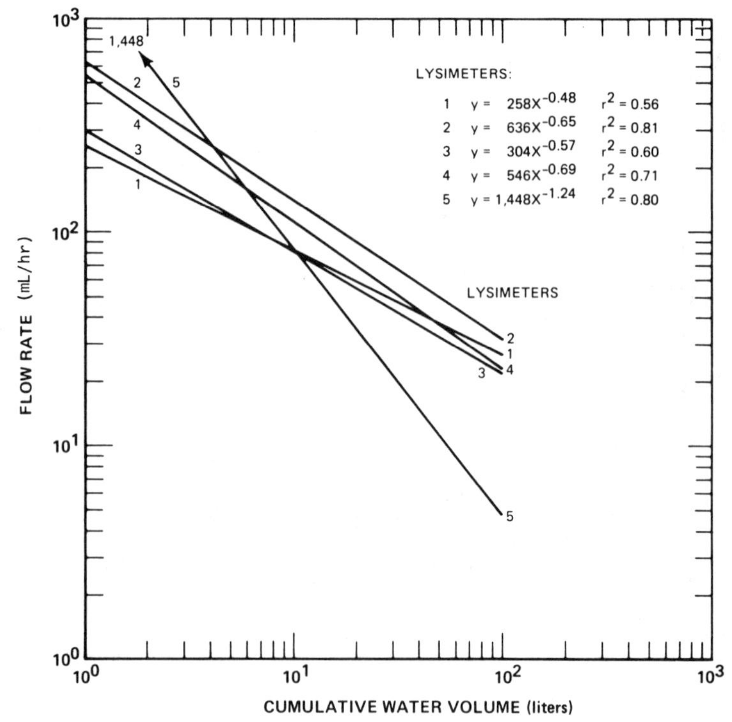


FIG. 8—*Coefficient of determination* ($r^2$) *for the lysimeter flow rate reduction.*

such a manner as to uniformly cover the testing area (see Fig. 10). Three tensiometers were installed in the test chamber with the porous tips of the tensiometers at the exact depth of the pore segments on each of the suction lysimeters. The seven leads from the extractor tubes and the one tube from the extractor plate were connected to an eight-lead manifold in turn connected to a 38-L (10-gal) vacuum reservoir and a vacuum pump. The test chamber was saturated by adding water until all three tensiometers indicated zero vacuum or saturation. Two days of adding water to the system were required to stabilize the tensiometers at zero suction.

At saturation, each of the three lysimeters was attached to a three-lead manifold and 80 centibars of suction was established. After 15 min of suction, samples were immediately retrieved from each of the lysimeters using 0.34 atm (5 psi) of pressure. Based upon the sample volume collected and the time frame used, a plot of the flow rates for each of the suction lysimeters was established and plotted (see Fig. 11).

A vacuum of 10 centibars was established on the 38-L (10-gal) reservoir connected to the extractor tubes and extractor plate. After one day, the soil suction within the test chamber (as evidenced by the three tensiometers) came up to a value of 5 centibars. As soil moisture was pulled from the test chamber, the vacuum in the 38-L (10-gal) reservoir dropped, and required occasional readjustment to maintain a vacuum on the whole system.

At 5 centibars of suction, all three lysimeters were evacuated with 80 centibars of suction. After 15 min of vacuum, each suction lysimeter was pressurized to 0.34 atm (5 psi) and the samples were immediately retrieved. The same procedure was repeated at increments of soil suction up to 60 centibars of suction. At each increment, a sample was obtained and the flow rate was determined for each lysimeter. The sampling interval ranged from 15 min at 5 centibars of suction up to 1

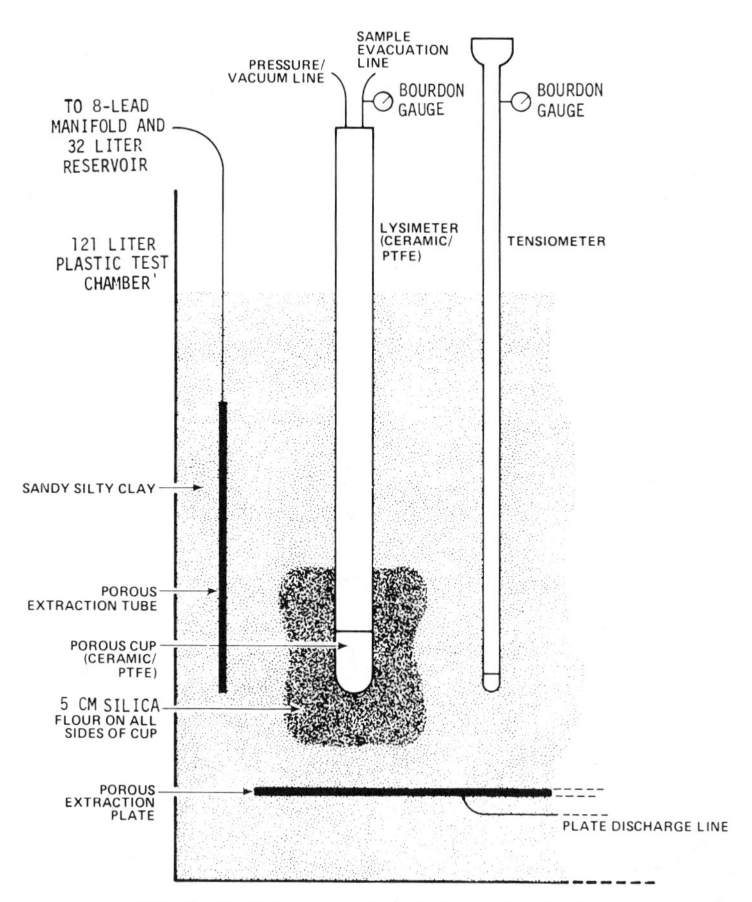


FIG. 9—*Partial cross section of lysimeter installation.*

week at 60 centibars of suction. Over 2 weeks were required to bring the soil suction in the test chamber from 50 centibars up to 60 centibars of suction. Even using a full vacuum on the 38-L (10-gal) reservoir, considerable time would have been required to bring the soil suction in the test chamber up from 60 centibars to 70 or 80 centibars of suction. The flow rate at 60 centibars of suction was so low that, for most practical purposes, the sample interval would have been too long.

*Dead Space in Lysimeters*

As a part of the experiments dealing with reduction in flow rates as a function of increased soil suction, it was determined that "dead" spaces may exist within suction lysimeters. As shown in Fig. 12, the ceramic cups are glued to the inner wall of a Schedule 20 PVC body tube. This results in a projection, or lip, on the inside of the suction lysimeter. As polyethylene tubes are pushed down or twisted through the two-holed stopper at the top of the lysimeter, they develop a characteristic twist in their length. (The tubing, in most cases, is delivered on a spool and tends to retain a residual bend.) The polyethylene tube may catch on the inside lip of the cup and the operator may conclude that the tube has reached the bottom of the ceramic cup. Since the bottom

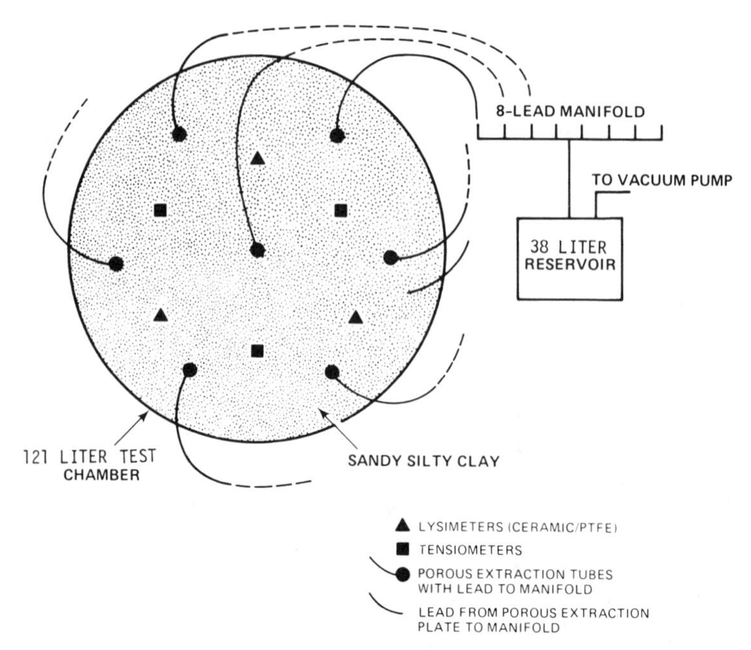


FIG. 10—*Plan view of test chamber for lysimeter operating range experiments.*

of the cup cannot be seen, it is difficult to determine whether the tube has actually reached the bottom of the cup. Even measuring with a tape rule may result in the tape rule hanging up on the edge of the cup, giving the impression that the depth to the bottom of the cup has been determined. As a result, an 80-mL (2.70 fl oz) error can occur in any rate determinations. This 80 mL (2.70 fl oz) of fluid accumulates in the cup and cannot be extracted through the discharge line.

In all-PTFE suction lysimeters, the discharge line is a rigid PTFE tube extending to the bottom of the PTFE cup. This design results in zero dead space in the all-PTFE lysimeters. However, the PTFE cups with a PVC body tube have been designed with a rigid interior tube which does not extend to the bottom of the PTFE cup. Since it is impossible to extend this rigidly fixed interior tube, the PTFE/PVC units used as a part of the preceding experiments had a constant dead space of 34 mL (1.149 fl oz). The authors are aware of numerous lysimeter investigations where rates of intake and volumes of samples have been reported. To date, most operators, including the manufacturers, have not been aware of the potential dead space within their lysimeters.

## Volatile Organics Loss Experiment

Fundamental chemical phase relationships indicate that under increasing negative pressures volatiles can be drawn from the liquid phase to the gas phase. Consequently, hazardous waste landfarm operators and environmental enforcement agencies have suspected that volatile organics would be lost when obtaining a sample in a suction lysimeter. Under controlled laboratory conditions, using selected volatile organics, basic chemical equilibrium laws can be demonstrated. The phenomena associated with the emissions of volatile organic compounds (VOCs) and other gaseous compounds from the land treatment of refinery oily sludges, however, are extremely complex [*37*]. The effects of sludge volatility, soil loading, soil type, soil moisture, air humidity, air temperature, soil temperature, wind speed, biodegradation, spreading technique, etc., are very

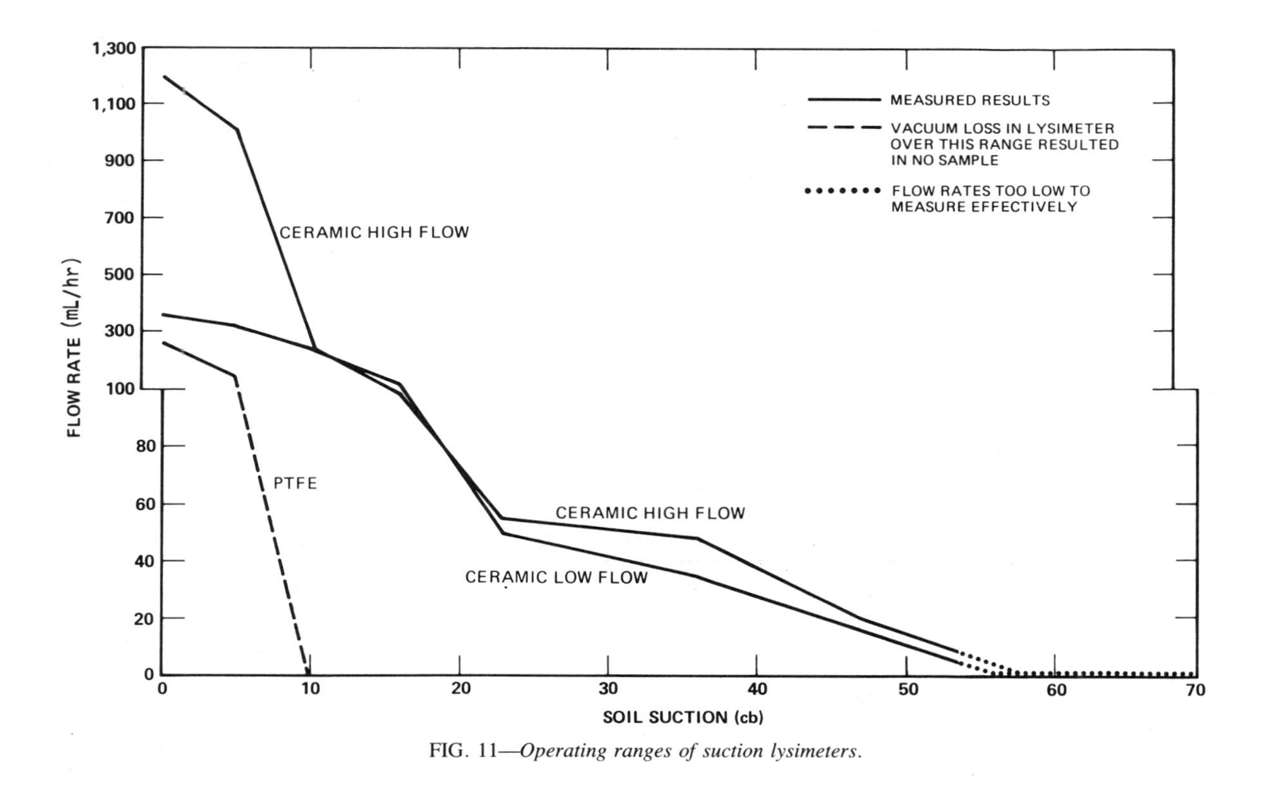


FIG. 11—*Operating ranges of suction lysimeters.*

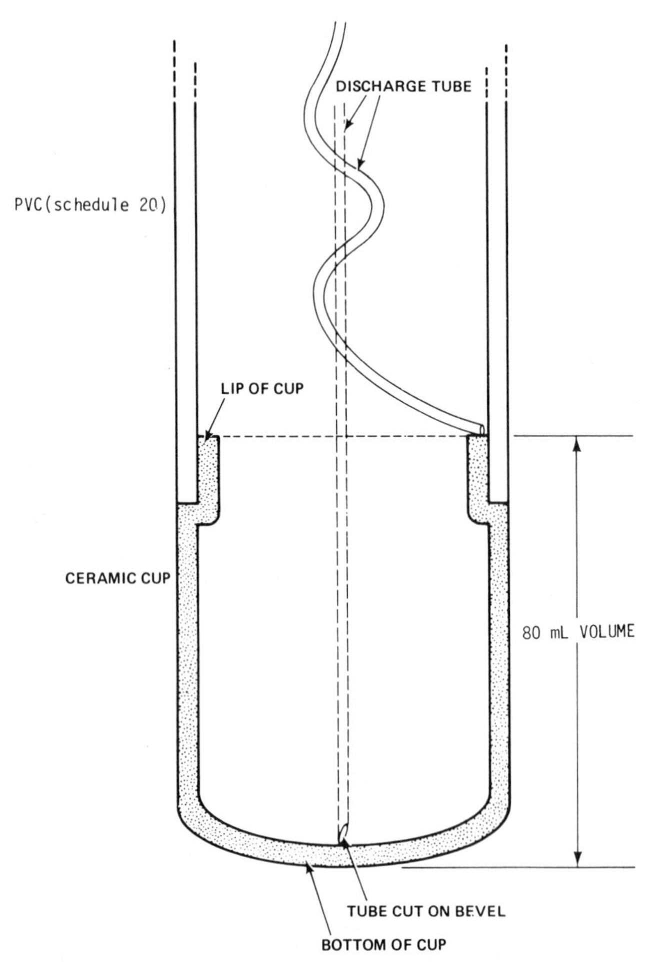


FIG. 12—*Location of potential dead space in suction lysimeters.*

difficult to quantify. By minimizing the majority of these variables, we anticipated that the affect of increasing negative pressures on volatile organics loss rates could be determined.

Although there are several sources of volatile organics at refineries (see Table 1), EPA requested that the laboratory tests be conducted on American Petroleum Institute (API) separator sludge. API separator sludge is typically characterized as having approximately 30 to 36% aromatics as a percentage weight in the oil phase. Permission was obtained from a refinery to obtain 19 L (5 gal) of API separator sludge typical of the material landfarmed at this facility.

Both ceramic cup lysimeters with PVC body tubes and all-PTFE lysimeters were used in the experiment. Each suction lysimeter was placed into a saturated sample of API separator sludge. Negative pressures of 20, 40, 60, and 80 centibars were pulled on each of the suction lysimeters. At each negative pressure, a sample was obtained from each of the ceramic/PVC lysimeters and the PTFE lysimeters. Each of the samples was analyzed for volatile organics to determine the volatile organics loss at higher vacuums.

One sample was initially analyzed for aromatics hydrocarbons by gas chromatograph/mass spectrometer (GC/MS) (Method 624). However, the relatively high concentration of aromatic and

TABLE 1—*Organics present in selected refinery wastes.*

| Type of Sludge | Sludge Oil Content, % weight | Oil Phase Hydrocarbon Fractions | | | | Phenols, ppm |
|---|---|---|---|---|---|---|
| | | Saturates | Olefins (% weight in oil phase) | Aromatics | Polar/ Asphaltenes | |
| Composite centrifuged oily sludge | 11.8[a] (5.6 to 18.4) | 57.1 (46.0 to 65.2) | 1.3 0 to 3.0 | 30.0 (24.8 to 35.3) | 11.6 | 1.7 |
| API separator sludge | 9.0 (4.6 to 17.6) | 53.5 (48.0 to 66.1) | 1.3 (0 to 2.3) | 33.3 (30.3 to 36.0) | 12.2 (8.5 to 19.9) | 2.8 |
| Composite oily sludge | 21.8 (3.2 to 41.0) | 59.6 (51.3 to 70.2) | 2.6 (0.9 to 2.9) | 30.2 (24.7 to 35.5) | 7.6 (5.9 to 12.8) | 6.7 (0.9 to 14.9) |
| Bio sludge | 0.9 (0.1 to 4.4) | 41.3 (32.8 to 49.8) | 0 | 28.2 (22.3 to 34.2) | 30.5 (27.9 to 33.0) | 0.6 (0.3 to 1.1) |
| DAF sludge | 8.4 (6.7 to 10.1) | 44.1 (39.2 to 49.0) | 0.7 (0.4 to 1.0) | 38.5 (35.4 to 41.7) | 16.3 (13.8 to 18.8) | . . . |
| Tank bottoms | 8.5 (4.8 to 12.3) | 50.7 (46.5 to 55.0) | 1.7 (0.5 to 3.0) | 36.0 (31.3 to 40.8) | 11.4 (10.7 to 12.2) | . . . |

[a] Mean value; data in parentheses represent the range of the available data [*36*].

aliphatic hydrocarbons made it necessary to dilute the samples into an acceptable range for the instrument for surrogate spike and internal standard resolution. The large dilutions required to analyze those samples by GC/MS would have resulted in elevated detection limits; consequently, we decided to analyze the samples by a Gas Chromatograph/Hall Photoionization Detector (GC/Hall Detector/PID) (EPA Methods 601/602). The GC/PID method is much more selective toward the aromatic compounds of interest. The interference issue and the elevated detection limits required to run the GC/MS method have created a fundamental problem not only with this particular experiment, but also with respect to analysis of oily samples throughout the nation. This fundamental analysis problem, which has been spearheaded by API and the Rocky Mountain Laboratory, is under intensive review by EPA at this time; however, a cleanup procedure for the method is not anticipated to be complete until the second quarter of 1986.

All samples were run using accepted quality-assurance/quality-control procedures. Chain of custody was maintained in addition to a complete set of standards, blanks, spiked samples, duplicates, and replicates for each of the analyses.

The raw sludge sample, as represented in Table 2, showed that the organics present were principally made up of trichloroethene, benzene, ethylbenzene, toluene, and *o, m,* and *p*-xylenes. The samples obtained at 20, 40, 60, and 80 centibars of suction for the ceramic and PTFE lysimeters are present in Table 2 as 2, 4, 6, 8 for ceramic (*C*), and PTFE (*T*), respectively. The difference between the levels found in the sludge and in the water sample is, understandably, substantial. The difference between the volatile organics present under increasing negative pressures, however, does not indicate a trend. As the negative pressure is increased from 20 to 80 centibars of suction, we would expect an increase in the loss of volatile organics. This trend is not demonstrated by the test results.

These results are consistent with the extensive laboratory investigations conducted by the Radian Corp. on hydrocarbon emissions from refinery landfarms [*37*]. In the Radian investigation, greater than 19 refinery sludges were evaluated for their volatility. The investigation showed that the volatility of the sludge "was found to be almost overwhelming in its effect" during the testing phases. Only by categorizing various sludges into low, medium, and high-volatility sludges could Radian control their experiments. Just as the intensive Radian investigation using 19 sludges could not develop an expression for estimating the rate of emissions as a function of elapsed time after sludge application from the laboratory experiments, this study showed that an expression for estimating the rate of volatile organics loss from a lysimeter as a function of increased negative pressure could not be developed for an API separator sludge sample.

Note, however, that during the laboratory experiments, the PTFE suction lysimeter continually lost its vacuum. The time required to obtain a sample at each of the negative pressures was held constant. The volatile organics were almost totally lost from the PTFE lysimeters. Although the possibility exists that the volatile organics were adsorbed onto the PTFE, a more reasonable explanation is that the repeated reestablishment of the vacuum in the PTFE lysimeters removed all of the volatiles within the liquid sample. The ceramic/PVC suction lysimeter, however, held its vacuum completely throughout each of the experiments. As a result, the liquid sample inside the lysimeter quickly came into equilibrium with the air phase in the suction lysimeter, and as such, further volatiles were not lost from the liquid sample.

**Results and Recommendations**

Although the aforementioned experiments represent the first phase of an extended testing program, the following definitive conclusions and recommendations can be made.

1. Prior to field installation, pressure techniques should be used to check all lysimeters for leaks.

2. The approximate bubbling pressure of ceramic low-flow cups is 2.38 atm (35 psi); for ceramic high-flow cups 1.224 atm (18 psi); and for PTFE cups 0.068 atm (1 psi).

3. Low-flow ceramic cups are capable of holding their vacuum for several months.

4. PTFE lysimeters must be used with a silica flour slurry.

5. The dead space in suction lysimeters must be determined prior to field installation or laboratory tests.

6. Suction lysimeters placed in most types of soil will experience a rapid drop-off in intake rate but will stabilize after about 15 L (4 gal) has been pulled through the porous segments.

7. Use of silica flour around the porous segments negates most plugging associated with finer particles in the soils.

8. The effective operating range of ceramic lysimeters is between 0 and 60 centibars of suction regardless of the use of silica flour.

9. The operating range of PTFE lysimeters without silica flour is extremely narrow, but with the use of silica flour is extended to about 7 centibars of suction.

10. An estimate of the rate of volatile organics loss with increasing vacuum in a suction lysimeter is difficult using a random API separator sludge sample.

11. Volatile organics can be obtained using a suction lysimeter where equilibrium is established and maintained.

12. Volatile organics are lost from suction lysimeters if the vacuum needs to be intermittently reestablished to draw sufficient sample.

The information herein will be used as a part of the ongoing experiments conducted by Kaman Tempo. The results are planned for inclusion in an EPA guideline document entitled "Unsaturated Zone Monitoring at Hazardous Waste Land Treatment Units."

*Acknowledgments*

This research was funded under EPA Contract No. 68-03-3090 with the Environmental Monitoring Systems Laboratory in Las Vegas, Nevada 89114. Special recognition is given to Dr. E. Keller, Mr. E. Xhao, Mr. P. Kieniewicz, professor and graduate research associates, respectively, Department of Geological Sciences, University of California at Santa Barbara.

Although the research described herein has been funded wholly or in part by the U.S. Environmental Protection Agency, it has not been subjected to Agency review and therefore does not necessarily reflect the views of the Agency, and no official endorsement should be inferred.

## References

[1] Everett, L. G. and Wilson, L. G., "Unsaturated Zone Monitoring for Hazardous Waste Land Treatment Units," U.S. Environmental Protection Agency, Las Vegas, NV, 1984.

[2] Tanner, C. B., Bourget, S. J., and Holmes, W. E., "Moisture Tension Plates Constructed from Alundum Filter Discs," *Soil Science Society of America Proceedings*, Vol. 18, 1954, pp. 222–223.

[3] Cole, D. W., "Alundum Tension Lysimeter," *Soil Science*, Vol. 85, 1958, pp. 293–296.

[4] Cole, D. W., Gessel, S. P., and Held, E. E., "Tension Lysimeter Studies of Ion Moisture Movement in Glacial Till and Coral Atoll Soils," *Soil Science Society of America Proceedings*, Vol. 25, 1961, pp. 321–325.

[5] Jackson, D. R., Brinkley, F. S., and Bondietti, E. A., "Extraction of Soil Water Using Cellulose-Acetate Hollow Fibers," *Soil Science Society of America Journal*, Vol. 40, 1976, pp. 327–329.

[6] Levin, M. J. and Jackson, D. R., "A Comparison of *In Situ* Extractors for Sampling Soil Water," *Soil Science Society of America Journal*, Vol. 41, 1977, pp. 435–536.

[7] Zimmermann, C. F., Price, M. T., and Montgomery, J. R., "A Comparison of Ceramic and Teflon *In*

TABLE 2—*Volatile organics analysis of API separator sludge.*

| Detection Limit | | (μg/kg) Raw Sludge | Water (μg/L) | | | | | | | | | |
|---|---|---|---|---|---|---|---|---|---|---|---|---|
| | | | 2C[a] | 2T[b] | 4C[c] | 4T[d] | 6C[e] | 6T[f] | 8C[g] | 8T[h] | 8CD[i] | 8TD[j] |
| 0.02 | chloromethane | ... | ... | ... | ... | ... | ... | ... | ... | ... | ... | ... |
| 0.06 | bromomethane | ... | ... | ... | ... | ... | ... | ... | ... | ... | ... | ... |
| N/D | dichlorodifluoromethane | ... | ... | ... | ... | ... | ... | ... | ... | ... | ... | ... |
| 0.05 | vinyl chloride | ... | ... | ... | ... | ... | ... | ... | ... | ... | ... | ... |
| 0.1 | chloroethane | ... | ... | ... | ... | ... | ... | ... | ... | ... | ... | ... |
| 0.02 | methylene chloride | ... | ... | ... | ... | ... | ... | ... | ... | ... | ... | ... |
| N/D | trichlorofluoromethane | ... | ... | ... | ... | ... | ... | ... | ... | ... | ... | ... |
| 0.07 | 1,1-dichloroethene | ... | ... | ... | ... | ... | ... | ... | ... | ... | ... | ... |
| 0.05 | 1,1-dichloroethane | 376 | 13.9 | ... | 18.5 | 4.0 | 15.7 | ... | 18.6 | ... | 18.3 | ... |
| 0.09 | trans-1,2-dichloroethene | ... | ... | ... | ... | ... | ... | ... | ... | ... | ... | ... |
| 0.05 | chloroform | ... | ... | ... | ... | ... | ... | ... | ... | ... | ... | ... |
| 0.07 | 1,2-dichloroethane | ... | ... | ... | ... | ... | ... | ... | ... | ... | ... | ... |
| 0.03 | 1,1,1-trichloroethane | 424 | ... | ... | ... | ... | ... | ... | ... | ... | ... | ... |
| 0.08 | carbon tetrachloride | ... | ... | ... | ... | ... | ... | ... | ... | ... | ... | ... |
| 0.04 | bromodichloromethane | ... | ... | ... | ... | ... | ... | ... | ... | ... | ... | ... |
| 0.03 | 1,2-dichloropropane | ... | ... | ... | 9.5 | 2.9 | ... | ... | 12.7 | ... | ... | ... |
| 0.11 | trans-1,3-dichloropropene | ... | ... | ... | ... | ... | ... | ... | ... | ... | ... | ... |
| 0.06 | trichloroethene | 2520 | 37.7 | ... | ... | 13.5 | 12.7 | ... | ... | ... | ... | ... |
| 0.2 | benzene | 15300 | 322 | ... | 436 | ... | 295 | ... | 379 | ... | 381 | ... |

| | | | | | | | | | | | | |
|---|---|---|---|---|---|---|---|---|---|---|---|---|
| 0.07 | dibromochloromethane | ... | ... | ... | ... | ... | ... | ... | ... | ... | ... | ... |
| 0.03 | 1,1,2-trichloroethane | ... | ... | ... | ... | ... | ... | ... | ... | ... | ... | ... |
| 0.07 | cis-1,3-dichloropropene | ... | ... | ... | ... | ... | ... | ... | ... | ... | ... | ... |
| 0.03 | 2-chloroethylvinyl ether | ... | ... | ... | ... | ... | ... | ... | ... | ... | ... | ... |
| 0.09 | bromoform | ... | ... | ... | ... | ... | ... | ... | ... | ... | ... | ... |
| 0.03 | 1,1,2,2-tetrachloroethane | ... | ... | ... | ... | ... | ... | ... | ... | ... | ... | ... |
| 0.03 | tetrachloroethene | 406 | ... | ... | ... | ... | ... | ... | ... | ... | ... | ... |
| 0.16 | chlorobenzene | ... | ... | ... | ... | ... | ... | ... | ... | ... | ... | ... |
| 0.4 | 1,3-dichlorobenzene | ... | ... | ... | ... | ... | ... | ... | ... | ... | ... | ... |
| 0.4 | 1,2-dichlorobenzene | ... | ... | ... | ... | ... | ... | ... | ... | ... | ... | ... |
| 0.6 | 1,4-dichlorobenzene | ... | ... | ... | ... | ... | ... | ... | ... | ... | ... | ... |
| 0.1 | ethylbenzene | 30700 | 83.3 | ... | 106 | ... | 80.6 | ... | 88.2 | ... | 91.7 | ... |
| 0.4 | toluene | 47400 | 196 | ... | 313 | ... | 248 | ... | 260 | ... | 264 | ... |
| 1.0 | *m*-xylene | 100000 | 166 | ... | 253 | ... | 183 | ... | 213 | ... | 222 | ... |
| 1.0 | *o,p*-xylenes | 109000 | 204 | ... | 272 | ... | 206 | ... | 230 | ... | 247 | ... |

[a] 2C = 20 centibars suction in ceramic lysimeter.
[b] 2T = 20 centibars suction in tetrafluoroethylene lysimeter.
[c] 4C = 40 centibars suction in ceramic lysimeter.
[d] 4T = 40 centibars suction in tetrafluoroethylene lysimeter.
[e] 6C = 60 centibars suction in ceramic lysimeter.
[f] 6T = 60 centibars suction in tetrafluoroethylene lysimeter.
[g] 8C = 80 centibars suction in ceramic lysimeter.
[h] 8T = 80 centibars suction in tetrafluoroethylene lysimeter.
[i] 8CD = 80 centibars suction in ceramic lysimeter duplicate.
[j] 8TD = 80 centibars suction in tetrafluoroethylene lysimeter duplicate.

*Situ* Samplers for Nutrient Pore Water Determinations,'' *Estuarine Coastal Marine Science,* Vol. 7, 1978, pp. 93–97.

[*8*] Morrison, R. D., ''A Modified Vacuum-Pressure Lysimeter for Soil Water Sampling, *Soil Science,* Vol. 134, No. 3, Sept. 23, 1982, pp. 206–210.

[*9*] Long, L. F., ''A Glass Filter Soil Solution Sampler,'' *Soil Science Society of America Journal.* Vol. 42, 1978, pp. 834–835.

[*10*] Silkworth, D. R. and Grigal, D. F., ''Field Comparison of Soil Solution Samplers,'' *Soil Science Society of America Journal,* Vol. 45, 1981, pp. 440–442.

[*11*] Nielsen, D. and Phillips, R., ''Small Fritted Glass Bead Plates for Determination of Moisture Retention,'' *Soil Science Society of America Proceedings,* Vol. 22, 1958, pp. 574–575.

[*12*] Chow, T., ''Fritted Glass Bead Materials as Tensiometers and Tension Plates,'' *Soil Science Society of America Journal,* Vol. 41, 1977, pp. 19–22.

[*13*] Stevenson, C. D., ''Simple Apparatus for Monitoring Land Disposal Systems by Sampling Percolating Soil Waters,'' *Environmental Science and Technology,* Vol. 12, 1978, pp. 329–331.

[*14*] Shimshi, D., ''Use of Ceramic Points for the Sampling of Soil Solution,'' *Soil Science,* Vol. 101, 1966, pp. 98–103.

[*15*] Tadros, V. T. and McGarity, J. W., ''A Method for Collecting Soil Percolate and Soil Solution in the Field,'' *Plant Soil,* Vol. 44, 1976, pp. 655–667.

[*16*] Quin B. and Forsythe, L., ''All-Plastic Suction Lysimeters for the Rapid Sampling of Percolating Soil Water,'' *New Zealand Journal of Science,* Vol. 19, 1976, pp. 145–148.

[*17*] Wagner, G., ''Use of Porous Ceramic Cups to Sample Soil Water Within the Profile,'' *Soil Science,* Vol. 94, 1962, pp. 379–386.

[*18*] Wood, W., ''A Technique Using Porous Cups for Water Sampling at Any Depth in the Unsaturated Zone,'' *Water Resources Research,* Vol. 9, 1973, pp. 486–488.

[*19*] Bell, R., ''Porous Ceramic Soil Moisture Samplers: An Application in Lysimeter Studies on Effluent Spray Irrigation,'' *New Zealand Journal of Experimental Agriculture,* Vol. 1, 1974, pp. 173–175.

[*20*] David, M. and Struchtemeyer, R., ''The Effects of Spraying Sewage Effluent on Forested Land at Sugarloaf Mountain, Maine,'' Life Science and Agricultural Experimental Station, No. 773, 1980, pp. 1–16.

[*21*] Johnson, T. M. and Cartwright, K., ''Monitoring of Leachate Migration in the Unsaturated Zone in the Vicinity of Sanitary Landfills,'' Illinois Geological Survey Division, Circular 154, Champagne, IL, 1980.

[*22*] Everett, L. G., Wilson, L. G., and Hoylman, E. W., *Vadose Zone Monitoring for Hazardous Waste Sites,* Noyes Data Corp., Park Ridge, NJ, 1984.

[*23*] Duke, H. R. and Haise, H. R., ''Vacuum Extractors to Assess Deep Percolation Losses and Chemical Constituents of Soil Water,'' *Soil Science Society of America Proceedings,* Vol. 37, 1973, pp. 963–964.

[*24*] Trout, T. J., Smith, J. L., and McWhorter, D. B., ''Environmental Effects of Land Application of Digested Municipal Sewage Sludge,'' Report submitted to the City of Boulder, Colorado, Department of Agricultural Engineering, Colorado State University, Ft. Collins, CO, 1975.

[*25*] Linden, D. R., ''Design, Installation and Use of Porous Ceramic Samplers for Monitoring Soil-Water Quality,'' U.S. Department of Agricultural Research Service Technical Bulletin, 1562, 1977.

[*26*] Harris, A. R. and Hansen, E. A., ''A New Ceramic Cup Soil-Water Sampler,'' *Soil Science Society of America Proceedings,* Vol. 39, 1975, pp. 157–158.

[*27*] DeJong, E., ''Inexpensive Micro-Soil Solution Sampler,'' *Canadian Journal of Soil Science,* Vol. 56, 1976, pp. 315–317.

[*28*] Yamasaki, S. and Kishita, A., ''Studies on Soil Solutions—A Historical Review,'' Research Bulletin, Hokkaido National Agricultural Experimental Station, Japan, Vol. 96, 1970, pp. 54–72.

[*29*] Everett, L. G., *Groundwater Monitoring,* General Electric Co., Business Growth Services, Schenectady, NY, 1980.

[*30*] Shuford, J., Fritton, D., and Baker, D., ''Nitrate Nitrogen and Chloride Movement through Undisturbed Field Soil,'' *Journal of Environmental Quality,* Vol. 6, 1977, pp. 255–259.

[*31*] Tyler, D. and Thomas, G., ''Lysimeter Measurements of Nitrate and Chloride Losses from Soil under Conventional Non-Tillage Corn,'' *Journal of Environmental Quality,* Vol. 6, 1977, pp. 63–66.

[*32*] Warrick, W. and Amoozegar-Ford, A., ''Soil Water Regimes Near Porous Cup Samplers,'' *Water Resources Research,* Vol. 13, 1977, pp. 213–217.

[*33*] van der Ploeg, R. and Beese, F., ''Model Calculations for the Extraction of Soil Water by Ceramic Cups and Plates,'' *Soil Science Society of America Journal,* Vol. 41, 1977, pp. 466–470.

[*34*] Barbarick, K., Sabey, B., and Klute, A., ''Comparison of Various Methods of Sampling Soil Water for Determining Ionic Salts, Sodium, and Calcium Content in Soil Columns,'' *Soil Science Society of America Journal,* Vol. 43, 1979, pp. 1053–1055.

[*35*] Everett, L. G., Wilson, L. G., Hoylman, E. W., and McMillion, L. G., "Constraints and Categories of Vadose Zone Monitoring Devices," *Ground-Water Monitoring Review,* Winter, 1984.
[*36*] Weldon, R., "Atmospheric Emissions for Oil Waste Landspreading," prepared for the American Petroleum Institute, Washington DC, 1980.
[*37*] Wetherold, R. B., Randall, J. L., and Williams, K. R., "Laboratory Assessment of Potential Hydrocarbon Emissions from Land Treatment of Refinery Oily Sludges," prepared by Radian Corp. for the American Petroleum Institute, Washington, DC, 1983.

# Laboratory and Field Analyses

*Harold W. Olsen,*[1] *Thomas L. Rice,*[1] *and Roger W. Nichols*[1]

# Measuring Effects of Permeant Composition on Pore-Fluid Movement in Soil

**REFERENCE:** Olsen, H. W., Rice, T. L., and Nichols, R. W., "**Measuring Effects of Permeant Composition on Pore-Fluid Movement in Soil,**" *Ground-Water Contamination: Field Methods,* ASTM STP 963, A. G. Collins and A. I. Johnson, Eds., American Society for Testing and Materials, Philadelphia, 1988, pp. 331–342.

**ABSTRACT:** A laboratory experimental system is described for measuring the effects of permeant composition on pore-fluid movement in a soil specimen mounted in a conventional triaxial cell. These effects include the variation of permeability that arises when the pore-fluid composition changes within a soil, and also the tendency for a permeant composition gradient to cause pore-fluid movement. The system is based on the previously reported capabilities and advantages of the flow-pump method for permeability measurements. The system also utilizes a permeant-control subsystem that facilitates bringing deaired solutions of arbitrary chemical composition from storage reservoirs to the ends of a test specimen.

**KEY WORDS:** permeability, osmosis, contamination, waste disposal, flow-pump method, permeant chemistry

The growing concern about contamination associated with waste disposal has generated interest in the need for improved laboratory methods to measure the effects of permeant composition on pore-fluid movement in soil. These effects include the variation of permeability that arises when the pore-fluid composition changes within a soil, and also the tendency for a permeant composition gradient to cause pore-fluid movement.

An experimental system is described which enables these effects to be measured on a test specimen in a standard triaxial cell under simulated *in situ* physical and chemical conditions. The system employs the flow-pump method for permeability testing which facilitates the avoidance or control of well-known difficulties with conventional constant-head and falling-head methods, including errors in the measurement of very low flow rates and seepage-induced consolidation and fabric changes [*1*]. The system also employs a permeant control subsystem that enables control of the air content and the chemistry of permeant solutions in contact with the ends of a test specimen.

The flow-pump method and the permeant control system described herein were introduced in fundamental permeability studies nearly two decades ago [*2–4*]. In those studies, emphasis was placed on conducting permeability measurements on rigidly confined specimens of idealized materials. Recent studies concerning soil permeability testing methods [*5–9*] have emphasized the practical need for testing natural materials under simulated in situ physical and chemical conditions. In this regard, the importance of using flexible wall permeameters has become widely recognized. The experimental system described herein utilizes the aforementioned flow-pump method and

[1] Research civil engineer, environmental engineer, and civil engineering technician, respectively, U.S. Geological Survey, Box 25046, M.S. 966, Denver, CO 80225. Mr. Rice's current address is: Consolidated Gas Transmission Corp., 445 Main St., Clarksburg, WV 26301.

FIG. 1—*Experimental system.*

permeant control system in conjunction with a conventional triaxial cell that serves as a flexible wall permeameter.

## Equipment

Figure 1 presents a photographic overview of the experimental system that consists of (*a*) a flexible wall permeameter, (*b*) a flow pump, (*c*) a differential pressure transducer, (*d*) a bank of permeant reservoirs, (*e*) a permeant exchange subsystem, and (*f*) a manifold that interconnects the above elements with the permeant lines leading to the base pedestal and top cap for a specimen.

The flexible wall permeameter (Fig. 2) is a conventional Wykeham Farrance triaxial cell that has been modified in two ways. First, the cell and the cell base are equipped with a stainless steel cylinder and Whitey ball valves, rather than the plastic cylinder and Klinger valves that were supplied with the equipment. Second, the base pedestal and the top cap were modified to facilitate changing the permeant solutions in contact with the ends of a test specimen. Figure 3 illustrates

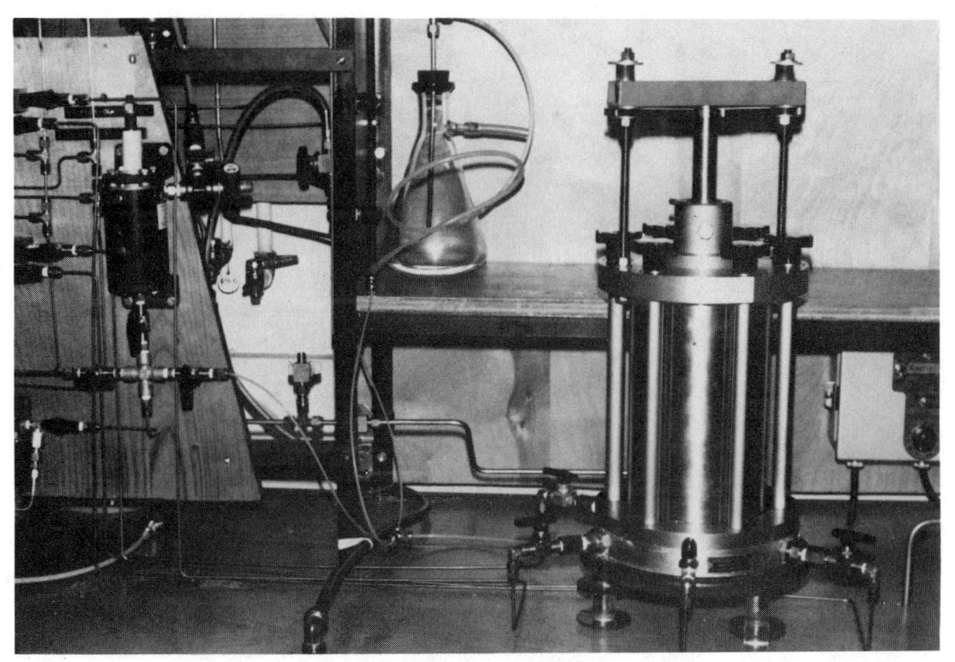

FIG. 2—*Triaxial cell.*

how this modification enables radial flushing of the porous disk between a center hole in the disk and a cavity in the disk container just outside the circumference of the disk. The center hole and the circumferential cavity are connected with the channels in the base pedestal and the top cap that transmit permeant to and from the ends of the specimen.

The flow pump (Fig. 4) consists of a stainless steel actuator mounted in a variable-speed drive. The actuator has the configuration of a syringe with a cylindrical barrel and a piston having a seal consisting of a pair of "O" rings. The variable-speed drive is a Harvard Apparatus Model 906 infusion-withdrawal pump. This pump controls the piston by means of a worm gear driven by a variable-speed d-c motor through a transmission box with twelve combinations of gears between

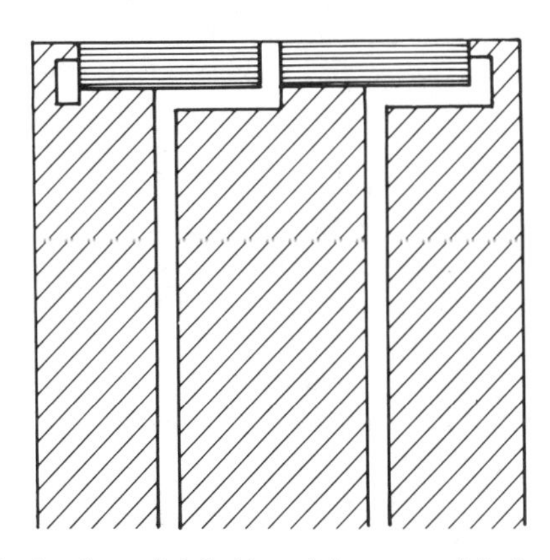

FIG. 3—*Scheme of the mechanism for radial flushing of the porous disks in the base pedestal and top cap.*

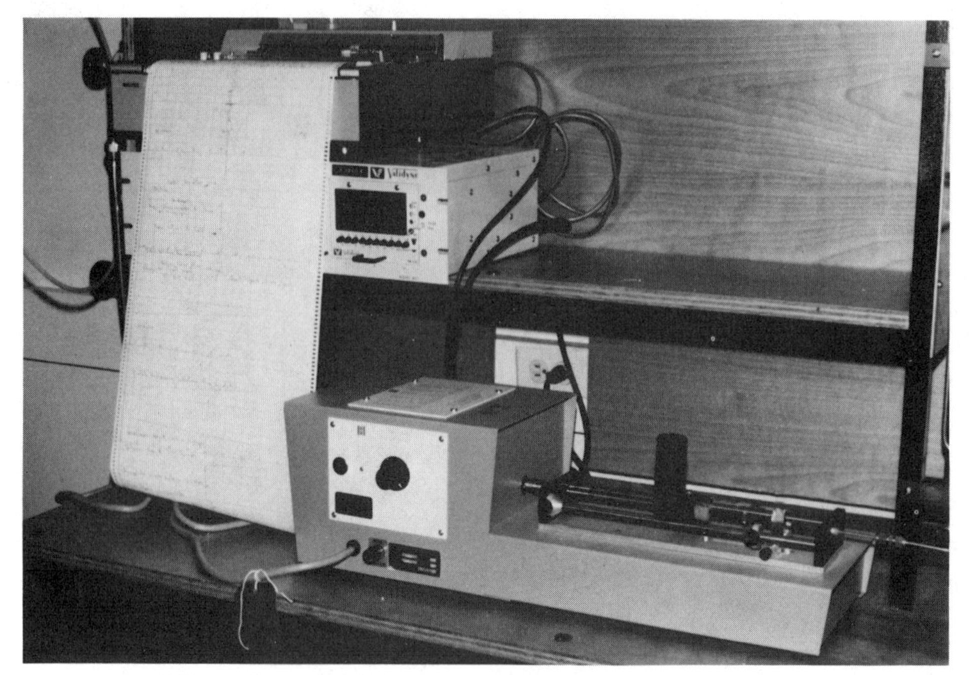

FIG. 4—*Flow pump and the readout system for the differential transducer.*

the worm gear and the motor. A controller on the d-c motor allows the worm gear rotation to be controlled at speeds intermediate between those determined by the gear selector. These features enable the piston to be advanced or withdrawn at any constant rate ranging from about $10^{-1}$ to less than $10^{-5}$ cm/s. The syringes used with this actuator have had capacities ranging from 20 $cm^3$ to 0.5 $cm^3$. Accordingly, this flow pump has produced flow rates ranging from about $10^{-1}$ $cm^3/s$ to $10^{-7}$ $cm^3/s$.

The differential pressure transducer (Fig. 5) is a Validyne variable reluctance Model P300D equipped with a $\pm$350-kN/$m^2$ diaphragm, and operated with a Validyne power and signal conditioning unit, Model MC1. The output of the transducer is monitored with the digital voltmeter and the Soltec model 210 strip-chart recorder shown in Fig. 4.

The bank of permeant reservoirs is mounted on a high shelf above and behind the triaxial cell (Fig. 6). The reservoirs are 19-L carboys with plumbing attached for storing the solutions under vacuum, and for transmitting solutions to the experimental system. Figure 7 shows a scheme of this plumbing, which includes a subsystem for elevating the pressure in the permeant before it is transmitted to the manifold. This subsystem utilizes a pair of Bellowfram cylinders (''Bellowfram'' is a registered trademark of Bellowfram Corp., Burlington, Massachusetts, for rolling seal diaphragms of its design and manufacture). While one Bellowfram cylinder is being filled with permeant flowing under gravity from a permeant reservoir, the other Bellowfram cylinder is discharging permeant into the manifolds, driven by elevated air pressure. When the first Bellowfram cylinder is full and the second is empty, their functions are interchanged with the four-way valves. Then, the second Bellowfram cylinder is filled with permeant from the reservoir while the first Bellowfram cylinder discharges permeant to the manifold. Thus, the permeant exchange subsystem provides a means to bring permeant from a reservoir at atmospheric pressure into the manifold under backpressure on a nearly continuous basis.

The manifold is organized around four identical elements that are arranged in pairs, as shown

FIG. 5—*Manifold including the upper and lower pair of manifold elements, the differential transducer beneath the manifold elements, and the permeant sink Bellowfram to the right of the manifold elements.*

in Fig. 5. The upper pair is connected to the two permeant lines to the top cap, and the lower pair to the two permeant lines to the base pedestal. Figure 8 shows that each manifold element interconnects a permeant line with the permeant reservoirs, the flow pump, the differential transducer, and the Bellowfram that serves as a downstream permeant sink.

## Leaks and Undissolved Air in the Permeant System

The permeant system is tested for leaks by determining whether it can retain elevated pressure or vacuum or both for substantial periods of time, say, on the order of a few days. Leaks are located by using the valves and the transducers to isolate and test zones within the system.

Before mounting a test specimen, the entire permeant system is filled with deaired permeant and tested for undissolved air, as follows. First, vacuum is applied simultaneously to a reservoir of permeant solution and to the entire permeant system, to remove dissolved air from the permeant solution and entrapped air from the permeant system. Second, while vacuum is maintained on both the reservoir of deaired solution and the permeant system, deaired permeant is allowed to flow under gravity from the permeant reservoir into the permeant system. Third, the vacuum is removed from, and elevated pressure is applied to, the permeant system to drive any remaining undissolved air into solution. Fourth, the permeant in the permeant system is replaced with fresh deaired solution from the reservoir, using the permeant exchange subsystem (Fig. 7) to maintain elevated backpressure in the permeant system during the exchange process.

## Equipment Capabilities

The following describes and illustrates the capabilities of the equipment for (*a*) simple and rapid performance checks on the equipment, (*b*) measuring the permeability of a test specimen and its

FIG. 6—*Bank of permeant reservoirs above and behind the triaxial cell.*

variation with changes in the chemistry of the specimen pore fluid caused by leaching, and (*c*) measuring the tendency of a permeant composition gradient to cause pore-fluid transport through the specimen.

*Performance Checks*

The configuration of the permeant system allows simple and rapid performance checks on the calibration of the differential transducer (including all the adjustments and settings in the associated signal conditioning and readout equipment), and also for the presence of undissolved air in the permeant system and the test specimen.

To conduct these checks, one side of the differential transducer is open to all four of the manifold elements illustrated in Fig. 8 while the other side is open to the Bellowfram that usually serves as a backpressured permeant sink, but which is isolated from the manifold for the purpose of these performance checks.

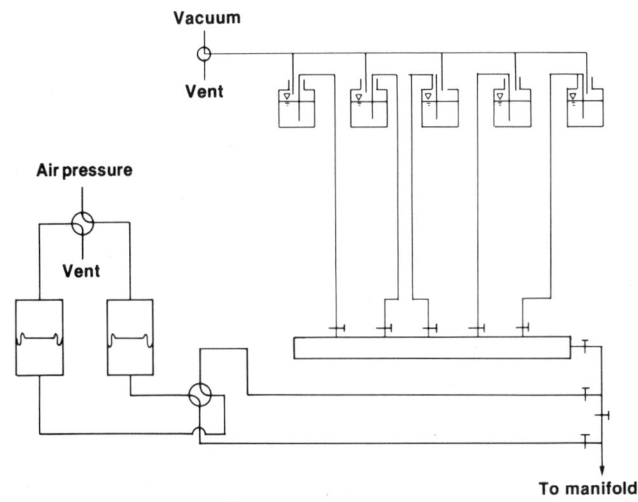


FIG. 7—*Scheme of the bank of permeant reservoirs and the permeant exchange subsystem.*

The "early" part of Fig. 9 shows the response of the differential transducer to externally controlled changes of $\pm 138$ kN/m$^2$ in the reference pressure in the permeant-sink Bellowfram. Clearly, the transducer output is fully consistent with the externally controlled pressure changes. Moreover, it is to be noted that this performance check was accomplished in about 2 min.

The "later" part of Fig. 9 shows the response of the differential transducer to externally controlled changes in the cell pressure acting on the test specimen, while the reference pressure in the permeant sink Bellowfram is maintained constant. The transducer response shows that the changes in pore pressure in the specimen ($\Delta u$) that were induced by changes in the total stress generated by the cell pressure ($\Delta\sigma_c$) are approximately equal. Hence, this performance check, which was also accomplished in a few minutes, indicates the test specimen and permeant system are fully saturated, at least to a first approximation, because this measurement is equivalent in principle to the B test in conventional triaxial procedures.

In Fig. 10, the transducer response was generated by using the flow pump to infuse and withdraw permeant from the manifold. During this experiment, the valves between the manifold and the triaxial cell were closed. Therefore, the data in Fig. 10 show the behavior of the permeant system by itself. These data allow the compliance of the permeant system ($S_{ps}$) to be defined in terms of the amount by which the volume of permenat contained within the system changes ($dV$) in response to a change of head ($\Delta h$) within the system, as follows:

$$S_{ps} = \frac{dV\,(cm^3)}{dh\,(cm\ \mathrm{H_2O})} = \frac{dV/dt\,(cm^3/s)}{dh/dt\,(cm/s)} \tag{1}$$

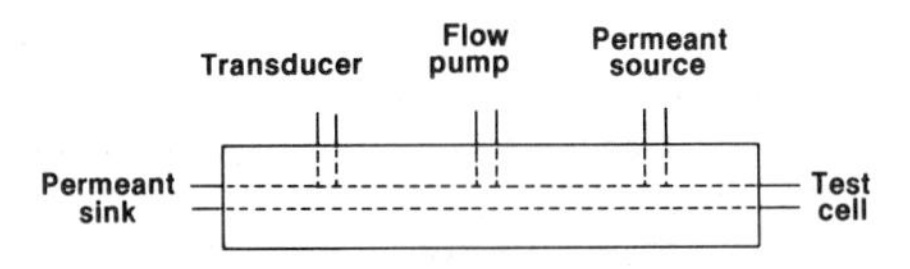


FIG. 8—*Scheme of a manifold element.*

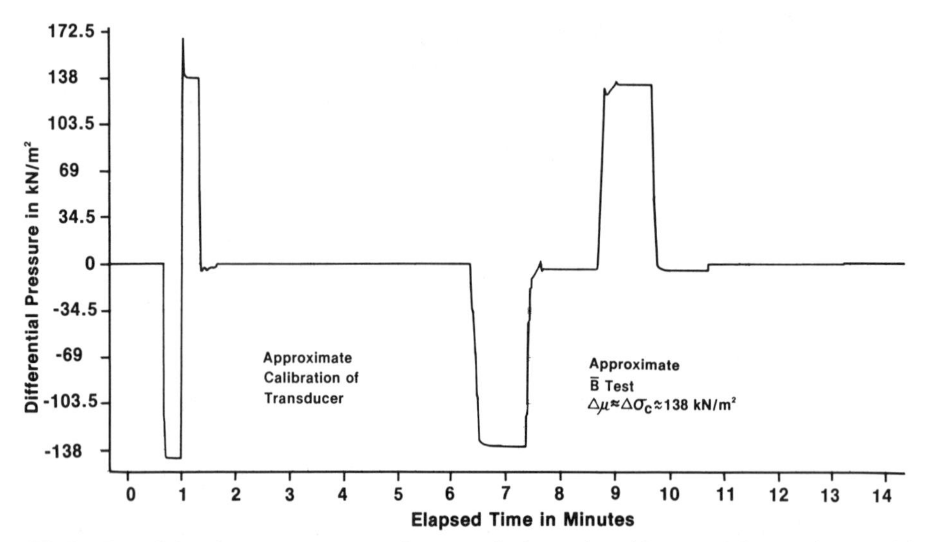


FIG. 9—*Typical data from approximate performance checks on the calibration of the transducer and the degree of saturation of the test specimen.*

where $dV/dt$ is the flow rate generated by the flow pump, and $dh/dt$ is proportional to the slope of the pressure versus time data in Fig. 10.

Such a compliance measurement enables the presence of undissolved air to be detected because the compressibility of undissolved air is very high compared with the compressibility of liquid. It follows that the compliance of a permeant system varies with the amount of undissolved air present, and that the compliance will be a minimum when no undissolved air is present. Comparison of the minimum compliance of a permeant system with a measurement such as that in Fig. 10 provides a measure of the amount of undissolved air in the system.

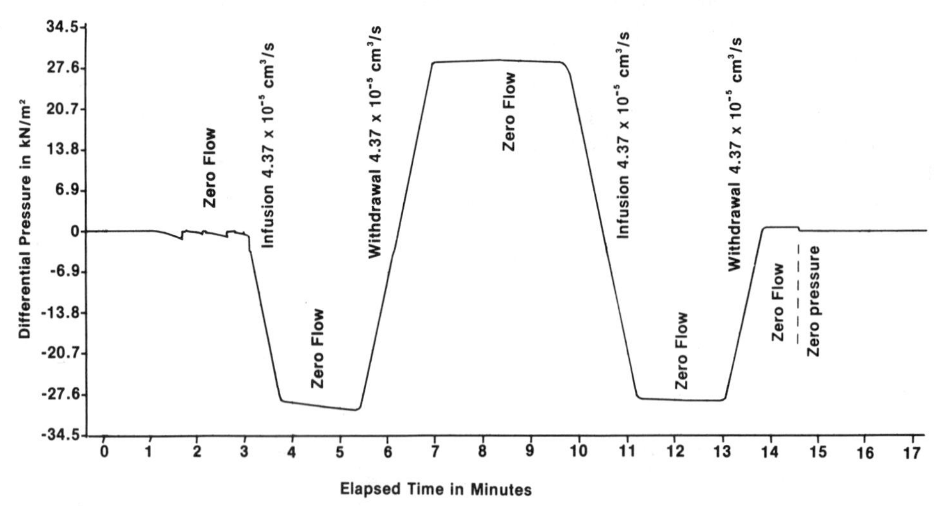


FIG. 10—*Typical data on the compliance of the permeant system.*

*Permeability Measurements*

The experimental system utilizes the flow-pump method for conducting permeability measurements, the advantages (compared with conventional constant-head and falling-head methods) of which are [*1*]: (*a*) Direct flow-rate measurements are avoided together with the associated errors that arise from the effects of contaminants on capillary menisci and the long periods of time involved in flow-rate measurements; (*b*) permeability measurements can be obtained much more rapidly and at substantially smaller gradients; and (*c*) errors from the small intercept in the otherwise linear flow-rate versus hydraulic-gradient relationship, and also from seepage-induced permeability changes, can easily be recognized and avoided or minimized.

In the flow-pump method, the pressure difference across a test specimen is measured with the differential pressure transducer while flow through the specimen is being generated with a flow pump, as illustrated in Fig. 11. When the flow pump is turned either on or off, the pressure difference across the specimen initially varies with time and gradually approaches a steady-state value. The magnitude of the steady-state pressure difference is directly related to the permeability of the specimen, according to Darcy's Law.

Figure 11 shows that the time needed for a permeability measurement is governed by the time required for the initial transient response. The initial response time, in turn, is caused, at least in part, by seepage-induced consolidation of the test specimen in response to the changes in effective stress generated by the applied flow rate. In consequence, the response time for a flow-pump permeability test varies with the permeability, compressibility, and length of a test specimen. In Fig. 11, the response time is short, consistent with the relatively high permeability of the test specimen. Data reported elsewhere [*1*] show that response times of a few hours are to be expected for conventional 3.5-cm-diameter by 7.1-cm-length specimens of materials having permeabilities on the order of $10^{-8}$ cm/s.

Figure 12 shows how the permeability-measurement approach illustrated in Fig. 11 can be utilized for measuring the variation of the permeability of a test specimen as the chemistry of its pore fluid is changed by leaching. The data points in Fig. 12 were derived from a continuous recording of the pressure difference across the specimen versus time, similar to that illustrated in Fig. 11. Simultaneously, the flow pump was withdrawing pore fluid from the base of the specimen while the permeant lines to the top of the specimen were connected to the permeant-exchange subsystem illustrated in Fig. 7. In addition, the permeant solution coming to the top of the specimen from the permeant-exchange subsystem was a 4% sodium metaphosphate solution, whereas the ends of the specimen had previously been exposed only to distilled water in the permeant system. Thus, the abscissa in Fig. 12 shows the extent to which the specimen was leached by the sodium metaphosphate solution, where the quantity of flow is expressed in terms of the volume of voids in the test specimen.

*Chemical Causes of Pore-Fluid Movement*

The permeant control system provides a means to control the permeant composition at each end of the test specimen independently. Therewith, arbitrary permeant composition gradients can be applied to a test specimen. The consequence of such a gradient concerning pore-fluid movement is that the flow rate versus hydraulic head relationship is displaced from the origin, as illustrated in Fig. 13. This figure shows data obtained with an experimental system identical in concept to that described herein except that the permeant control system was attached to a one-dimensional consolidometer rather than to a triaxial cell [*3*]. When the flow rate is equal to zero, the head difference across the specimen is the osmotic pressure, that is, the hydraulic pressure difference needed to counteract the tendency of a chemical gradient to cause pore-fluid movement through the specimen. The experimental system described herein allows such osmotic pressure measurements

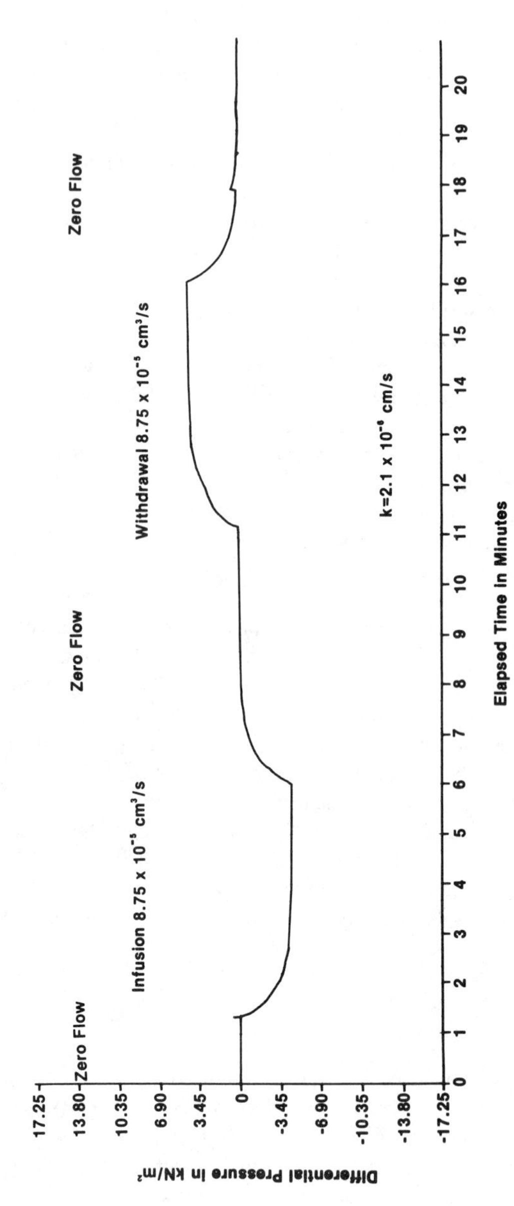


FIG. 11—*Typical data from a flow-pump permeability test.*

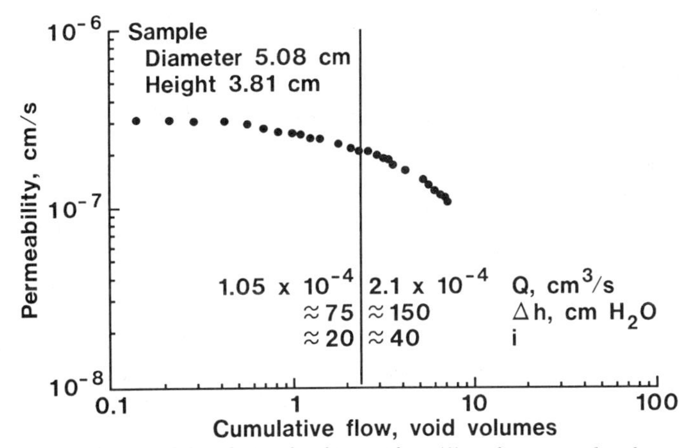


FIG. 12—*Variation of permeability during leaching with a 4% sodium metaphosphate solution, where leaching was controlled at a constant rate with a flow pump while the permeability was monitored continuously by measuring the pressure difference across the specimen with the differential pressure transducer.*

to be conducted on specimens of natural soils under simulated *in situ* stress and chemical conditions in response to chemical gradients associated either with naturally occurring geochemical processes or with waste-management activities.

## Summary and Conclusions

The growing concern about contamination associated with waste disposal has focused attention on the need for improved laboratory methods to measure the effects of permeant composition on

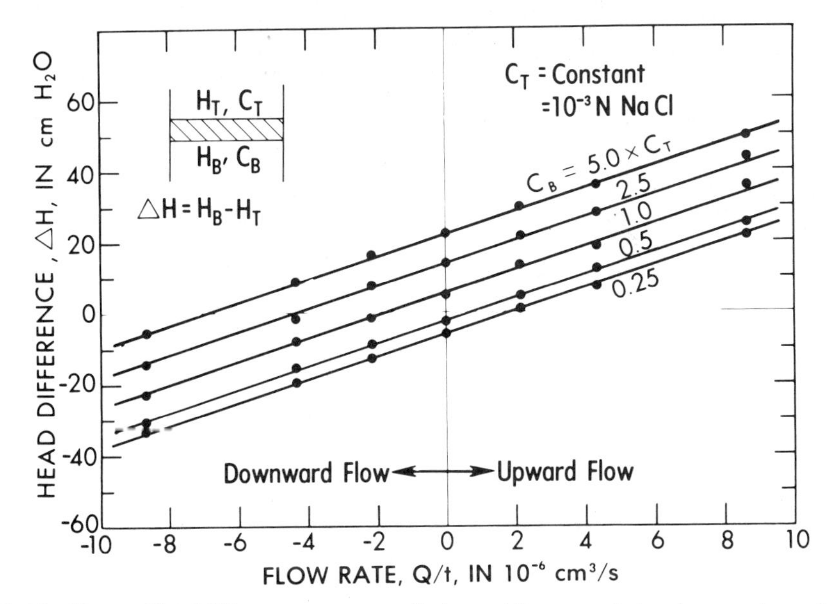


FIG. 13—*Measured head difference versus externally imposed flow rate relations for a specimen of sodium kaolinite. The electrolyte concentration in the permeant at the top of the specimen ($C_T$) is the same for all the relations. The displacement of the relations from the origin varies with the electrolyte concentration in the permeant at the bottom of the specimen ($C_B$).*

pore-fluid movement in soil. These effects include the variation of permeability that arises when the pore-fluid composition changes within a soil, and also the tendency for permeant composition gradients to cause pore-fluid movement.

An experimental system is described that enables these permeant composition effects to be measured on saturated soil specimens in a conventional triaxial cell. The system is based on the previously reported capabilities and advantages of the flow-pump method for permeability measurements. It also utilizes a previously developed permeant control system that facilitates bringing deaired solutions of arbitrary compositions from storage reservoirs to the ends of a test specimen. The tendency for a composition gradient to induce pore-fluid movement is determined by preventing flow into or out of one end of the specimen, and measuring the induced pressure difference across the specimen with a differential pressure transducer. The variation of permeability that arises when the pore fluid composition changes within a soil is determined by measuring the head difference across the specimen continuously as the specimen is leached by withdrawing pore fluid from the base of the specimen with the flow pump and allowing liquid from a selected permeant solution reservoir to be drawn into the top of the specimen.

## References

[1] Olsen, H. W., Nichols, R. W., and Rice, T. L., "Low Gradient Permeability Measurements in a Triaxial System," *Geotechnique*, Vol. 35, No. 2, 1985, pp. 145–147.

[2] Olsen, H. W., "Darcy's Law in Saturated Kaolinite," *Water Resources Research*, Vol. 2, No. 6, 1966, pp. 287–295.

[3] Olsen, H. W., "Simultaneous Fluxes of Liquid and Charge Through Saturated Kaolinite," *Proceedings*, Soil Science Society of America, Vol. 33, No. 3, 1969, pp. 338–344.

[4] Olsen, H. W., "Liquid Movement through Kaolinite under Hydraulic, Electric, and Osmotic Gradients," *Bulletin of the American Association of Petroleum Geologists*, Vol. 56, 1972, pp. 2022–2028.

[5] Olson, R. E. and Daniel, D. E., "Measurement of Hydraulic Conductivity of Fined-Grained Soils," *Permeability and Ground Water Contaminant Transport, ASTM STP 746*, R. F. Zimmie and C. O. Riggs, Eds., American Society for Testing and Materials, Philadelphia, 1981, pp. 18–64.

[6] Tavenas, F., Leblond, P., Jean, P., and Leroueil, S., "The Permeability of Natural Soft Clays, Part 1, Methods of Laboratory Measurement," *Canadian Geotechnical Journal*, Vol. 20, No. 4, 1983, pp. 629–644.

[7] Daniel, D. E., Trautwein, S. J., Boynton, S. S., and Foreman, D. E., "Permeability Testing with Flexible-Wall Permeameters," *Geotechnical Testing Journal*, Vol. 7, No. 3, 1984, pp. 113–122.

[8] Dunn, R. J. and Mitchell, J. K., "Fluid Conductivity Testing of Fine-Grained Soils," *Journal of Geotechnical Engineering*, Vol. 110, No. 121, 1984, pp. 1648–1665.

[9] Boynton, S. S. and Daniel, D. E., "Hydraulic Conductivity Tests on Compacted Clay," *Journal of Geotechnical Engineering*, Vol. 111, No. 4, 1985, pp. 465–478.

*Charles P. Gerba*[1]

# Methods for Virus Sampling and Analysis of Ground Water

**REFERENCE:** Gerba, C. P., "**Methods for Virus Sampling and Analysis of Ground Water,**" *Ground-Water Contamination: Field Methods, ASTM STP 963*, A. G. Collins and A. I. Johnson, Eds., American Society for Testing and Materials, Philadelphia, 1988, pp. 343–348.

**ABSTRACT:** The consumption of contaminated ground water is responsible for almost half of the reported waterborne disease outbreaks each year in the United States. Enteric viruses continue to be a significant cause of such waterborne disease. Methods for virus detection in ground water are needed to assess the safety of ground-water supplies. Viruses in ground water may originate from septic tanks, leaking sewer lines, domestic and sludge landfills, land disposal of sewage effluent, leaking sewage ponds, etc. Because of the potential health significance of low numbers of viruses in water, it is necessary to sample large volumes of water (40 to 1000 L). The method most commonly used is the microporous filter adsorption/elution technique. This involves passing water through a filter to which the viruses adsorb and subsequently eluting (deadsorbing) the viruses off the filter using a 1 to 2-L suspension of 3% beef extract. This eluate is further reconcentrated to a volume of 20 to 30 mL before assay. Currently, this concentrate is then assayed by animal cell culture. Newer techniques for virus detection should dramatically reduce the time and cost of virus detection in ground water.

**KEY WORDS:** viruses, ground water, gene probe, enteroviruses, virus concentration

There are presently over 120 different types of enteric viruses excreted in human feces and likely to be present in domestic wastes. Enteric viruses can cause a wide variety of diseases in man, some of which are listed in Table 1. While obvious sewage contamination often results in outbreaks of disease, low numbers of enteric viruses in water are also believed capable of causing significant illness and mortality. It has been estimated that as little as one hepatitis virus per 1000 L of drinking water can result in a lifetime risk of mortality of $5 \times 10^{-4}$ [*1*]. The possible high risk that may be associated with enteric viruses in water has led the U.S. Environmental Protection Agency (EPA), Office of Drinking Water to propose a recommended maximum contaminate level (RMCL) of zero for drinking water [*2*].

Viruses in ground water represent a very real threat to the health of the people consuming the water. From 1971 to 1979, an estimated 57974 persons in the United States were involved in waterborne disease outbreaks [*3*]. The consumption of contaminated ground water was responsible for over one half of all reported waterborne outbreaks and 45% of all reported cases of waterborne disease in the United States. It has been estimated that as much as 65% of the documented outbreaks of waterborne disease in the United States could be attributed to illness of probable viral etiology [*4*].

Methods for virus detection in ground water are needed to assess the safety of ground-water supplies. Viruses in ground water may originate from septic tanks, leaking sewer lines, domestic

[1] Professor, Departments of Nutrition and Food Science, and Microbiology and Immunology, University of Arizona, Tucson, Arizona 85721.

TABLE 1—*Human enteric viruses which may be present in sewage.*

| Viruses | Type | Disease Causes |
|---|---|---|
| Enteroviruses | | |
| Poliovirus | 3 | meningitis, paralysis, fever |
| Echovirus | 31 | meningitis, diarrhea, rash, fever respiratory disease |
| Coxsackievirus A | 23 | meningitis, herpangina, fever respiratory disease |
| Coxsackievirus B | 6 | myocarditis, congenital heart anomalies, pleurodynia, respiratory disease, fever, rash meningitis |
| New enteroviruses (Types 68-71) | 4 | meningitis, encaphalitis, acute hemorrhagic conjunctivities, fever, respiratory disease |
| Hepatitis Type A (enterovirus 72) | 1 | infectious hepatitis |
| Norwalk virus | 1 | diarrhea, vomiting, fever |
| Calicivirus | 1 | gastroenteritis |
| Astrovirus | 1 | gastroenteritis |
| Reovirus | 3 | not clearly established |
| Adenovirus | 40 | respiratory disease, eye infections gastroenteritis |
| Rotavirus | 4 | diarrhea |
| Snow-Mountain Agent | unknown | gastroenteritis |

waste and sludge landfills, land disposal of sewage, leaking sewage ponds, etc. [*4*]. Once they have entered the ground water they may travel hundreds of meters [*4*]. Several studies have shown that coliform bacteria, currently used to judge the microbial safety of drinking water, do not always indicate the presence of enteric viruses [*5,6*]. Few major field studies on virus occurrence in ground water have been conducted. In the most extensive field study to date, Marzouk et al. [*5*] isolated enteroviruses in 20% of 99 samples of drinking water wells located in areas where septic tanks were present. Recently, the Office of Technology Assessment suggested that viral contamination of drinking water (including ground-water sources) may be of greater significance than has been recognized [*7*].

This paper reviews the most commonly used methods today for virus concentration and detection in ground water.

## Concentration of Viruses from Ground Water

### *Filter Systems*

Because of the potential health significance of low numbers of viruses in water, it is necessary to sample large volumes of water (40 to 1000 L). To accomplish this, viruses are concentrated from water by adsorption onto microporous filters [*8*]. This involves passing water through a filter to which the viruses adsorb and subsequently eluting (deadsorbing) the viruses off the filter using a 1 to 2-L suspension of 3% beef extract.

Two types of filtering systems are used and both have advantages and disadvantages. The electronegative filters have been shown to have a greater capacity for virus adsorption in waters with high turbidities and organic matter, but require the addition of aluminum trichloride ($AlCl_3$) or magnesium dichloride ($MgCl_2$) and acidification of the water to pH 3.5 to get maximum adsorption of the viruses to the filter. This can be cumbersome in the field as it requires modifying the water sample prior to filtering (addition of $AlCl_3$) and additional materials and equipment (pH meter). It also requires extensive trained and experienced personnel for proper use.

The most commonly used negatively charged filter in the United States is the Filterite fiberglass Duo-Fine brand filters (Filterite, Timonium, Maryland). They are marketed as pleated cartridge units, although disk filters can be purchased on special order. Usually, a 25-cm-length cartridge filter is used for virus sampling. The pleated cartridge configuration allows a large filter surface area in a small unit which in turn allows larger volumes of turbid water to be sampled before filter clogging. However, ground waters are generally of low turbidity, and a disk filter usually can be successfully used to sample 400 to 1000 L. Our laboratory uses a 293-mm-diameter filter contained in a stainless steel holder manufactured by the Millipore Corporation (Bedford, Massachusetts). Although maximum sampling flow rate is reduced from 40 L/min for the cartridge filter to 20 to 25 L/min for the disk filter, the cost of the disk filter is one twentieth the cost of the cartridge filter.

Electropositive filters offer a major advantage over the electronegative filters in that the water being sampled does not have to be preconditioned if the pH of the water is 8.5 or less. This reduces the needed sampling equipment to only a filter housing and a flowmeter. The 1 MDS Virozorb is a positively charged microporous filter especially manufactured by CUNO (Meriden, Connecticut) for virus concentration from water. It is available as a 25.4-cm pleated filter cartridge or as disk filters of various sizes. Again, sampling of large volumes of ground water is possible with a 293-mm-diameter disk filter. Little training of collection personnel is necessary for use of the 1-MDS filters, and the necessary sampling equipment and filters can be sent to remote sites in a medium-size ice chest through the mail.

*Field Methods for Virus Concentration from Ground Water*

The following is a description of the procedures necessary for concentration of viruses from ground water using either the electronegative or electropositive filters. It is provided so that an appreciation of the general procedure and equipment required can be gained by those not familiar with virus monitoring of ground water.

*Electronegative Filters*—The collection and filtering procedures for electronegative filters are as follows:

1. Collect sample in a large plastic(s) container; usually a 200 to 400-L container is used.
2. Dechlorinate sample, if necessary, by addition of sodium thiosulfate.
3. Adjust pH of the water to 3.5 with 1-N hydrochloric acid (HCl) while mixing the water in the container with a pumping system.
4. Add 1-M $AlCl_3$ solution to obtain a final concentration of 0.0005 M $AlCl_3$ to the water being sampled. Although more complex, in-line injectors may be used for addition of sodium thiosulfate, acid, and salts, this requires some skill for proper use but can avoid the need for a large-volume plastic container(s). The Derma-type unit used by Payment and Trudel [*9*] is useful and inexpensive.
5. Place the presterilized Filterite (electronegative filter) cartridge (should be sterilized by autoclaving) filter in the holder, secure, and connect tubing. The water may be pumped up to 40 L/min through the filter.
6. After the desired volume has been filtered or after the filter reaches capacity and clogs, freeze immediately and ship on dry ice for processing [*10*]. The filter may be also eluted in the field with 1 to 2 L of 3% beef extract (pH 9.5 to 10.0), neutralized (adjust to pH 7.0), frozen, and shipped. Samples should be neutralized with 1 N HCl. In no case should higher concentrations of acid be used in the neutralization as significant virus inactivation may take place. Concentrates may be also stored at 4°C for 2 to 3 days without significant viral inactivation occurring.

*Electropositive Filters*—Electropositive filters are easier to use than the negatively charge filters since no preconditioning of the water is necessary.

1. Connect filter housing directly to a faucet or pump.
2. Add sodium thiosulfate, to dechlorinate if necessary, with a Dema [9] in-line injector or metering pump. The water sample may be also collected in large plastic containers and dechlorinated as described for the electronegative filters.
3. Place the 1 MDS Virozorb (electropositive) cartridge filter in the housing, secure, and connect all tubing.
4. Begin pumping at a flow rate of up to 40 L/min.
5. After the desired volume has been filtered or the filter clogs, place filter immediately at 4°C or on ice and ship for processing or elution.

*Filter Transportation*—The filter can be left in the filter holder or can be placed in a plastic Ziplock bag for shipping. Electropositive filters (1 MDS Virozorb) must be shipped at 4°C to prevent significant die-off of the viruses [*11*]. The electropositive filters should be eluted within 48 to 72 h after collection. The electronegative filters must be frozen and shipped on dry ice [*10*]. If the filters are removed from their housings, they should be handled only with disposable plastic gloves to prevent contamination.

**Required Equipment and Materials for Field Sampling**

*Cartridge filters (most commonly used)*—Filterite (25.4-cm pleated cartridge filters, electronegative: Filterite Corp. Timinium, Maryland, pore size: 3.0 μm [prefilter] and 0.45 μm set up in series in two separate holders).

1 MDS Virozorb (25.4 cm pleated cartridge filters, electropositive, CUNO, Meriden, Connecticut, nominal pore size 0.2 μm).

*Pumps*—Gasoline or electric water pump (the Homelite P100 waterbug (Homelite Textron, Charlotte, North Carolina gasoline pump is a lightweight, inexpensive pump which can be used for this purpose).

*Fittings and Hose*—Snap-Tite (Snap-Tite Inc., Union City, Pennsylvania) or other type quick-disconnect fittings are the most convenient to use. A pressure hose is needed on the intake before the filters, because of pressure buildup if the filter clogs. Garden-type hoses can be used on the outlet side of the filter.

*Flowmeter*—Water flowmeter.

*Cartridge Filter Holders*—Lightweight clear plastic cartridge filter holders available from several manufacturers can be used (CUNO Meriden, Connecticut, Model 3MS1 cartridge filter holder will hold both 1 MDS Virozorb and Filterite filters).

*Other*—Sodium thiosulfate ($Na_2S_2O_3$) for chlorinated waters.

400-L containers or several plastic garbage cans, in-line injector for addition of sodium thiosulfate (not needed if the sample is not chlorinated).

*Additional Materials Needed for Use with Electronegative Filters*—pH meter; in-line injectors (Model 202-P, Dema Engineering Co., St. Louis, Missouri); aluminum chloride (1 M $AlCl_3$); and, 1 N HCl.

**Laboratory Procedures for Isolating Viruses**

Once a water sample has been filtered and the filter is brought into the laboratory, further processing of the virus sample is accomplished through elution, reconcentration, clarification, and assay on cell culture, although the elution step can be carried out in the field. Laboratory analysis may take anywhere from two to four weeks to complete, utilizing tissue culture, and from 48 to 72 h for gene probes.

*Filter Elution*

Residual water is removed from the filter while in the holder. One litre of 3% beef extract, with 0.05 M glycine at a pH of 9.5 is added to the filter holder and passed through the filter by applying air pressure. The viruses are eluted off the filter, and the eluate is immediately adjusted to a neutral pH with 1 N HCl. The pH of the eluate must be adjusted immediately after elution to prevent virus inactivation due to the high pH of the eluent.

*Reconcentration and Clarification*

The 1 to 2-L eluate is reconcentrated by precipitation of the proteins and virus with acid followed by centrifugation [*12*]. The pellet from centrifugation is resuspended in 20 to 30 mL of buffer at a high pH. The bacteria are removed through low-speed centrifugation and treated with antibiotics, if necessary. The pH of the final sample is brought to neutral.

*Assay on Cell Culture*

The standard cell line used to assay environmental samples for enteroviruses is the BGM monkey kidney cell line (BGM). The cells are grown in confluent monolayers in 75-$cm^2$ surface area plastic flasks and then innoculated with the water or wastewater filter concentrates [*13*]. The cells are examined for 14 days for cell destruction (CPE, cytopathic effect) caused by the virus. Any positive flask is confirmed by passage into fresh cells and subsequent CPE. This procedure is repeated until all the sample volume has been assayed. Enteroviruses are usually quantitated by a Most Probable Number (MPN) method similar to coliform bacteria or by the Plaque Forming Unit (PFU) method. In the PFU method, the cell monolayer is covered with agar, and then the cell monolayer is exposed to a dye that stains only living cells [*14*]. Cells killed by viruses produce clear zones or plaques in the monolayer. For detection of rotaviruses and hepatitis A virus, additional cell lines (MA-104) and fluroescent and radioactively labeled antibodies are required [*14,15*].

*Enterovirus Detection by Gene Probes*

To reduce the cost and time necessary to detect enteric viruses we have recently applied gene probes to their detection in ground water. Gene probes are small pieces of nucleic acid (either DNA or RNA) that hybridize to the complementary base pairs of specific genes and can be used to identify the genetic information of any organism. The cDNA or cRNA probes that hybridize to viral nucleic acids now make it possible to identify the presence of small amounts of genetic information of any organism. Nucleic acid probes are at least 1000 times more sensitive than serological tests and do not require cultivation of the organisms [*16*]. Recently in our laboratory, we have developed a cDNA probe which is labeled with both $^{32}P$ dATP and dCTP [*17*]. It is capable of detecting as little as 1 femtrogram of nucleic acid or one tissue culture infectious dose of poliovirus or hepatitis A virus. This is the most sensitive nontissue culture technique ever developed for enteroviruses ]*16*]. An advantage of gene probes is that using autoradiography results can be obtained within 72 h rather than 2 to 3 the weeks with tissue culture. Another advantage of the gene probes is that they will detect closely related viruses. Thus, a probe developed for poliovirus will also detect other enteroviruses.

Currently, gene probes cannot be used to determine if a virus is infectious; thus they can be used only in the analysis of undisinfected water. Positive results by gene probe should be confirmed in tissue culture. Current direct costs for enterovirus concentration and detection in cell culture run from $300 to $350 per sample. The same analysis using gene probes is less than $100 per sample.

## Discussion

Methods involving virus adsorption-elution from microporous filters continue to be the most promising and useful methods for concentrating viruses from large volumes of ground water. The recent development of positively charged microporous filter media offers further simplification of current methodology. Samples can now be collected by personnel with a minimum of training. Recent development of gene probes for virus detection in environmental samples has the potential of greatly reducing the cost and time necessary for the laboratory analysis [*16*]. Current costs for virus analysis of ground water, including collection and laboratory analysis, run from $250 to $750 per sample [*1*]. Total costs using gene probes in our laboratory run from $70 to $150 per sample. Further research should continue to reduce costs for virus detection in ground water.

## References

[*1*] Gerba, C. P., "Strategies for the Control of Viruses in Drinking Water," American Association for the Advancement of Science, Washington, DC, 1984.

[*2*] *Federal Register,* Microbiological RMCLS, Vol. 50, 1985, pp. 46951–46957.

[*3*] Craun, G. F., *Waterborne Diseases in the United States,* CRC Press, Boca Raton, FL, 1986.

[*4*] Keswick, B. H. and Gerba, C. P., "Viruses in Groundwater," *Environmental Science and Technology,* Vol. 14, 1980, pp. 1290–1297.

[*5*] Marzouk, Y., Goyal, S. M., and Gerba, C. P., "Prevalence of Enteroviruses in Ground Water in Israel," *Ground Water,* Vol. 17, 1979, pp. 487–491.

[*6*] Slade, J. S., "Viruses and Bacteria in a Chalk Well," *Water Science and Technology,* Vol. 17, 1985, pp. 111–125.

[*7*] "Protecting the Nation's Groundwater from Contamination," Office of Technology Assessment, Vol. 1, U.S. Congress, Washington, DC, 1984.

[*8*] *Standard Methods for the Examination of Water and Wastewater,* American Public Health Association, 16th ed., Washington, DC, 1985.

[*9*] Payment, P. and Trudel, M., "Improved Method for the Use of Proportioning Injectors to Condition Large Volumes of Water for Virological Analysis," *Canadian Journal of Microbiology,* Vol. 27, 1981, pp. 455–457.

[*10*] Dahling, D. R. and Wright, B. A., "Processing and Transport of Environmental Virus Samples," *Applied Environmental Microbiology,* Vol. 47, 1984, pp. 1272–1276.

[*11*] Sobsey, M. D. and Glass, J. S., "Improved Electropositive Filters for Concentrating Viruses from Large Volumes of Water" in *Viruses and Wastewater Treatment,* M. Goddard and M. J. Butler, Eds., Pergamon Press, New York, 1982.

[*12*] Katzenelson, E., Fattal, B., and Hostovesky, T., "Organic Flocculation: An Efficient Second-Step Concentration Method for the Detection of Viruses in Tap Water," *Applied Environmental Microbiology,* Vol. 32, 1976, pp. 638–639.

[*13*] Payment, P. and Trudel, M., "Influence of Innoculum Size, Incubation Temperature, and Cell Culture Density on Virus Detection in Environmental Samples, *Canadian Journal of Microbiology,* Vol. 31, 1985, pp. 977–980.

[*14*] Smith, E. M. and Gerba, C. P. "Development of a Method for the Detection of Human Rotavirus in Water," *Applied Environmental Microbiology,* Vol. 43, 1982, pp. 1440–1450.

[*15*] Lemon, S. M., Binn, L. N., and Marchwicki, R. H., "Radioimmunofocus Assay for Quantitation of Hepatitis A Virus in Cell Cultures," *Journal of Clinical Microbiology,* Vol. 17, 1983, pp. 834–839.

[*16*] Kulski, J. K. and Norval, M., "Nucleic Acid Probes in Diagnosis of Viral Diseases of Man: Brief Review," *Archives of Virology,* Vol. 83, 1985, pp. 3–15.

[*17*] Margolin, A. B., Martinez, M. J., and Gerba, C. P., "Use of cDNA Dot-Blot Hybridization Technique for Detection Enteric Viruses in Water," Abstract, Annual Meeting of the American Society of Microbiologists, 1985, p. 270.

*Sharon S. Lindsay*[1] *and Mary Jo Baedecker*[2]

# Determination of Aqueous Sulfide in Contaminated and Natural Water Using the Methylene Blue Method

**REFERENCE:** Lindsay, S. S. and Baedecker, M. J., **"Determination of Aqueous Sulfide in Contaminated and Natural Water Using the Methylene Blue Method,"** *Ground-Water Contamination: Field Methods, ASTM STP 963*, A. G. Collins and A. I. Johnson, Eds., American Society for Testing and Materials, Philadelphia, 1988, pp. 349–357.

**ABSTRACT:** The methylene blue method for the colorimetric determination of aqueous sulfide was modified and evaluated for field use in the concentration range of 0.3 to 1500 μM sulfide. Aqueous sulfide ($H_2S$, $HS^-$, and $S^=$) reacts with *N,N*-dimethyl-ρ-phenylenediamine sulfate and ferric chloride in an acidic solution, forming a methylene blue complex which is measured spectrophotometrically at 670 nm. Modifications to the method permit rapid and repetitive sampling, produce reproducible standard curves, and define the limitations on stability of reagents and of the color-complex. Field measurements are made easily after the reagents have been standardized in the laboratory. The method has a relative standard deviation of ±3.8% at 80 μM $S^=$ and ±0.84% at 420 μM $S^=$.

The method was laboratory tested for possible interferences from phenols, polynuclear aromatic hydrocarbons and salts. Sulfide concentrations up to 70 μM were measured in ground water contaminated with creosote waste products including phenols and aromatic hydrocarbons. No interferences were found from these constituents.

**KEY WORDS:** aqueous sulfide, methylene blue, sulfide in ground water, sulfide in contaminated water

Hydrogen sulfide is produced biologically under anoxic conditions in diverse environments. Sulfide species are major contributors to the total anions in some reducing environments and are important constituents in oxidation-reduction reactions. Although numerous methods are reported in the literature for the determination of aqueous sulfide species, a satisfactory field method that is easy to use and gives reproducible and accurate measurements is not available. Methods that precipitate sulfides as metallic sulfides or convert the sulfides to sulfate and precipitate the sulfate as a salt are not satisfactory for measuring low levels of sulfides because quantification on small samples is difficult. The measurement of sulfide with a sulfide electrode [*1,2*] is subject to interferences, requires long equilibration times at low concentrations, and is difficult to manage in the field. Other methods such as ion chromatography [*3*] and gas chromatography [*4*] are not suited for field use. Several colorimetric methods are used, such as reaction of sulfide with 2,2′-dipyridyl disulfide [*5*], with ρ-phenylenediamine [*6*], and with *N,N*-dimethyl-ρ-phenylenediamine sulfate and ferric chloride [*7–11*]. The latter method, the methylene blue procedure, is sensitive, specific for sulfides, and generally free of interferences from other constituents that may be dissolved in natural or contaminated water.

[1] Chemist, U.S. Geological Survey, 431 National Center, Reston, VA 22092; present address, Department of Chemistry, San Jose State University, San Jose, CA 95192.

[2] Research chemist, U.S. Geological Survey, 431 National Center, Reston, VA 22092.

The methylene blue procedure described here, modified from that of Cline [*11*], provides a method for the determination of aqueous sulfide for field use that is fast, reproducible, and suitable for small sample volumes (1 mL). A procedure is given to obtain a reproducible standard curve, which is a problem with other methods.

## Summary of Method

The formation of the methylene blue color complex occurs when sulfide ($H_2S$, $HS^-$ and $S^=$) reacts with *N,N*-dimethyl-ρ-phenylenediamine sulfate (diamine) and ferric chloride in an acidic solution [*9*]. The methylene blue complex is measured spectrophotometrically at 670 nm.

$2(CH_3)_2$-N · HCl, $NH_2$ — $Fe^{3+}$, $H^+$ oxidation → [ $(CH_3)_2$-N⁺ =, =N–, N-$(CH_3)_2$ · HCl ] $Cl^- + NH_4^+ + H^+$

$H_2S$ reduction ↓

[ $(CH_3)_2N$⁺ =, =N–, S, $N(CH_3)_2$ · HCl ] $Cl^- + 4H^+$

Because the sulfide stock solution is stable for only a few hours it is prepared just prior to use and standardized using an iodometric titration. The diamine reagent is stable for up to nine months if kept in the dark and in the refrigerator and the ferric chloride solution is stable indefinitely. A standard curve (absorbance as a function of concentration) is prepared with the standard sulfide solution and the diamine and ferric chloride reagents. These reagents, the standard curve, and the same spectrophotometer are used in subsequent determinations of aqueous sulfide concentrations. For field use, the diamine and ferric chloride reagents are mixed prior to sampling in disposable syringes and water samples are introduced through a four-way stopcock. The samples are diluted if necessary and can be stored in an ice chest for one or two days before determining the absorbance. The concentration is calculated from the standard curve.

## Experimental Methods

### *Apparatus*[3]

Glaspak or plastic disposable syringes (Becton-Dickinson) 3, 10, 50 mL.
Polycarbonate 4-way stopcocks (Propper Manufacturing).
Disposable glove bag ($I^2R$).
Glassware for iodometric titration.
Spectrophotometer, single beam with 1-cm cells (Spec 100, Bausch & Lomb).

### *Reagents*

*Hydrochloric Acid (HC1)*—Concentrated reagent grade.

*Purged Water*—To prepare oxygen-free water, boil double deionized water and purge with nitrogen ($N_2$) for 30 min.

[3] The use of brand names does not imply endorsement by the U.S. Geological Survey.

TABLE 1—*Reagent concentrations and volumes for the determination of aqueous sulfide.*

| Reagent | A | B1 | B2 | C1 | C2 |
|---|---|---|---|---|---|
| Range | | | | | |
| sulfide range, μM | 0.312 to 31.2 | 31.2 to 312 | 31.2 to 312 | 312 to 1560 | 312 to 1560 |
| sample size, mL | 8 | 1 | 8 | 1 | 8 |
| Standard $S^=$ solution | | | | | |
| approximate concentration of stock $S^=$ solution, μM | 31 | 310 | 310 | 1560 | 1560 |
| $(NH_4)_2S$[a], mL | 0.01 | 0.1 | 0.1 | 0.5 | 0.5 |
| Standardization | | | | | |
| $K10_3$[b], mL | 2 | 5 | 5 | 10 | 10 |
| $H_2O$, mL | 8 | 5 | 5 | 5 | 5 |
| KI, mg | 20 | 20 | 20 | 20 | 20 |
| approximate amount of titrant[c] | 1.9 | 3.8 | 3.8 | 3.8 | 3.8 |
| $S^=$ determination | | | | | |
| diamine reagent, g | 1.15 | 2.89 | 5.76 | 14.43 | 28.85 |
| ferric chloride reagent, g | 1.68 | 4.22 | 8.41 | 21.06 | 42.10 |
| syringe size, mL | 10 | 3 | 10 | 3 | 10 |
| amount of mixed reagent, mL | 2 | 0.5 | 2 | 0.5 | 2 |
| sample size, mL | 8 | 1 | 8 | 1 | 8 |
| spectrophotometer reading[d] | no dilution | dilution to 10 mL | dilution to 100 mL | dilution to 50 mL | dilution 3 mL to 100 mL |

[a] Added to 1 L purged water.
[b] 1.668 mM.
[c] 0.010 N $Na_2S_2O_3$ standardized solution.
[d] Dilute contents of syringe with deionized water.

*Sulfide Stock Solution*—In a $N_2$-filled glove bag, pipet the appropriate amount (Table 1) of 22 to 24% ammonium sulfide solution [$(NH_4)_2S$] (Mallinckrodt, 3524) into 1 L of purged water.

*Iodate Solution*—To prepare a 1.668 mM potassium iodate ($KIO_3$) solution dissolve 0.3567 g of dried (180°C for 2 h) American Chemical Society (ACS) certified $KIO_3$ in purged water to make a 1-L solution.

*Potassium Iodide Crystals*—Test crystals to determine that they are iodate free by dissolving about 0.1 g in 5 mL water and adding two drops of concentrated sulfuric acid ($H_2SO_4$) and a small amount of thyodene (a soluble starch indicator). Appearance of blue color indicates the presence of iodate contamination.

*Sodium Thiosulfate Standard Solution*—To prepare a 0.01-N solution, dissolve 2.482 g of $Na_2S_2O_3 \cdot 5H_2O$ in purged water to make a 1-L solution. To standardize this solution, prepare an iodate standard solution [*12*] by pipetting 25.0-mL $KIO_3$ solution into a 250-mL Erlenmeyer flask. Add 75-mL double deionized water, 0.5-g potassium iodide (KI), and 10-mL concentrated HCl. Stopper the flask and place in the dark for 5 min to permit iodine ($I_2$) formation. Titrate the $I_2$ with the $Na_2S_2O_3$ solution prepared above. Add thyodene when the solution is straw-colored and titrate to the disappearance of the blue color. From Eqs 1 and 3, 6-mol $S_2O_3^=$ react with 1-mol $IO_3^-$, and thus the normality of the $Na_2S_2O_3$ solution is calculated as follows

$$N\,(Na_2S_2O_3) = \frac{N\,(KIO_3) \times mL\,(KIO_3) \times 6}{mL\,(Na_2S_2O_3)}$$

*Diamine*—Dissolve the appropriate amount (Table 1) of *N,N*-dimethyl-p-phenylenediamine sulfate (Kodak, 1333) in a 250-mL solution of 50% (volume/volume) reagent-grade HCl and purged water (cooled). Store in a dark bottle in a refrigerator.

Ferric Chloride—Dissolve the appropriate amount (Table 1) of $FeC1_3{\cdot}6H_2O$ in 250-mL 50% (volume/volume) reagent grade HCl and purged water.

Thyodene—Solid—Use as an end-point indicator.

*Standardization Procedure*

The sulfide stock solution is standardized by the iodometric titration method [*12,13*]. Sulfide is reacted with an excess of iodine in acid solution, and the excess iodine is titrated with a standard thiosulfate solution.

$$IO_3^- + 5I^- + 6H^+ \longrightarrow 3I_2 + 3H_2O \quad (1)$$

$$I_2 + S^= \xrightarrow{H^+} S^0 + 2I^- \quad (2)$$

$$I_2 + 2S_2O_3^= \xrightarrow{H^+} S_4O_6^= + 2I^- \quad (3)$$

Use the appropriate amounts of reagents (Table 1) according to the concentration range of sulfide and the sample size. Pipet the $KIO_3$ solution into a beaker and draw it into a 50-mL plastic syringe. Transfer the appropriate volume of purged water into the beaker with 20 mg of KI and draw this solution into the syringe. Next, draw in 5 mL of concentrated HC1 and exclude all air from the syringe (without losing any solution). The iodate is converted quantitatively to $I_2$ in the syringe.

In the $N_2$-filled glove bag draw 20 mL of the sulfide solution into the solution. Reproducible results are obtained only when transfers of the sulfide solution are handled in a glove bag. Remove the syringe from the glove bag, transfer the contents to a 50-mL Erlenmeyer flask, and immediately titrate the remaining iodine with the standard $Na_2S_2O_3$ solution (approximate volume in Table 1), using thyodene as the end-point indicator. Calculate the concentration of the sulfide solution as follows:

$$S^= (\mu M) = \frac{\text{mmoles } I_2 \text{ (in syringe)} - \text{mmoles } I_2 \text{ (titrated)}}{\text{mL } S^= \text{ solution} \times 10^{-6}}$$

The mmoles $I_2$ (in syringe) is calculated from the amount of standard $KIO_3$ solution added to the syringe (Table 1). The mmoles $I_2$ (titrated) is equal to 0.5 $\times$ mL of titrant ($Na_2S_2O_3$) $\times$ N of titrant.

*Preparation of Standard Curve*

The standard curves are prepared over three concentration ranges to keep within the response limit of the spectrophotometer. Range A is 0.31 to 31 $\mu$M $S^=$, Range B (1 and 2) is 31 to 310 $\mu$M $S^=$, and Range C (1 and 2) is 310 to 1560 $\mu$M $S^=$. The following procedure is given for Range A sulfide concentrations. For other sulfide ranges and samples sizes, use Table 1 to determine the correct amounts of reagents. Mix equal amounts of diamine and ferric chloride reagents and draw 2-mL mixed reagent into a 10-mL plastic syringe. In a $N_2$-filled glove bag draw 8 mL of the sulfide stock solution into the syringe. Place the syringe in the dark for 20 min for full color to develop. Then, using this solution as the most concentrated standard (31 $\mu$M), make nine serial dilutions in 10-mL volumetric flasks, to the least concentrated standard (0.31 μM). Dilute with a blank solution, which is mixed reagent and purged water in the same ratio (1:4) as in the concentrated standard. This maintains a constant pH for all the standards, which is essential to obtain a valid standard curve. Read the standards on a spectrophotometer at 670 nm using the blank solution to zero the instrument. The same spectrophotometer and cells are used for all subsequent determinations. An absorbance-versus-concentration plot is prepared using the determined value of the sulfide stock solution. The standard curve (Fig. 1, solid line, for Range A) conforms to Beer's law and is linear.

For $S^=$ concentrations in ranges B1, B2, C1, and C2 the standards are diluted a second time with double deionized water (Table 1) before the solutions are read on the spectrophotometer. This procedure is followed so that, in the field, samples can be diluted with double deionized water rather than the acid solution. The resulting solutions are all at the same pH, even though

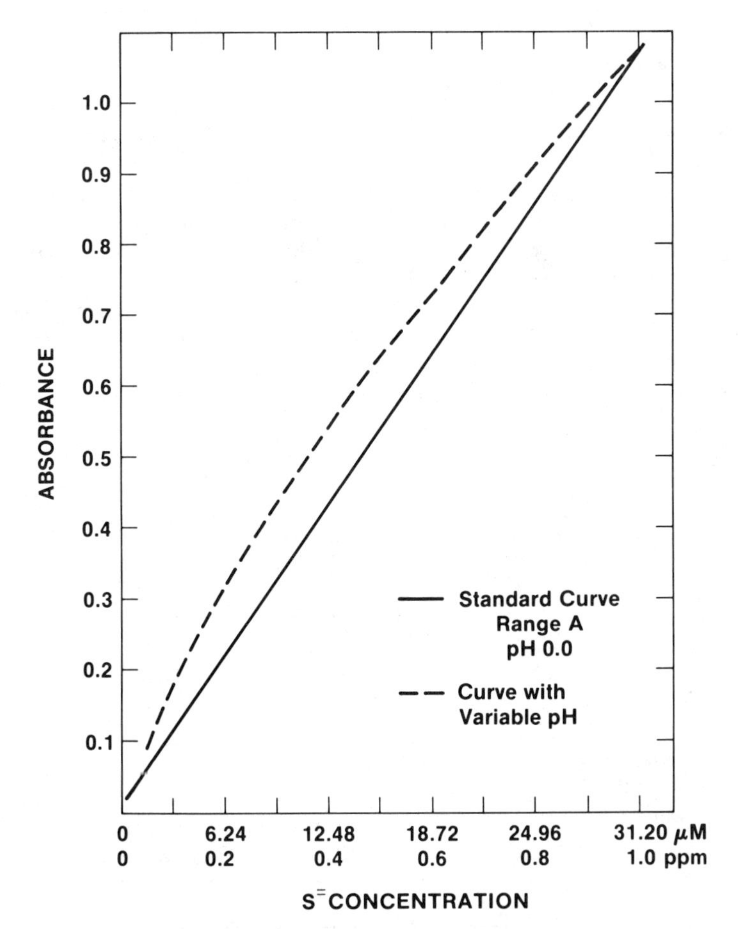


FIG. 1—*Standard curves for Range A (0.312 to 31.2 μM) sulfide at a pH of 0.0. Solid line determined by the following two methods which gave identical results, (1) dilution of concentrated standard with solution blank after formation of the color complex and (2) formation of the color complex on sulfide solutions prepared at different concentrations. Dashed line, which is not linear, determined by dilution of concentrated standard with deionized water without maintaining constant pH.*

they are not at the optimum pH of 0.35 [9]. Although the standard curve deviates slightly from Beer's law, the standard curves are reproducible.

*Field Procedure*

On the day samples are collected the diamine and ferric chloride reagents are mixed in equal amounts. Plastic syringes with four-way stopcocks are filled with 2 mL of mixed reagent (Range A). They are stored on ice and in the dark and can be left for up to 12 h. Water samples (8 mL) containing sulfide are collected directly into the syringes by way of the stopcock and the syringes immediately put on ice. Dilutions, if needed, are done after the color is developed (about 10 min). Then, the samples can be stored on ice for one or two days before reading on the spectrophotometer. Dilutions are made with double deionized water. The samples are brought to ambient temperature before measuring the absorbance. After determining the absorbances, the concentrations are calculated from the standard curve.

## Results and Discussion

Several methods of preparing a standard curve were compared. The method described above is dilution of the most concentrated standard with solution blank after the sulfide complex is developed. A second method is to develop the color complex on sulfide solutions prepared at different concentrations. These two methods, tested in the A and B ranges, maintain a constant pH near 0.0 and give identical results (Fig. 1, solid line). A third method of diluting the most concentrated standard with deionized water does not yield the same results (Fig. 1, dashed line) because the molar absorptivity of methylene blue is pH dependent [*14*].

The mole ratio of the diamine reagent to the sulfide solution in the syringe is critical. The absorbance varies with the mole ratio unless the ratio is greater than 20 (Fig. 2). Another diamine reagent, *N,N*-diethyl-ρ-phenylenediamine oxalate was tested. The ethylene blue complex is less sensitive than the methylene blue complex and the mole ratio does not reach a plateau.

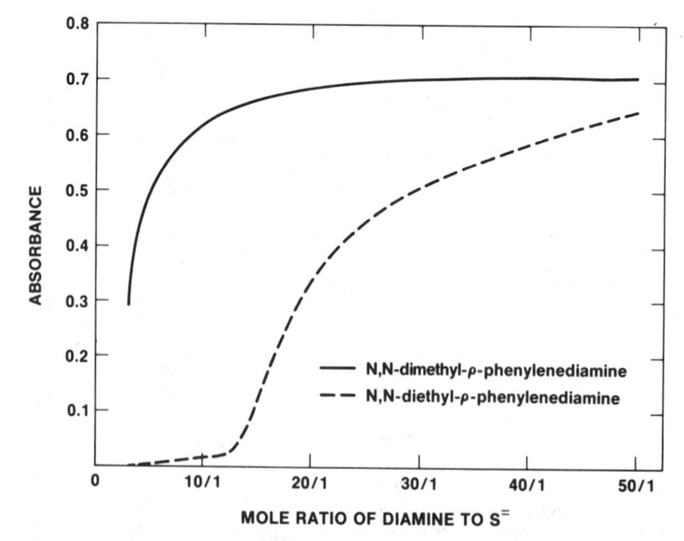


FIG. 2—*Absorbance versus mole ratio of diamine to sulfide for two diamine reagents. The* N,N-*diethyl-ρ-phenylenediamine is less sensitive and does not reach a plateau.*

TABLE 2—*Precision data for methylene blue method.*

| | Absorbance for $S^=$ Concentrations of | | |
|---|---|---|---|
| Sample No. | 1.25 μM | 81.1 μM | 415 μM |
| 1 | 0.042 | 0.198 | 0.737 |
| 2 | 0.047 | 0.199 | 0.736 |
| 3 | 0.041 | 0.183 | 0.753 |
| 4 | 0.045 | 0.192 | 0.744 |
| 5 | 0.039 | 0.181 | 0.747 |
| 6 | 0.038 | 0.199 | 0.738 |
| 7 | 0.040 | 0.185 | 0.734 |
| 8 | 0.039 | 0.192 | 0.744 |
| 9 | 0.038 | 0.199 | 0.749 |
| 10 | 0.040 | 0.203 | 0.736 |
| 11 | 0.049 | 0.199 | 0.745 |
| 12 | 0.046 | 0.189 | |
| RSD[a] (%) | ±9.0 | ±3.8 | ±0.84 |

[a] Relative standard deviation.

The accuracy of the methylene blue method was tested by comparison to determinations by iodometric titration of a sodium sulfide solution. At a sulfide concentration of 840 μM, the results agree within ±2.4% (relative standard deviation). The precision of the method was determined at three sulfide concentrations (Table 2). The relative standard deviation was largest (±9.0%) at low sulfide concentration (1.25 μM). The relative standard deviation was ±3.8 and ±0.84% at sulfide concentrations of 81 and 415 μM, respectively. The detection limit was 0.3 μM. The precision is improved by using larger cells to measure the absorbance. One-centimetre cells were used routinely for the convenience of field measurements and because the method was designed for use on small samples from a variety of environments. Although large samples can be obtained from ground water, the yield of pore fluids from sediments is generally small.

The effects of interferences from organic compounds and salt were examined in several experiments. A sulfide stock solution was prepared in a 1-mM phenol solution and a standard curve determined. The standard curve reproduced the original curve for that range. No interference was found with a sulfide stock solution prepared in a 3.7 eq/kg sodium chloride solution. Samples from the field that were turbid did interfere with the color complex. This was avoided by using an in-line filter (0.4 μm). Because small sample volumes were used, most of the water from a pumping well was diverted by a splitter and a small portion of the water passed through a small in-line filter and into the syringe with no exposure to air. Highly colored water may cause interference. However, such samples were not encountered in this work.

The adaptability of the method for field use was determined at a waste disposal site in a coastal plain aquifer [*15*]. Ground water contaminated with phenols and polynuclear aromatic hydrocarbons—components of creosote—was analyzed for sulfide along the contaminant plume. Sulfide concentrations ranged from below the detection limit (0.3 μM) to 70 μM. The dissolved organic carbon concentration of the most contaminated water sample was 400 mg/L. Ground water was withdrawn by a peristaltic pump and part of the water (a splitter was used to divert most of the water) was filtered through a 0.4-μm in-line filter to remove particulate material. Sulfide was detected (above 0.3 μM) at 14 sampling sites and the determinations were done in triplicate. The average percent difference from the median for samples with sulfide above 3 μM (0.1 mg/L) is < 5% (Fig. 3). The average percent difference from the median for samples with sulfide at or below 3 μM is 18%.

The methylene blue method modified for field use is easy to use for reproducible measurements

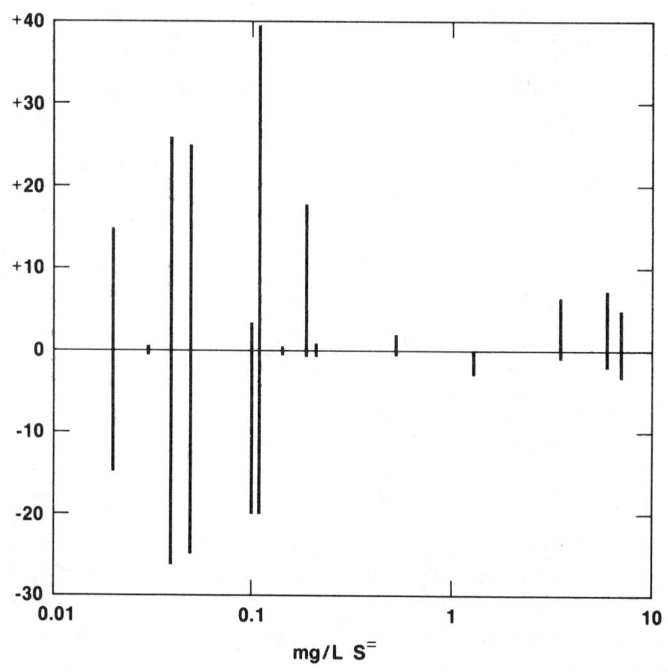


FIG. 3—*Percent difference from the median versus sulfide concentration for water samples from a contaminated aquifer.*

of sulfide in natural and contaminated water. This method eliminates some of the problems in obtaining a reproducible standard curve because of unstable sulfide compounds and interference from oxygen.

*Acknowledgments*

The cooperation and assistance of Jessica Hopple, U.S. Geological Survey, is gratefully appreciated. We appreciate the review comments and helpful suggestions of Isabelle Cozzarelli, Marge Kennedy, and Dave Vivit, U.S. Geological Survey, and two anonymous reviewers.

## References

[1] "Instruction Manual—Sulfide Ion Electrode, Silver Ion Electrode Model 94-16," Orion Research, Inc., Cambridge, MA, 1979.

[2] Vivit, D., Ball, J., and Jenne E., "Specific-Ion Electrode Determination of Sulfide Preconcentration from San Francisco Bay Waters," *Environmental and Geological Water Science*, Vol. 6, No. 2, 1984, pp. 79–90.

[3] Okada, T. and Kuwamoto, T., "Potassium Hydroxide Eluent for Nonsuppressed Anion Chromatography of Cyanide, Sulfide, Arsenite, and Other Weak Acids," *Analytical Chemistry*, Vol. 57, 1985, pp. 829–833.

[4] Funazo, K., Hirashima, T., Tanaka, M., and Toshiyuki, S., "Determination of Sulphide Ion at Trace Levels by Ethylation and Gas Chromatography," *Fresenius Zeitschrift fuer Analytische Chem.*, Vol. 311, No. 1, 1982, pp. 27–29.

[5] Svenson, A., "A Rapid and Sensitive Spectrophotometric Method for Determination of Hydrogen Sulfide with 2,2′-Dipyridyl Disulfide," *Analytical Chemistry*, Vol. 107, 1980, pp. 51–55.

[*6*] Strickland, J. D. H. and Parsons, T. R., "A Manual of Seawater Analysis," *Bulletin of the Fisheries Research Board of Canada*, Vol. 167, 1968, p. 311.
[*7*] Almy, L. H., "A Method for the Estimation of Hydrogen Sulfide in Proteinaceous Food Products," *American Chemical Society*, Vol. 47, 1925, pp. 1381–1390.
[*8*] Sheppard, S. E. and Hudson, J. H., "Determination of Labile Sulfur in Gelatin and Proteins," *Industrial and Engineering Chemistry. Analytical Edition,* Vol. 2, 1930, pp. 73–75.
[*9*] Patterson, G. D., "Sulfur," *Colorimetric Determination of Nonmetals*, D. F. Baltz, Ed., Interscience, New York, 1958.
[*10*] Siegel, L. M., "A Direct Microdetermination for Sulfide," *Analytical Biochemistry*, Vol. 11, No. 1, 1965, pp. 126–132.
[*11*] Cline, J. D., "Spectrophotometric Determination of Hydrogen Sulfide in Natural Waters," *Limnology and Oceanography*, Vol. 14, 1969, pp. 454–458.
[*12*] "Methods for Determination of Inorganic Substances in Water and Fluvial Sediments," *Techniques of Water Resources Investigations of the U.S. Geological Survey*, M. W. Skougstad, M. J. Fishman, L. C. Friedman, D. E. Erdmann, and S. S. Duncan, Eds., Book 5, Chapter A1, 1979, p. 619.
[*13*] *Standard Methods for the Examination of Water and Wastewater*, 16th ed., American Public Health Association, Washington, DC, 1985, pp. 476–477.
[*14*] Sands, A. E., Gratius, M. A., Wainwright, H. W., and Wilson, M. W., "The Determination of Low Concentrations of Hydrogen Sulfide in Gas by the Methylene Blue Method," U.S. Bureau of Mines Report Investigation No. 4547, 1949, pp. 1–18.
[*15*] Baedecker, M. J. and Lindsay, S. S., "Chapter B—Distribution of Unstable Constituents in Ground Water near a Creosote Works, Pensacola, Florida," U.S. Geological Survey Open-File Report 84-466, 1984, pp. 13–22.

*M. E. Crocker*[1] *and L. M. Marchin*[1]

# Adsorption and Degradation of Enhanced Oil Recovery Chemicals

**REFERENCE:** Crocker, M. E. and Marchin, L. M., "**Adsorption and Degradation of Enhanced Oil Recovery Chemicals,**" *Ground-Water Contamination: Field Methods, ASTM STP 963,* A. G. Collins and A. I. Johnson, Eds., American Society for Testing and Materials, Philadelphia, 1988, pp. 358–369.

**ABSTRACT:** The possible contamination of ground waters by injection of chemicals used for enhanced oil recovery was evaluated at the National Institute for Petroleum and Energy Research. The objective of the study was to delineate the nature of the reservoir rock/injected chemical interactions and the potential for ground-water contamination.

As a preliminary step in the investigation, a method for determining the cation exchange behavior of consolidated sandstone specimens at *in situ* flow-through conditions as compared to a crushed specimen method at ambient conditions was developed. The flow-through method was tested in the laboratory on small core specimens (0.038 by 0.102 m) and was used as an indicator of the potential of chemicals to interact with the rock matrix.

Results obtained provided individual concentration profiles for major and minor amounts of cations removed from the core after exchanging with lanthanum. The information obtained is valuable in assessing the potential compatibility of injected fluids with a reservoir formation. Upon comparing the cation exchange capacity (CEC) for specimen C071 (Parkman formation, sandstone) the values were 13.9 meq/kg for the crushed method versus 6.11 meq/kg for the flow-through method. This was a consistent finding for all the specimens tested; CECs determined on the crushed core specimens were higher than those determined by the lanthanum flood on the whole core. These displacement tests could also indicate prior chemical contamination by observation of the cations displaced by the flow-through method. This option is not available using the crushed core method.

The whole core flow-through method was adapted to a larger (0.155 by 0.762 m) closed core system to evaluate the stability/degradability of enhanced oil recovery (EOR) chemicals: three surfactants, a co-surfactant, and a polymer. Seven separate tests were conducted with test periods ranging from 17 to 135 days. The tests were conducted at simulated reservoir conditions of temperature, pressure, and fluid flow rate. Specimens were taken periodically and analyzed using gas chromatography, liquid chromatography, ultraviolet spectroscopy, or colorimetric techniques of both. The chemicals were found to be stable for the conditions studied during the testing periods with no detectable amounts of degradation products. These tests indicate the stability of these EOR chemicals over extended time periods at *in situ* conditions and the potential of the chemicals to contaminate ground-water sources over a long time period.

**KEY WORDS:** ion exchange, enhanced oil recovery, cation exchange capacity, aquifer, stability, degradation, toxicity, core tests

Geologic formations containing oil are often located near fresh water aquifers; therefore, knowledge of the fate of chemicals injected into these formations is essential for the preservation of fresh water quality. Possible mechanisms whereby these fresh water aquifers could be contaminated include (1) a spill of potentially harmful chemicals followed by migration of the

[1] Project leader and associate chemist, respectively, National Institute for Petroleum and Energy Research, Division of IIT Research Institute, P.O. Box 2128, Bartlesville, OK 74005.

chemical into the ground water, or (2) a communication of the chemical from the petroleum reservoir to the fresh water aquifer through incompetent well casing, incompetent cementing of the well, or a fractured aquiclude [*1*].

Some enhanced oil recovery (EOR) processes, for example micellar-polymer flooding, involve injection of large quantities of chemicals into subterranean geologic formations [*2–4*]. The purpose of the injected chemicals is to increase displacement of oil from the geologic formations to adjacent producing wells. Two of the environmental impacts associated with EOR are possible contamination of surface and ground water and possible contamination of agricultural land.

EOR technologies which could cause contamination of land, potable surface water, and potable ground water include: (1) micellar-polymer (surfactant) flooding, (2) polymer flooding, (3) microbial enhanced oil recovery (MEOR), (4) alkaline flooding, (5) carbon dioxide flooding, (6) *in situ* combustion, (7) steam flooding, and (8) gas miscible flooding. Any EOR technique has the potential to cause environmental problems.

The effect of these chemicals on the underground environment is a matter of concern. An assessment of the toxicological nature of EOR chemicals, and the possible constraints to EOR commercialization resulting from the toxicity of these substances has been previously reported [*5*]. In addition to the specific EOR chemicals being used, the possible degradation products of those chemicals must also be considered. Under reservoir conditions of high temperature and pressure, EOR chemicals known to be environmentally safe could degrade to hazardous products.

This research at the National Institute for Petroleum and Energy Research (NIPER) was designed to evaluate the possible contamination of ground waters that could occur when certain EOR chemicals are injected into geologic formations. A quantitative method of evaluating the potential of geologic material to interact with these chemicals is described. Cation exchange capacities (CEC) of several geologic formations are evaluated as an indicator of the potential of these formations to react with injected chemicals. Although the literature provides several methods for determining the CEC of geologic samples, all of these methods use a crushed specimen instead of a consolidated specimen and therefore do not accurately evaluate the actual flow path followed by the reservoir fluids [*6–9*].

Predictions of the degradation of particular substances are often made on the basis of their physical and chemical properties, but few degradation studies of specific EOR chemicals have been made [*10*]. This second area of research was performed to determine the degradability/stability, at reservoir conditions, of some chemicals being used or considered for EOR. The research was designed to determine: (1) under what conditions (if any) specific EOR chemicals would degrade, (2) the rate of degradation, and (3) if degradation occurred, the identity and concentrations of the products. Only chemical degradation was considered in this study even though biological degradation may also be important.

## Experimental Procedure

### *Fluid Flow Apparatus*

A diagram of the fluid flow apparatus used in this investigation is shown in Fig. 1. The system resembles a high-pressure liquid chromatograph except that the chromatographic column has been replaced by a geologic core specimen.

To simulate reservoir pressure conditions, the core specimen is encased in a high-pressure stainless steel cell. The cell is filled with fluid (usually water), which provides the means of simulating an overburden pressure on the core specimen. The core is physically separated from the overburden fluid by means of the rubber Hassler sleeve. A heating mantle encompasses the cell, providing the desired operating controls to simulate reservoir temperature. Experimental fluids are driven through the core by a high-pressure constant-rate pump and the internal pressure of the system is controlled by an adjustable pressure relief valve.

Figure 1 shows that specimens eluting from the core may be collected via an automatic fraction collector, or may be diverted to an ultraviolet (UV) spectrophotometer, to a refractometer, or to both for on-stream analysis. This arrangement allows the analysis of several different types of simulated chemical floods, with the method of detection dependent upon the particular type of chemical being analyzed. Generally, specimens collected directly at the exit port are those which cannot be detected in the UV region of the electromagnetic spectrum or by a change in refractive index. Brine fractions, for example, are collected for subsequent analysis by inductively coupled plasma emission spectroscopy (ICP/AES), and complex surfactant solutions also are collected for analysis by high-pressure liquid chromatography.

*High-Pressure Fluid Pumping System*

To duplicate the linear flow rates found in oil reservoirs, the high-pressure pumping system must provide a constant flow rate, approximately $3.33 \times 10^{-9}$ $m^3/s$, with negligible pulsation at operating pressures approaching 250 atm. The pump should have an inlet manifold capable of switching from one fluid to another without disruption or mixing of the flow regime. This is accomplished by assuring that the connections of the inlet manifold have zero dead volume. Several different types of high-pressure pumps are commercially available which meet these requirements.

*Overburden Pressure System*

The pressure applied to the outside of the geologic core specimen contained in the rubber sleeve both simulates the overburden pressure on an aquifer system and prevents leakage of the fluids being pumped through the core as the internal pressure of the system is increased.

The overburden pressure is simulated using a standard hydraulic pump with a pressure capability of at least 340 atm. Control valves (Fig. 1) provide for release of pressure that is generated as the core specimen is being brought up to operating temperature.

*Fluid Flow Lines*

The flow lines shown in Fig. 1 are assembled using high-quality stainless steel tubing, and the tubing connections are made with 316 stainless steel Swagelok fittings. This design permits the analysis of reactive materials such as brine solutions and corrosive chemical compounds. Capillary tubing is used where possible to minimize dead volume. The total internal volume of the system shown is $3.86 \times 10^{-6}$ $m^3$.

*Ion Exchange Test Cell*

The stainless steel test cell is shown in Fig. 2. The cell, machined from 321 stainless steel tubing stock with an inside diameter of 0.064 m and an outside diameter of 0.076 m, has been tested to a pressure of 680 atm, which is more than twice the normal operating pressure. A cutaway drawing of the containment of the core material within the test cell is shown in Fig. 3.

*Stability/Degradability Test Cell*

The mode used for the stability/degradation apparatus is shown in Fig. 1 as indicated by the dotted lines. It consists of a 0.762-m-long, 0.038-m-diameter sandstone core centered and sealed inside a 0.813-m stainless steel cylinder. The cylinder has an inside diameter of 0.051 m and an outside diameter of 0.070 m. The sealant is a lead/tin alloy which binds to the core and cylinder

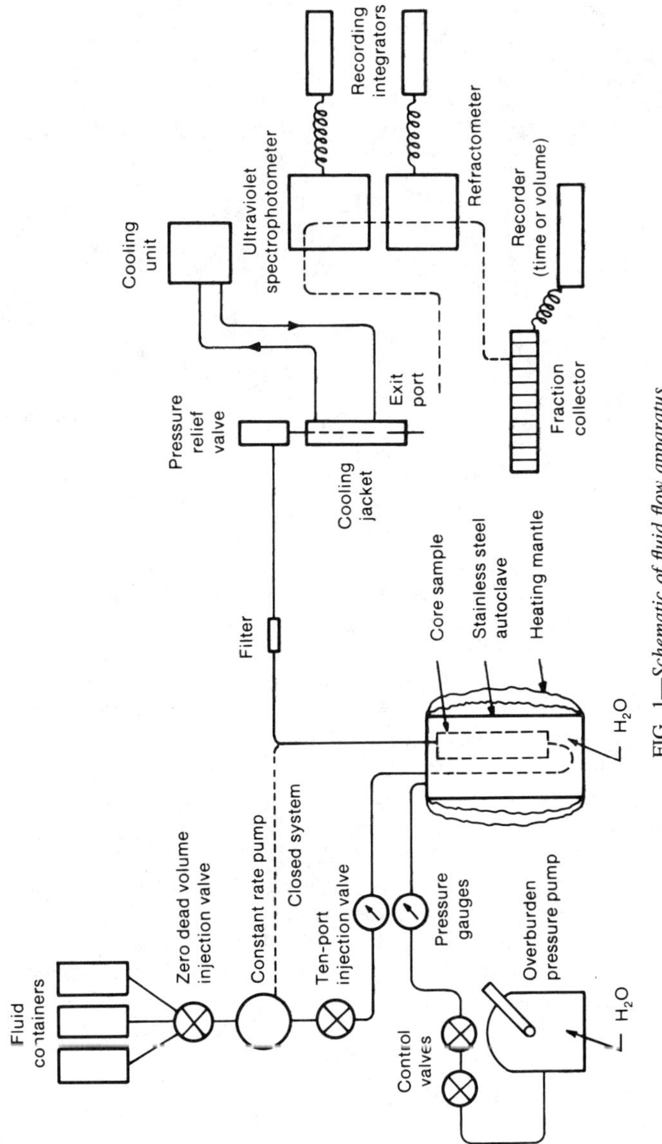


FIG. 1—*Schematic of fluid flow apparatus.*

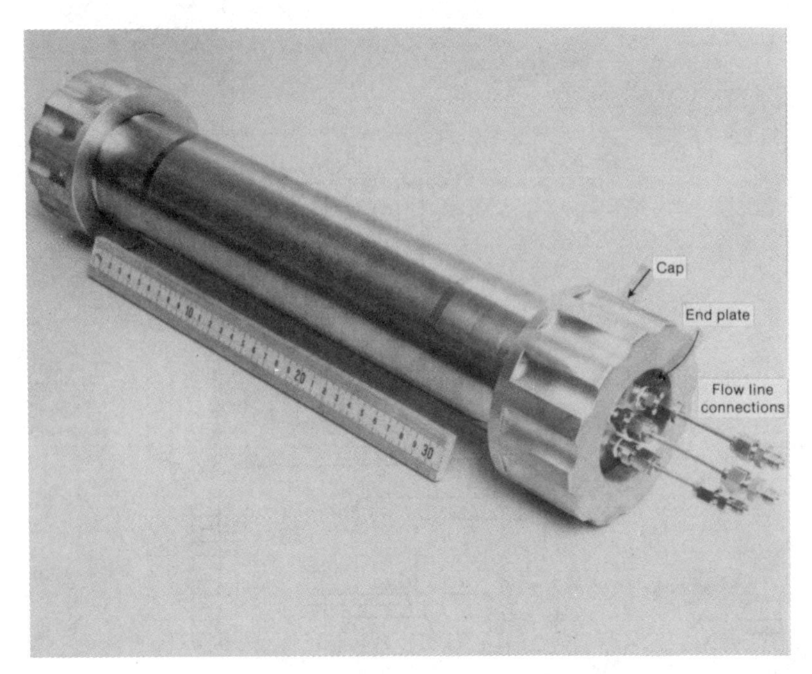


FIG. 2—*Stainless steel test cell.*

which assures that fluid flow will occur only through the core itself. Specially designed stainless steel end-plugs sandwich the core in the cylinder.

A thermistor wire is taped to the outside of the cylinder, and the cylinder is wrapped with a heat tape. The wrapped cylinder is then centered inside a 0.102-m-diameter, 0.787-m length polyvinyl chloride (PVC) pipe. Several holes (0.0064 m diameter) are drilled in the pipe. Polycel foam is injected through the drilled holes to fill the void space in the pipe. The foam cures and hardens overnight. In this manner, the core is centered, stabilized, and insulated.

The thermistor wire is connected to a digital thermometer, and the heating tape is plugged into a variable auto-transformer. This allows for control and monitoring of temperature. A liquid chromatography pump is used to pump fluid through the system. An adjustable, low-volume pressure-relief valve is placed downstream from the outlet of the core to maintain fluid-flow pressure at a set constant value. In this way, reservoir overburden pressure is simulated. The pressure gage on the pump provides a readout of the flowing pressure of the system. A stainless steel sampling cylinder ($3 \times 10^{-6}$ $m^3$ capacity) is located downstream from the outlet of the pressure relief valve to facilitate specimen removal.

### *Core Specimens*

The core specimens used for experimentation were selected from local geologic outcrop formations having similar physical characteristics to the rock matrices found in subsurface reservoirs. A portable coring unit was used to obtain representative core specimens.

### *Fluid Flow Systems*

The geologic core specimen is mounted in the cell, as previously described, the constant rate pump is primed, and a simulated overburden pressure (about 20.4 atm above internal pressure) is

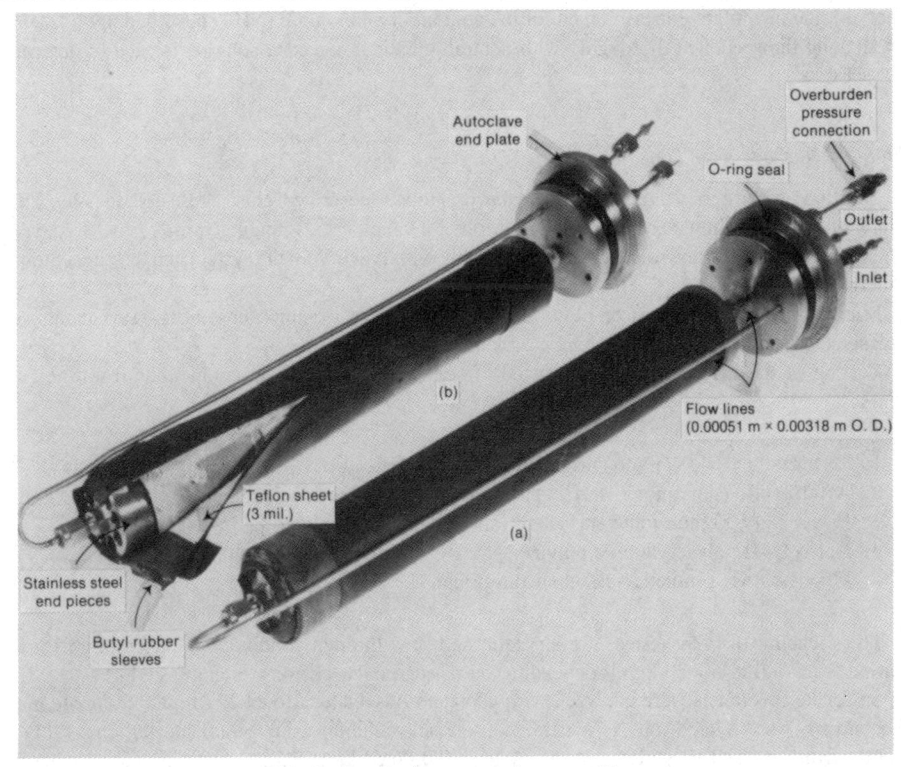


FIG. 3—*Detail of core mounting assembly.*

applied around the core. The flowing pressure of the system, controlled by the spring setting of the pressure relief valve, is set to the desired pressure, and the initial solution is pumped through the core. During this period, the temperature of the cell is carefully increased to the desired temperature. When the predetermined temperature and pressure settings reach equilibrium, the specimen is injected into the core by means of the injection valve; simultaneously, the fraction collector and recorder are activated.

## Experimental Procedures

### *Ion Exchange Displacements*

Examination of results of the water and lanthanum nitrate solution floods of whole cores used for cation exchange studies can reveal basic qualities of the reservoir formation that should be considered before chemicals are injected to stimulate oil production. In the procedure described in this report, water-soluble salts present in the cores are displaced by allowing two to three pore volumes of deionized water to flow through the core. Specimens of the effluent are collected for analysis by the inductively coupled plasma, atomic emission spectrophotometer (ICP/AES). Following the deionized waterflood, the core is flooded with 67 M/m$^3$ lanthanum nitrate [$La(NO_3)_3$]. As with the waterflood of the core, fractional volumes of the effluent of the $La(NO_3)_3$ flood are collected via an automated fraction collector. The individual fractions are analyzed by ICP/AES for specific cations and their concentrations determined. In this research, the most common cations that were found include calcium, magnesium, sodium, and potassium. For comparison of results, Grim's [*9*] method for CEC of geologic specimen was used for the crushed specimens. In this method, the quantitative measurement of the amount of ammonium ($NH_4^+$) absorbed upon leaching or shaking the specimen thoroughly with neutral ammonium acetate solution is used to determine the CEC.

*Stability/Degradability Tests*

Solutions to be tested were placed in the previously described closed test cell in which they were subjected to simulated reservoir conditions for time periods ranging from 17 to 135 days. A small specimen of the test solution was periodically withdrawn from the system and the concentration determined.

Micellar-polymer flooding employs different classes of organic substances—surfactants, co-surfactants, and polymers—to enhance oil recovery.

Testing of the EOR chemicals consisted of selecting the following compounds from the basic classes:

1. Tertiary-butyl alcohol; co-surfactant [$(CH_3)_3COH$].
2. AA-10 sulfonate; anionic-surfactant [sodium dodecylbenzene sulfonate].
3. Igepal CO-630; non-ionic surfactant [alkylphenoxypolyethanol].
4. Nalco Q41; polyacrylamide-polymer [15.1% proprietary co-surfactant].
5. Floodaid 141; synthetic surfactant [proprietary].

These chemicals were tested at both static and flow-through conditions. Concentrations levels varied from 500 to 50000 kg/m$^3$ depending on the particular chemical being tested.

Sulfonate specimens were analyzed with a Waters Associates Model 201 liquid chromatograph. Specimens were eluted on a Waters AX/Corasil column using gradient elution from 1:1 tetrahydrofuran (THF)/distilled water to 1:1 THF/0.06 *M* $NaH_2PO_4$. Separation was complete in 10 to 12 min with a flow rate of $3.3 \times 10^{-8}$ m$^3$/s and a gradient elution time of 4 min from 100% A/0% B to 20% A/80% B.

Solutions of Igepal were analyzed on a Beckman Model 26 UV/Visible spectrophotometer. Spectra were observed over the 200 to $300 \times 10^{-9}$ m wavelength range. Peaks at $280 \times 10^{-9}$ m and $210 \times 10^{-8}$ m were used to quantify results.

Specimens of *t*-butyl alcohol were analyzed using a Hewlett-Packard model 5730A gas chromatograph: 2.1 m $\times$ 0.0032 m SS, Poropak-QS, 80–100 mesh, 415 K, He $6.7 \times 10^{-7}$ m$^3$/s.

The specimens of polyacrylamide were analyzed by the starch-triiodide method [*11*].

## Results and Discussion

*Ion Exchange*

The physical and chemical behavior of the reservoir rocks can be significantly affected by ion exchange with injected chemicals if large amounts of clays are present. These potential interactions of clays with injected chemicals can result in an ion exchange, adsorption or precipitation. Therefore, this method for quantifying not only the total CECs of a core specimen, but also the individual amounts of different cations that can be exchanged, is important in defining the overall potential chemical interactions that can occur in a reservoir formation. This method is also useful in defining the impact that the interactions can have on ground-water quality [*12*].

The results of the water displacement and the $La(NO_3)_3$ displacement for the whole core test are presented in Table 1. The salts displaced during the waterflood are due to salts precipitated from the formation brine. This water displacement gives a good indication of divalent cations that could induce precipitation if chemicals are injected into the formation. For example, calcium

TABLE 1—*Ion exchange capacities of sandstones determined by flow-through whole core* (a) *versus crushed core method* (b).

| | | Meq/kg | | | | | |
|---|---|---|---|---|---|---|---|
| Specimen | Method | Na | K | Mg | Ca | Sr | Total |
| 216 | a. whole core | 4.59 | 0.41 | 0.72 | 3.14 | 0.02 | 8.88 |
| | b. crushed core | | | | | | 56.10 |
| 222 | a. whole core | 0.00 | 0.00 | 0.00 | 0.00 | 0.00 | 0.00 |
| | b. crushed core | | | | | | 14.40 |
| B100 | a. whole core | 2.56 | 0.08 | 1.05 | 1.79 | 0.06 | 5.54 |
| | b. crushed core | | | | | | 41.90 |
| B101 | a. whole core | 0.14 | 0.00 | 0.00 | 0.47 | 0.00 | 0.61 |
| | b. crushed core | | | | | | 5.00 |
| B110 | a. whole core | 0.12 | 0.01 | 0.00 | 0.10 | 0.00 | 0.23 |
| | b. crushed core | | | | | | 0.00 |
| B693 | a. whole core | 3.95 | 0.15 | 0.49 | 3.60 | 0.00 | 8.19 |
| | b. crushed core | | | | | | 12.30 |
| B976 | a. whole core | 0.58 | 0.00 | 0.00 | 1.97 | 0.00 | 2.55 |
| | b. crushed core | | | | | | 8.40 |
| C034 | a. whole core | 0.00 | 0.04 | 0.80 | 1.45 | 0.00 | 2.29 |
| | b. crushed core | | | | | | 6.44 |
| C071 | a. whole core | 0.68 | 0.15 | 0.86 | 4.39 | 0.03 | 6.11 |
| | b. crushed core | | | | | | 13.90 |
| W100 | a. whole core | 0.31 | 0.00 | 0.00 | 0.60 | 0.00 | 0.91 |
| | b. crushed core | | | | | | 0.00 |

sulfonates may precipitate if large amounts of divalent cations are present. The lanthanum nitrate displacement, on the other hand, indicates the capacity of the rock to undergo ion-exchange and the potential of chemicals to react within the reservoir matrix (Fig. 4).

All of the core specimens tested for CECs by the flow-through method consisted of consolidated sandstone specimens. As seen in Fig. 4, at the initiation of the lanthanum flood the most prevalent cations are the first to be displaced ($Na^+$ and $CA^{++}$). At approximately one pore volume the cations of lower concentrations are being displaced ($K^+$, $Mg^{++}$, and $Sr^{++}$). The lanthanum cation is first seen at approximately 1.25 pore volumes and approaches its initial concentration (67 M/$m^3$) as the other cations are approaching zero values. This was the case for all cores tested.

Upon comparing the CECs for Specimen C071 (crushed versus flow-through) the values are 13.9 and 6.11 meq/kg, respectively. This was a consistent finding of all sandstone core specimens tested; CECs determined on the crushed specimens of cores were higher than those determined by the lanthanum flood on the whole core. The results of ten sandstones analyzed by both methods are given in Table 1. One plausible reason for these differences is that crushing the specimen exposes additional sites for ion exchange to occur, whereas the cation-exchange as determined by the fluid flow method is determined on the permeability system found within each specific reservoir matrix. These results concur with research done by Campos and Hilchie on the effects of specimen grinding on CEC measurements [*13*]. In their work the CECs were found to increase by as much as five times depending upon the grain size. Smaller grain sizes (325 mesh) resulted in larger CEC values for 20 specimens tested. Therefore, the CEC as determined by the flow-through method should indicate more realistically the tendency for EOR fluids to interact (adsorb, ion exchange or precipitate) with the reservoir matrix.

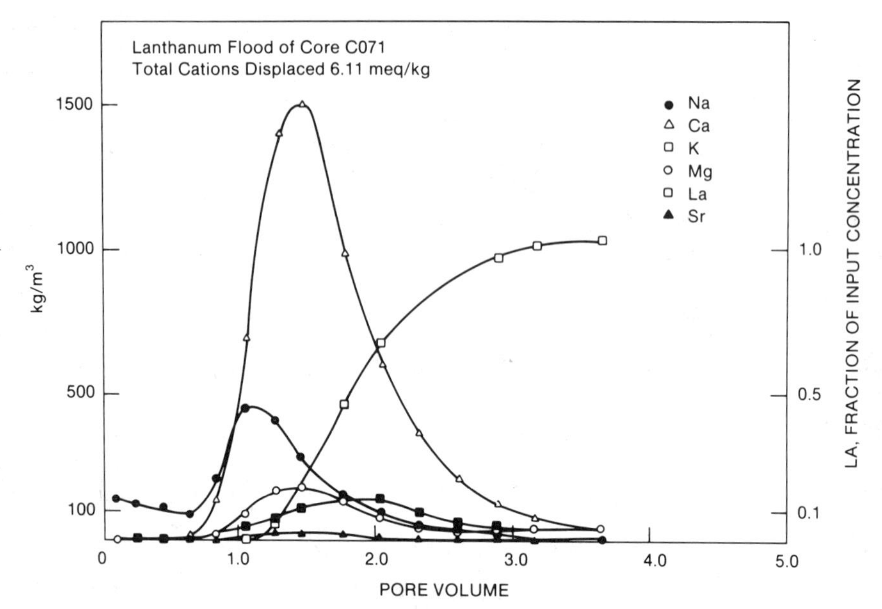


FIG. 4—*Lanthanum nitrate displacement for Core C071.*

*Stability/Degradability*

Enhanced oil recovery is an emerging technology, and understandably there are several areas of this technology lacking in fundamental chemical and physical engineering data [*5*,*10*]. The issue of environmental damage that may result from use of large quantities of various chemical compounds to enhance oil recovery is one example. Engineers are unable to predict the magnitude and duration of environmental hazards that may result from the intentional underground injection of large quantities of chemicals or accidental surface spillage because scientific data on several vital natural related phenomena are not available.

To our knowledge, very little research has been done in this field. The issue was addressed at a forum on ground-water contamination. The results of the forum indicated that little is known of the fate and toxicity of organic compounds entering the underground environment. It was noted that some organic solvents are known to be persistent in the soil for years [*14*].

The results of two flow-through and one static test of EOR chemicals tested for stability/degradation are presented in Figs. 5 through 7. As the graphs indicate, neither of these EOR chemicals (*t*-butyl alcohol and AA-10 Sulfonate) gave evidence of degradation under the conditions of temperature and pressure and over the time periods tested. The other chemicals (Igepal, Nalco, and Floodaid) were equally stable over the test periods. Tests were conducted at temperatures well beyond temperatures that would be expected in a typical underground aquifer.

A co-surfactant, *t*-butyl alcohol, was found to be stable at reservoir conditions (338 K and 170 atm) when a 5% solution of *t*-butyl alcohol in brine was pumped at 0.61 m/day through a 0.76-m-long Berea sandstone core. Although the alcohol concentration initially decreased indicating adsorption, there was no evidence that any degradation products were formed over a 30-day testing period (Fig. 5). The stability of *t*-butyl alcohol solution was confirmed at more extreme conditions of temperature and pressure in a separate no-flow experiment. In this experiment, the alcohol solution was mixed with crushed sandstone and placed in a closed, pressurized, and heated reaction vessel. The concentration did not change over a 20-week testing period. During this period the

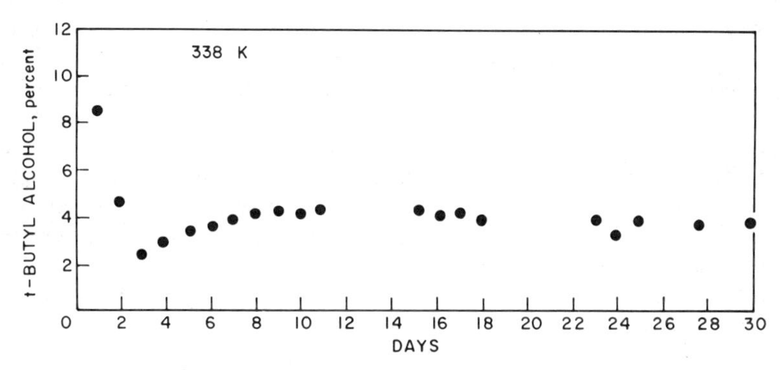


FIG. 5—*Flow-through stability test for* t-*butyl alcohol.*

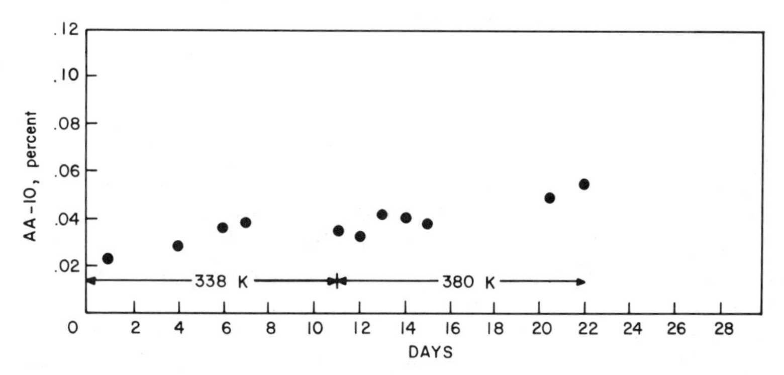


FIG. 6—*Flow-through stability test for AA-10 sulfonate.*

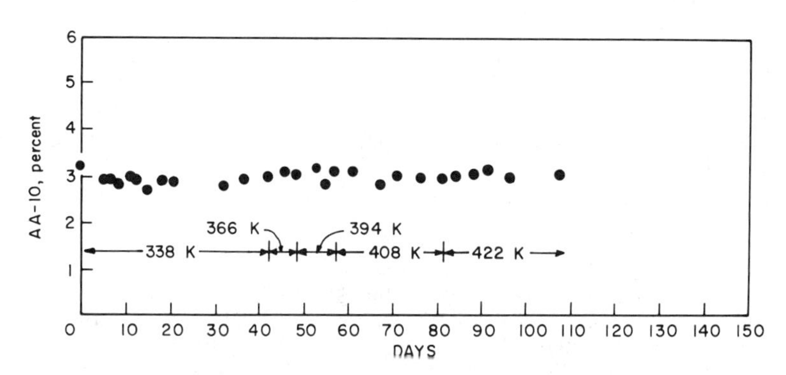


FIG. 7—*Static stability test for AA-10 sulfonate.*

pressure was maintained at 170 atm, and the temperature was increased three times from an initial 338 K to a final 394 K.

Two stability experiments run on AA-10 sodium showed it to be chemically stable in solution under reservoir conditions. When a 500-kg/m$^3$ solution of AA-10 was recycled at 1.36 atm through a 0.76-m core at 0.61 m/day for 22 days, there was no evidence of degradation (Fig. 6).

Temperature was maintained at 338 K the first 12 days, and 380 K the last 10 days. In a no-flow experiment involving a 3% solution of AA-10 and crushed sandstone, the sulfonate was stable over a 15-week period in which the temperature was increased four times from an initial of 338 K to a final of 422 K (Fig. 7).

A nonionic type surfactant, Igepal C0-360, was tested at static conditions. This surfactant adsorbed on the sandstone: $0.45 \times 10^{-3}$ kg per kg of sandstone at ambient temperature and static conditions. However, this surfactant was found to be stable (concentration 600 kg/m$^3$) and did not degrade. This experiment was run 25 days, and the temperature was increased three times beginning at ambient temperature and ending at 394 K.

Floodaid 141, which was tested under flow-through conditions, was stable at the end of the test period (33 days) at a final temperature of 394 K. In this test a 600 kg/m$^3$ solution was continuously injected at 0.61 m/day through Berea sandstone.

Finally, Nalco Q41F, a polyacrylamide polymer, was tested under flow-through conditions. The temperature was increased from an initial 338 to 394 K over a 19 day test period. This chemical indicated an adsorption/temperature dependency, but there was no indication of degradation during the 19 day test period.

## Conclusions

1. A method was developed for determining the cation exchange capacities of consolidated specimens (whole core) at reservoir conditions. Individual amounts of displaced cations may be monitored and potential reactions between the cations and injected chemicals can be identified.
2. The flow-through method resulted in CEC values up to eight times lower than the crushed method; therefore, the flow-through method should be more representative of interactions between EOR chemicals and the reservoir matrix.
3. EOR chemicals tested were found to be stable up to 422 K and for as long as 135 days, indicating their potential to have a long-term contamination effect on the subsurface environment.
4. Not only were the EOR chemicals stable, but they also displayed little adsorption at reservoir conditions. These chemicals and homologs therefore have the ability to be transported long distances and thereby possibly contaminate ground-water sources.

*Acknowledgment*

This work was performed for the U.S. Department of Energy under Cooperative Agreement DE-FC01-83FE60149.

## References

[*1*] Collins, A. G., *Geochemistry of Oilfield Waters,* Elsevier, New York, 1975, p. 434.

[*2*] Trantham, J. C., Patterson, H. L., and Boneau, D. F., "The North Burbank Unit, Tract 97 Surfactant/Polymer Pilot-Operation and Control," *Journal of Petroleum Technology,* July 1978, pp. 1068–1074.

[*3*] Mack, J. C. and Warren, J., "Performance and Operation of a Crosslinked Polymer Flood at Sage Creek Unit A, Natrona County, Wyoming," *Journal of Petroleum Technology,* July 1984, pp. 1145–1163.

[*4*] Gogarty, W. B., "Enhanced Oil Recovery through the Use of Chemicals—Part 1," *Journal of Petroleum Technology,* Sept. 1983, pp. 1581–1590.

[*5*] Silvestro, E. and Desmarais, A. M., "Toxicity of Chemical Compounds Used for Enhanced Oil Recovery," U.S. Department of Energy Report DOE/BC/10014-5, Feb. 1980.

[*6*] Mehlich, A., "Use of Triethanolamine Acetate-Barium Hydroxide Buffer for the Determination of Some Base Exchange Properties and Lime Requirement of Soil," *Soil Science Society Proceedings,* 1938, pp. 162–166.

[*7*] Worthington, A. E., "An Automated Method for Measurement of Cation Exchange Capacity in Rocks," *Geophysics,* Feb. 1973, pp. 140–153.

[8] Nevins, M. J. and Weintritt, D. J., "Determination of Cation Exchange Capacity by Methylene Blue Adsorption," *Ceramic Bulletin,* 1967, Vol. 46, No. 6, pp. 587–592.
[9] Grim, R. E., *Clay Mineralogy,* McGraw-Hill, New York, 1968.
[10] Howard, P. H., Saxena, J., and Sikka, H., "Determining the Fate of Chemicals," *Environmental Science and Technology,* Vol. 12, No. 4, April 1978, pp. 398–407.
[11] Scoggins, M. W. and Miller, J. W., "Determination of Water-Soluble Polymers Containing Primary Amide Groups Using the Starch-Triiodide Method," *Society of Petroleum Engineers Journal,* Vol. 19, No. 3, June 1979, pp. 151–154.
[12] Celik, M. S., Maneu, E. D., and Somasundaran, P., "Sulfonate Precipitation-Redissolution-Precipitation in Inorganic Electrolytes," *American Institute of Chemical Engineers Symposium Series,* Vol. 78, 1982, pp. 86–96.
[13] Campos, J. C. and Hilchie, D. W., "The Effects of Sample Grinding on Cation Exchange Capacity Measurements," presented at the Society of Professional Well Log Analysts 21st Annual Logging Symposium, July 8–11, 1980.
[14] National Research Council, *Groundwater Contamination,* National Academy Press, Washington, DC, 1984.

*Stanley M. Klainer,*[1] *John D. Koutsandreas,*[2] *and Lawrence Eccles*[3]

# Monitoring Ground-Water and Soil Contamination by Remote Fiber Spectroscopy

**REFERENCE:** Klainer, S. M., Koutsandreas, J. D., and Eccles, L., **"Monitoring Ground-Water and Soil Contamination by Remote Fiber Spectroscopy,"** *Ground-Water Contamination: Field Methods, ASTM STP 963,* A. G. Collins and A. I. Johnson, Eds., American Society for Testing and Materials, Philadelphia, 1988, pp. 370–380.

**ABSTRACT:** Fiber optics, lasers, chemistry, fiber optic chemical sensors (FOCS), optics, and spectroscopy have been integrated to form the new technology of remote fiber spectroscopy (RFS). This method permits the development of ductile probes to detect and monitor ground-water contaminants. The *key* to this concept is the FOCS, a fiber termination with preselected chemical and physical properties. This is attached to the distal end of the fiber so that specific, sensitive analyses of ground-water constituents can be made. A single fiber is used for both excitation and for collection of the return signal, thus keeping the sensor small and optically simple.

The first FOCS being developed is to be used for both "early warning" and long-term monitoring of organic chloride. The chemical basis for this FOCS is a modified Fujiwara reaction. The FOCS, in conjunction with a field fluorimeter, have been used to make *in situ* measurements in a chloroform-contaminated well. Preliminary data indicate good agreement between the FOCS data and independent gas chromatography analysis of collected samples.

**KEY WORDS:** fiber optics, ground-water contamination, soil gas, remote fiber spectroscopy

The importance of safe and plentiful ground-water supplies to the nation's future cannot be overstated. Yet domestic water quality is being threatened in many areas by the intrusion of toxic contaminants into the soil and the ground water. Underground aquifers are the source of drinking and agricultural water in over one half of the United States. However, the explosive growth of synthetic chemicals in the past 30 years has resulted in a problem of vast but unspecified magnitude. In recent years it has become clear that these chemicals have made their way into the nation's soil and water supply through agricultural runoff of pesticides and herbicides; industrial discharge into lakes and rivers; and, perhaps most serious from the standpoint of public health, into the ground water from solid waste sites (landfills, storage lagoons, and waste piles).

The contamination of soil and ground water has been called a "subterranean time bomb," and it is likely to grow worse at the very time that water quality is becoming an increasingly sensitive issue in many communities. The contamination of ground water is an insidious process with plumes of chemicals diffusing into the soils and slowly streaming into aquifers and contaminating the water. Unlike surface contaminants, which are quickly diluted, chemicals in the soil and ground water often remain highly concentrated both underground and in the water which flows

[1] President, ST&E, Inc., 1214 Concannon Blvd., Livermore, CA 94550.

[2] Chief, Water and Waste Management Monitoring and Research Staff, U.S. Environmental Protection Agency, 401 M Street, S.W., Washington, DC 20460.

[3] Hydrologist, U.S. Environmental Protection Agency, 944 E. Harmon Ave., Las Vegas, NV 89109.

from the faucet. The magnitude of the problem is illustrated by the fact that the U.S. Environmental Protection Agency (EPA) has identified thousands of industrial sites containing potentially hazardous wastes which have no safeguards to prevent seepage. In addition, there are over 275 000 Subtitle D facilities (municipal and industrial sites) which may contain dangerous materials. It is, therefore, essential that an economical, practical monitoring system be in place as soon as possible.

## Background

It has been well established through diagnostic investigations, using the most up-to-date and exotic analytical equipment, that highly toxic chemical contaminants have entered underground water supplies from hazardous waste sources in many areas of the country. Many of these toxic contaminants which have been identified and quantified include heavy metal ions; soluble salts, such as sulfates and nitrates; a wide variety of organic compounds ranging from chloro- and bromo-hydrocarbons to phenols to organophosphates; and in some cases, radioactive wastes. In addition, many organic industrial chemicals find their way into the soil and ground water, and some of these are known or suspected to be extremely dangerous.

In order to provide adequate protection of water sources, methods of detecting low concentrations of toxic contaminants are urgently needed. The public health, as well as the public's confidence in domestic water supplies, requires an early warning system so that prompt action may be taken to track down the sources of the contamination and to take appropriate steps to protect the public.

In order to assure soil and water quality, the contaminants must first be identified, their method of getting into the ecological system established, and their toxic limit determined. This has resulted in a long-term exploratory investigation during which each of these areas has been addressed. Inasmuch as diagnostic research and development is not based on any assumptions, the analytical equipment used must be capable of measuring both the anticipated as well as the unexpected species which may be present. This has led to the use of sophisticated state-of-the-art equipment and has required that wells be drilled for proper access to the vadose zone and ground water. Typically, gas chromatography, mass and atomic (absorption and emission) spectroscopy have been used in conjunction with special pumps and samplers to collect the soil and water to be analyzed. In many instances the diagnostic needs have also resulted in upgrading existing equipment and have been conducive to developing new ideas, methodologies, and instrumentation. Unfortunately, problem areas that have not been sufficiently addressed are the contamination of samples by well construction materials and the fact that most sampling techniques cause degradation of sample integrity. This could result in questionable data and make enforcement difficult.

The problem has thus become one of *in situ* monitoring, which obviates the questions associated with collected samples and thus assures that violations do not occur and our natural resources are protected.

## Monitoring

The ideal monitoring system for ground water would provide *in situ* determination of the levels of potential contaminants at very low concentrations. The techniques and instrumentation would have to be inexpensive to install and maintain, be capable of automatic operation, and give reliable results when used by operators with only modest levels of technical training.

It is important to recognize that there is a major conceptual transition from diagnosing to monitoring. In the first case the objective is to characterize and define unknown contamination sources, whereas in the second case the objective is to closely watch the behavior of a predefined system. Unfortunately, the tendency has been to use tried and true diagnostic methods to accomplish the monitoring task. This approach has impeded the development of new concepts specifically directed toward monitoring. Diagnostic techniques are generally suited to making a broad range of measurements with limited repetition. The need for analytical versatility overshadows the cost and complexity of the data gathering process. On the other hand, monitoring requires frequent selected repetitive measurements. This means that analytical versatility can be traded for a dedicated, inexpensive, simple, easy-to-operate device.

Spectroscopy has become a very popular technology on which potential monitoring systems are being based. This approach is popular because of the sensitivity, specificity, and versatility of the spectroscopic methodologies. Unfortunately, the very characteristics that make spectroscopy desirable also make it a costly and relatively complex system. Remote fiber spectroscopy (RFS), using fiber optic chemical sensors (FOCS), overcomes these objections by making the FOCS primarily responsible for the specificity of the measurement and partially accountable for its sensitivity. Furthermore, since FOCS are specific to a particular species, the spectrometer can be greatly simplified. It is projected that for RFS systems which operate in the fluorescent mode, individual FOCS will cost less than $25 and that an automatic, direct-reading spectrometer (fluorimeter) will be in the $3000 range. Minimum-capability fluorimeters should be available between $500 and $1000. It is because of the RFS-FOCS system's potential for good monitoring capabilities at a reasonable price that EPA has chosen to support development of this technology.

The use of fiber optics in conjunction with spectroscopy is not new. Several researchers are using fiber optics as an interface between a sample to be analyzed and the spectrometer, but only a few are exploring the FOCS concept, that is, a chemical or physical transducer at the distal end of the fiber.

At present, active FOCS programs exist at Lawrence Livermore National Laboratory, Livermore, California (Milanovich, Angel, and Klainer); University of Tennessee, Knoxville, Tennessee, (Sepaniak et al.); Tufts University, Medford, Massachusetts (Walt); Sperry Research Center, Sudbury, Massachusetts (Spillman); National Institutes of Health, Bethesda, Maryland, (Peterson et al.); University of California at San Francisco, California (Feustel); University of New Hampshire, Durham, New Hampshire (Seitz); and University of Graz, Graz, Austria (Wolfbeis). In addition, Cardiovascular Devices, Inc., Irvine, California and Kelsius, Inc. San Carlos, California are trying to develop FOCS for medical applications.

Tufts University (Chudyk et al.) and DSI, McClean, Virginia (Einzig) are trying to do in-ground measurements using fiber optics as simple light pipes. ST&E, Inc. and the University of Graz (Klainer and Wolfbeis) and Emory University, (Norman and Patoray), on the other hand, are using plain fiber optic cables coupled to a fluorimeter with a multidimensional correlation package to identify and quantify specific chemical species in complex mixtures.

## Instrumentation

### *Remote Fiber Spectroscopy (RFS)*

The RFS concept is shown schematically in Fig. 1. The key components of the system are the spectrometer, an optical coupler, and the FOCS. Here, excitation light of the appropriate wavelength is focused into single-strand optical fibers. The optical fiber transmits this excitation to the sampling region and, hence, to the FOCS. The interaction of the FOCS with the target molecule can result in changes in fluorescence, reflection, or absorption. The preference is for FOCS which utilize fluorescent reactions. Under these conditions the excitation light subsequently produces fluorescence in the FOCS. A small amount of this fluorescent light is collected by the optical fiber and returned to an optical coupler that has been specifically designed to separate the excitation light from the returning fluorescent light. The fluorescence is finally directed into a spectrometer for spectral analysis. The intensity of signal in a specific wavelength band (or bands) is related to the chemical information being sought.

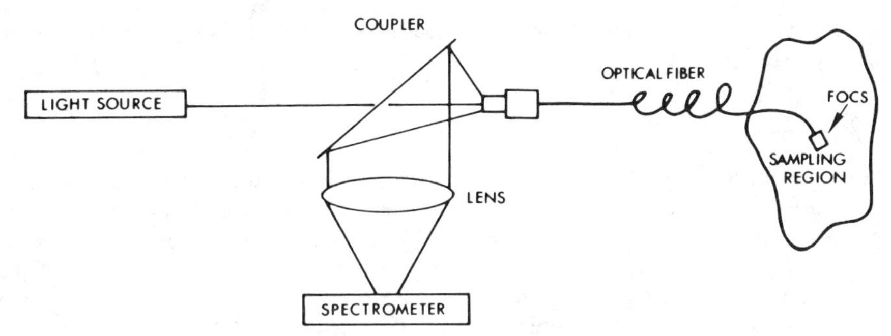


FIG. 1—*Remote fiber spectroscopy-fiber optic chemical sensor concept.*

*Field Fiber Spectrometer*

Figure 2 is a schematic representation of the spectrometer that was used for the initial fieldwork. The excitation is an air-cooled, argon-ion laser. The laser beam (514.5 nm) is expanded and then reduced in intensity (typically to 1 to 10 μW) by a series of neutral density filters. It is then directed through a dichroic mirror and focused into an optical fiber. The returning fluorescence is reflected at high efficiency (greater than 85%) by the dichroic mirror and directed into one of the three photomultiplier tubes (PMTs) selected for a particular analysis. Spectral sorting is performed by bandpass filters that are placed in front of the PMTs. The output of the PMTs is electronically conditioned and subsequently recorded on a strip chart. An automatic shutter exposes the FOCS

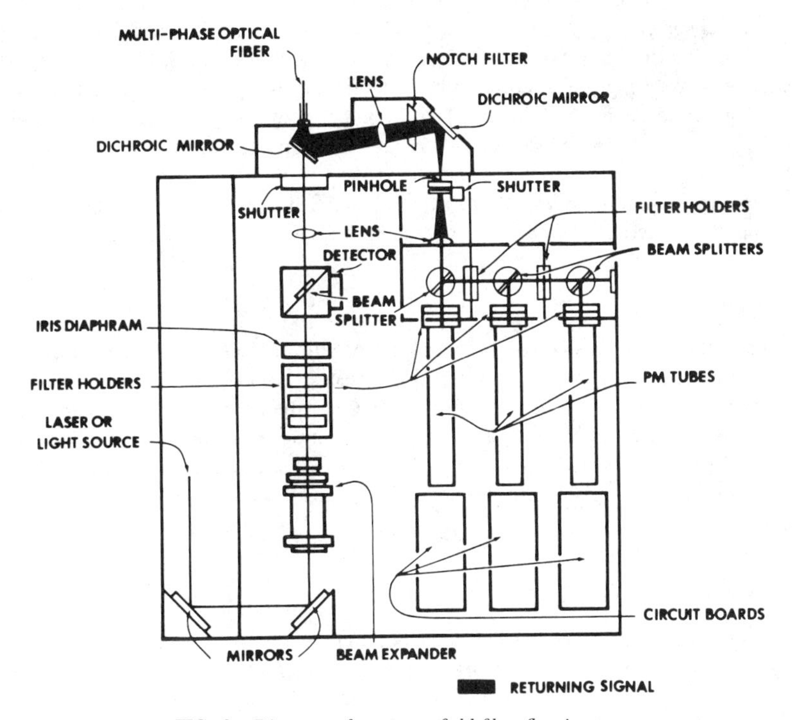


FIG. 2—*Diagram of a remote field fiber fluorimeter.*

to the laser excitation for a preset duration and frequency. To date this system has been operated using only a single PMT.

### *Portable Fluorimeter*

Figure 3 is the first portable fluorimeter. There are several innovations which have resulted in the fluorimeter being reduced in size and weight. These are: (1) the use of an incandescent lamp, instead of a laser, as the illumination source, and (2) the use of a photodiode detector in place of the photomultiplier tube. Furthermore, the optical system is internally connected using 630-μm fibers to minimize alignment problems. A 400-μm fiber is used to connect the spectrometer to the FOCS. Sensors of 100-μm diameter or larger can be used with this device. Three $x,y,z$ translation stages are provided for optical alignment. In addition, a fiber switch is furnished which permits an external light source, that is, a laser, to be used, if desired.

In operation, the light from the incandescent lamp is conditioned by the illuminator and sent to the fiber switch which directs it into the optical splitter. An electronic shutter is used to control analysis time and thus reduces or eliminates photobleaching of the fluorophore. The light entering the optical splitter is filtered so that only the green light passes through. This is then divided into two beams by a dichroic mirror set at a 45-deg angle to the incident beam. The dichroic mirror is designed to reflect the green light and pass light at 600 nm. Thus, when the entering green light hits the dichroic mirror it is reflected into the FOCS. The dichroic, however, is not perfect and a small amount of the green light passes through it into the reference channel where it is measured by a photodiode. The fluorescent signal, returning back through the fiber, from the sample FOCS (in this case, organic chloride) is at 600 nm and passes through the dichroic into the signal channel. Here it is purified by a narrow-band filter and detected by a photodiode. Provisions are made for measuring and displaying the individual reference and signal channel outputs, as well as the ratio of the two. The fluorimeter is 60 cm by 35 cm by 15 cm and weighs 10.8 kg.

An external power supply is required to run the present fluorimeter. It contains the electronics necessary to generate stable 15 V d-c and 5 V d-c output power to the fluorimeter from either 110 V a-c or 12 V d-c input. The power supply is about 35 cm by 15 cm by 15 cm and weighs about 2.25 kg.

A new system, currently under design, should be 45 cm by 20 cm by 10 cm, complete with power supply and batteries, and should weigh about 4.5 kg.

### *Organic Chloride FOCS*

The organic chloride FOCS is based on the work of Fujiwara [*1–4*], who demonstrated that the absorbance of basic pyridine changes when it is exposed to various organic chlorides. This change in absorbance was shown to be a quantitative measure of the presence of these compounds. The absorbance has been shown to be caused by the formation of a chromophore (Fig. 4). In the organic chloride FOCS, fluorescence of the chromophore is used to determine how much absorbing product is formed. In fact, fluorescence intensities can be directly related to organic chloride concentrations.

### *Pyridine/Potassium Hydroxide (KOH) FOCS*

The Fujiwara reaction has been evaluated as both one- and two-phase systems [*5–7*]. The two-phase reaction utilizes pyridine and KOH. Using this chemical system there is little difference between the two approaches in the laboratory. For a FOCS, however, the two-phase approach originally was felt to have two major benefits: (1) it overcomes the fact that pyridine and water (needed for the reaction to work and as a solvent for the potassium hydroxide) are immiscible, and

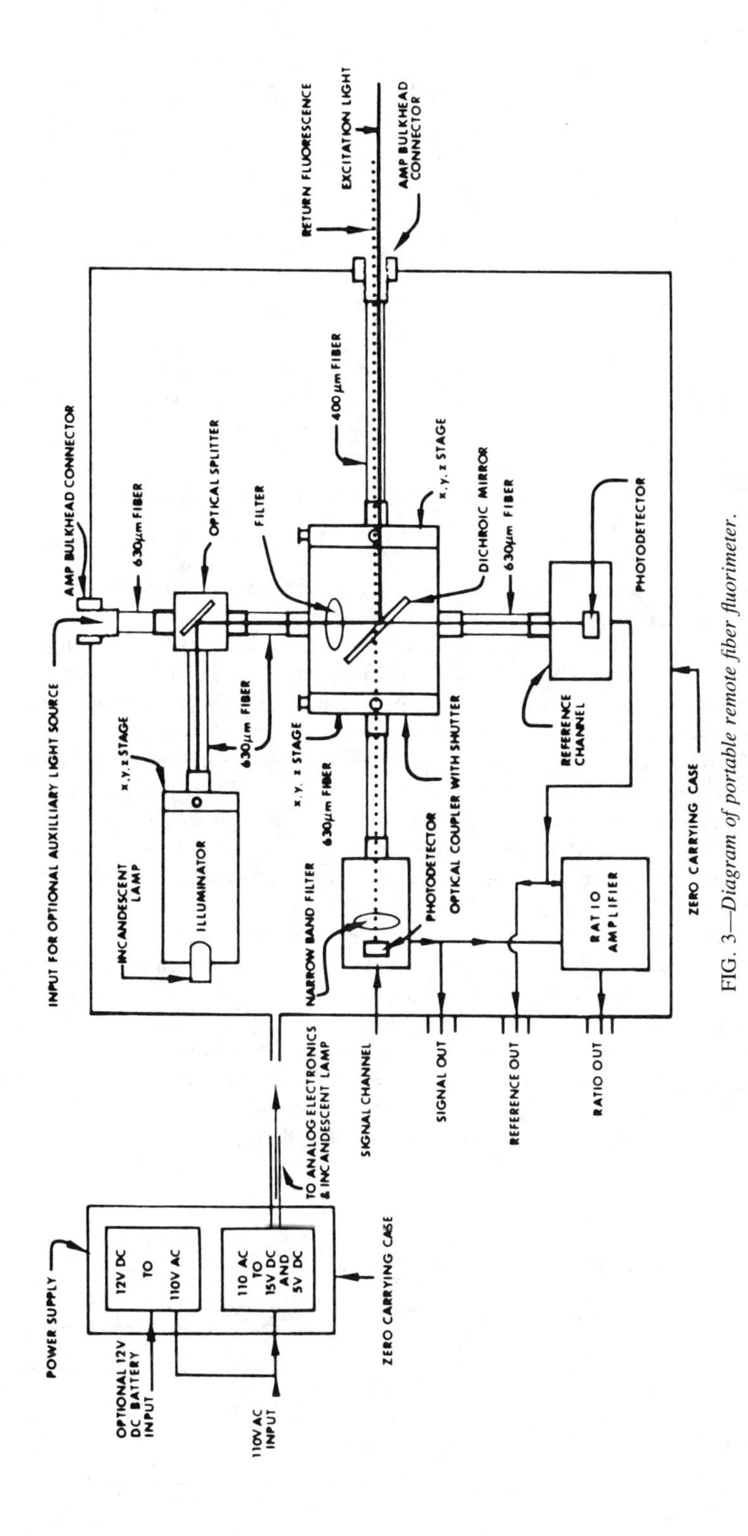


FIG. 3—*Diagram of portable remote fiber fluorimeter.*

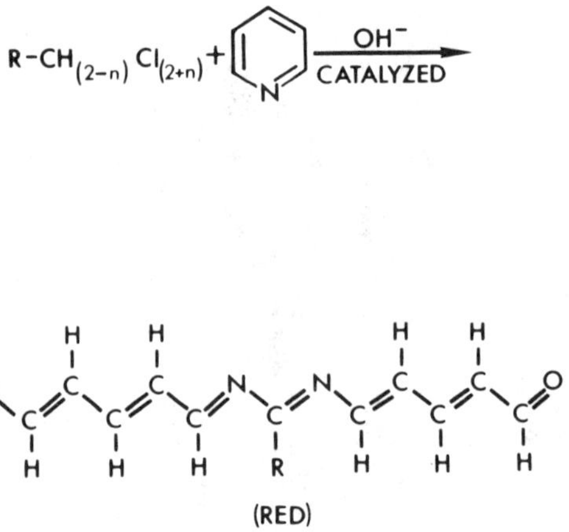

$R-CH_{(2-n)}Cl_{(2+n)}$ + pyridine $\xrightarrow[\text{CATALYZED}]{OH^-}$

(RED)

FIG. 4—*Fujiwara reaction with organic chloride compounds.*

(2) it provides a mechanism for replenishing the reactants on a continual basis. Figure 5 is a schematic of a two-phase pyridine/KOH FOCS.

The Fujiwara reaction is very dependent on the concentration of the constituents, that is, pyridine, KOH, and water. The pyridine is used pure while the KOH is made into a 10.75 *M* aqueous solution. The KOH concentration is critical as it controls the water content of the system. 10.75-*M* KOH keeps the water concentration at the optimum 7%. If the KOH molarity is too low, the spectral absorbance at 535 nm increases until it reaches the maximum and then decreases because of the increasing water concentration in the pyridine phase. On the other hand, if the amount of KOH is too high, a maximum absorbance is never attained because there is an insufficient amount of water, and thus KOH, present in the pyridine phase. The organic chloride FOCS that have been tested successfully have had KOH concentrations between 10 and 11 *M*. This is in agreement with the findings of Lugg [*5*].

The Fujiwara [*1*] reaction, as originally studied, demonstrated the ability to measure chloroform. It was later shown to work for several multiple chloride compounds such as carbon tetrachloride, tetrachloroethane, and trichloroethylene by Daroga et al. [*6*] and Hunold and Schuhlein [*7*]. More recently, Lugg [*5*] detected 17 halides using this technique. The results to date indicate that (1) the presence of multiple chlorines is not a sufficient criterion for reaction and that some compounds do not respond; (2) those compounds which do respond may do so with varying sensitivities; (3) a two-phase liquid system was impractical for field use because of problems with separation and mixing; and (4) it is desirous to continue using the modified Fujiwara reaction because of its direct

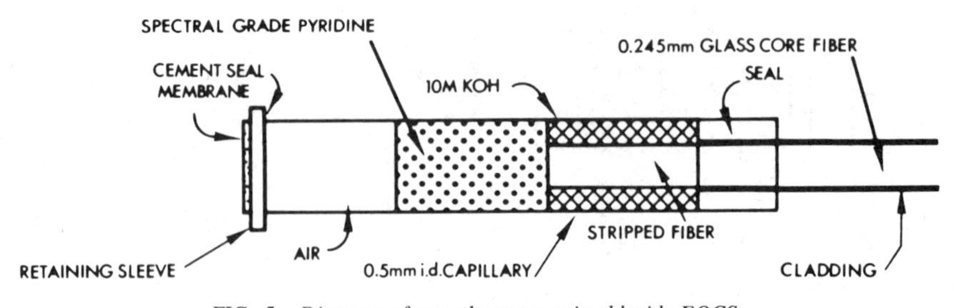


FIG. 5—*Diagram of two-phase organic chloride FOCS.*

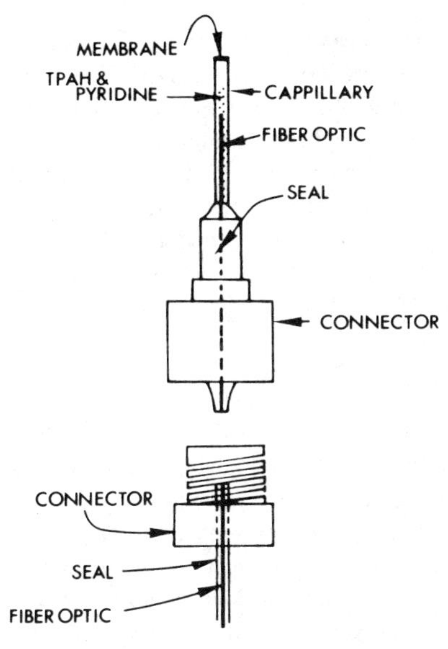


FIG. 6—*Diagram of single-phase organic chloride FOCS.*

sensitivity to volatile organic chlorides such as chloroform and trichloroethylene, which are of particular interest to EPA.

### *Pyridine/TPAH FOCS*

In order to overcome the deficiencies with the two-phase approach it was necessary to come up with a single-phase Fujiwara-type reaction which detects volatile organic chlorides with the same efficiency. To do this required finding a strong base which was miscible with pyridine and which had low volatility. The compound which meets this requirement is tetra(*n*-propyl)ammonium hydroxide {TPAH, $[(n\text{-}C_3H_7)_4N^+\ OH^-]$}. TPAH is purchased as either a 10 or 20% aqueous solution. The best recipe is 10% by volume of 10% aqueous TPAH in 90% by volume pyridine. This represents 0.05 *M* TPAH.

Figure 6 is a diagram of the FOCS that is being used with the single-phase system. This is *very close* to what a production model could look like.

The organic chloride FOCS *is not* a reversible chemical system. It integrates the amount of chloride versus time. The system can, however, be reset to "zero" time by bleaching the fluorescent indicator. This is done by raising the laser power by a factor of 100 to 1000 and exposing the dye.

## Results

The results of initial field tests of the organic chloride FOCS are most encouraging. Both the two- and single-phase organic chloride FOCS were tested in Henderson, NV in wells that were known to contain chloroform. These wells are continuously monitored using "grabbed" samples and gas chromatography/mass spectrometry. This provided an independent reference for the *in situ* measurements.

For each type of FOCS, detailed experiments are run to characterize the FOCS and to obtain a calibration curve. In the two-phase case, this was done using precalibrated aqueous chloroform solutions of 100, 10, and 1 ppm concentrations. The single-phase system was calibrated using 10, 2, and 1 ppm. Both FOCS are characterized by an initial time lag, and then the response (count rate) as a function of time is linear. Since the organic chloride FOCS are integrating devices, this means that the slope of the response curve can be directly related to sample concentrations. Abrupt changes in concentration would be indicated by a definitive slope shift.

Two distinctive pieces of data have been selected for the two-phase FOCS. Figure 7 shows the response of two FOCS placed in a chloroform-contaminated well. In this case, the data from both FOCS sensors fell on the same curve. The results of many measurements indicate that a plot of count rate versus time is parallel for all FOCS, but that they may be shifted on the intensity scale by as much as 10%.

Figure 8 shows data taken in the same well and the vadoze zone just adjacent to that well. Here the data are plotted against data taken at several wells in the area using "grabbed" samples and gas chromatography. It is important to note that in every case the *in situ* results are higher than those obtained using a sampler and laboratory analysis. This has been attributed to sample losses both during well pumping and sample storage prior to analysis.

The single-phase FOCS measurements represented two advances in the organic chloride FOCS technology: (1) the use of a single-phase basic pyridine system and (2) the elimination of the membrane; that is, it was decided to put the FOCS into the well *without* using a membrane to protect it from liquid water. The initial results were very encouraging. The system operated properly for several hours, its response time was about two seconds, and several different FOCS gave essentially the same response curve. Figure 9 summarizes the results from several *different* one-hour runs. Here all of the experimental points fall on the average curve within about 3%. This is exceptionally good performance for a monitor. Figure 10 shows the effect of pumping a

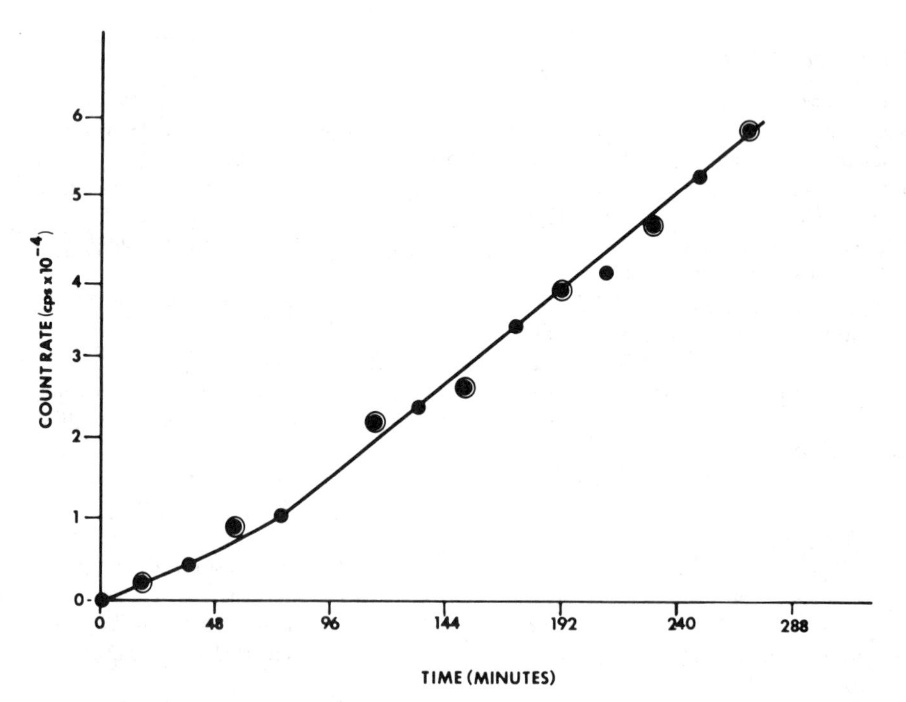


FIG. 7—*Performance of two different two-phase FOCS in a chloroform-contaminated well at equilibrium (slope: 10 ppm = 9.33, 1 ppm = 0.93).*

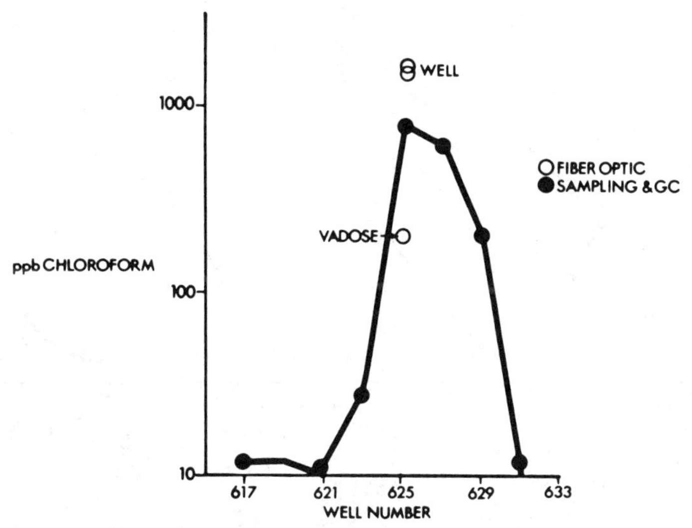


FIG. 8—*Comparison of the performance of a two-phase FOCS to gas chromatography/mass spectrometric analysis of a chloroform-contaminated well and the adjacent vadoze zone.*

well. The curves in this figure represent a series of one-hour measurements started immediately after the well was sampled. It should be noted that the data in Figs. 9 and 10 have not been corrected for the difference in temperature between the well water and the calibration solutions. This means the measured well concentrations may be slightly low. Comparing the data in Figs. 9 and 10 it is apparent that a time span of more than three hours was required until the chloroform concentration returned to the level it was at before pumping. It is evident, though insufficiently substantiated at this time, that present sampling techniques introduce an error into the concentration measurements. The fact that the *in situ* data consistently show higher concentrations in the well than the ''grabbed'' samples, and that these same *in situ* measurements show volatile depletion in a well after sampling, makes it a priority item to collect enough data to equate the two measurement approaches.

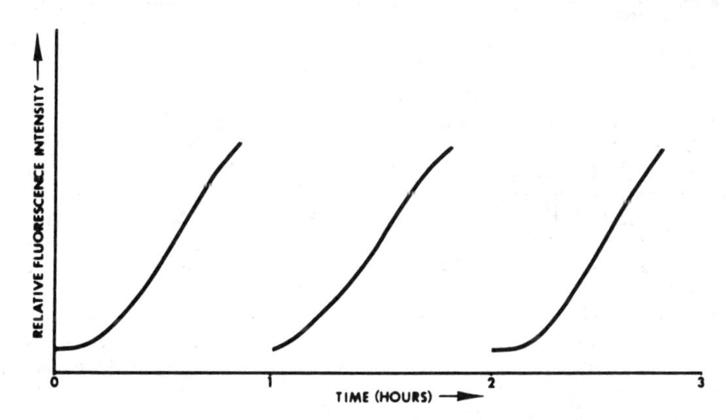


FIG. 9—*Performance of a single-phase FOCS in a chloroform-contaminated well at equilibrium (slope: 2 ppm = 4, 1 ppm = 2).*

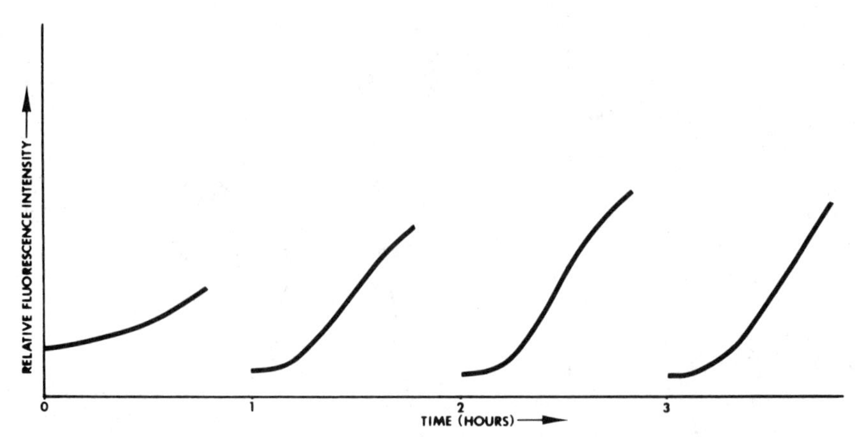


FIG. 10—*Performance of a single-phase FOCS in a chloroform-contaminated well immediately after pumping (slope: 2 ppm = 4, 1 ppm = 2).*

## Conclusions

Based on current data, it appears as though the RFS-FOCS methodology is a viable approach to solving the problem of developing a practical pollutant monitor. This is so because it represents a combination of technologies which will yield a sensitive, specific, easy-to-use, affordable system. The key to this system is an inexpensive, disposable FOCS which provides specificity to a preselected pollutant and which also contributes to system sensitivity. The spectrometer when used with a FOCS can be greatly simplified because of the great amount of responsibility placed on the FOCS. Projections are that the RFS-FOCS system will, eventually, be self-contained and fit in a coat pocket.

The only limitation to the RFS-FOCS approach is the need for a different FOCS for each pollutant to be monitored. This requires that new, or modified, chemistry be developed for each species of interest. Although there are no evident pitfalls in this requisite, it does take time and money to develop several of these sensors. On the other hand, once this has been accomplished, many of these 100- to 200-μm fibers can be bundled into a very small package which can still be placed in the soil or ground water using a punched, or drilled, small hole, that is, 1.27 cm (½ in.) or less in diameter, rather than a costly, outfitted minimum 10-cm (4 in.) well.

## References

[*1*] Fujiwara, K., *Sitzungsber Abh. Naturforsch Ges. Rostock,* Vol. 3, No. 33, 1916.

[*2*] Hirschfeld, T., Deaton, T., Milanovich, F., and Klainer, S., "The Feasibility of Using Fiber Optics For Monitoring Ground Water Contaminants," Technical Report UCID-19774, Vol. 1, Lawrence Livermore National Laboratory, Livermore, CA May 1983.

[*3*] Milanovich, F., Hirschfeld, T., Miller, H., Garvis, D., Anderson, W., Miller, F., and Klainer, S., "The Feasibility of Using Fiber Optics for Monitoring Ground Water Contaminants. II. Organic Chloride Optrodes," Technical Report UCID-19774, Vol. II, Lawrence Livermore National Laboratory, Livermore, CA March 1985.

[*4*] Milanovich, F., Anderson, W., Garvis, D., Angel, M., and Klainer, S., "The Feasibility of Using Fiber Optics For Monitoring Ground-Water Contaminants. III. Preliminary Field Test Results, Organic Chloride FOCS," Technical Report UCID-19774, Vol. III, Lawrence Livermore National Laboratory, Livermore, CA Sept. 1985.

[*5*] Lugg, G. A., *Analytical Chemistry,* Vol. 38, No. 1532, 1966.

[*6*] Daroga, R. P. and Pollard, A. G., *Journal of the Society of Industry,* Vol. 60, No. 218, 1941.

[*7*] Hunold, G. A. and Schuhlein, B. L., *Zeitschrift für Analytische Chemie,* Vol. 179, No. 81, 1961.

*Kent J. Voorhees,*[1] *Michael J. Malley,*[2] *James C. Hickey,*[2] *Ronald W. Klusman,*[1] *and William W. Bath*[3]

# Application of a New Technique for the Detection and Analysis of Low Concentrations of Contaminants in Soil

**REFERENCE:** Voorhees, K. J., Malley, M. J., Hickey, J. C., Klusman, R. W., and Bath, W. W., **"Application of a New Technique for the Detection and Analysis of Low Concentrations of Contaminants in Soil,"** *Ground-Water Contamination: Field Methods, ASTM STP 963,* A. G. Collins and A. I. Johnson, Eds., American Society for Testing and Materials, Philadelphia, 1988, pp. 381–396.

**ABSTRACT:** A direct method for trapping and detecting volatile organic compounds emanating from contaminated ground water or soil has been successfully demonstrated. The method consists of a static trapping device which is placed just below the soil surface and left for seven to thirty days to insure time for integrated collection of the soil gas. Analysis of the trapped compounds is performed by Curie-point desorption mass spectrometry. Results from two studies conducted in separate Denver industrial complexes showed that the technique is a powerful tool for delineating surface contaminations such as trichloroethylene (TCE), tetrachloroethylene (PCE), toluene, and hydrocarbons. The studies also showed the flux data are useful for inferring the extent and direction of contamination where limited hydrological data are available.

**KEY WORDS:** soil contamination, ground-water contamination, vadose zone, static trapping, Curie-point desorption mass spectrometry, analysis, hydrocarbons, trichloroethylene, tetrachloroethylene, toluene

Commonly used methods for monitoring ground-water or vadose-zone contamination or both include gas chromatography/mass spectroscopy (GC/MS) analysis of water or soil samples, a solvent soil extraction, or remote portable detectors such as gas chromatographs or photoionization detectors. The GC/MS methods have been successful, but drilling and collection, preparation, and analysis of samples are expensive. The application of portable detectors is economical, but in some cases does not provide for adequate species differentiation. A direct detection method utilizing a time-integrated static collection device for trapping trace volatile organic compounds (VOCS), followed by analysis with mass spectrometry, has demonstrated that it overcomes many of the problems inherent in the methods previously described.

Static collectors have been used extensively to monitor contamination in the workplace [*1*] but until recently had not been utilized for the collection of potential contaminants in the soil or emanating from ground water [*2*]. By means of a specially constructed static collector combined with analyses by mass spectrometry, VOC's originating from contaminated soils and ground water can commonly be detected just below the soil surface.

[1] Professor of chemistry and professor of geochemistry, respectively, Department of Chemistry and Geochemistry, Colorado School of Mines, Golden, CO 80401.

[2] Petrex, 605 Parfet St., Suite #100, Lakewood, CO 80228.

[3] Martin Marietta Environmental Systems, 9200 Rumsey Road, Columbia, MD 21045.

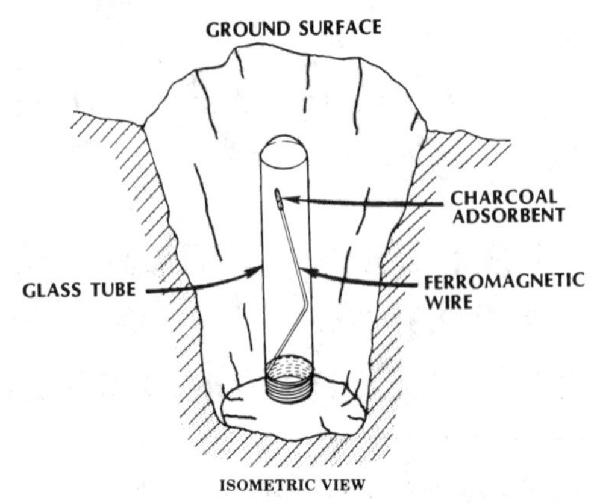


FIG. 1—*Schematic diagram of static collector used for trapping volatile organic compounds.*

The static collector illustrated in Fig. 1 is used for trapping the VOC's. It is prepared by applying presieved-sized activated charcoal to the tip (1 cm) of a ferromagnetic wire. After construction, the collectors are cleaned by heating to 358°C in a high vacuum system followed by placement into a sealed culture tube under an inert atmosphere.

At the field site, the static collectors are placed in augered holes to a depth of about 45 cm, backfilled, and left to equilibrate with soil vapors for a period of seven to thirty days. After the integrative collection period, the collectors are retrieved from the field and returned to the laboratory for analyses by Curie-point desorption mass spectrometry. Previous research has shown that, compared with instantaneous collection methods, time-integrated collection techniques yield statistically superior results when measuring flux rates [*3,4*].

The Curie-point method allows for a clean reproducible desorption of compounds from the static trap. This system, when coupled with a mass spectrometer, allows for the detection and identification of compounds to mass 240. The compounds reported in this range can be interpreted by mapping the relative ion count (flux) of the identified compounds. The pattern developed as a result of correlating flux data over multiple sample points (for example, a grid) is used to determine the extent and apparent direction of migration of the contaminant plume.

This paper describes the application of the technique to measure near-surface volatile organic compounds originating from two areas of vadose zone, soil, and ground-water contamination in two different hydrogeologic settings.

## Experiment

### *Study Areas*

Two industrial areas (A and B) were used in the study. Study Area A is near Denver, CO where several sources of both soil and ground-water contamination had been documented. A chemical storage facility, an abandoned petroleum refinery, a chemical waste disposal area, and a landfill disposal were all located within the study area (approximately 1.7 km$^2$). Figure 2 illustrates site location and sample distribution. The refinery had been operated until 1965, when a fire destroyed the facility and released an estimated 182 m$^3$ (48000 gal) of petroleum products onto the site. The chemical storage facility had operated for a number of years and had been cited frequently for poor operating procedures. The chemical disposal site had received waste from a pesticide manufacturing operation for several years and had disposed of the waste in rubber-lined pits. The

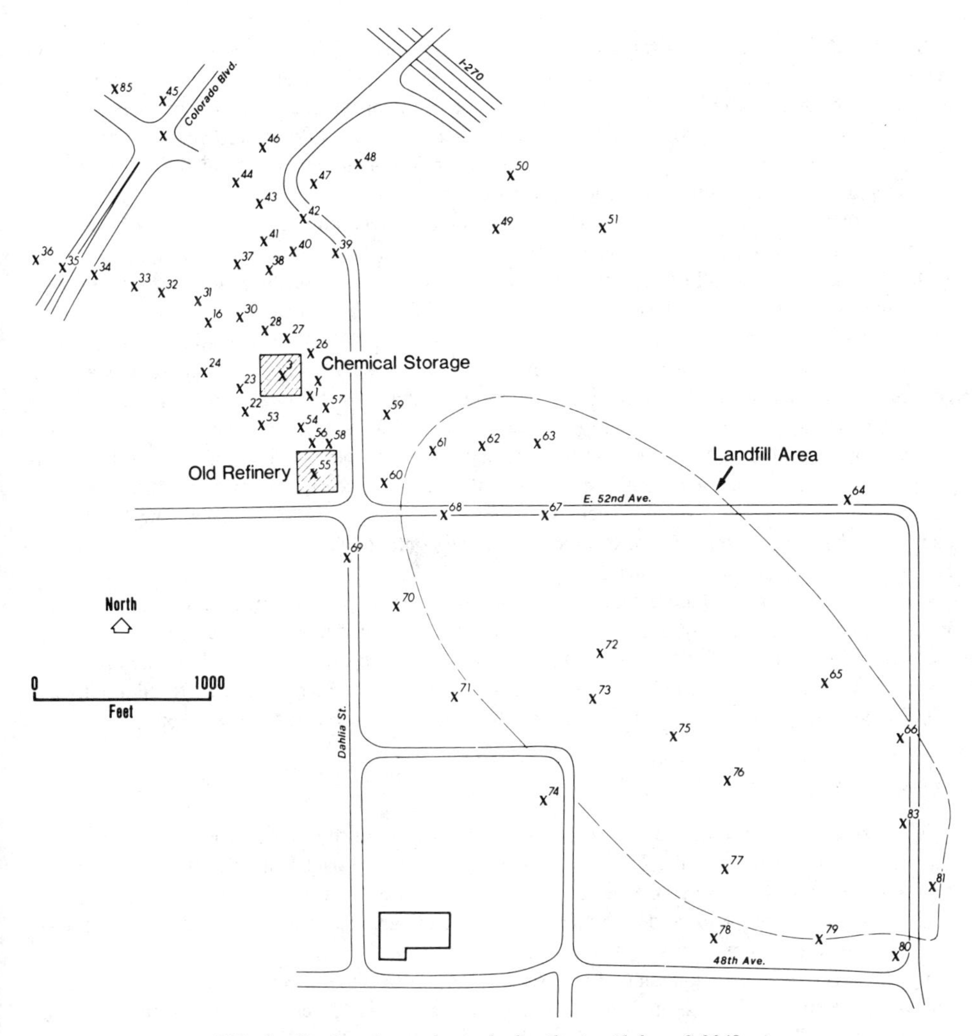


FIG. 2—*Site location and sample distribution (1 ft = 0.3048 m).*

refuse landfill was operated for about 20 years prior to 1973. Several major surface spills had been documented in this area over the past 20 years, which led to the selection of the site as a study area.

The survey area sample site, located on Holocene and Quaternary alluvium and aeolian deposits, is on the south side of a stream. The alluvium and aeolian deposits range in thickness from 6 to 60 m. The alluvium deposits are underlain by the Tertiary-Cretaceous Denver Formation, which is composed of interbedded shale, claystone, siltstone, and sandstone. A probable paleochannel of the stream underlies the landfill as is evidenced by the bedrock elevations.

The sandy and lenticular nature of the alluvial deposits results in high permeability and high hydraulic conductivities. Water wells in the alluvium of the area have moderate to high yields. The piezometric surface indicates ground-water movement to the north or northeast toward the channel of the stream. Organic and inorganic contaminates at the sites can be expected to move vertically and horizontally in the alluvial deposits of the area.

Study Area B is on the extreme west flank of the Denver Basin near the base of the Front Range. A large industrial complex, which had used various industrial solvents, had been at the location since the 1940's. The site is underlain by the Pennsylvanian-age Fountain Formation, which consists of reddish-brown sandstone with minor shale. These strata dip about 50 deg to the east in the site vicinity, and differential erosion of the Fountain and overlying sedimentary units has created a series of northward-trending hogbacks and valleys.

The Fountain Formation is overlain by up to 9 m of Quaternary alluvium consisting predominantly of reddish-brown gravelly, clayey sands. The alluvial veneer is continuous over large areas of the valley bottom, but is locally interrupted by small spires and ridges of the Fountain Formation.

Drilling within the geological setting described previously, but outside of the immediate study area, indicates that ground water should be encountered in the Fountain Formation at all locations (given sufficiently deep holes), but that ground water is only locally present in the alluvial overburden. Ground-water is more likely to be encountered in the valley bottoms than in the intervalley areas, presumably because the overburden thickness is greatest near the center of the valleys. Well-yield data from nearby areas indicate that the permeability of the overburden is appreciably greater than that of the Fountain Formation. Consequently, the two units are considered as separate water-bearing zones. Where saturated conditions are present in both the "shallow" and "deep" systems, contaminant transport rates and contaminant concentrations are expected to be greatest in the shallow overburden system.

To define the extent of contamination in the study area, overburden monitoring wells were installed prior to and concurrently with emplacement of the static collectors. The locations of these wells are shown in Fig. 3. Wells in the lower part of the valley were dry; wells in the upper part generally were not dry. Ground water drawn from wells in the upper valley had trichloroethylene (TCE) concentrations ranging from 5 parts per billion (ppb) to greater than 2000 ppb.

*Sample Collector*

The static collection device illustrated in Fig. 1 was used in this study. The description of the fabrication and cleaning of the device has been previously described in detail [*2,5*]. Integration time for Field Study A was from one day to one week. Integration time for Study B was 18 days. The times for integration were determined based on loadings observed in the analysis of duplicate wires which were removed at selected times.

Quality assurance was maintained by providing ten unexposed collectors which were carried to and from the survey site and analyzed as transportation blanks. In addition, duplicate collectors, representing 10% of the survey total, were placed at randomly selected sites and analyzed to test the reproducibility of flux measurements.

*Mass Spectrometer Analysis*

The adsorbed compounds were desorbed and analyzed with a Fisher Curie-point pyrolyzer (1.5 kW, 1.1 MHz) in tandem with an Extrel SpectrEL quadrupole mass spectrometer (Fig. 4). Low-energy electron ionization was used to minimize fragmentation and to preserve the molecular ion of the desorbed compounds [*6*]. A scan rate of 1250 atomic mass units/second (amu/s) and a scan range of 10 to 240 amu were used for all analyses. Analysis time was two minutes per sample with no sample preparation required. The Curie-point wire acts as a heating element for desorbing volatiles into the mass spectrometer inlet system. Quality assurance was maintained by tuning the mass spectrometer with the correct mass assignment and peak resolution. In addition, the mass spectrometer was monitored every 25 samples to assure low-level background readings.

In addition, identification of individual species in Study Area A was made on GC/MS analysis of a limited number of duplicate samples. A Curie-point inlet [*7*] connected to the GC/MS enabled

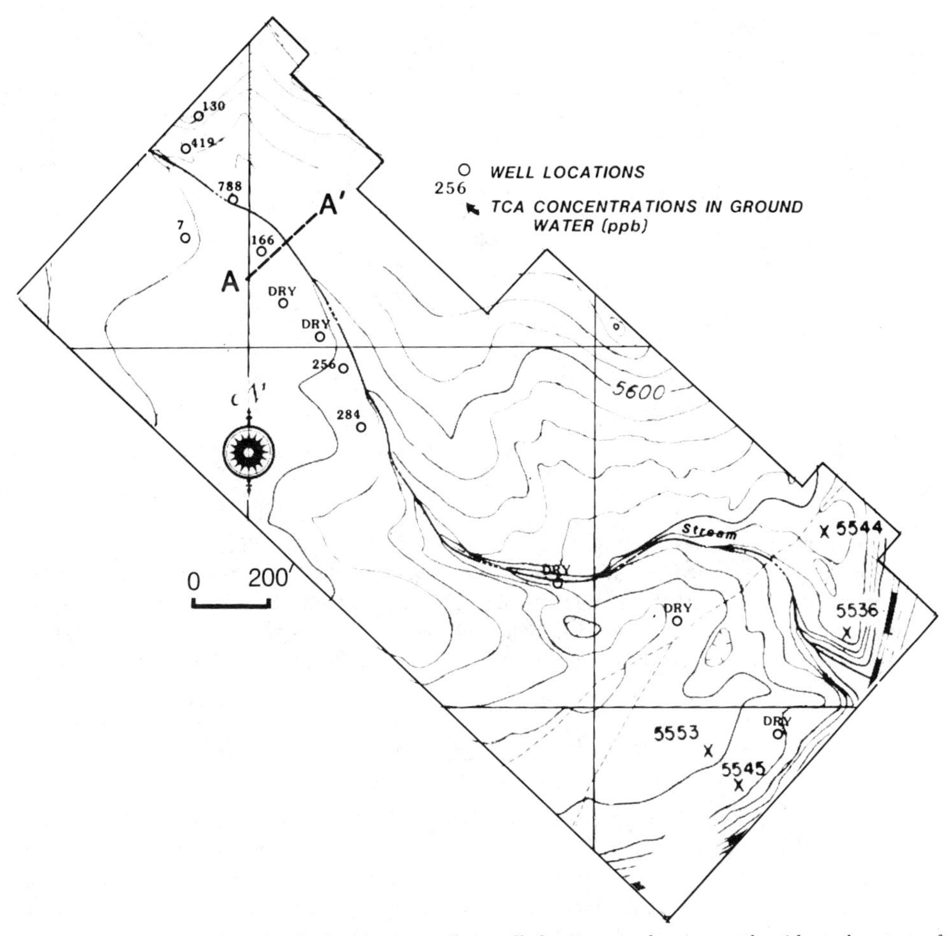


FIG. 3—*Locations of overburden monitoring wells installed prior to and concurrently with emplacement of Petrex static collectors.*

introduction of the sample into the system. A 30-m DB-5 fused silica column programmed between 50 and 250°C at 8°C/min was used for compound separation. An Extrel Simulscan GC/MS system using electron ionization at 70 eV was used in all analyses. Compound identification was made by comparison with the National Institute of Health (NIH) mass spectral library [*8*].

### *Compound Identification: Curie-point Desorption-Mass Spectrometry*

Identification of compounds was done by comparing mass spectra of the survey samples with a reference library of spectra of the pure compounds suspected of being present within the survey area. Library spectra of the 32 volatile priority pollutants were generated by exposing the sample collectors to the individual pure compounds in a headspace environment followed by analyses by mass spectrometry. Resultant data were stored on disk, and raw data and hard copies of the spectra were produced.

The compounds detected in both study areas were identified by plotting the mass-to-charge ratio (m/z) versus the relative intensity of all the major ions identified and comparing those with the

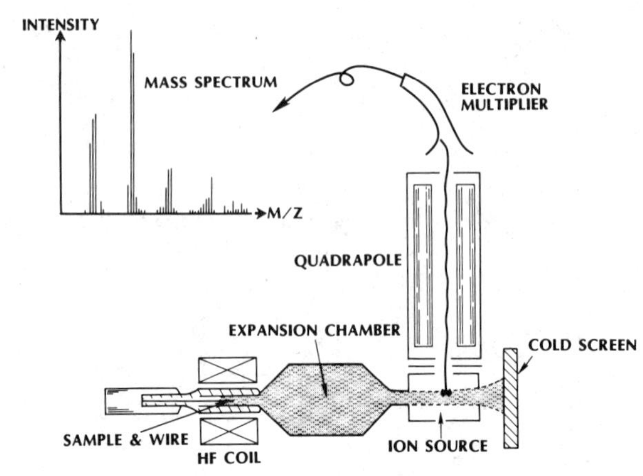


FIG. 4—*Schematic diagram of the Curie-point desorption mass spectrometry system.*

library mass spectra. Since several of the contaminants of interest were chlorinated hydrocarbons, identification was performed by utilizing the isotopic characteristics of a chlorine atom which has occurrences of major isotopes at m/z 35 and 37. The method of identification allows for determination of the proper isotopic distribution of masses (m/z) in each mass spectrum [*9*]. In addition, each mass spectrum was compared with the listing of the major Environmental Protection Agency (EPA) listed ions [*10*].

### *Data Presentation: Flux Maps*

The process of mapping fluxes (relative ion count) of the compounds identified in the survey area was completely computerized. The sample sites were digitized from base maps of the study areas and were reported as *X-Y* coordinates. The *X-Y* data were then paired with the relative ion counts of the corresponding parent ion of the compounds identified. Parent ions for compounds identified are listed in Table 1. A file was constructed which was subsequently run through a plotting routine to produce relative flux maps for each compound.

The indicated flux values are not direct concentration values, but represent the ion count values from the spectra and are proportional to the vapor phase emanation rate at a given site. The relative flux values cannot be related to absolute concentrations in soil or ground water, but can be used to determine the extent of the plume and infer the direction of the plume movement by analyzing the pattern of flux detected at the surface. The indirect relationship is a result of a multitude of physical and chemical parameters, some of which include atmospheric heterogeneity, partitioning

TABLE 1—*Parent ions for compounds of interest.*

| Compound(s) | Parent Ion(s), m/z |
|---|---|
| Tetrachloroethylene (PCE) | 164 |
| Trichloroethylene (TCE) | 130 |
| Toluene | 92 |
| Hydrocarbons | 57, 71, 85, 99 |

effects, compound modification during transport, biological activity, soil types, and hydrogeologic characteristics.

## Results and Discussion

Prior to the survey, contaminant species in the Field Study A were not known. Therefore, GC/MS analysis was employed to determine potential contamination in each area. The results of the GC/MS analysis conducted on the duplicate wires are summarized in Table 2. Documentation from the chemical storage facility indicated a wide range of organic compounds had been used, including pesticides stored at the site. Due to the strong association of tetrachloroethylene (PCE) with the pesticides, it was used as a pathfinder to define migration routes from the area. Analysis of samples collected from the chemical storage area produced data that clearly defined tetrachloroethylene as the major volatile species identified. Most pesticides have vapor pressures less than $10^{-4}$ torr and, based on previous laboratory exposure results, are not efficiently collected on the charcoal wires. In this study the effect of trapping a low-vapor-pressure compound in the presence of a compound with a high-vapor pressure clearly indicated that only the high-vapor-pressure compounds would be detected. The tetrachloroethylene in this situation totally overwhelmed any low-vapor-pressure compound present in the area. As indicated in Table 1, the GC/MS analysis of wires collected in the refinery area showed primarily hydrocarbons, while toluene was the major species identified as emanating from the landfill. No volatile species could be identified from the chemical disposal area.

Mass spectral peaks from the Curie-point desorption MS data for tetrachloroethylene (mass/charge ratio/m/z 164), toluene (m/z 92), and the hydrocarbons as a compound class (m/z 57,71,85,99) were used for the mapping procedures. Figure 5 shows the map for the tetrachloroethylene flux. The indicated flux values are not direct concentration values, but are represented as the ion current values from the spectra and are proportional to the emanation rate at a given site. The area showing the highest flux for the PCE falls primarily around the chemical storage facility. A strong migration pathway is observed originating at the storage facility and traveling west northwest. The extent of the diffusion of the plume cannot be defined because of the lack of sample stations to the north and south of the main sample line. The ground water in the area flows

TABLE 2—*GC/MS results from duplicate wires for Site A.*

| Area | Compounds Identified |
|---|---|
| Refinery | heptane<br>hexane<br>PCE |
| Chemical storage | heptane<br>hexane<br>PCE<br>TCE (trace)<br>toluene |
| Refuse landfill | toluene<br>benzene (trace)<br>heptane<br>PCE |
| Chemical waste disposal areas | PCE<br>all other compound levels extremely low |

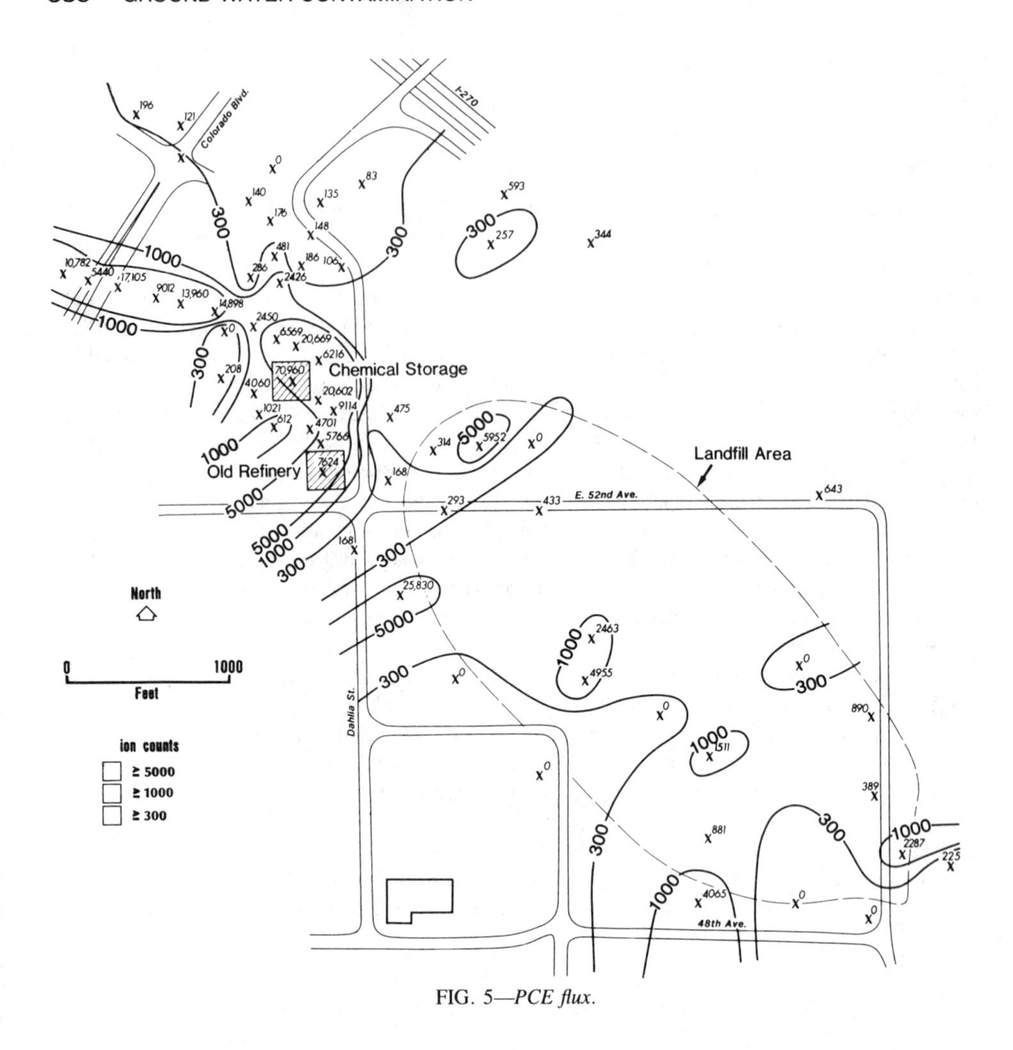


FIG. 5—*PCE flux.*

north northeast. The westward migration of the PCE has probably resulted because of the natural surface drainage to the west northwest. Because of the extremely high values for the PCE flux, it is likely that significant amounts of surface spillage has occurred in the area.

Based on the results of this study and other studies, tetrachloroethylene appears to be a mobile species in both the ground water and the vadose zone. The mobile property of this molecule allows for its use as a pathfinder for determining future migration pathways for less mobile compounds. Future monitoring in the areas with high PCE values should be a high priority.

Figure 6 indicates the distribution of the hydrocarbon plume associated with the refinery incident. The highest flux values are located to the south, north, and northwest. Again, the definition of plume around the chemical storage facility is hampered by the sample design. The refinery is less of a point source than the storage facility. During the fire, the spill covered a large surface area and resulted in hydrocarbon migration from topographic highs to the western portion of the study area. The surface drainage to the west prevented the hydrocarbon migration from contaminating the eastern half of the study area. The southern contamination is in an area where the topography

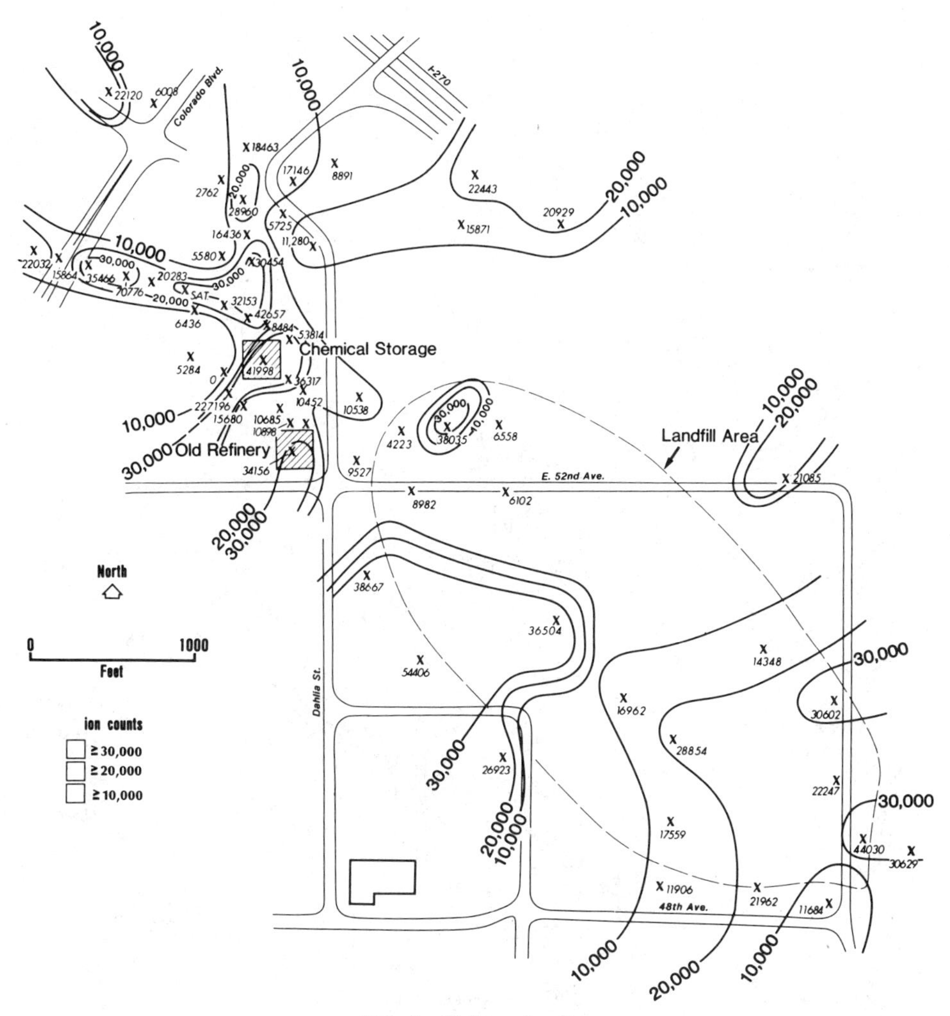


FIG. 6—*Hydrocarbon flux.*

shows no strong drainage pattern. In addition, at present there is a trucking operation occupying the site which has fueling facilities.

The distribution of toluene detected in the landfill is shown in Fig. 7. The toluene plume extends toward the northwest and it is speculated that possible drum disposal could be the toluene source. As was the case for the hydrocarbon spill, the migration of the toluene was extensive enough to result in contamination of the area west of the chemical storage area.

Each of the areas has been mapped with an individual compound to show general plume migration pathways as well as possible contamination. Because of the discriminatory capabilities of the method, species differentiation was attained and allows for mapping distribution of individual species and compound classes. In order to compare the results of the data represented in Figs. 5–7 with a nondiscriminating method, a map representing the total ion current for all sample locations has been prepared (Fig. 8). This map is a sum of all individual compounds trapped on the static

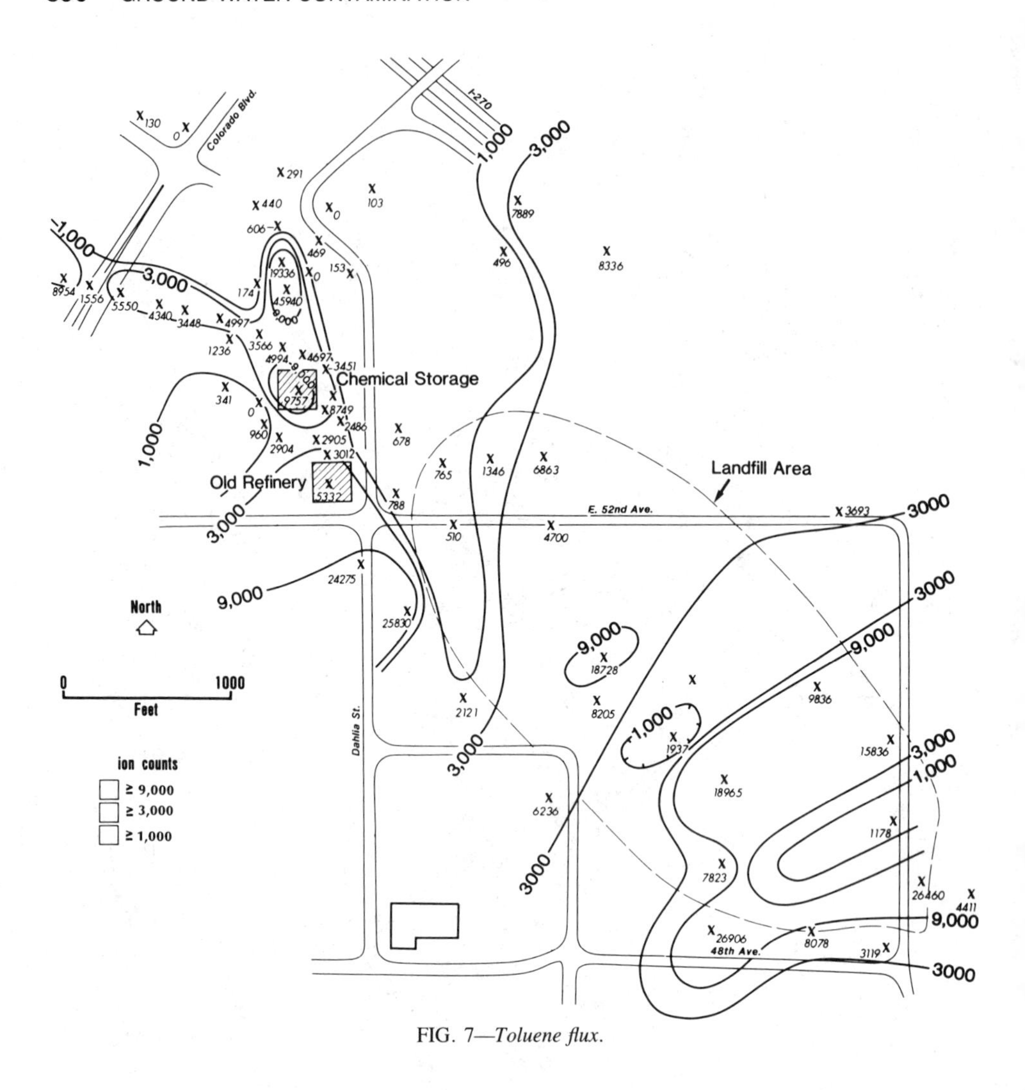


FIG. 7—*Toluene flux.*

collector. Examination of this figure and Figs. 5–7 shows that the overall contamination of the area is defined as well as the migration trend to the west northwest; however, due to the sample design, it was not possible to apportion the various known sources.

The survey in Study Area B consisted of a total of 280 static collectors placed within a grid pattern on 32-m centers (Fig. 9). Since previous soil gas studies had not been conducted within the survey area, the determination of adequate spacing density could not be statistically calculated. In lieu of a statistical evaluation, a sample density was selected such that there would be a minimum of three data points per unit width of the estimated anomaly. Existing hydrogeologic data indicated the estimated plume width ranged from 97 to 170 m; consequently 32-m spacing was chosen.

Figure 10 indicates the relative flux values of TCE vapors measured at near-surface sample sites. Contours were drawn with each interval indicating an increase in relative ion counts measured.

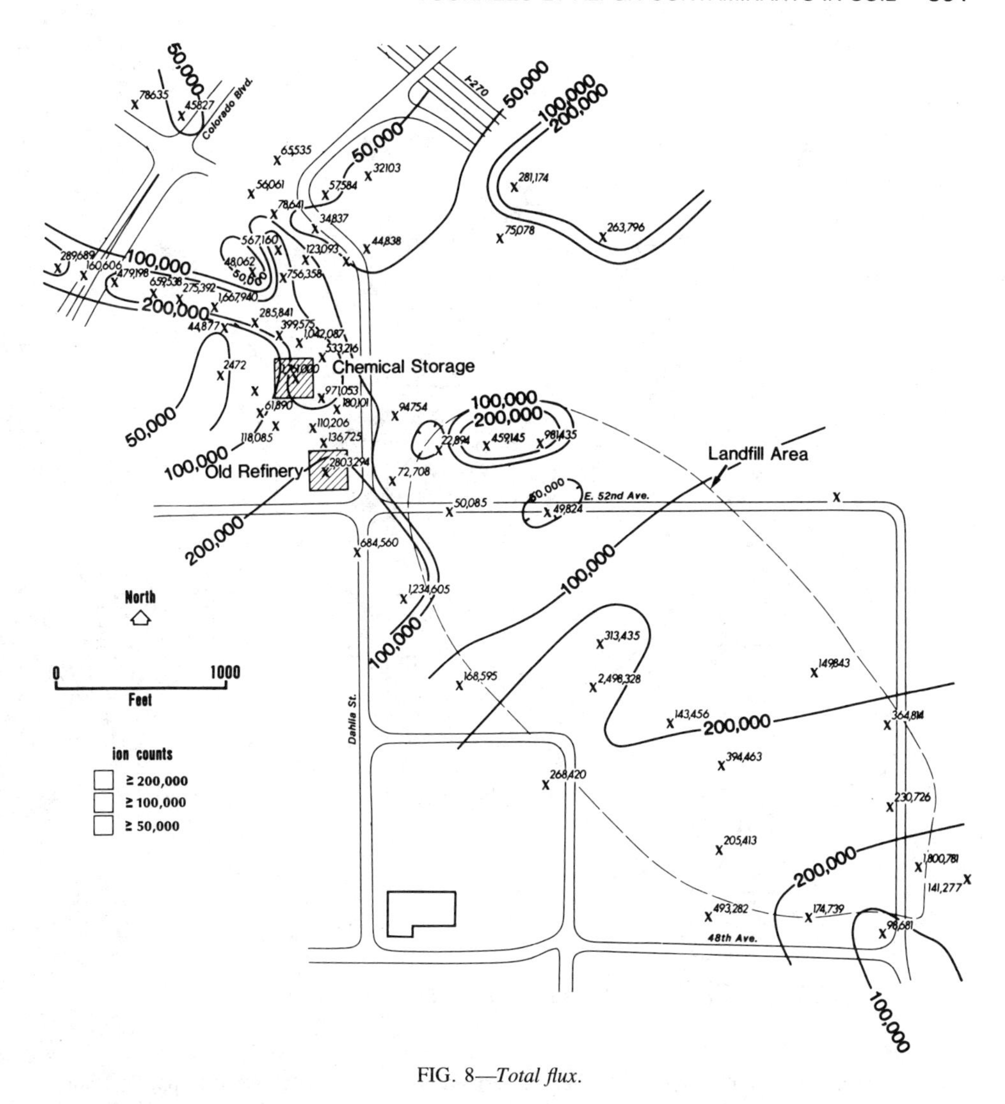


FIG. 8—*Total flux.*

Figure 11 shows the relationship of the TCE flux map and the TCE plume defined by previous ground-water GC/MS analysis results. As a result of the high correlation between TCE concentrations in ground water and the TCE flux data detected in near-surface soils, the flux data were used to infer the distribution of contaminated ground water outside the area defined by hydrogeologic and water quality data.

In order to confirm the presence of TCE contamination outside the area previously defined by monitoring wells, several additional wells were drilled in areas exhibiting anomalies (Fig. 11). The additional wells all encountered shallow ground water and TCE concentrations which ranged from 15 to 419 ppb. The dry wells installed concurrently with the survey were generally outside of or at the fringe of the anomalies. This indicates that ground water is only locally present in the shallow alluvial system and that, where present, it typically exhibits trace concentrations of TCE.

To further examine the localized flow system and to provide the basis for designing a ground-

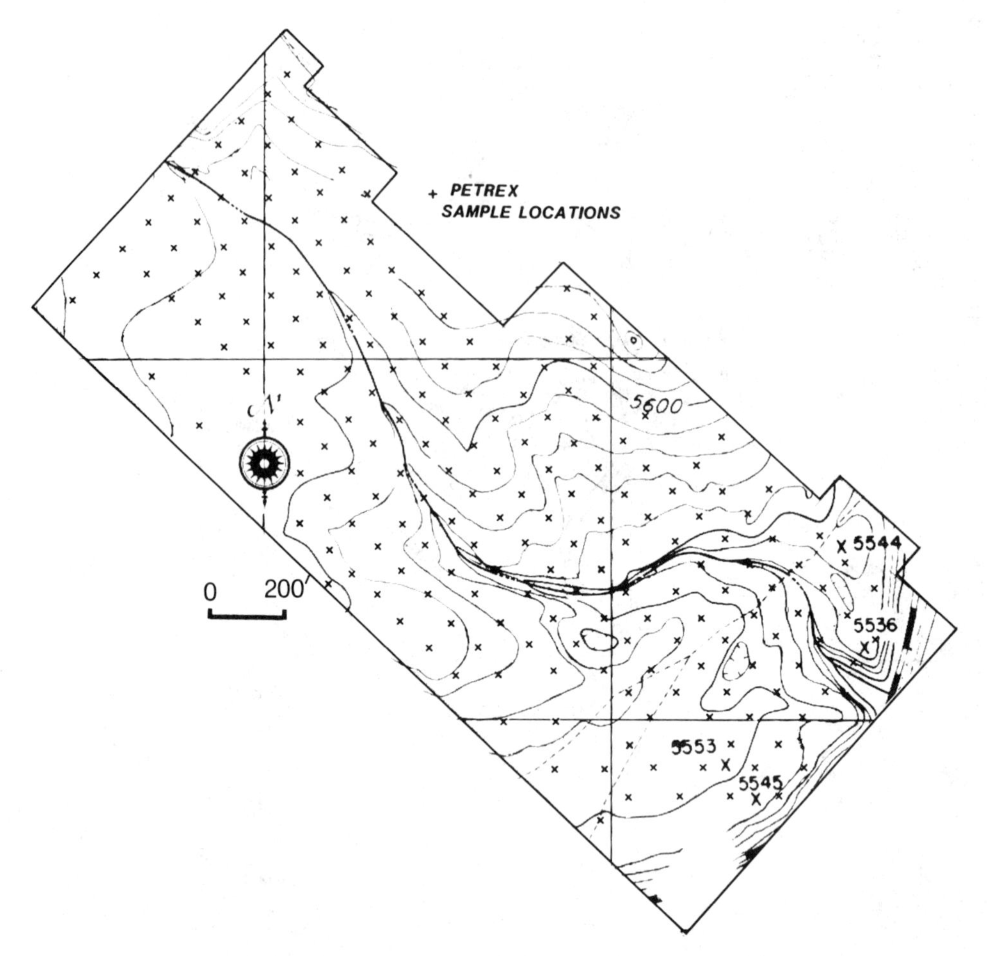


FIG. 9—*Survey design showing sample locations of Petrex static collector.*

water containment/treatment system, wells were drilled on 3-m centers across the apparent plume. This section, shown in Fig. 12, is approximately parallel to the dip direction on the Fountain Formation and perpendicular to ground-water plume axis.

Figure 12 shows that the contaminant plume consists of three separate flow zones which are separated high points on the bedrock surface. Assuming that the buried bedrock highs represent erosionally resistant strata, they should be relatively continuous along strike and effectively separate shallow flow zones. The water quality data shown in Fig. 12 support the interpretation that three discrete flow zones exist within the shallow system. This underscores the need for an extremely detailed hydrogeologic study in contaminant investigations. If water quality and water level data collected from the three flow systems identified in the study area were presumed to represent a single water-bearing zone, some erroneous interpretations might have been made. Although the discontinuous nature of the shallow flow systems could have been identified by drilling alone, the survey, supplemented with well drilling and water sampling, expedited delineation of the flow system as well as the extent of contamination.

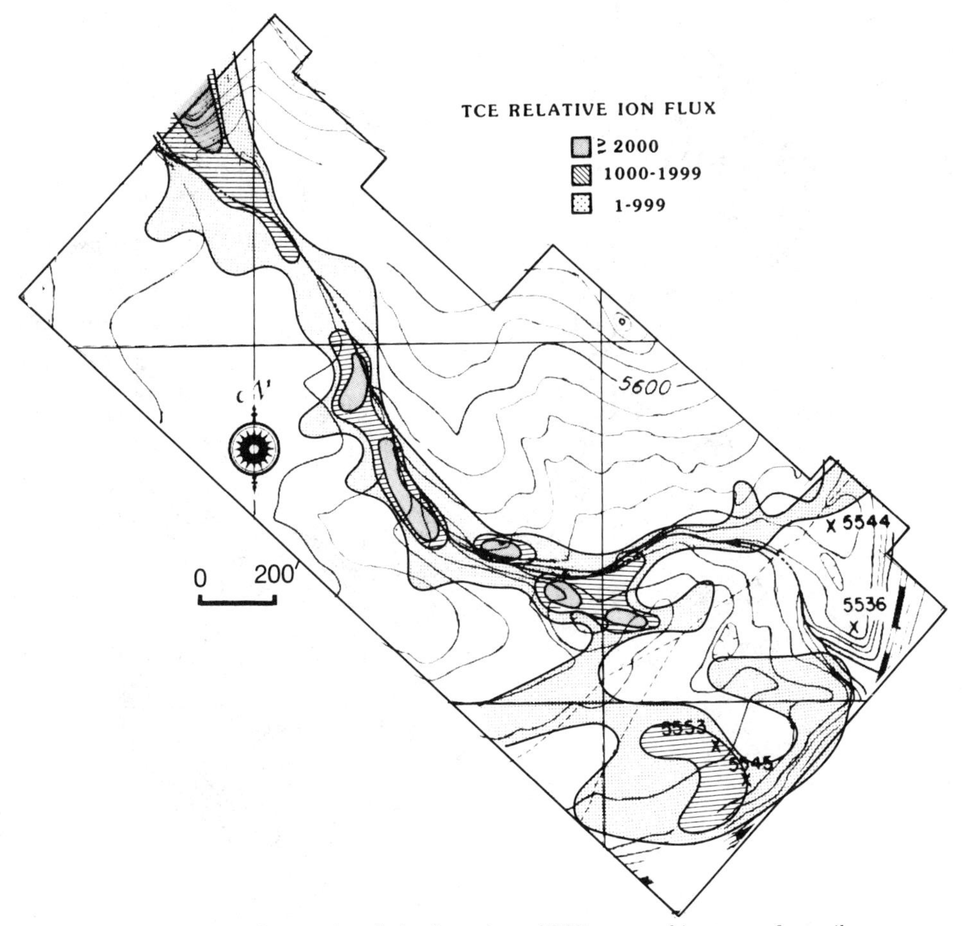


FIG. 10—*Map indicating the relative flux values of TCE measured in near-surface soils.*

As shown in Fig. 12, there is a strong correspondence between the TCE ground-water plume and TCE flux values measured at the surface; however, the results did not delineate the very narrow unsaturated zones along bedrock highs. The widths of the unsaturated zones are considerably less than the 30-m sample density. This probably accounts for the occasional lack of agreement between monitoring well data and surface flux anomalies detected by the static collection technique.

In order to assess the reproducibility of the system, a site was chosen with known PCE contamination about 500 m northwest of Area A. Forty double-wire tubes were installed at predetermined sample locations. These tubes were utilized in two ways. At 30 sample locations, a double-wire tube was placed approximately 0.3 m from the location of a single-wire tube. This provided a total of three adsorption wires per duplicate site. Also, ten double-wire tubes were placed at separate sample locations. The flux data for a given site were then ratioed (low/high value) according to two classification schemes: intrasampler ratios (between two wires in the same tube) and intersampler ratios (between two values). The mean standard deviation and relative standard deviation (RSD) for each class are listed in Table 3.

The previous examples illustrate that surface trapping can be used for the remote detection of

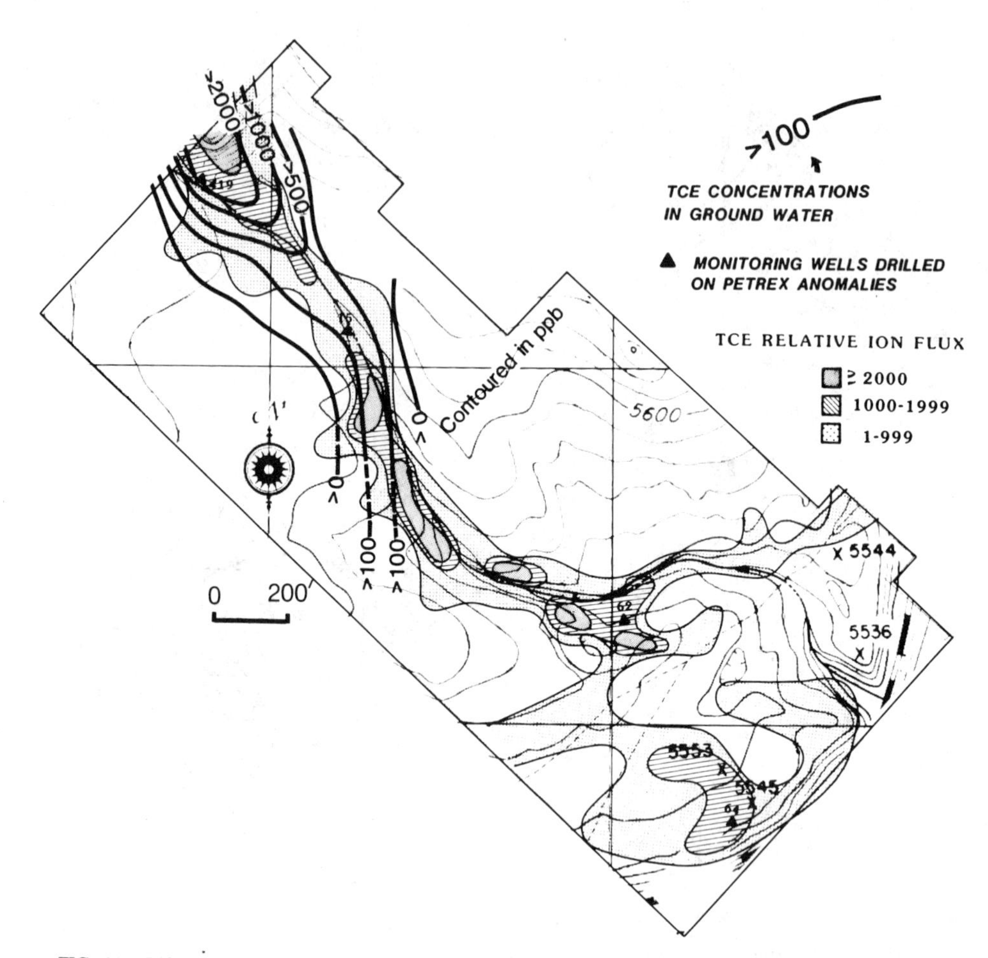


FIG. 11—*Relationship of TCE flux map and TCE ground-water contamination levels determined by previous ground-water monitoring.*

VOC's in the vadose zone originating from ground-water and soil contamination. The use of mass spectrometry in the analysis procedure provides an important capability to perform source discrimination by using selected unique compounds from a particular source. The variation of the geological variables in the study areas had no measurable effect on the flux data. With the use of GC/MS to analyze duplicate wires, a high degree of confidence can be placed on the Py-MS mapping procedure.

## References

[1] Lewis, R. G., Mulik, J. D., Wooten, G. W., and McMillin, C.R., *Analytical Chemistry,* Vol. 57, 1985, 214–219.

[2] Voorhees, K. J., Hickey, J. C., and Klusman, R. W., *Analytical Chemistry,* Vol. 56, 1984, pp. 2602–2604.

[3] Klusman, R. W., Voorhees, K. J., Hickey, J. C., and Malley, M. J., "Unconventional Methods in Exploration for Petroleum and Natural Gas-IV," Southern Methodist University Press, Dallas, TX, 1986, pp. 219–243.

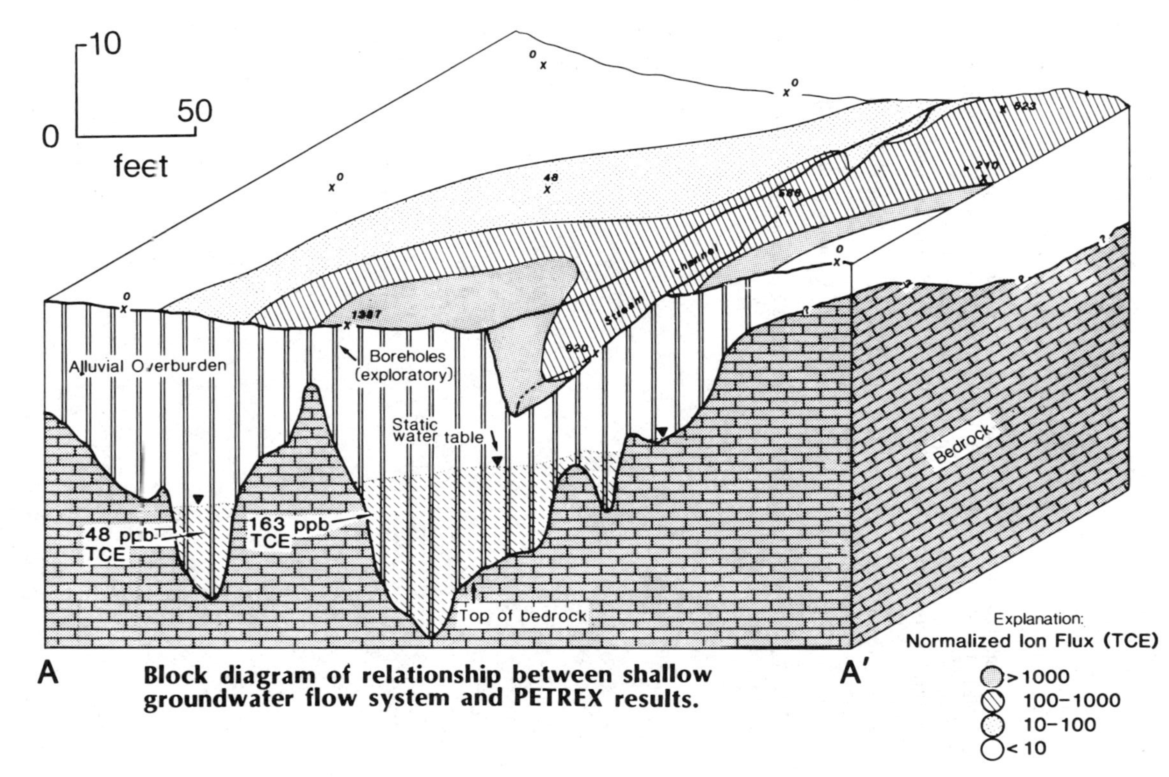


FIG. 12—*Block diagram along Section A-A′ showing relationship between shallow ground-water flow system and Petrex results.*

TABLE 3—*Replicate sample statistics—PCE flux.*

| Classification | Mean | Standard Deviation | RSD |
|---|---|---|---|
| Intrasampler[a] | 0.74 | 0.19 | 0.26 |
| Intersampler[b] | 0.48 | 0.31 | 0.64 |

[a] On 29 values.
[b] On 44 values.


4] Hickey, J. C., Voorhees, K. J., and Klusman, R. W., *American Association of Petroleum Geologists Bulletin,* in press.

[5] Colenutt, B. A. and Thornburn, S., *Chromatographia,* Vol. 12, 1979, pp. 12–16.

[6] Meuzelaar, H. L. C., Haverkamp, J., and Hileman, F. D., *Pyrolysis Mass Spectrometry of Recent and Fossil Biomaterial: An Atlas and Compendium,* Elsevier, Amsterdam, 1982.

[7] deLeeuu, J. W., Maters, W. L., Meent, D., and Boon, J. J., *Analytical Chemistry,* Vol. 47, 1977, pp. 1881–1882.

[8] Environmental Protection Agency/National Institute of Health Mass Spectral Data Base, NSRDS-NBS 63, U.S. Government Printing Office, Washington, DC, 1978.

[9] McLafferty, F. W., *Interpretation of Mass Spectra,* University Science Books, Mill Valley, CA, 1980, p. 303.

[10] Middleditch, B. S., Missler, S. R., and Hines, H. B., *Mass Spectrometry of Priority Pollutants,* Plenum Press, New York, 1981. p. 308.

*Hsai-Yang Fang*[1] *and Jeffrey C. Evans*[2]

# Long-Term Permeability Tests Using Leachate on a Compacted Clayey Liner Material

**REFERENCE:** Fang, H.-Y. and Evans, J. C., **"Long-Term Permeability Tests Using Leachate on a Compacted Clayey Liner Material,"** *Ground-Water Contamination: Field Methods, ASTM STP 963,* A. G. Collins and A. J. Johnson, Eds., American Society for Testing and Materials, Philadelphia, 1988. pp. 397–404.

**ABSTRACT:** This paper presents the results of laboratory tests on a compacted clayey liner material using landfill leachate. Laboratory permeability tests were conducted with continuous permeation for a period from three to to six months in duration in specially designed triaxial cell permeameters. The soil samples were obtained from a borrow pit in eastern Pennsylvania and proposed for use as a compacted liner material. Routine soil tests, including Atterberg Limits, gradation, compressive strength, and compaction tests were conducted on samples exposed to both tap water and leachate. Limited tests were conducted on bentonite-clay mixtures to illustrate the behavioral difference between high-swelling and non-high-swelling clays.

It was found that permeation of the clayey liner material with landfill leachate did not significantly alter permeability of the material. Further, the physical properties remained relatively unchanged. In contrast, test results conducted on bentonite-clay mixtures resulted in more significant changes. It is concluded that leachate permeation of natural silty clays of low activity, such as the material investigated herein, can result in inconsequential change in engineering properties. In contrast, leachate permeation of a high-swelling sodium montmorillonitic soil could result in significant changes in physical and engineering properties of the soil.

**KEY WORDS:** ground water, laboratory testing, hydraulic conductivity, leachate, bentonite

The use of natural clayey soils to form barriers of low permeability has recently been called into question. Permeation of clay materials with *concentrated* organic fluids can result in significant increases in the hydraulic conductivity of the clay liner material [*1*]. The results of the studies described herein, coupled with the results of recent studies of soil-bentonite cutoff walls [2] and other clay barrier materials [*3*], indicate significant increases in permeability would not be expected if the permeant is a relatively *dilute* leachate from a solid waste landfill. This is particularly true if clay liner material mineralogy is such that the clay minerals are those of relatively low activity.

To supplement the long-term permeability tests, additional soil tests were conducted. These included Atterberg Limits, compaction, gradation, and compressive shear strength tests. These tests were performed on samples exposed to both tap water and leachate permeants. In addition, leachate fluids were used on bentonite-clay mixtures to compare the effects found in the natural clayey materials with those of a high-swelling sodium montmorillonitic clay.

[1] Professor of civil engineering, Lehigh University, Bethlehem, PA 18015.
[2] Associate professor of civil engineering, Bucknell University, Lewisburg, PA 17837.

TABLE 1—*Typical test results for proposed clayey liner material.*

| Type of Test | Results | |
|---|---|---|
| | S-1 | S-2 |
| Liquid Limit (ASTM D 4318) | 19 | 24 |
| Plastic Limit (ASTM D 4318) | 15 | 18 |
| Plasticity Index (ASTM D 43180) | 4 | 6 |
| Specific Gravity (ASTM D 854) | 2.73 | . . . |
| Unified Soil Classification (ASTM D 2487) | ML–CL | |
| Maximum Dry Density (ASTM D 698) | see Fig. 2 | |
| Optimum Moisture Content (ASTM D 698) | see Fig. 2 | |
| Particle-Size Analysis (ASTM D 422) | see Fig. 1 | |
| Unconfined Compressive Strength (ASTM D 2166) | see Fig. 5 | |

## Experimental Study

### *Materials*

*Clayey Soil*—Brown sandy silty clay/clayey silt was obtained from the borrow pit to be used for the proposed liner for controlling leachate migration from a solid waste facility. Laboratory test results are presented in Table 1. As shown, the material is classified as a sandy, silty clay/clayey silt of low plasticity.

*Bentonite*—The bentonite was an untreated commercially available clay material. The basic chemical composition is:

| | |
|---|---|
| $SiO_2$ | 56.5% |
| $A1_2O_3$ | 20.0% |
| $Fe_2O_3$ | 4.8% |
| $MgO$ | 4.0% |
| $Na_2O$ | 3.5% |
| $CaO$ | 1.8% |
| $K_2O$ | 0.7% |

*Pore Fluid*—Two types of pore fluid were used, tap water and landfill leachate. The leachate was obtained from an urban landfill site. The results of a chemical analysis of the leachate are presented in Table 2. A comparison of the test results shown in Table 2 with those of characteristics of landfill leachates compiled by others [*3*] shows that the concentrations are in the range expected of solid waste landfill leachate. It is essentially neutral in pH and has relatively high concentrations of several anions and cations in solution. These include iron, magnesium, and manganese as well as chloride and sulfate. The tap water was not analyzed but originated as ground water in eastern Pennsylvania. These and previous studies [*2*] have demonstrated that the tap water has an inconsequential effect upon the long-term hydraulic conductivity of clayey soils.

## Test Equipment and Procedures

The primary geotechnical property of concern for this investigation is the hydraulic conductivity. This was determined in a permeability test utilizing a triaxial cell permeameter developed at Lehigh

TABLE 2—*Physico-chemical properties of landfill leachate.*

| Parameter | Leachate Concentration, mg/L |
|---|---|
| Biochemical oxygen demand (BOD) | 7 300 |
| Chemical oxygen demand (COD) | 11 000 |
| Total organic carbon (TOC) | 3900 |
| Chloride | 898 |
| Magnesium | 300 |
| Sulfate | 400 |
| Total organic nitrogen | 220 |
| Manganese | 48 |
| Aluminum | 2.1 |
| Total iron | 278 |
| Nitrite | 0.21 |
| Nitrate | 0.89 |
| Fluoride | 0.41 |
| Oils and greases | 43 |
| Color | 4957 color units |

University and described elsewhere [*4*]. Samples were compacted at optimum water content using Standard Proctor Energy extruded and trimmed to 70 mm in diameter by 140 mm in length. The samples were capsulated in a membrane and consolidated to an isotropic state of stress. The permeant percolates through the sample from the bottom to the top. A back pressure of 345 kPa was applied to aid saturation and was maintained throughout the test. Ports are provided to permit sampling of the permeant just as it goes into and just as it exits the sample. Details regarding the permeability test system are discussed elsewhere [*4,5*]. In addition, unconfined compression tests were conducted on the samples compacted with both the leachate and the water. Atterberg Limit tests were conducted in accordance with ASTM procedures and equipment on both clay liner materials and bentonite-clay mixtures utilizing both water and a mixture of water and leachate. Hydrometer analyses were likewise conducted utilizing water and a combination of water and leachate. These tests were conducted in accordance with the standard ASTM methods [*6*].

## Test Results and Discussion

The routine soil classification data of the liner material is summarized as given in Table 1. Based on these data, the liner soil is described as sandy silty clay/clayey silt. The Unified Soil Classification is CL-ML. The gradation curve (Fig. 1) indicates that there is approximately 35% sand and about 65% silt or clay. Figure 2 presents a comparison of hydrometer test results with liner materials utilizing water in accordance with the ASTM procedures [*6*] and a combinaton of 50% water and 50% leachate. As shown, with any particle size, there appears to be a 5 to 50% difference in the results. Given a relatively high total dissolved solids in leachate, one might expect the materials exposed to leachate to flocculate and settle out more quickly than those in just water. The opposite, however, was observed. It is possible that interactions between the dispersing agent and the leachate contributed to the observed behavior.

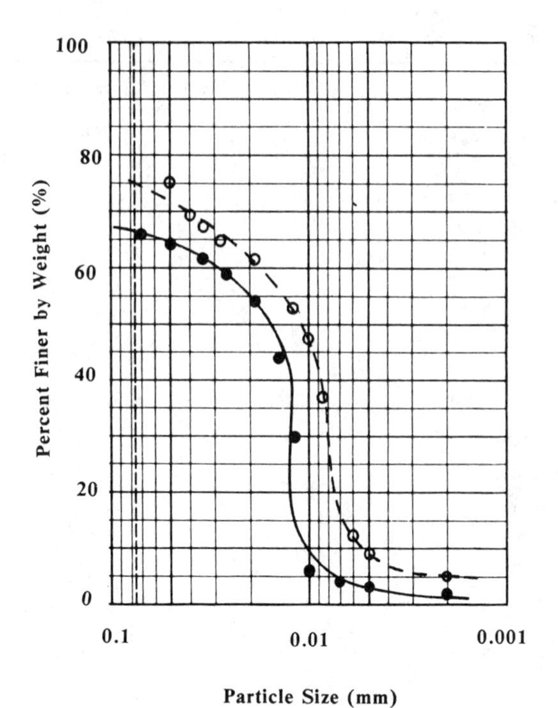


FIG. 1—*Particle size distribution: hydrometer test results.*

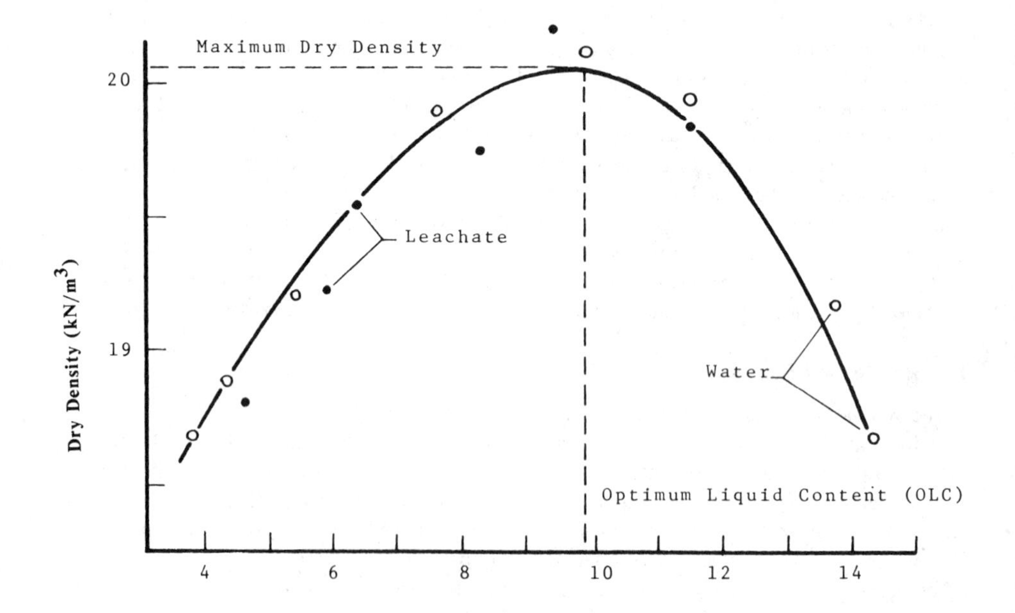


FIG. 2—*Standard Proctor compaction tests results.*

TABLE 3—*Summary of permeability tests on liner material.*

| Sample No. | 1 | 2 | 3 | 4 | 5 |
|---|---|---|---|---|---|
| Equilibrium Average | $6 \times 10^{-9}$ | $2 \times 10^{-8}$ | $3 \times 10^{-8}$ | $3 \times 10^{-8}$ | $4 \times 10^{-8}$ |
| Permeant type | leachate[b] | water | leachate[b,c] | leachate[b] | water |
| Time | 23 days | 8 days | 150 days | 27 days | 95 days |

[a] Geotechnical properties of clay liner material (see Table 1).
[b] Physio-chemical properties of leachate (see Table 2).
[c] Detailed test results of Sample No. 3 (see Figs. 3, 4, and 5).

Figure 1 shows the standard compaction curve on liner material. It indicates that there is no significant difference between the compaction results using water and leachate so far as compaction is concerned.

The results of the long-term permeability tests on the clay liner material are presented in Table 3. A total of five samples were tested, three utilizing both water and leachate as a permeant and two utilizing tap water. The test duration varied from 8 to 150 days. As shown in Table 3, the equilibrium hydraulic conductivity, $K$ varied from approximately $4 \times 10^{-8}$ to $6 \times 10^{-9}$ cm/s.

The results indicate no difference between the permeability of the clay liner when permeated with water as compared to the leachate. Shown in Fig. 3 is a relationship between the hydraulic conductivity and time in terms of pore volume displacement for Sample No. 3. The test was begun with water and then leachate was introduced to the sample. As shown, the permeability is approximately $3 \times 10^{-8}$ cm/s for both water and leachate. If any trend in permeability is apparent, there appears to be some slight decrease in permeability with time. This slight decrease might be attributed to further reductions in void ratio due to secondary compression, plugging by suspended solids in the leachate or precipitation of dissolved solids in the leachate. Note that these tests were conducted for 150 days. As noted, shown in Fig. 3 are the hydraulic conductivity results presented as a function of pore volume displacement. The pore volume displacement has exceeded seven

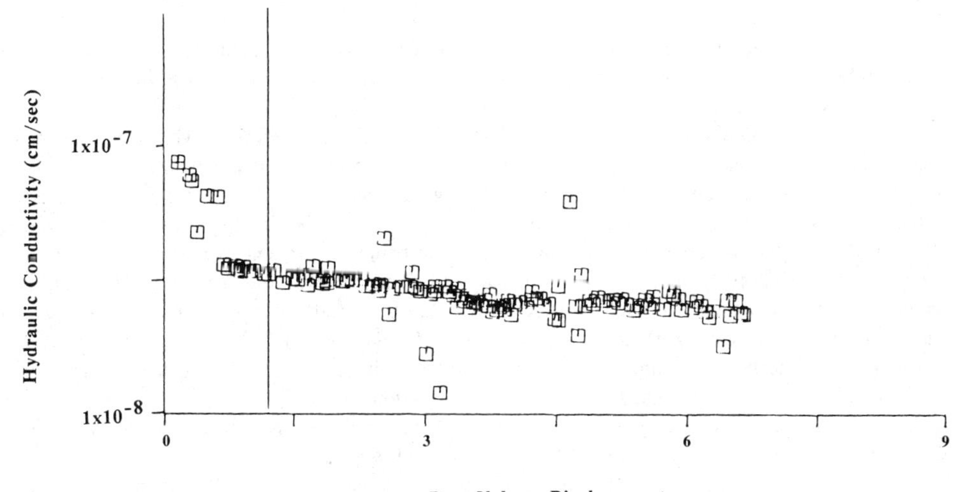


FIG. 3—*Hydraulic conductivity versus pore volume displacement.*

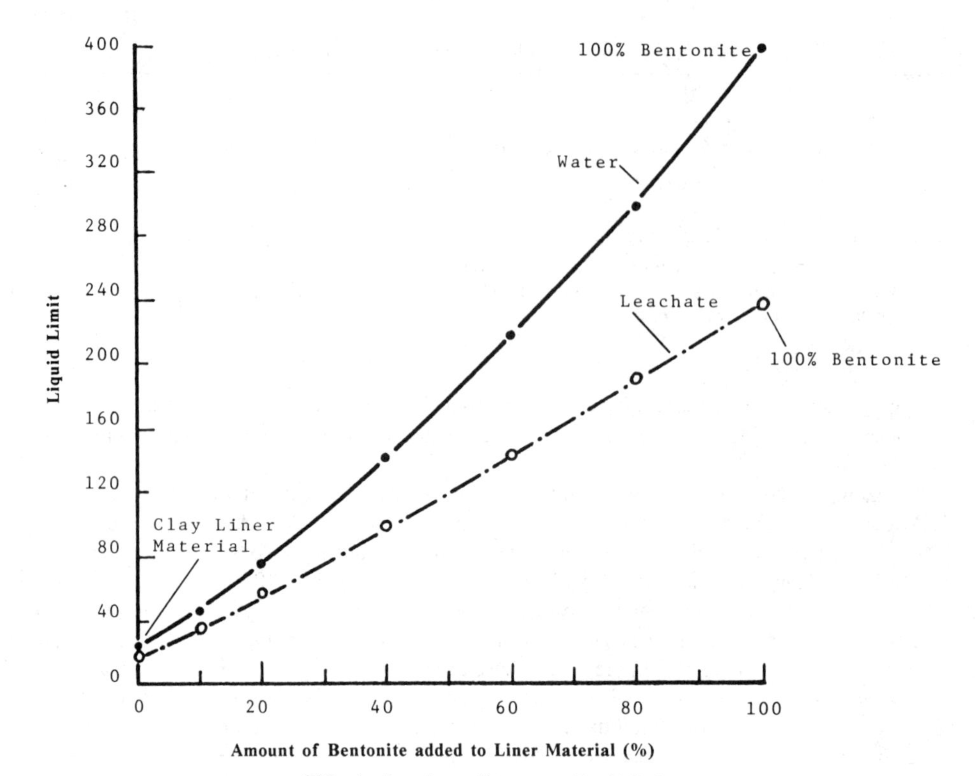


FIG. 4—*Leachate effects upon liquid limit.*

for this testing period. Similar results were obtained for the other two samples. The hydraulic conductivity in the clay liner material is essentially the same for water and for the leachate. Long-term permeation with the leachate did not cause any degradation in the laboratory value of the hydraulic conductivity.

The effect of the leachate upon the Atterberg Limits was also investigated. The soil was found to have a liquid limit of 22 for water and 20 for leachate. This observed difference in liquid limit is relatively small.

To further investigate leachate effects, samples of liner material were mixed with bentonite and tested. Presented in Fig. 4 is a plot of liquid limit as a function of bentonite content added to clay for both water and leachate. The difference becomes greater as more and more bentonite is added to the liner material. Since bentonite is primarily sodium montmorillonitic high-swelling clay, the effect of the leachate becomes more pronounced as the bentonite content increases. In addition, bentonite has a relatively high liquid limit and the liquid limit of the bentonite-soil mixture increases with increasing bentonite as expected. It is clear that at 100% bentonite, the leachate has a considerably greater effect than it does for a 100% liner material. This would be expected considering the interlayer swelling which occurs with bentonitic particles, whereas the low activity clays would be less subject to these diffuse ion layer phenomena [*1*,*5*].

Presented in Fig. 5 is a comparison of unconfined compressive strength test results for both water and leachate compacted samples. The shear strength appears less for the leachate-prepared samples than for the water-prepared samples. The effect of the dissolved solids upon the diffuse ion layer appears to reduce the strength of the materials. There is, however, insufficient data to strongly conclude that this effect is universal.

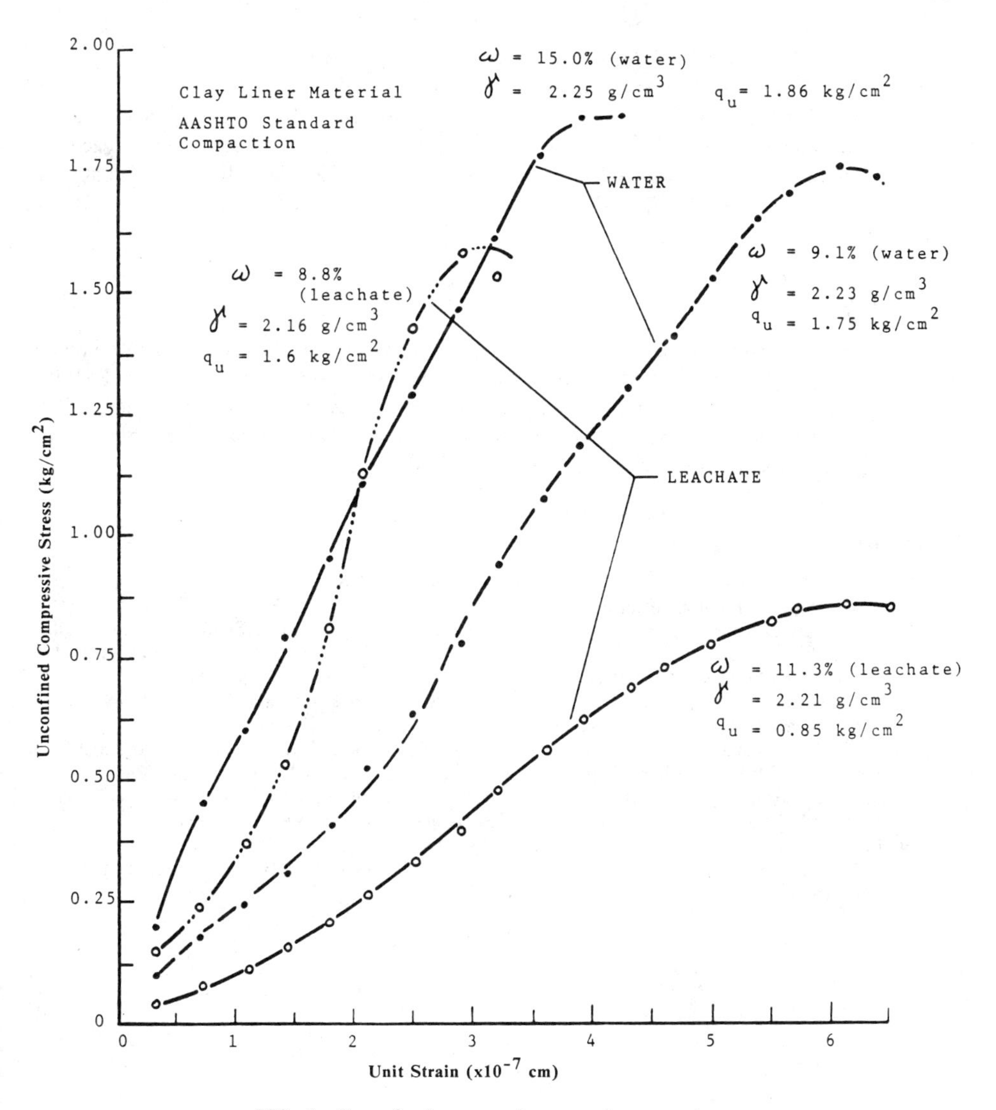


FIG. 5—*Unconfined compressive strength test results.*

## Conclusions and Recommendations

It was found that permeation of a compacted clayey liner material with landfill leachate results in hydraulic conductivities approximately the same as those measured with tap water. Further, no degradation in long-term permeability tests was observed. Physical property testing also indicates little change in property behavior for these silty clayey liner materials. The leachate did have an effect upon the Atterberg Limits of sodium montmorillonitic (high-swelling) clays.

Based upon the above findings, it is concluded that the use of silty clays of low plasticity for barrier applications offers greater containment resistance than the use of high-swelling clays. Although high-swelling clays are frequently proposed because of their ability to significantly reduce the hydraulic conductivity permeability at low clay contents, it is this high swelling which may be affected by leachates. Hence, for liner materials the use of swelling materials incur the greatest risk of permeability increases due to permeation with a landfill leachate. While it has been demonstrated that concentrated organic fluids can have detrimental effects on clay liner materials [*1*], these and other studies [*2,3*] suggest that this is not the case for dilute landfill leachates. These results are consistent with other findings that aqueous phase landfill leachate containing 50000 mg/L or less of organic compounds behave the same as water in terms of effect on permeability [*3*].

*Acknowledgments*

The financial support for these studies was partially provided by Envirotronics Corporation, International, and the United States Environmental Protection Agency under Grant No. R810992 to Lehigh University. Thanks are extended to Dr. George Mikroudis for his review of the manuscript. The opinions, findings, and conclusions expressed in this paper are those of the authors and not necessarily those of the project sponsors.

**References**

[*1*] Brown, K. W. and Anderson, D. C., "Effects of Organic Solvents on the Permeability of Clay Soils," EPA-600/2-83-016, Environmental Protection Agency, March 1983, p. 153.

[*2*] Evans, J. C., Kugelman, I. J., and Fang, H.-Y., "Organic Fluid Effect on the Strength, Compressibility, and Permeability of Soil-Bentonite Slurry Walls," *Toxic and Hazardous Wastes*, I. J. Kugelman, Ed., Technomic Publishing Co., Lancaster, PA, 1985, pp. 275–291.

[*3*] Daniel, D. E. and Liljestrand, H. M., "Effects of Landfill Leachates on Natural Liner Systems," report to the Chemical Manufacturers Association, Jan. 1984. p. 86.

[*4*] Evans, J. C. and Fang, H.-Y., "Triaxial Equipment for Permeability Testing with Hazardous and Toxic Permeants," *ASTM Geotechnical Testing Journal,* Vol. 9, No. 3, Sept. 1986, pp. 126–132.

[*5*] Evans, J. C., Fang, H.-Y., and Kugelman, I. J., "Influence of Hazardous and Toxic Wastes on the Engineering Behavior of Soils," *Management of Toxic and Hazardous Wastes, Chapter 21,* H. G. Bhatt et al., Lewis Publishers, Chelsea, MI, 1985, pp. 237–264.

[*6*] *Annual Book of Standards, Soil, and Rock*, Vol. 04.08, American Society for Testing and Materials, Philadelphia, 1986.

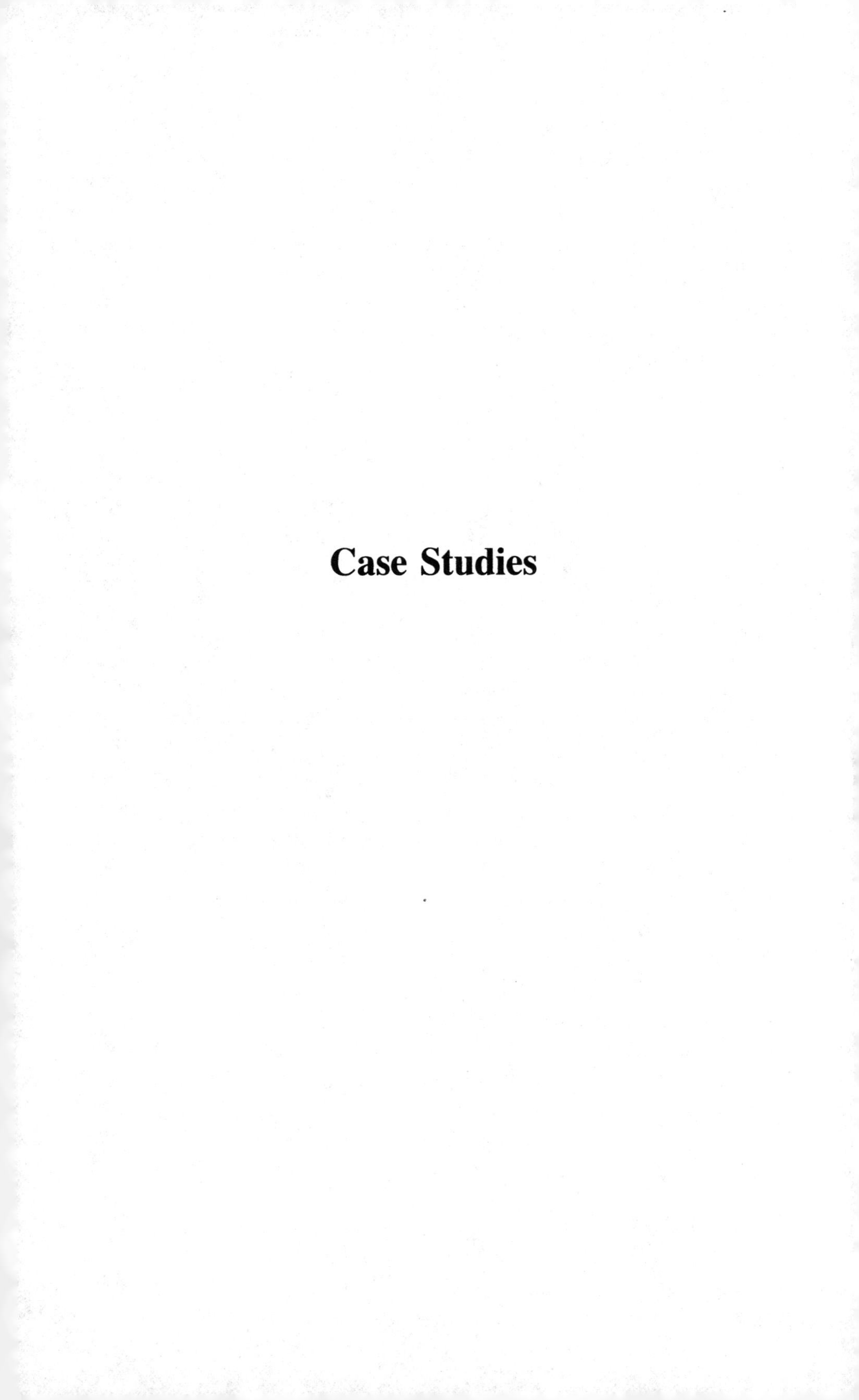

# Case Studies

*Joel G. Melville,*[1] *Fred J. Molz,*[1] *and Oktay Güven*[1]

# Field Experimental Methods in Stratified Aquifers

**REFERENCE:** Melville, J. G., Molz, F. G., and Güven, O., **"Field Experimental Methods in Stratified Aquifers,"** *Ground-Water Contamination: Field Methods, ASTM STP 963,* A. G. Collins and A. I. Johnson, Eds., American Society for Testing and Materials, Philadelphia, 1988, pp. 407–415.

**ABSTRACT:** Tracer injection and observation experiments are the best field methods available to characterize advection and dispersion in aquifers. Near Mobile, Alabama, single and two-well tracer experiments have been conducted in a confined sandy aquifer. Multilevel observation wells were designed, constructed, and operated in eleven different experiments.

Data have been selected from all the results which demonstrate observation well behavior. Isolation of sampling zones is shown to be necessary. Drilling mud remnants in or near the observation well are shown to invalidate measurements. In an experiment where an inflatable packer system suddenly failed, the loss of vertical integrity is shown in the electrical conductivity signals. Finally, mixing within sampling zones, sampling zone lengths, well installation, vertical integrity, and vertical integrity testing are discussed with an emphasis on desirable design characteristics.

**KEY WORDS:** ground water, tracer, field experiment, observation well, multilevel

## Nomenclature

- $Q$ Injection flow rate (gpm), $m^3/min$
- $V$ Injected volume, $m^3$
- $C_0$ Maximum injection concentration, mg/L
- $R$ Radius from injection to observation well in the single-well experiments, m
- $t_t$ Total duration of experiment, h
- $t_c$ Initial phase of injection when tracer is being added to the injection water, h
- $t_i$ Duration of the injection phase, h
- $L$ Spacing between injection and recovery wells in the two-well experiments, m

Existing or potential contamination problems and regulatory pressure for predictive capability have generated the need to measure dispersive behavior in ground water. Recent theoretical studies and the revival of established concepts [*1*] have indicated that the vertical variation of the horizontal hydraulic conductivity, referred to as $K(z)$, is a significant parameter influencing dispersion in saturated porous media. Field experiments [*2–4*] have verified the importance of $K(z)$. More generally the hydraulic conductivity may also depend on horizontal position, $K = K(x,y,z)$. Practical predictive modeling depends on field data describing $K(x,y,z)$. The major question posed is how the variation can be measured and to what detail should it be measured.

Tracer tests, because they are controlled contaminant transport simulations, are the most reliable field experiments to describe advection and dispersion in ground-water flow. Eleven tracer tests have been completed in a confined, unconsolidated aquifer in an effort to determine the necessary

[1] Associate professor, professor and associate professor, respectively, Department of Civil Engineering, Auburn University, Alabama 36849.

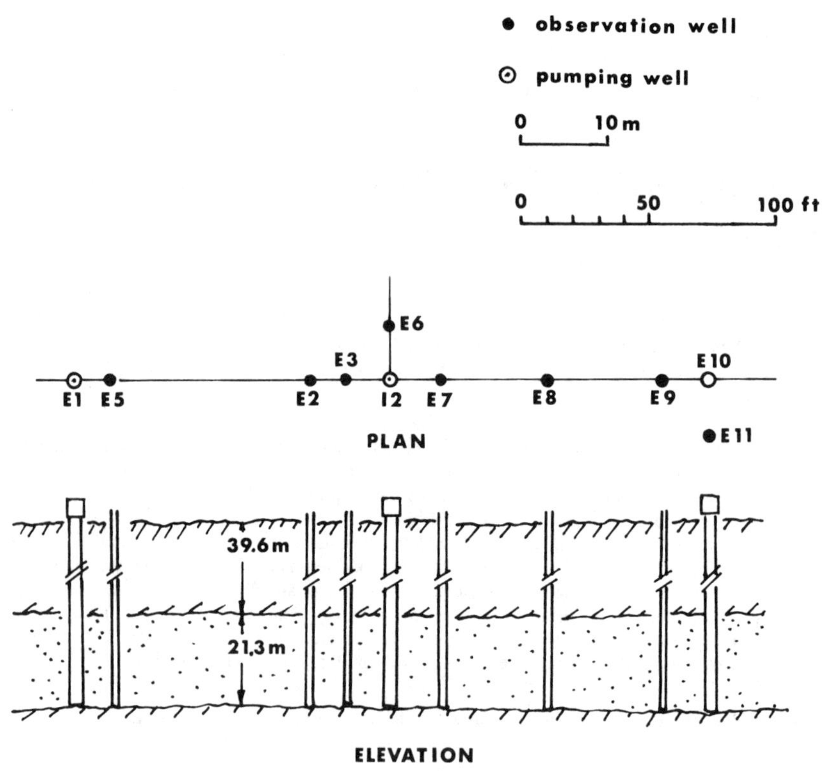


FIG. 1—*Tracer test well field.*

measurements and to evaluate measurement techniques. The purpose of this paper is to describe the design and performance of multilevel sampling wells used in the tracer tests. Results which characterize the observation well behavior are presented and discussed.

## Description of Experiments

Two types of experiments were conducted, single-well and double-well. In the single-well experiments, water was injected into the confined aquifer at a constant flow rate, $Q$, creating a radial, steady-state flow field. For an initial phase of the injection, sodium bromide tracer was mixed with the injected water. A multilevel observation well located a distance $R$ from the injection well then recorded the tracer arrival. The $K(z)$ variation was indicated by different travel times for different elevations in the aquifer. Figure 1 shows the tracer test well field; the series of single-well experiments are summarized in Table 1.

Double-well tests were conducted on a larger scale than the single-well tests. In these tests an injection well and a recovery or withdrawal well, separated by distance $L$, were pumped simultaneously at the same flow rate, $Q$. Again, a slug of tracer was added to the injected water. Tracer arrivals at multilevel observation wells between the two pumping wells were recorded along with the dispersed tracer at the recovery well. The purpose of the double-well tests was to verify the importance of the $K(z)$ parameter (measured in the single-well test) to describe dispersive behavior in the double-well flow field [*4*].

TABLE 1—*Single-well experiments.*

| Experiment No. Date | $Q$, gpm, $m^3$/min | $V$, $m^3$ | $C_0$, mg/L | Injection Well | Observation Well | $R$, m | $t_t$, h | $t_c$, h | $t_i$, h | Measurements, Comments |
|---|---|---|---|---|---|---|---|---|---|---|
| 1 3/7/84 | 133 (0.50) | 1368 | 300 | I2 | E3 | 5.57 | 117.5 | 17.5 | 46.0 | conductivity, E3 open well bore, I2 partial penetration, 7 levels |
| 2 5/22/84 | 139 (0.53) | 5587 | 340 | I2 | E3 | 5.57 | 328.5 | 17.5 | 177.3 | conductivity, E3 packers, no mixing I2 partial penetration, 7 levels |
| 3 6/18/84 | 242 (0.92) | 3309 | 250 | I2 | E3 | 5.57 | 175.5 | 36.0 | 69.6 | conductivity, concentration, E3 packers no mixing, 7 levels |
| 4 7/20/84 | 242 (0.92) | 2962 | 242 | I2 | E3 | 5.57 | 141.0 | 32.0 | 53.0 | conductivity, concentration, E3 packers mixing, 7 levels |
| 5 8/13/84 | 245 (0.93) | 3894 | 250 | E1 | E5 | 5.36 | 190.0 | 37.0 | 70.0 | concentration, conductivity, E5 packers, mixing, 7 levels, drilling mud invalidated E5 |
| 7 4/10/85 | 245 (0.93) | 3950 | 140 | E1 | E5 | 5.36 | 71.0[a] | 58.0 | 71.0[a] | concentration, E5 packers, mixing, 7 levels, E5 cleaned and new insert |
| 8 6/17/85 | 248 (0.94) | 5350 | 108 | I2 | E6 | 6.17 | 95.0 | 67.3 | 95.0 | concentration, direction perpendicular to all other experiments, 7 levels, E6 packers and mixing |
| 9 7/8/85 | 249 (0.94) | 5223 | 144 | I2 | E7 | 6.07 | 92.5 | 56.5 | 92.5 | concentration, E7 packers, mixing, 7 levels, mirror image of Experiment 4 |
| 10 7/29/85 | 253 (0.96) | 5411 | 138 | I2 | E9 | 5.97 | 94.0 | 62.5 | 94.0 | concentration, E9 packers, mixing, 14 levels, packer failure $t$ = 25 to 30 h |

[a] No recovery phase.

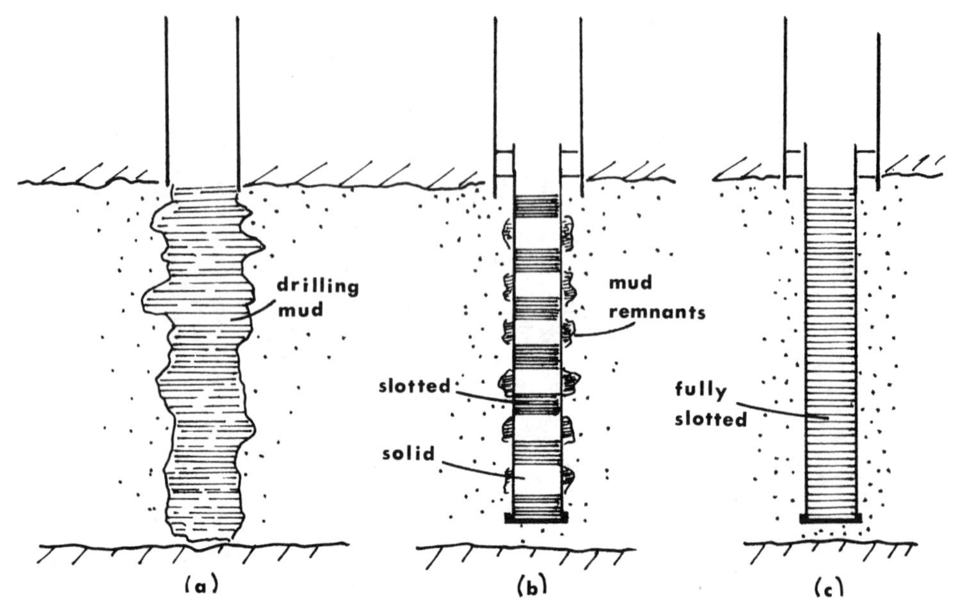


FIG. 2—*Well bore and casing:* (a) *well bore after removal of drilling equipment,* (b) *segmented casing,* (c) *full slotted casing.*

**Observation Wells**

The purpose of the observation wells was to allow water samples to be taken from different elevations in the confined aquifer. An approach to observation well design is to attempt minimal disturbance to the natural structure of the aquifer. Recognizing that the drilling and construction process must disturb the structure, the design approach is more reasonably directed at minimal interference with or even enhancement of the particular measurements to be taken.

The drilling and casing implacement procedure is shown in Fig. 2. Having removed the drilling equipment, the drilling mud and disturbed aquifer material are significantly disarranged as shown in Fig. 2. Two types of Schedule 40 polyvinyl chloride (PVC), 10-cm (4 in.) casing construction were completed. As shown in Fig. 2, segments of slotted pipe and solid casing were installed with the slotted segments at elevations where samples were to be taken. Later, in an effort to simplify construction and to allow for sampling at more elevations, a fully slotted pipe was installed as shown in Fig. 2.

The cleaning and development procedure then was to pump and surge the wells until the water came clear without drilling mud and fine material. In Case (*b*), Fig. 2, this procedure probably left some drilling mud adjacent to the solid casing segments and a disturbed (perhaps more permeable) aquifer material near the slotted segments. In Case (*c*), Fig. 2, a more uniform distribution of disturbed aquifer material was likely to result.

As shown in Fig. 3, seven levels of 0.9-m (3-ft) sampling zones were constructed in observation wells E2 to E7. In observation wells E8, E9, and E11, there were 14 levels of 0.6-m (2-ft) sampling zones. As shown in Fig. 3, the sampling levels were numbered from the top to bottom of the aquifer.

The actual sampling zones were created with a 5-cm (2-in.) Schedule 40 PVC insert composed of slotted and solid segments to match the 10-cm (4-in.) casing. The insert system, Fig. 3, consisting of vacuum tubing, electrodes, silicon plugs, and inflatable packers was constructed at

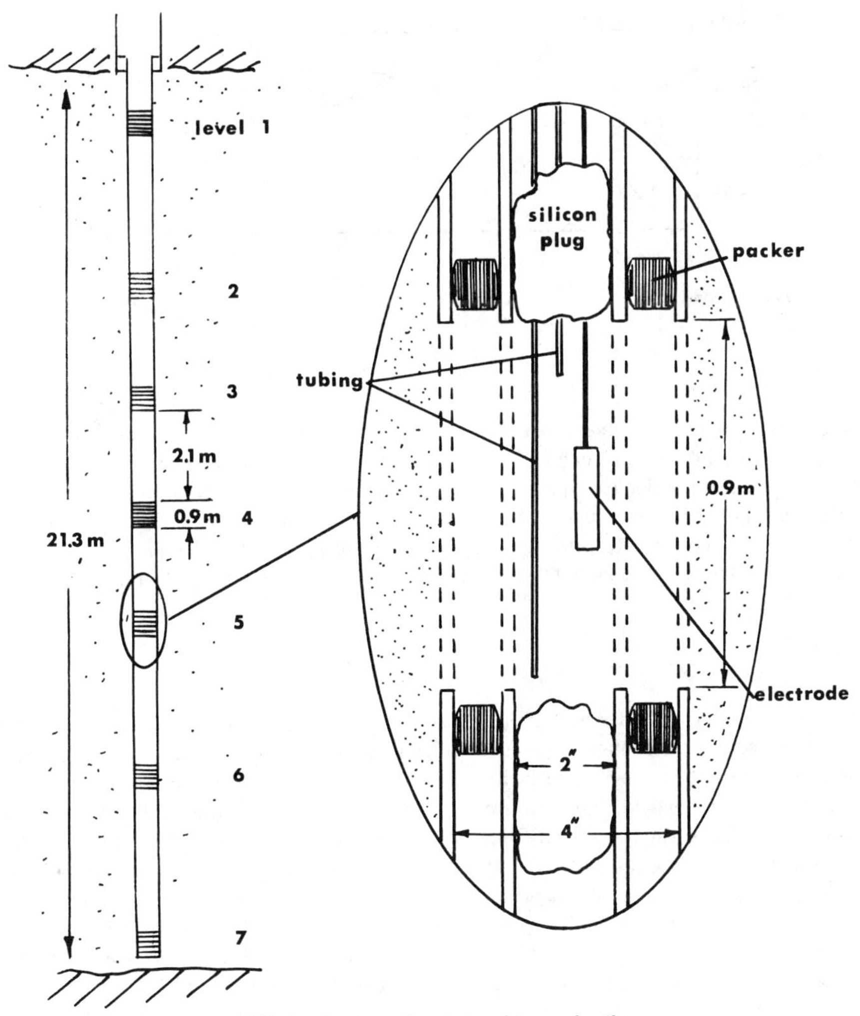


FIG. 3—*Segmented casing and insert detail.*

the ground surface and then lowered into the well. After implacement, the packers were inflated with water with approximately 207-kPa (30-psi) additional pressure imposed at the surface, where the pressure could be monitored.

The vacuum tubing, connected to peristaltic pumps at the surface, was used for two purposes. It provided a means to collect water samples which could eventually be tested for tracer concentration in the laboratory. Also, with two tubes in each zone, mixing within the sampling zone was accomplished by continuous circulation of water from the zone, through the tubing, and back into the sampling zone. Temporarily interrupting the mixing circulation, samples were obtained from the injection tube at the surface. The purpose of mixing was to eliminate any concentration gradients within the sampling zone.

The conductivity electrode shown in the sampling zone in Fig. 3 was alternatively installed in the recirculation system at the surface for some of the observation wells. At the surface the

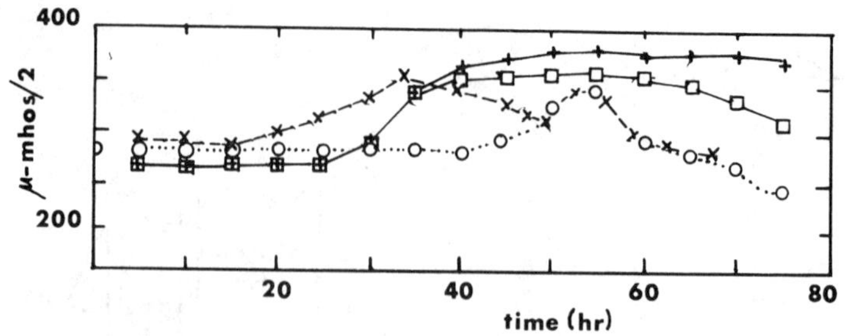


FIG. 4—*Level 7 conductivity response at Well E5: Partial penetration, E1:* × = *Experiment 1, open well bore;* ○ = *Experiment 2, packers. Full penetration, E1:* + = *Experiment 3, packers;* □ = *Experiment 4, packers and mixing.*

electrodes were more accessible, but subject to temperature fluctuations not present down in the well. Because of background conductivity differences between native and injected water and because of inconsistent electrode performance, conductivity data were used only qualitatively to alert the operators of tracer arrival at observation levels so that sampling rates could be increased. At this stage of experience with these field tests, there seems to be no simple, dependable, and accurate substitute for the tedious task of collection, labeling, and precise laboratory analysis of water samples.

## Selected Results

The primary purpose of the field project was to improve understanding of the transport of a conservative tracer in the confined aquifer. Detailed presentation and analyses of the measurements in regard to transport may be found in Refs *4* and *5*. The following results were selected from several of the experiments to demonstrate observation well behavior, successful and unsuccessful.

Complete results of single-well experiments (Experiments 1, 2, 3, and 4) are presented in Ref *3*. A partially penetrating injection zone (upper 44% of the aquifer) was used in Experiments 1 and 2 to impose vertical head gradients larger than any which would exist in all the later experiments, where fully penetrating injection wells were used. If the vertical gradients imposed with the partially penetrating injection did not cause interference between the sampling zones, then it was assumed there would not be interference with the fully penetrating tests.

In Experiment 1, the electrodes were simply suspended in the 10-cm (4-in.) casing without the packers and insert installed. All levels exhibited the behavior as shown for Level 7 in Fig. 4. All breakthroughs began at approximately 35 h. This uniform response could be explained as a result of homogeneous aquifer properties or mixing in the well bore. Later experimental results indicated that the aquifer was not homogeneous and that rapid vertical mixing had occurred in the open well bore. In Experiment 2 the insert and packer system was installed. Breakthroughs occurred at Levels 2, 4, and even slightly at Level 7 as shown in Fig. 4. It was concluded that the packer system did isolate the sampling zones. Also, a complex flow field had apparently been established with the partial penetration of the injection well.

In Experiments 3 and 4, the injection well was altered from partial to full penetration. In these experiments distinct breakthroughs occurred at all levels. Level 7 results are shown as an example in Fig. 4. The difference between Experiments 3 and 4 was that mixing within the sampling zone was accomplished in Experiment 4 by continuous circulation through the sampling tubes through the experiment.

Single-well experiments, Nos. 5 and 7, were conducted at the same injection (E1) and observation (E5) wells (see Table 1). The reason for the repetition was that incomplete concentration

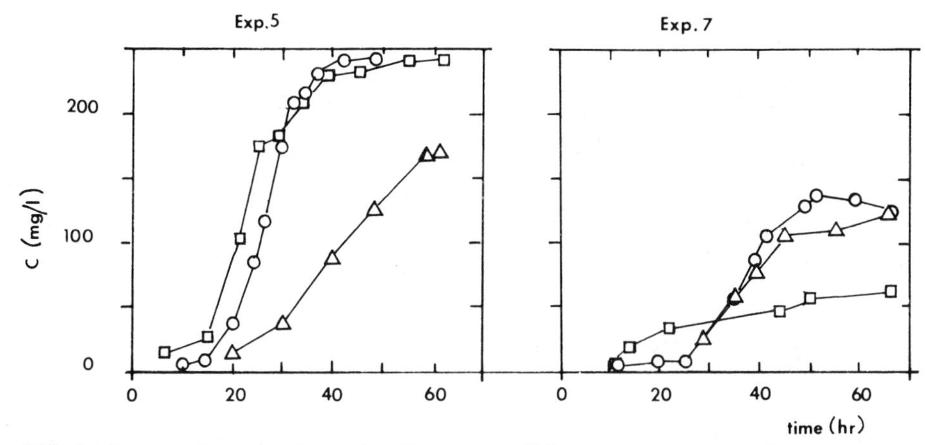


FIG. 5—*Concentration on breakthrough at observation well for Experiments 5 and 7:* ○ = *Level 1;* △ = *Level 3;* □ = *Level 5.*

breakthroughs were observed in Experiment 5 at some levels as shown for Level 3 in Fig. 5. There was more complete behavior at Levels 1 and 5 as shown where the concentration quickly rose to equal the injection concentration. The behavior at Level 3 indicated restricted flow to the sampling region. The low flow rate could be explained by a natural low permeability lens at that level or could also be explained by drilling mud in the observation well or in adjacent formation.

After termination of Experiment 5, the 5-cm (2-in.) PVC insert from the observation well E5 was removed. Significant amounts of drilling mud were found adhering to the insert and inflatable packers. The well E5 [segmented casing of solid and slotted 10-cm (4-in.) PVC] was then flushed using compressed air injection, and a new insert was installed before conducting Experiment 7. With the exception of Level 5, all levels in the repeated experiment showed complete breakthrough as shown in Fig. 5. Levels 1 and 3 are shown as examples. This change indicated that drilling mud indeed was the problem in Experiment 5. The behavior of Level 5 in Experiment 7 was not consistent and can only be attributed to some unknown experimental error. These data, of course, were not used in any analyses described in Refs *4* and *5*. The fact that an expensive experiment had to be repeated indicates how adequate cleaning is important.

Experiment 10 was a single-well experiment with an average injection concentration $C_0$ = 138 mg/L. In this experiment some of the packers failed at $t$ = 25 h. Although the injection well was fully penetrating, unlike Experiment 2, and strong vertical head gradients would not be expected, there was remarkable response at the time of packer failure. As shown for the high-velocity zone, Fig. 6, Level 2, the breakthrough concentration which occurred before the failure was well defined and increased to the injection concentration as expected. However, in a lower velocity zone, Level 1, the breakthrough concentration started to increase but then suddenly decreased and fluctuated after $t$ = 25 h. This behavior, coinciding with loss of packer pressure, can only be attributed to vertical mixing in the well or to flooding of the sampling region with water leakage from the packers.

## Discussion

Although limited data are available based on these experiments, it appears that mixing within an isolated sampling zone is desirable. Conductivity measurements were slightly more erratic when mixing was not imposed in Experiment 3. From a general overview, mixing also seems desirable.

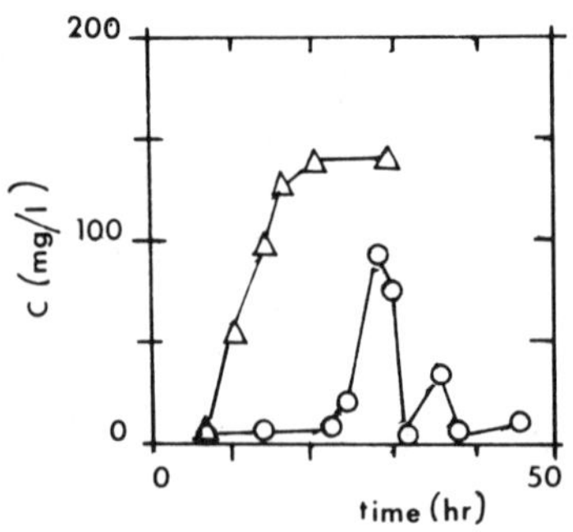


FIG. 6—*Concentration breakthrough for Experiment 10, packer failure at* t = *25 to 30 h in observation well E9:* ○ = *Level 1;* △ = *Level 2.*

For a sampling zone of finite length it is possible for the tracer to enter the zone anywhere along the slotted length and then be recorded depending on unknown natural mixing and probe position. Imposed mixing forces an integration effect known to represent the entire length of the sampling zone. Without imposed mixing, the integration length is really unknown. Migration away from the probe of buoyant or dense tracers could also be reduced with imposed mixing.

Sampling zone lengths for observation wells should be chosen to reflect the parameter variation in the aquifer. Since the variation is really unknown prior to experimentation, detail of experimental design does involve judgment. Usually the number of sampling zones is limited by data acquisition facilities at the surface. It is probably better to err with sampling zones that are so long they disperse fast-moving tracer fronts, rather than with short zones, which could miss fast-moving fronts entirely.

Observation well installation does create considerable disturbance to the natural aquifer structure. For the two types of observation wells constructed at the site, the aquifer-well interface is probably approximated by the two drawings of Fig. 7. For the segmented casing, Fig. 7*a,* there was likely to be drilling mud remnant adjacent to the solid casing. This barrier to vertical flow is very desirable. For the fully slotted casing, very little mud remains after development and a disturbed aquifer material of possibly higher permeability would be adjacent to the casing. As shown in Fig. 7*b,* this disturbed layer may offer an opportunity for significant interference between sampling zones. For aquifers, particularly water table aquifers, with any vertical head gradients this layer would act like an open conduit for vertical flow and completely invalidate any attempt to measure vertical stratified parameters in the natural system. Elimination of vertical flow is an important objective to observation well design.

To insure vertical integrity, some testing to be conducted after well installation is probably desirable and should be considered in the design phase of observation well development. With vertical head gradients larger than those occurring naturally at the site, there would likely be vertical seepage, especially with the fully slotted casing and insert. If there is any question about vertical integrity during monitoring, then the multilevel data will be questionable.

*Acknowledgments*

This work was supported in part by the U.S. Environmental Protection Agency under Assistance Agreement CR 810704-01-0 to Auburn University. It has not been subjected to the agency's required peer and administrative review. Therefore, it does not necessarily reflect the views of the agency and no official endorsement should be inferred. Special thanks are due to field engineers Steve Nohrstedt and Jon Overholtzer for their efforts. At Auburn University the project was administered by the Water Resources Research Institute.

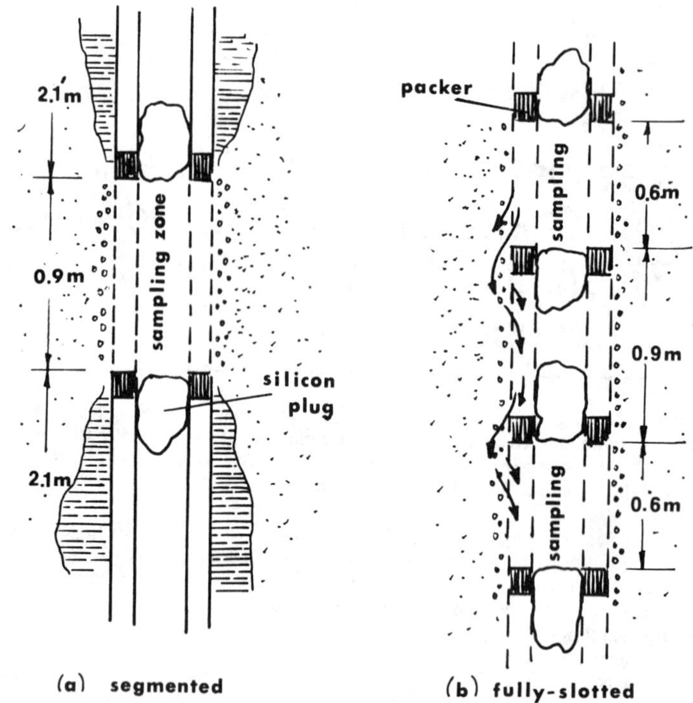


FIG. 7—*Detail of aquifer casing insert for observation well:* (a) *casing and insert segmented-slotted an solid pipe, 3-ft sampling zones with 7-ft intervals;* (b) *casing and insert fully slotted, 2-ft sampling zones wi 3-ft intervals (1 ft = 0.3048 m).*

## References

[*1*] Güven, O., Molz, F. J., and Melville, J. G., "An Analysis of Dispersion in a Stratified Aquifer," *Resources Research,* Vol. 20, No. 10. 1984, pp. 1337–1354.

[2] Pickens, J. F. and Grisak, G. E., "Scale Dependent Dispersion in a Stratified Granular Aquifer," *Resources Research,* Vol. 17, No. 4, 1981, pp. 1191–1211.

[*3*] Molz, F. J., Melville, J. G., Güven, O., Crocker, R. C., and Matteson, K. T., "Design and Perf of Single-Well Tracer Tests at the Mobile Site," *Water Resources Research,* Vol. 21, No. 10, 1 1497–1502.

[*4*] Molz, F. J., Güven, O., Melville, J. G., Crocker, R. C., and Matteson, K. T., "Performance, and Simulation of a Two-Well Tracer Test at the Mobile Site," *Water Resources Research,* V 7, 1986, pp. 1031–1037.

[*5*] Güven, O., Falta, R. W., Molz, F. J., and Melville, J. G., "Analysis and Interpretation of Tracer Tests in Stratified Aquifers," *Water Resources Research,* Vol. 21, No. 5, 1985, pp.

*Kevin A. Morin*[1] *and John A. Cherry*[2]

# Field Investigation of a Small-Diameter, Cylindrical, Contaminated Ground-Water Plume Emanating from a Pyritic Uranium-Tailings Impoundment

**REFERENCE:** Morin, K. A. and Cherry, J. A., **"Field Investigation of a Small-Diameter, Cylindrical, Contaminated Ground-Water Plume Emanating from a Pyritic Uranium-Tailings Impoundment,"** *Ground-Water Contamination: Field Methods, ASTM STP 963,* A. G. Collins and A. I. Johnson, Eds., American Society for Testing and Materials, Philadelphia, 1988, pp. 416–429.

**ABSTRACT:** As part of a larger study on the geochemical behavior and computer simulation of subsurface seepage from pyritic tailings impoundments, a detailed field study of such seepage has been conducted. At the field site, the tailings lie over a glaciofluvial sand aquifer. Near the southeastern corner of the tailings, acidic contaminated water (pH 4) seeps through the base of the tailings into the sand aquifer and flows laterally to the south as a distinct, roughly cylindrical plume. This plume contains up to 6000 ppm iron, up to 14 000 ppm sulfate ($SO_4$), and elevated levels of other contaminants. The acidic seepage eventually encounters carbonate minerals in the aquifer, pH is then neutralized, and contaminants are attenuated through chemical precipitation, co-precipitation, and adsorption over a distance of several metres.

At the edge of the impoundment, the ground-water plume has a cross-sectional area of about 3 $m^2$. Although the cross-sectional area suggests the plume is rather unimportant, the ground-water velocity is about 440 m/year, resulting in a lateral flux of more than $10^6$ L of contaminant-laden water each year. The multilevel bundle piezometers installed to define this small-diameter plume are described and the installation techniques used to minimize disturbance of the plume are also addressed. Each bundle contains up to twelve individual piezometers with a vertical screen spacing of as small as 0.5 m. Horizontal distance between bundles is as small as 2.5 m.

The high total dissolved solids (TDS) water in the plume presents many problems for field and laboratory geochemical measurements. High partial pressures of carbon dioxide ($CO_2$) in the water result in rapid degassing and subsequent pH fluctuation upon sampling. Because of the high degree of temperature-sensitive aqueous complexing in the water, temperature variations can affect pH measurements. Mixing of air with the water initiates iron oxidation. To acidify and stabilize a water sample, 30 mL of concentrated hydrochloric acid (HCl) must be added to each litre of sample, lowering the pH to about 1.5. Laboratory analysis of these samples often requires significant dilution, and the addition of neutral-pH and alkaline reagents to the sample causes ferric hydroxide precipitation and subsequent instrument clogging. Techniques are described to circumvent many of these problems and to collect reliable physical and chemical hydrogeologic data on this site.

**KEY WORDS:** ground water, water contamination, piezometers, water sampling, uranium tailings

n order to define the physical and chemical hydrogeology in and around uranium-tailings oundments, the Institute for Groundwater Research at the University of Waterloo initiated field lies in the Elliot Lake uranium district of Ontario. The main field site was the pyritic Nordic n impoundment (Figs. 1 and 2), which lies on a glaciofluvial sand aquifer (Fig. 3). An initial

orwijk Enterprises, Suite 1706L, Laurier House, 1600 Beach Avenue, Vancouver, British Columbia, da V66 1Y6.

stitute for Groundwater Research, University of Waterloo, Waterloo, Ontario, Canada N2L 3G1.

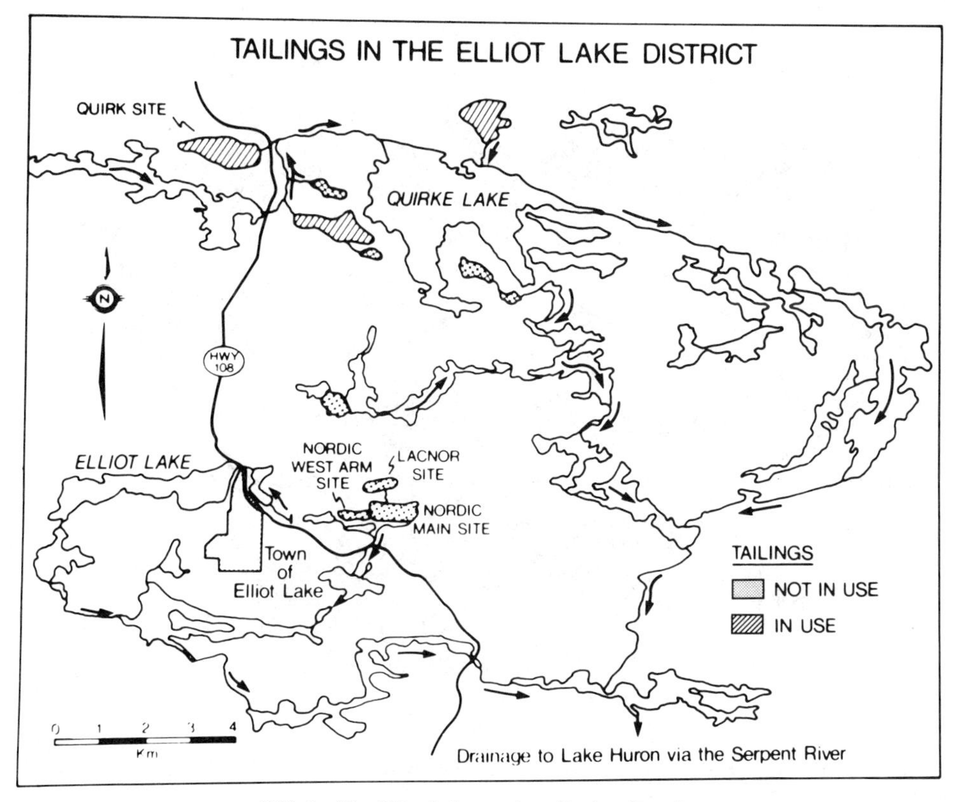


FIG. 1—*The Elliot Lake uranium district, Ontario.*

ground-water survey in 1978 indicated acidic, contaminated seepage was entering the aquifer and flowing southward in the region labeled "Seepage Area A" on Fig. 2 [*1*]. Beginning in 1979, Area A was instrumented with piezometers and a roughly cylindrical, acidic plume of small diameter was defined near the impoundment dam. Through an iterative approach, the seepage area was further instrumented and monitored until late 1984. This temporal and spatial monitoring, in addition to information from other tailings sites, allowed the definition of contaminant migration, the delineation of processes controlling contaminant concentrations, the formation of a general conceptual model of behavior, and the computer simulation of past, present, and future migration [*1–7*].

Because of the unusual nature of the seepage—for example there are several thousands of ppm iron at neutral pH—installation and monitoring were carefully defined and implemented while emphasis was placed on relatively low-cost materials. Through an iterative approach, the techniques were modified until good-quality data were obtained.

This paper presents the requirements identified at the site for installation and monitoring, the techniques employed to fulfill the requirements, and the rationale for the techniques. Also briefly presented are techniques used for monitoring the tailings.

## Piezometer Design and Installation

Because the plume has a cross-sectional area of about 3 m² near the impoundment dam in Seepage Area A, a closely spaced three-dimensional network of piezometers was required.

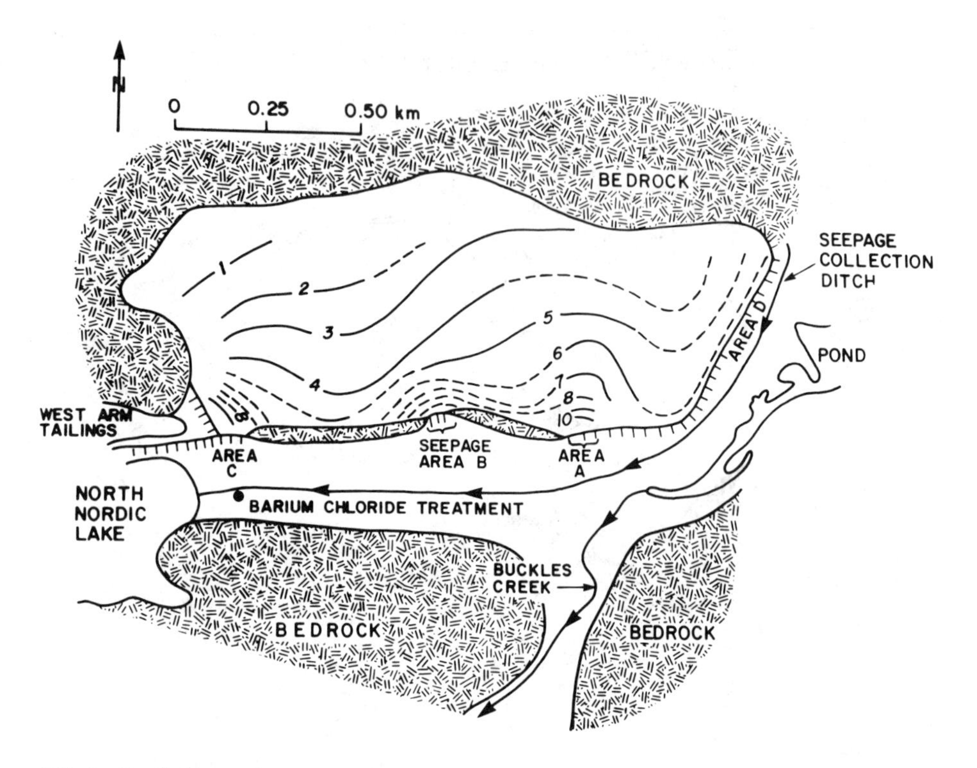


FIG. 2—*Nordic Main tailings impoundment area. Contours in the tailings indicate depth to water table in meters.*

Additionally, short screens and small diameters were desired in order to obtain depth-specific samples and minimize standing water in the pipe. Consequently, multilevel bundle piezometers were chosen (Fig. 4) [*8*]. The bundle piezometers at the field site, marked "M" on Fig. 5, consist of up to twelve individual minipiezometers, terminating at different depths, attached to a polyvinyl chloride (PVC) centerpipe for support. The limit of twelve is the consequence of the inside diameter (approximately 8 cm) of the hollow-stem augers discussed below. Each individual minipiezometer in a bundle consists of flexible 0.9-cm-inside-diameter polyethylene tubing with a

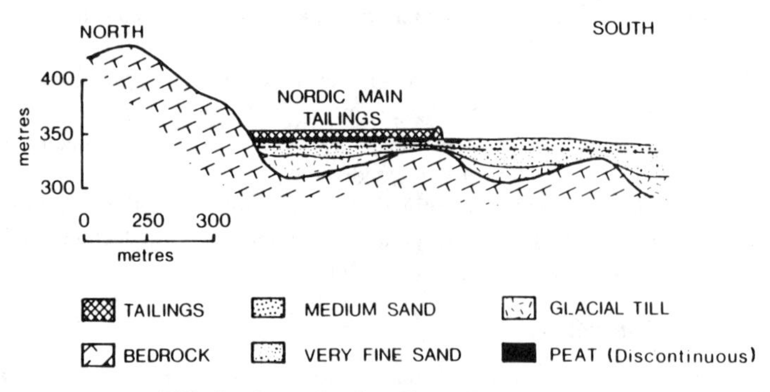


FIG. 3—*Generalized north-south cross section.*

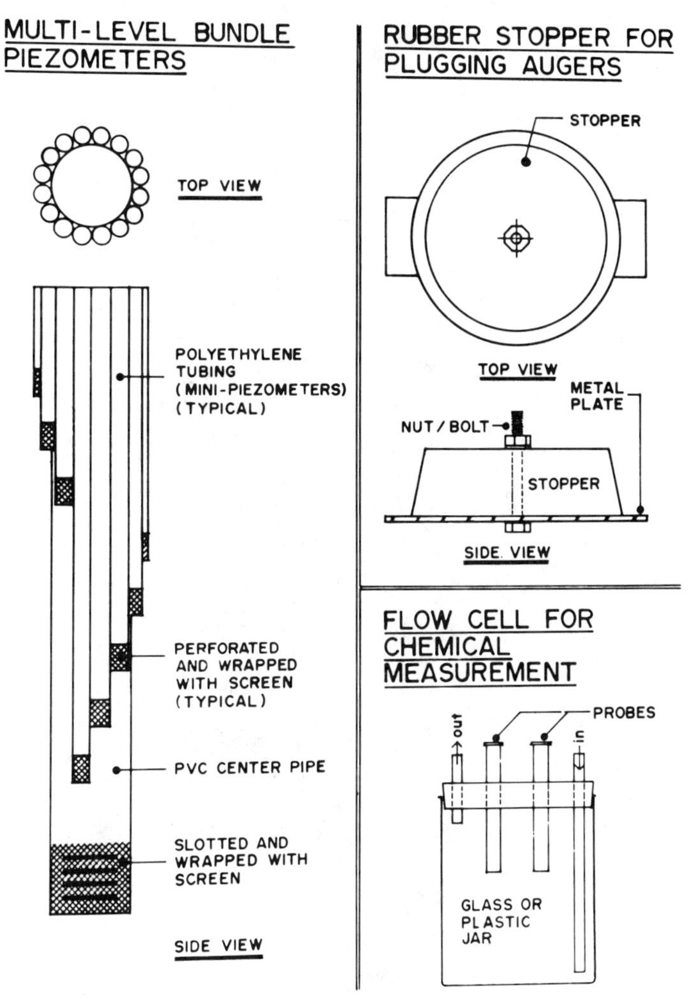


FIG. 4—*Schematic diagrams of bundles, rubber stopper, and chemical flow cell.*

10-cm-long, 0.9-cm-inside-diameter polyethylene screen at the base. The screen was made by plugging the end of the tube and drilling holes through the polyethylene tube. The perforated tube was then wrapped in synthetic screen material; initial screen wrap was probably nylon while later wrap was polyethylene. The base of the PVC centerpipe was similarly plugged, drilled, and wrapped. The screens, tubing, and PVC pipe were brought to the field site unassembled.

The glaciofluvial sand in Seepage Area A is cohesionless below the water table, at a depth of about 1 to 3 m. As a result, retrieval of a relatively undisturbed sample was extremely difficult (See later section on solid sampling). In order to provide a stable opening for the installation of piezometers, either casing or hollow-stem augers were required. Because the sand was capable of flowing upward into either of these alternatives and nearly filling them within several seconds, a bottom plug was required. The faster method of hollow-stem augers was chosen and a laboratory-grade rubber stopper plugged the bottom of the augers. A long narrow metal plate was attached across the face of the stopper and held in place with a stainless-steel bolt through the center of

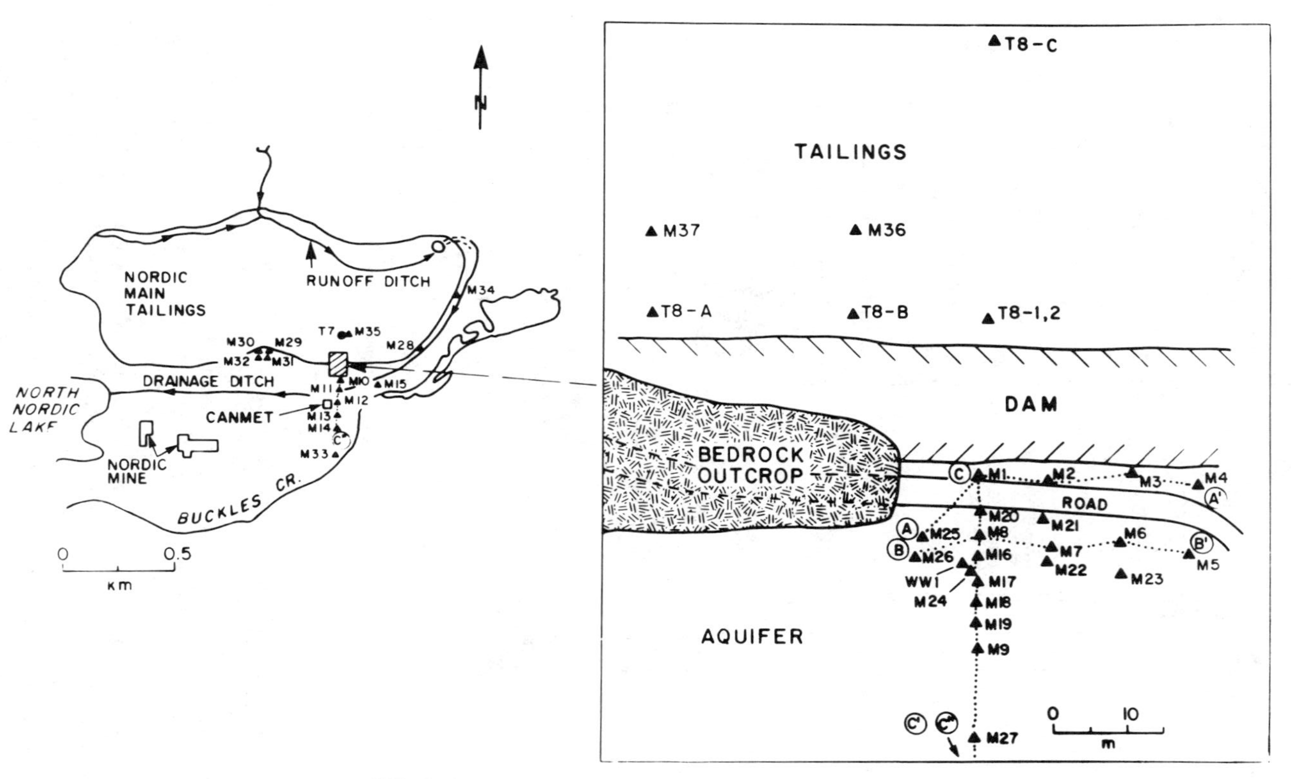


FIG. 5—*Location of bundle piezometers in Seepage Area A.*

the stopper (Fig. 4) in order to prevent it from moving up into the augers during drilling. Rotation of the augers was minimized so that minimal disturbance of the sand occurred. During augering, the depth of each minipiezometer was decided with vertical spacings ranging from 0.5 to 3 m, the materials assembled, and the tubing attached to the PVC centerpipe with nylon-reinforced tape. When the desired total depth of the centerpipe was reached, the bundle was inserted into the center of the auger string. As the string was pulled upwards by a cable and winch, the bundle was repeatedly slammed down onto the stopper until the stopper popped out. The augers were then pulled up around the bundle and the sand quickly and tightly collapsed around the piezometers. No sand-gravel packs and seals were required. Because no foreign fluids were used for augering, typical well development was not necessary.

A PVC well of about 4 cm diameter and 6 m length was installed at WW1 (Fig. 5). The lower 3 m of well was screened with a slot size of about $d_{10}$ of the sand. Installation techniques were those used for the bundle piezometers.

## Collection of Physical Hydrogeologic Data

Bundle piezometer installations were allowed to stabilize for at least one week prior to data collection. Because the sand collapsed tightly around the bundles, no open hydraulic connections were likely created along piezometer pipes between various levels in the aquifer. This conclusion is supported by water chemistry (below) which shows differences in certain locations over a vertical distance of 0.5 m. Physical, as well as chemical, data are compiled in Refs *2* and *5*.

Water levels were measured with a narrow coaxial cable attached to a meter, buzzer, or light. Water levels indicated that flow in the sand was essentially horizontal toward the south with a westward component of flow as water flowed around the bedrock outcrop (Fig. 5). This information was essential for determining the position of flow lines (line C-C′ on Fig. 5) along which chemical changes and contaminant migration could be followed. Finite-element simulation and sensitivity study of flow along the vertical cross section of C-C′ were performed [*5*].

Hydraulic conductivity of the sand was measured by several techniques. A pump test was performed 1.5 km to the west of Seepage Area A, yielding $K = 1.4$ to $2.2 \times 10^{-4}$ m/s; recovery of an observation well yielded $1.2 \times 10^{-4}$ m/s, and permeameter tests on disturbed samples from Area A yielded $2.1 \times 10^{-4}$ m/s for the upper portion of the sand where the plume is located [*1*]. In order to obtain an undisturbed conductivity, a borehole dilution test was performed to obtain an undisturbed velocity from which conductivity was calculated using measured gradients and porosities.

The borehole dilution apparatus (Fig. 6) employed a double packer assembly which was used to isolate a 0.3-m section of piezometer WW1. The circulation tubing was of small diameter and short length to minimize the volume in the circuit. The intended injection solution was distilled water, but it was believed this would not provide a sufficient electrical conductivity contrast. As a result, a concentrated sodium chloride (NaCl) solution was injected. The results of the test [*5*] indicated a velocity of $1.4 \times 10^{-5}$ m/s (440 m/year) which led to a hydraulic conductivity of $1.0 \times 10^{-4}$ m/s in excellent agreement with other values. The velocity measurement indicated that, although the plume cross section is only about 3 m$^2$, the plume represents a volume flow of more than $10^6$ L of water each year.

Grain-size analyses of the sand were also used to calculate hydraulic conductivities by the method of Masch and Denny [*9*], which includes $d_5$, $d_{16}$, $d_{84}$, and $d_{95}$ in the calculation. The mean of four analyses was $1.2 \times 10^{-4}$ m/s.

## Collection of Chemical Hydrogeologic Data

There appears to be no artificial vertical hydraulic connection created in the aquifer by piezometer installation based on observed sand collapse and sharp changes in chemical concentrations between

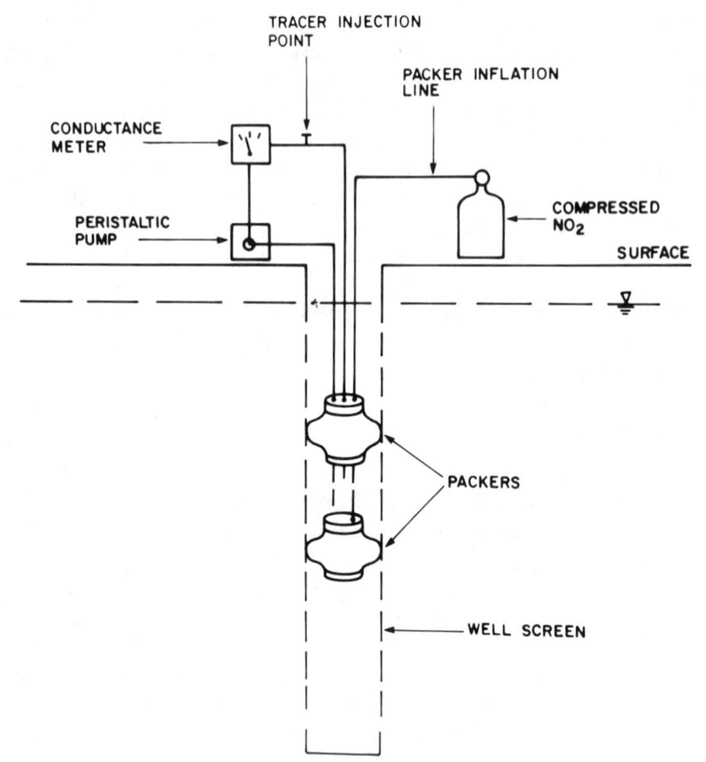


FIG. 6—*Borehole dilution equipment.*

vertically adjacent screens. Measured chemical parameters and observed ranges of concentrations in Seepage Area A are listed in Table 1. Detailed data are compiled in Refs. *2* and *5*.

The general approach for collecting data followed the recommendations of Environment Canada [*10*]. Environment Canada recommendations are partially based on those of the U.S. Environmental Protection Agency.

Because the water levels in piezometers located in Seepage Area A are within 3 m of the surface, water was pumped directly from the piezometers by simply attaching a pump to the top of the pipe. In this way, the piezometers also represented dedicated sampling tubing, thereby minimizing cross-contamination of samples.

Contamination of water samples by the pump mechanism and atmosphere was minimized by the use of a peristaltic pump. The pump was connected to piezometer tops through silicone tubing and polyethylene connectors. In most cases, an airtight seal was obtained at the piezometer top and the ground water thus remained isolated from the atmosphere as it passed through the piezometer screen, tubing, and pump.

Duration of pumping and pumping rate from any one piezometer in a bundle was minimized so that (1) degassing by vacuum of the high $CO_2$ levels was minimized and (2) water was not drawn from a higher or lower level in the aquifer; otherwise the sample would not be representative of conditions at the pumped piezometer. In past years, this effect was noted during pH measurements after 10 to 15 min at a pumping rate of about 0.5 L/min. The lowering of the water table around a bundle has also been noted when shallow piezometers are pumped (Davé and Lim, personal communication). As a general rule, no piezometer in Area A was pumped for more than 10 min

TABLE 1—*Parameters and concentration ranges measured in Seepage Area A (Concentration in ppm unless noted).*

| Parameter | Concentrations | Parameter | Concentrations |
|---|---|---|---|
| Field pH | 3.20—6.81 pH units | Cl | 170—5.5 |
| Lab pH | 2.4—6.9 pH units | F | 33—<0.1 |
| Field Eh | +114—+444 mV | $NO_3$ | 18—<0.4 |
| Field temperature | 10.5±1.5°C | $NH_3$ | 55.9—0.33 (as N) |
| | | $PO_4$ | <0.04 |
| Field conductivity | 11500—380 μS/cm | V | <1 |
| Fe | 5930—8 | Se | <1 |
| Ca | 860—40 | | |
| Mg | 702—12 | DOC | 8.0—22 (as C) |
| Na | 152—1.2 | Ra-223 | 83—6 pCi/L |
| Mn | 56—0.91 | Ra-226 | 213—1 pCi/L |
| Al | 206—<0.02 | Pb-210 | 110—0.1 pCi/L |
| Cu | 0.44—<0.001 | Th-227 | 600—6 pCi/L |
| Co | 5.9—0.01 | Th-228 | 9.7—<0.5 pCi/L |
| Zn | 0.50—0.004 | Th-230 | 48—<0.5 pCi/L |
| Pb | 0.9—<0.003 | Th-232 | 11—<0.5 pCi/L |
| Ni | 6.0—0.015 | Total Th | 96—11 μg/L |
| Cr | 0.14—<0.01 | Ac-227 | 319—16 pCi/L |
| Cd | <0.005 | U-234 | 204.5—14.9 pCi/L |
| | | U-238 | 166.5—3.8 pCi/L |
| As | 0.011—<0.001 | U-234/U-238 | 1.2—3.4 |
| $SO_4$ | 14420—111 | | |
| Field alkalinity | 290—1 (as $HCO_3$) | Total U | 8500—0.4 μg/L |
| DIC | 86.4—21.6 (as C) | O-18 | −11.9 to −10.2 0/00 SMOW |
| $SiO_2$ | 77.0—<1 | H-2 | −84 to −69 0/00 SMOW |
| | | C-13 | −20.8 to −19.2 0/00 PDB |

at 0.5 L/min, resulting in the removal of 5 to 40 piezometer volumes, depending on the depth of the screen. Because pH values occasionally required more than 5 min to stabilize during pumping, the field study was broken into two independent tasks: collection of pH and other data, and collection of water samples.

Field pH was measured by continuously directing pumped water into a covered 150-mL container ("flow cell") containing a combination pH electrode which was inserted through a hole in the cover (Fig. 4). Water passed from the flow cell by flowing through an opening or exit tube in the cover. In this way, water in the flow cell was constantly being replaced and refreshed, and $CO_2$ degassing and atmospheric disturbances of pH measurements were minimized. The flow cell was kept in a water bath, consisting of the water that continuously flowed from the cell, to minimize effects of temperature fluctuation on pH measurement. Temperature fluctuations were usually less than 3°C.

The pH probe and meter were calibrated in nominal pH 7 and pH 4 buffers, which were replaced at least once every 24 h and were kept in the water bath. Buffers at ground-water temperatures minimize temperature-equilibration time of electrodes. At ground-water temperatures, the pH 7 buffer was about 7.06 and that of the pH 4 buffer about 4.00. For the first day of measurements in each field season, the pH meter was calibrated with buffers about once an hour. After the stability of the equipment was established, calibration checks were made about once every 2 to 5 h. Calibration checks often did not deviate by more than 0.04 pH units from the buffer value. In one extreme case, the pH 7.06 buffer read 7.18 after five pH measurements and 30 min time; the meter was readjusted to 7.06 and the pH 4.00 buffer then read 4.00. This indicated the preceeding five ground-water pH values could probably be adjusted by a constant value ($-0.12$ pH units in this case) rather than a pH-dependent value (see Ref *11* for further discussion of pH meter behavior) although it was not clear when the shift of 0.12 units occurred over the 30 min. For all pH measurements, water was pumped at a rate of about 0.5 L/min through the flow cell until a drift of less than 0.02 pH units/min was noted and the value was then recorded as field pH. Approximately 5 min of pumping was often required before the pH stabilized. The pH probe and cell were rinsed with distilled water between each measurement.

Eh was measured with a combination platinum electrode having a 1.3-$cm^2$ platinum button. The electrode was standardized with Zobell's solution and inserted into the flow cell with the pH electrode. The reliability of the resulting Eh measurements was evaluated following the suggestions in Ref *12*.

Electrical conductivity of the water was measured by taking a sample of about 100 mL in a small container after pH and Eh had been recorded. Because the cell constant of the conductivity cell varied through the field study, the cell constant was measured at each bundle with potassium chloride (KCl) standards. Conductivity was later corrected to 25°C and recorded as $\mu$S/cm. However, because conductivity measurements are insensitive to neutral-charge aqueous complexes and because up to half of the aqueous ions in the plume water exist as neutral complexes, electrical conductivity was not considered a reliable quantitative parameter in this study.

The second phase of the field study, which began after all pH, Eh, and conductivity measurements had been taken, involved water sampling and field alkalinity titrations. Fifty to 100 piezometers were selected each year for water sampling and subsequent laboratory analysis. Water was pumped via a peristaltic pump through a 142-mm disk filter assembly, similar to that described in Ref *13*, containing 0.45-$\mu$m nitrocellulosic filter paper. New filter paper was used for each sample, and adsorption of metals onto the filter following initial flushing was considered negligible when compared with concentrations (Table 1). After pumping the piezometer for a few minutes and after flushing the system with about 1 L of ground water to wet and stabilize the filter and to rinse the sample container, the filtered water was directed into a 4-L polyethylene container. After collection the sample was acidified with 30 mL concentrated HCl/L of sample, lowering the pH of the sample to about 1.5. This relatively large volume of acid was required to stabilize the high concentrations and to overcome $SO_4$-$HSO_4$ pH buffering around pH 2. This volume also highlights the weakness of procedures which specify a volume of acid/L sample (for example, 2 mL [*10*]). HCl was chosen over the other acids because (1) its strength minimized acid addition, (2) other acids would cause supersaturation and mineral precipitation by anion addition, and (3) HCl minimized artificial redox reactions in the sample. A separate, unacidified 100-mL sample was collected for chloride analysis only, because chloride is not significantly adsorbed by many precipitated minerals above pH 2. The filter assembly was rinsed with distilled water and a new filter was installed before the next sample was taken.

While the water was filling a sample container, a smaller container was rinsed with distilled water, rinsed with filtered ground water, and then filled with about 50 mL of ground water. This water was titrated for alkalinity using standardized HC1. Hydrochloric acid was again used for the same aforementioned reasons. Although alkalinity is not wholly attributed to $HCO_3^-$, measured dissolved inorganic carbon (DIC) in Area A waters compared well with calculated DIC by assuming the concentration of $HCO_3^-$ was equal to alkalinity [*5*]. Samples for DIC analysis were collected directly from pump tubing in glass syringes. The tips of the syringes were sealed immediately and returned to the laboratory for analysis. Reagents were added directly to the syringes and the resulting gas was directly injected into gas-chromatography equipment. In this way, atmospheric contamination was minimized.

In 1984, five of the 50 sampled piezometers were selected for replicate sampling. Because over 10 L of ground water had to be pumped from each of these five piezometers, water quality was expected to change during sampling as water moved from a higher or lower elevation in the aquifer into the screen. This problem had two major implications. First, because representative water may not be obtained and the plume could be disturbed, only nonstrategic piezometers were chosen. Second, because water quality could change during pumping, once large sample had to be taken and homogenized before splitting into replicates; otherwise comparison of "replicate" analysis could be invalid. Therefore, the following technique was used for each of the five piezometers. A 15-L plastic bucket was rinsed with distilled water, rinsed with diluted HCl (pH 1), rinsed with distilled water, shaken dry, and covered with a sheet of thick styrofoam. Tygon tubing carried water from the filter assembly into the bucket through a hole in the styrofoam. As the water was filling the bucket, the bucket was occasionally agitated to assist in mixing of water, and concentrated HCl was added (30 mL HCl/1 L water). When the bucket was filled, it was gently agitated; then the water was poured into seven 4-L polyethylene collapsible containers approximately 1 L at a time. The containers were marked with various piezometer numbers so that all appeared as an integral part of the sampling program from different locations. Sample distribution for the seven containers was: three to the main analytical laboratory and two each to two other laboratories. Table 2 contains the letter requesting analysis to each of the laboratories, following the recommendations in Ref *14*.

As a general indicator of variation in water quality during pumping of the large 15-L sample, two alkalinity titrations were made during pumping—at the beginning and the end. For the three samples with pH > 5 (alkalinity > 30 ppm $HCO_3$), variation in alkalinity was less than 10% or within analytical accuracy; for the two samples with pH < 5 (alkalinity < 30 ppm $HCO_3$), variation was between 20 and 25%, which is considered the general limit of accuracy at low concentrations. This indicates that there was probably no major variation in water quality during pumping. Nevertheless, the 15-L sample was homogenized before partitioning. Results of this data quality study are given in Ref *2*.

Most samples were delivered to the laboratory within 48 h of sampling and were stored at ambient temperature in the laboratory.

TABLE 2—*Letter requesting analyses.*

The accompanying large bottles contain water samples acidified with HCl to a pH of about 1.5. Please analyze for the following. The samples in the small bottles are unacidified and are for Cl analysis only. [The numbers in brackets are approximate concentrations in ppm.]

Ca,Mg,K,Na [400,20,80,200]
Fe,Mn,Co,Cu,Zn,Ni,Pb [3000,20,2,0.04,0.1,1,0.2]
$SiO_2$ [20]
$SO_4$ [9000]
Cl [10]
$NH_3$ [20]
$NO_3$ [2]
Total uranium [0.003 to 1]
Total thorium (X-ray method preferably) [0.04]
Radium-226 [3 to 150 pCi/L]

Please enclose with the results of analyses a brief description of the method used to analyze each species (one or two sentences only), the frequency of calibration for each method, and the precision for each method.

Send results and details on methods to:

Kevin Morin
Morwijk Enterprises

### Laboratory Analysis, Data Quality, and Speciation

Chemical analysis of Seepage Area A waters was found to be difficult. High concentrations of iron and sulfate and elevated levels of other metals produced interferences. Analysis of anions by column-exchange methods was problematic, because dilution through the addition of pH-neutral reagents initiated iron precipitation, which subsequently coated the resin. Thus, iron had to be removed before analysis. Because of the high levels of sulfate, sulfate was often analyzed by gravimetric methods, probably accounting for the relatively high observed analytical error.

The reliability of overall analysis was evaluated through cation-anion balances:

$$\text{Error} = [(\text{CAT} - \text{AN})/0.5(\text{CAT} + \text{AN})] \times 100\%$$

where CAT is the sum of cations in equivalents per litre and AN the sum of anions in equivalents per litre.

Mean errors for 1980, 1982, and 1984 were +7.98%, +1.10%, and +6.18%, respectively. However, because iron and sulfate are dominant in concentrations, the error is sensitive only to the reliability of their analysis. In order to assess the accuracy and precision of all aqueous constituents, the 1984 study included replicate sampling described in the previous section [2].

Water analyses indicated that significant geochemical processes were operating as the plume water flowed southward from the tailings impoundment. Most contaminant concentrations decreased and pH increased along the flowpath defined by water level data (see earlier section on collection of hydrogeologic data). Because chemical precipitation-dissolution, including pH-neutralizing carbonate-mineral dissolution, was the likely cause of change, water analyses were speciated [*15*]. The speciation program, WATEQ2 [*16*], was used for this study and saturation indices calculated by WATEQ2 indicated which minerals were likely involved in precipitation-dissolution.

### Solid Sampling

Because saturation indices indicated that mineral precipitation-dissolution was probably occurring, undisturbed-unoxidized solid samples were required for analysis. Because mass balance calculations indicated that most reactive minerals would be at concentrations of less than 1 weight %, detection limits of most current methods did not allow identification of reactive minerals [*17*].

Initial disturbed sand samples for grain-size analysis and thin-section examination were obtained from auger flights. Iron-staining on these samples indicated aqueous iron had oxidized and precipitated, thereby creating acidity which could dissolve up to 0.2 weight % of the calcite identified in thin-section examination and through the method of Barker and Chatten [*18*]. Also, iron oxidation suggested that the minor amount of siderite expected to occur probably oxidized to ferric hydroxide [*17*]. Therefore, attempts were made to obtain an undisturbed sample and maintain its isolation from the atmosphere.

Standard split-spoon sampling failed because the cohesionless sand would flow upward into the hollow-stem augers whenever the centerpole was pulled. A split-spoon sample could not be taken from inside the augers because the sampler could jam within the augers. A variation on the standard method, employing the rubber stopper described earlier, failed because sand flowed into the augers upon popping the stopper and, on the first and only attempt, nearly jammed the sampler below the base of the augers.

Other methods were tried. A steel plug with a swing gate was placed on the base of the augers. According to theory, as augering proceeded, the gate remained closed and no sand entered the augers. When desired depth was reached, a sampler was lowered from the surface which completely covered the gate aperture. The auger string was then turned slightly in the reverse direction, causing the gate to open, and sand theoretically flowed into the sampler. This method failed because the coarse matrix sheared the gate off during augering. Driving of long (6 m) thin-walled aluminum pipe with various sample catchers at the base failed because the cohesionless sand would flow past the sample catchers.

In despair, this driving of long pipe was continued without sample catchers until some of the cored sand jammed inside the pipe on one occasion. This core was sealed tightly, returned for analysis, and opened in a nitrogen-filled glove box. Because most relevant analytical methods required long periods of time for preparation or special preparation techniques, thereby allowing sample oxidation, rapid X-ray diffraction scans of the sand were made [*5,17*]. Results were inconclusive because of low concentrations.

## Tailings Installations

Installation and monitoring techniques employed in the tailings were similar to those for the sand aquifer [*4,19*]. However, because of the low hydraulic conductivity, small grain size, and deep water table at some locations, the piezometers were standard open standpipes made of 3.2-cm-inside-diameter PVC rather than multi-level bundles. Screens were of a nonstandard design: perforated PVC wrapped with Vyon (registered tradename of Povair Ltd., United Kingdom, also known as Sinterpore), which is made of polyethylene, has a 50-μm pore size to prevent tailings from entering the piezometer, and has a relatively high conductivity to allow rapid water influx and single-well response tests. The base of the screen was tapered (Fig. 7) to allow percussion driving of the piezometers into the tailings.

Water sampling from the piezometers involved small-diameter polyethylene tubing and a peristaltic pump [*19*]. Where the water table was less than 7 m deep, a steady low volume of water was pumped into the flow cell (see earlier section on collection of hydrogeologic data) and

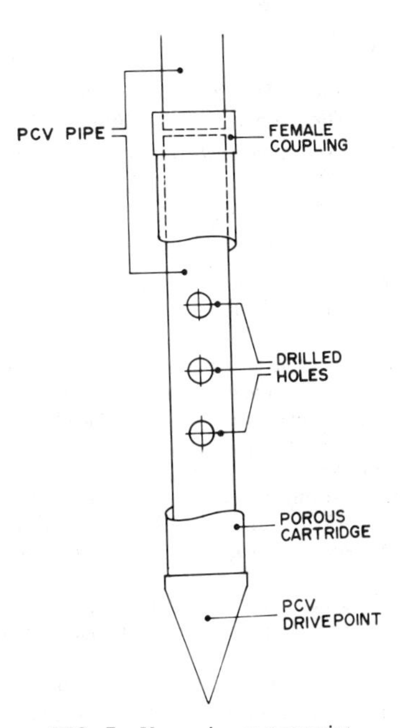


FIG. 7—*Vyon piezometer point.*

sample container. Where the water table was deeper, the polyethylene tubing was used as a bailer by turning the pump on high speed, lowering the tubing to the base of the piezometer, allowing the water to rise as high as possible in the tubing, then pulling the tubing to the surface. This method minimized contact with the atmosphere. Downhole syringe sampling was also used, but offered no significant advantage [*19*]. As with solid sampling in Seepage Area A, tailings were sampled with long, thin-walled aluminum pipe which was driven into the peat layer beneath the tailings, thereby plugging the pipe and allowing sample recovery.

## Summary

The presence of a cylindrical ground-water plume of small diameter in acidic tailings seepage required a detailed monitoring network consisting of multilevel bundles composed of up to twelve minipiezometers in each bundle. Vertical spacing between adjacent screens in a bundle was as small as 0.5 m and horizontal spacing between bundles was as small as 2.5 m. Installation techniques were designed to minimize the physical and chemical disturbance of sand and plume.

The plume water is sensitive to many conditions, such as temperature variation, mixing with the atmosphere, and degassing of $CO_2$. Techniques to minimize these problems during data collection were discussed. Because of the large amount of pH buffering in the water, an unusually large amount of acid was added to samples for proper stabilization. Laboratory analysis of water was difficult because of the addition of pH-neutral reagents, which initiated chemical precipitation, which in turn plugged equipment. Also, less-accurate analytical methods were sometimes substituted to avoid machine problems.

Collection of undisturbed sand samples was difficult. The driving of long, thin-walled aluminum pipe eventually provided a sample.

Techniques for and problems with the tailings were similar. Piezometer screens included a Vyon cover to prevent tailings from entering the piezometer.

*Acknowledgments*

Funds for this study were provided by the Natural Science and Engineering Research Council (Canada) Strategic Grant G0679, The Department of Energy, Mines, and Resources (Canada), Rio Algom Ltd., the Institute for Groundwater Research, and Morwijk Enterprises.

## References

[*1*] Blair, R. D., "Hydrogeochemistry of An Inactive Pyritic Uranium Tailings Basin, Nordic Mine, Elliot Lake, Ontario," M.Sc. thesis, Department of Earth Sciences, University of Waterloo, Waterloo, Ontario, Canada, 1981.

[*2*] Morin, K. A., "1984 Geochemical Study of the Acidic Contaminant Plume Near the Nordic Main Uranium-Tailings Impoundment, Elliot Lake, Ontario," Research Report OSQ84-00229, National Uranium Tailings Program; Energy, Mines and Resources Canada, April 1985.

[*3*] Cherry, J. A., Morin, K. A., and Dubrovsky, N. M., "Modelling of Contaminant Migration In Acidic Groundwater Plumes at Uranium Tailings Impoundments: ADNEUT3," Research Report OST83-00284, National Uranium Tailings Program; Energy, Mines, and Resources Canada, June 1984.

[*4*] Dubrovsky, N. M., Morin, K. A., Cherry, J. A., and Smythe, D. J. A., "Uranium Tailings Acidification and Subsurface Contaminant Migration in a Sand Aquifer," *Water Pollution Research Journal of Canada,* Vol. 19, No. 2, 1984, pp. 55–89.

[*5*] Morin, K. A., "Prediction of Subsurface Contaminant Transport in Acidic Seepage From Uranium Tailings Impoundments," Ph.D. thesis, Department of Earth Sciences, University of Waterloo, Waterloo, Ontario, Canada, 1983.

[*6*] Morin, K. A., Cherry, J. A., Lim, T. P., and Vivyurka, A. J., "Contaminant Migration in a Sand Aquifer Near an Inactive Tailings Impoundment, Elliot Lake, Ontario," *Canadian Geotechnical Journal,* Vol. 19, 1982, pp. 49–62.

[7] Cherry, J. A., Shepherd, T. A., and Morin, K. A., ''Chemical Composition and Geochemical Behavior and Contaminated Groundwater at Uranium Tailings Impoundments,'' Preprint No. 82-114, Annual Meeting, Society of Mining Engineers of the American Institute of Mining, Metallurgical, and Petroleum Engineers, Dallas, TX, Feb. 14–18, 1982.
[8] Cherry, J. A., Gillham, R. W., Anderson, F. G., and Johnson, P. E., ''Migration of Contaminants in Groundwater at a Landfill: A Case Study. Part 2: Groundwater Monitoring Devices,'' *Journal of Hydrology,* Vol. 63, 1983, pp. 31–49.
[9] Masch, F. D. and Denny, K. J., ''Grain-Size Distribution and Its Effect on the Permeability of Unconsolidated Sands,'' *Water Resources Research,* Vol, 2, 1966, pp. 655–677.
[10] ''Sampling for Water Quality,'' Environment Canada, Catalog No. En 37-64/1983E, Minister of Supply and Services Canada, 1983.
[11] Linnet, N., ''pH Measurements in Theory And Practice,'' Radiometer A/S, Copenhagen, 1970.
[12] Morin, K. A., ''Validity of Redox Measurements in Hydrogeologic Studies'' *Proceedings,* Third Canadian Hydrogeological Conference, Saskatoon, Saskatchewan, 21–23 April 1986.
[13] Kennedy, V. C., Jenne, E. A., and Burchard, J. M., ''Backflushing Filters for Field Processing of Water Samples,'' USGS Open-File Report 76-126, U.S. Geological Survey, 1976.
[14] Keith, S. J., Frank, M. T., McCarty, G., and Mossman, G., ''Dealing with the Problem of Obtaining Accurate Ground Water Quality Analytical Results'' in *Proceedings,* Third National Symposium on Aquifer Rehabilitation and Ground Water Monitoring, National Water Well Association, Columbus, OH, 25–27 May 1983.
[15] Morin, K. A., ''Simplified Explanations and Examples of Computerized Methods for Calculating Chemical Equilibrium in Water,'' *Computers & Geosciences,* Vol. 11, No. 4, 1985, pp. 409–416.
[16] Ball, J. W., Jenne, E. A., and Nordstrom, D. K., ''WATEQ2—A Computerized Chemical Model for Trace and Major Element Speciation and Mineral Equilibria of Natural Waters'' in *Chemical Modeling In Aqueous Systems,* E. A. Jenne, Ed., American Chemical Society Symposium Series 93, 1979, pp. 815–835.
[17] Morin, K. A. and Cherry, J. A., ''Trace Amounts of Siderite Near a Uranium-Tailings Impoundment, Elliot Lake, Ontario, and Its Implication in Controlling Contaminant Migration in a Sand Aquifer,'' *Chemical Geology,* Vol. 6, No. 1–2, 1986, pp. 117–134.
[18] Barker, J. F. and Chatten, S., ''A Technique for Determining Low Concentrations of Total Carbonate in Geologic Materials,'' *Chemical Geology,* Vol. 36, 1982, pp. 317–323.
[19] Dubrovsky, N. M., ''Geochemical Evolution of Inactive Pyritic Tailings in the Elliot Lake Uranium District,'' Ph.D. thesis, Department of Earth Sciences, University of Waterloo, Waterloo, Ontario, Canada, 1986.

*Clark Gregory Kimball*[1]

# Ground-Water Monitoring Techniques for Non-Point-Source Pollution Studies

**REFERENCE:** Kimball, C. G., "**Ground-Water Monitoring Techniques for Non-Point-Source Pollution Studies,**" *Ground-Water Contamination: Field Methods, ASTM STP 963,* A. G. Collins and A. I. Johnson, Eds., American Society for Testing and Materials, Philadelphia, 1988, pp. 430–441.

**ABSTRACT:** The results of a ground-water monitoring project are based on samples collected from monitoring wells. The wells must be properly installed, constructed, and placed in accordance with project objectives and goals. Ground-water sampling must be conducted such that a representative sample is consistently obtained. Ground-water monitoring system design considerations include project goals and objectives, geology, hydrologic characteristics, chemical parameters to be monitored, and analysis and evaluation plans. To achieve a high-quality monitoring project with the capability of meeting project goals and objectives, the integration of a number of disciplines is necessary. Hydrogeologists, geologists, chemists, statisticians, engineers, and trained and experienced technical field personnel all provide valuable input.

A Comprehensive Monitoring and Evaluation project designed to detect changes in nitrogen and pesticide concentrations from agricultural activities is underway in eastern South Dakota. The geology and sampling situations encountered vary considerably throughout the project area. Experience gained through this project and techniques and methodologies applicable to ground-water monitoring are presented.

**KEY WORDS:** hydrogeology, ground water, monitoring, wells, well completion, sampling

Ground-water monitoring requires careful planning in order to accomplish the project goals. Those designing a monitoring project must first determine what is to be monitored, what parameters will be analyzed, and what the data analysis or evaluation plans will be. Considerations in proper monitoring system design include how the well is installed, how and of what materials the well is constructed, and where the well is placed. Equally important is proper sampling equipment and techniques.

The South Dakota Department of Water and Natural Resources, in cooperation with the United States Department of Agriculture, is currently involved in a Comprehensive Monitoring and Evaluation project (CM&E). The goal of the CM&E project is to determine non-point-source pollution impacts to ground water from agricultural activities and if the implementation of certain best management practices (BMP's), such as fertilizer and pesticide management and conservation tillage, can reduce nitrogen and pesticide inputs to the ground-water system.

The approach is to monitor ten farm fields (sites) 81000 to 324000 m$^2$ (20 to 80 acres) in size: six to seven working farm fields that employ BMP's; two to three working farm fields that do not employ BMP's; and one unfarmed field. The sites fall into two geologic settings, glacial outwash and glacial till. The outwash is stratified sand and gravel deposited by the meltwater of a stagnant

[1] Project hydrogeologist, South Dakota Department of Water and Natural Resources, Brookings, SD 57007.

or receding glacier, and the till is nonsorted, unstratified clay, silt, sand, and gravel. The project area till is dominated by clay and silt particle-size fractions.

Well installation and construction techniques are dependent upon the geology and need to be flexible with the differing geologies encountered. Finished wells recharge at different rates depending upon the geology. Also, in combination with the range of weather conditions, collecting ground-water samples throughout the year requires a number of sampling techniques.

Ground-water monitoring analysis and evaluation is based on samples collected from monitoring wells, which once installed become fixed sampling points. Improper installation, construction, or placement of these fixed sampling points, or incorrect sampling procedures, compounds the problem of interpreting data already complicated by the natural variability inherent in a ground-water system. No matter how sophisticated the method of evaluation, there is no substitute for reliable data.

The CM&E project goals require strict adherence to high-quality monitoring techniques. Throughout this paper methodologies learned and developed as a result of these requirements are discussed. Figure 1 is a summary flow chart of the activities involved during a non-point-source ground-water monitoring project as they are presented in this paper.

## Premonitoring Information Gathering

After the project goals and objectives have been determined, the initial step is to gather available information regarding the area to be monitored. In non-point-source pollution studies, the land area under consideration is usually large. To adequately monitor a very large area with the detail necessary to resolve project goals, data collection and associated instrumentation requirements would be prohibitive. Smaller areas that can be described and monitored properly must be selected from the original area. Geology should be determined from available literature sources. Determination of the geology of the area cannot be overstated, for it dictates the ground-water flow system and the most likely zones of contamination. In some cases, information in the literature regarding the geology of an area may be limited, as it was for the CM&E project. Therefore, a preliminary drilling program will be needed. From the preliminary drilling, potential sites that are geologically favorable (for example, shallow ground water, sand and gravel, representative soils or specific site qualities needed to meet project objectives) can be chosen from the overall area. The next step is to obtain permission from, and tentative cooperation of, the landowner whose site appears to be potentially favorable.

Once a particular field may become part of the monitoring project, site-specific information must be collected to assure suitability to monitoring goals and to design the monitoring scheme. The site geology needs to be ascertained, geologic cross sections developed, and a triangular system of wells installed to determine depth to ground water and direction of ground-water flow. At this stage, beware of drilling without detailed sampling and logging. Thin layers of contrasting material can play an important role in contaminant migration. Geophysical borehole logging (may be expensive and needs samples for calibration) or split spoon sampling can yield the desired detail. Split spoon or relatively undisturbed samples provide a sample in hand which lends itself to detailed description and is then available for particle-size distribution analysis. Experience gained during the CM&E project has shown split spoon sampling through hollow flight augers to be invaluable in determining site-specific geology with the necessary degree of resolution. One site in particular was extremely confusing based on rotary drilling logs supplemented with only three hollow flight auger (with split spoon) logs. When the remainder of the drilling was finished using only hollow flight auger and split spoon sampling, the geology became clear, and the lack of resolution in the rotary drilling logs, which has been noted by others [*1*], became apparent. When a layer of sand has been penetrated during rotary drilling, sand material continued to come to the surface with the cuttings used for logging, and subsequent thin layers of sand were masked.

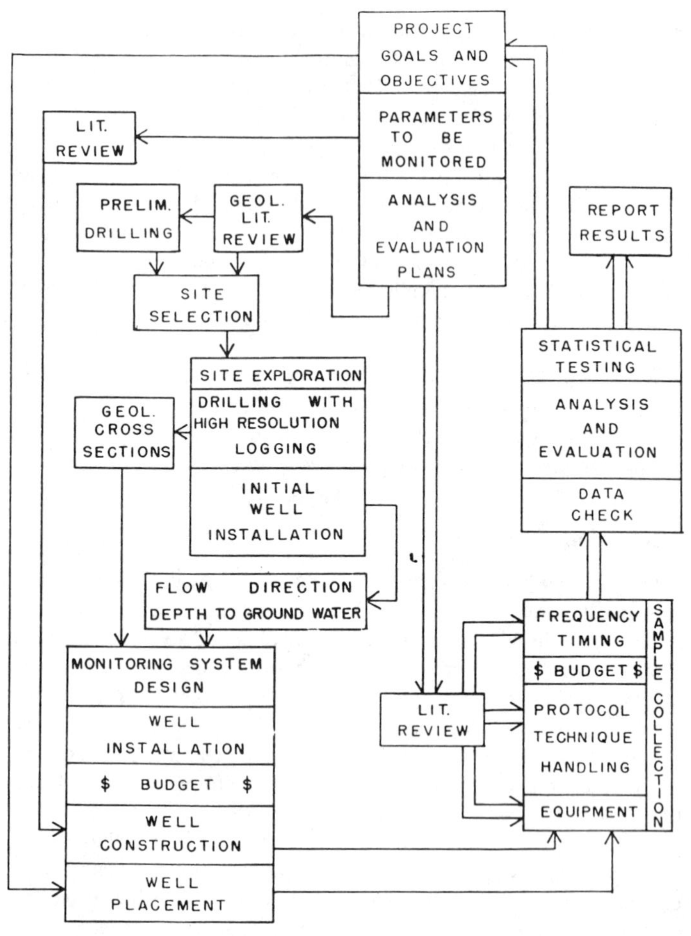


FIG. 1—*Summary flow chart of non-point-source ground-water monitoring project design. Beginning with goals and objectives the flow chart moves once through activities connected by a single line and as many times as necessary through activities connected by double lines.*

## Monitoring System Design

With site-specific geology, depths to ground water, and preliminary direction of ground-water flow information known, the monitoring system design can begin. A literature review of the parameters to be monitored should be done at this time or during the premonitoring information-gathering stage. Determine the current state of knowledge regarding expected concentrations, modes of transport through the ground-water system, the most likely zones of contamination, and what biases or interferences are associated with the parameter. The major concerns of the monitoring system design are well installation, construction, and placement.

### *Installation*

Well installation necessitates a choice of drilling methods. Three principal methods are employed in monitoring well installation: direct rotary, standard flight auger, and hollow flight auger. Of the three, rotary is probably the fastest, but this method requires the use of a drilling fluid to remove the cuttings, cool the bit, and form a mud cake on the walls of the borehole to keep it from collapsing. The drilling fluid is typically a mud, either clay (mostly bentonite) or an organic mud. A study showed that chemical oxygen demand was three to ten times higher in wells drilled with mud than in adjacent wells not drilled with a fluid, even after careful well development and purging prior to sampling. The elevated chemical oxygen demands lasted for 20 to 120 days [2]. Rotary drilling with clear water instead of mud may not keep the hole open. Additionally, the resolution of the logging of a rotary drillhole without additional samples may not be sufficient. Unless there is a reason to drill to substantial depths (not as likely in non-point-source pollution studies) or to penetrate consolidated formations, direct rotary drilling with mud is not recommended. An alternative in consolidated formations might be air rotary, which eliminates the need for mud.

Standard flight auger drilling has the advantage of not needing any drilling fluid and is usually quite mobile. The disadvantage is that often the hole does not stay open for well installation in unconsolidated, noncohesive formations. The resolution of the logging with a standard flight auger rig can be even less than with the rotary rig. In some formations no cuttings come to the surface at all, making logging rather difficult. Because it can drill quite quickly, the standard flight auger rig can be helpful as a supplement to a high-resolution logging drill rig, but for monitoring well installation the standard flight auger rig has some serious limitations.

The hollow flight auger, as the name implies, is an auger with a hollow center that needs no drilling fluid. Well installation can be accomplished through the center of the hollow auger, allowing a nearly exact placement of the well screen. The auger itself does not allow very high resolution logging, but the hollow center allows the introduction of split spoon or continuous samplers with which samples can be brought to the surface for inspection and logging. The CM&E project has been utilizing this method of drilling with a lot of success and a few problems.

A plug or bit plate is inserted in the hollow flight during drilling to keep material from entering the auger. When the plug is removed for a sample to be taken or a well installed, a problem can occur. A differential head exists between the inside and outside of the auger when the auger has penetrated any distance below the water table in a saturated formation of fine to medium sand. When the auger plug is removed, the heads inside and outside the auger equilibrate, and formation material is carried inside the auger. At one drilling location for the CM&E project, the auger plug was removed for sampling, and 5.2 m (17 ft) of material boiled into the auger stem. As a result of material in the auger, samplers either cannot be pushed through, or if they are, cannot be retrieved. Likewise, it may not be possible to install wells through the material, to the desired depth. There are three solutions to this problem:

1. Drill the material out of the hollow flight with clear water (which means rotary capabilities must be available).
2. Introduce water above the plug prior to pulling the plug to create a head that will equalize the head outside the auger.
3. Use a bit plate, which allows formation water to enter the auger but screens out the formation material [3].

The CM&E project has been using Method 2, which introduces a minimal amount of foreign water to the formation. Without having actually tried Method 3, it does seem to be the most desirable since nothing outside of the formation is introduced.

Another problem encountered has been differential collapse of the formation when the auger is removed from around the well pipe during well installation. The auger hole may collapse preferentially on one side if the auger is removed too quickly. This may result in a pipe that is bent, especially at a coupler where pipe sections have been joined. The pipe may have enough curvature that sampling equipment does not fit. The solution seems to be simply to lift the auger slowly enough to allow even collapse of the formation around the pipe.

*Construction*

Well casing materials are the first consideration in well construction and are based on the parameters to be monitored. The materials typically considered are trifluoroethylene, stainless steel, polyvinyl chloride (PVC), and a new material, fiberglass-reinforced epoxy-resin. Trifluoroethylene is the most expensive and PVC the least expensive material. Trifluoroethylene is one of the least reactive and has the least potential for bias or interference, followed by stainless steel and PVC [2]. Recent information on the fiberglass-reinforced epoxy-resin pipe indicates it may be the least reactive material, but all the evidence is not yet in. In many instances PVC is a chemically noninterfering material suitable for monitoring well construction, but in the case of organic compounds PVC may be involved in adsorptive uptake or release (field documentation of this is very limited [2]). It is recommended that PVC wells be constructed without glue, with sections joined with threaded couplers, because the solvent or PVC glue has been shown to contribute significant quantities of various organic compounds to the water [2].

The use of manufactured PVC well screens is also recommended, rather than hand-slotted PVC pipe (which increases freshly exposed surface area), to reduce the risk of organic compound interferences [2]. The CM&E project is monitoring for nitrogen in ground water with PVC wells joined with threaded couplings and manufactured screens. Wells constructed of PVC can be installed at a cost that allows more well installations than trifluoroethylene or stainless steel materials. But since the possibility exists that the wells may be interfering with the concentration of pesticides (organic compounds) in collected samples, a second well of fiberglass epoxy-resin (appropriate for organic compound monitoring and less expensive than trifluoroethylene) will be placed adjacent to some selected wells to verify if the PVC is in fact interfering, and if possible, to quantify to what extent.

The diameter of the well casing should be decided upon concurrently with the anticipated types of sampling equipment. Though many sizes are available, 0.05-m (2 in.) diameter is appropriate. A 0.05-m (2 in.) casing is large enough to accept most of the sampling equipment being produced or developed today while being small enough to allow standing water in the well to be relatively easy to purge prior to sampling.

Gravel pack (sand or gravel materials) is placed along the length of the screen between the well and the borehole, or natural materials (if the natural materials are sand/gravel) are allowed to collapse around the well screen. Whether the gravel pack is emplaced or natural, it is important in improving the intake efficiency of a well. When choosing a gravel pack material the hydraulic conductivity of the gravel pack should not be the limiting factor when hydraulic conductivity testing is performed. In the case of the CM&E project, some of the formations that were screened had zones of very coarse, well sorted gravel. In these cases, pea gravel was used to avoid limiting the hydraulic conductivity of the formation.

The screened area of the well must receive water only from the formation in which it is screened. A seal is placed in the borehole to isolate the screened portion of the well from waters that may migrate along the disturbed materials in the borehole. The material typically used for a seal is bentonite, a high-swell clay largely composed of montmorillonite.

Two schools of thought exist regarding seal placement in a well:

1. Seal for 0.91 to 1.22 m (3 to 4 ft) just above the screen (lower seal) and for 0.91 to 1.22 m (3 to 4 ft) at the surface (upper seal) with spoil materials in between.
2. Start the seal just above the screen and continue it to the surface. Placement of the seal 0.3 m (1 ft) above the screen is a good rule of thumb so that if the gravel pack settles, bentonite is still kept away from the well screen.

The CM&E project uses both methods. In constructing wells in glacial till, the borehole stays open during emplacement of the gravel pack and the seal, and seals are typically started 0.3 m (1 ft) above the screen and continued to the surface. A good seal is particularly important in the till wells because the permeability in the borehole is orders of magnitude higher than in the surrounding undisturbed material.

A seal just above the well screen has been difficult to accurately place in wells in glacial outwash where the sand and gravel collapses around the well screen as the hollow flight auger is withdrawn. The problem with seal placement has been the narrow annular space between the well and the inside of the hollow flight auger. Bentonite powder cannot be used since there is water in the auger and the powder (even the granular which the project uses extensively) will float. Bentonite pellets have been tried. But, even though the diameter of the pellets indicates that they should fit, they become caught and bridge upon reaching the water. A system is now being developed where a bentonite slurry can be pressure-injected through a narrow tube to the depth desired for the seal.

Do not consider a shallow well in sand and gravel unsuitable for ground-water monitoring because the lower seal is absent. A sand and gravel well should not be subject to receiving water outside the desired limits of the screen if it does not penetrate the water table very far, and the pumping rate during sampling is low. A 0.91-to-1.22-m-thick (3 to 4 ft) upper seal should be placed at the surface to preclude the entry of surface water into the well via the borehole.

After the well is in place, it should be developed to remove fine-grained materials in the gravel pack and adjacent formation that could clog the screen or cause turbidity in the water samples. Developing the gravel pack, especially natural gravel packs, increases the intake efficiency of the well. A well may be developed by overpumping (high rate of pumping), mechanical surging in the well screen with a surge block, air or water jetting within the screen, or any combination of these methods. The CM&E experience has been that development of a till well has little value except for removal of some of the fine-grained material from the gravel pack, and possibly for flushing clean some soil pores clogged during drilling and well installation.

*Placement*

Well placement is the final stage in the monitoring design and is based on the ground-water flow direction, the geology, the effects to be monitored, and the parameters to be monitored. If the monitoring program objectives are to gather baseline data over a large area, then placement of the wells should be representative of the predominant geologic strata and land uses with some consideration of potential stratification of contaminants. Some foreknowledge of potentially anomalous monitoring situations such as areas of contamination due to spills or waste disposal is necessary. Locations are typically chosen to be accessible, either in a right of way or where permission can be gained.

Well placement is more critical when the monitoring objectives are more directed as in the determination of effects or trends due to a particular activity. The CM&E project objective is to determine if an effect to the ground-water system, due to the use of BMPs, can be detected. Wells must be placed where causal effects can be detected and explained, and a number of factors have to be considered. The direction of ground-water flow determines whether a well is in an upgradient or downgradient position. Placement of wells in both positions is important. Downgradient wells include those wells which are located within the site. Upgradient wells yield background, ground-water chemistry data. Flow gradients should be determined prior to full site instrumentation so that the downgradient wells constitute the majority of the installations. It is through these downgradient wells, supplemented by upgradient wells, that the effect being monitored and quantification of causal relationships will be determined.

Wells need to be placed throughout the site rather than just at the upgradient and downgradient positions. The perimeter of a site may be instrumented such that parameters are monitored as they enter a site and then as they leave. But those evaluating the results will find it very difficult to ascertain gradients of change across the site and equally difficult to apply what has been learned to somewhat different situations.

Wells completed at different depths at a location are necessary to be able to discern stratification of a contaminant in the ground-water system. Shallow, medium, and deep wells (absolute depth is relative and differs depending on a particular site) need to be installed at one location either as a well nest, or multiple port sampler. Wells may be installed in different holes adjacent to each other or multiple wells in a single borehole. A good seal should be placed between multiple wells in a single borehole to ensure that isolation of the screens is maintained.

Vertical gradients can be determined only with multiple-level instrumentation at a location such as nested wells. Vertical gradients can have important ramifications with regard to the source of the ground water as it relates to observed chemistries. Some topographic influence on vertical gradients at the CM&E project sites located in till has been noted. Though the statistical correlation is not strong, the chemistries appear to be affected by the vertical gradients. Thus, in some cases, topography may be a factor in determining well placement at a site.

The decision on which depth is the most important will depend on the source of the contaminant, surface or subsurface, and the manner in which the contaminant moves through the ground-water system; that is, soluble, insoluble, denser or, less dense than water. In most non-point-source pollution studies, contaminants move from the surface to the subsurface, and shallow-depth wells constitute the bulk of the wells. To study the effect of a farming operation on the ground-water system, it is important to have wells located in the field. Wells placed in road ditches and drainage ways may reflect decreased concentrations of contaminants due to increased infiltration of lower-concentration water [*4,5*].

The question of how many wells are needed is project-specific and is based on the variability of the geology, the variability of the parameter to be monitored, the budget, and the analysis plans as they relate to proposed statistical testing. It should be apparent that the geology of the site controls many facets of the final system design. Ground-water flow and the chemical characteristics of the parameters to be monitored are also extremely important. Ground-water monitoring system design needs the input of hydrogeologists, geologists, chemists, engineers, statisticians, and personnel with field experience on a drill rig installing monitoring wells. Even after the best possible design has been determined, expect that changes will be necessary during the implementation of the design in the field.

## Ground-Water Sampling

The object of a sampling program is to collect a sample that is representative of the ground-water system being monitored. Reevaluate the parameters to be sampled and determine desired detection limits at this time. If available, search other data sources for typical concentrations and numerical variances of the parameters of interest to the project. Estimate the number of samples necessary to either describe the population within certain errors of estimation, or to detect a certain amount of change at a particular significance level. A statistician should provide assistance regarding the type of statistical testing envisioned in the analysis and evaluation stage, the legitimacy of the experimental design, and how to best meet the assumptions built into most statistical tests.

At some point during the monitoring or the evaluation of data, there will be a need to know the hydraulic conductivity at the well. Hydraulic conductivities determined prior to sampling can be helpful in determining the number of well volumes to be removed when obtaining a sample. Rather than a conventional pump test, the single-well baildown or slug test is more appropriate to monitoring projects. Pump tests result in hydraulic conductivities that represent an average over a large area. The distribution of hydraulic conductivities over a site and vertically within formations is important in evaluating the distribution of a contaminant and predicting where it might be.

The baildown and slug tests presented here make use of the equations developed by Hvorslev [*6*]. The assumptions of the baildown or slug test include instantaneous removal (or addition) of water from the well, homogeneous, isotropic, incompressible, and infinite extent of the medium [*7*]. Homogeneous and isotropic conditions are rarely met in the real world. Instantaneous removal of water is met in wells screened in till or slowly recharging material where bailing or pumping the well is essentially an instantaneous removal. In cases where part of the screen is in unsaturated material the baildown test is superior because the added water in a slug test might be filling the gravel pack. Care must be taken in till wells to assure that the gravel pack has been emptied during the water removal stage so that calculated hydraulic conductivities are based on the rate of water entering the well from the formation and not the draining gravel pack.

Difficulties arise with determination of hydraulic conductivity in rapidly recharging wells [hydraulic conductivity greater than 0.0010 cm/s (0.000397 in./s)]. Problems in depressing the water level instantaneously and recording a sufficient number of readings in a short time during recharge create a low confidence in the data used to calculate the hydraulic conductivity. The CM&E project is using a new technique [*8*]. A calibrated pressure transducer is placed at the bottom of the well, a packer is positioned and inflated to isolate the screened portion of the well from the rest of the casing, and the water is removed from above the packer. The pressure transducer is connected to a data acquisition system that records and stores three readings per second. As the data acquisition system is started, the packer is deflated and removed, instantaneous removal of water is achieved, and sufficient readings can be obtained. Three replications at wells that completely recovered in as little as eight seconds showed excellent reproducibility in calculated hydraulic conductivities.

Sample collection personnel are a key link to gaining reliable, high-quality data. Trained personnel should be used to collect the samples. There is a need for consistency of sampling methods either by maintaining consistency in the personnel or developing explicit, project-specific, standard operating procedures.

## Sample Collection Equipment

The sampling equipment must not alter the sample chemistry any more than necessary. Sampling equipment is chosen dependent upon several criteria:

1. Parameters to be monitored.
2. Potential effects or biases by the materials comprising the equipment.
3. Potential chemical alterations due to the manner in which the equipment delivers the sample to the surface.
4. Physical capability of the well to accept the equipment.
5. Accessibility of the well and associated portability of the equipment.
6. Ease with which the equipment can be maintained and cleaned in the field.
7. Applicability of the equipment to anticipated weather conditions.

The types of materials that comprise the sampling equipment are very important. The order of materials that have a low potential for introducing contamination, bias, or interferences is (least potential first): trifluoroethylene; stainless steel, PVC; lo-carbon, galvanized, and carbon steel for rigid materials; trifluoroethylene; polypropylene; PVC, vinylidine fluoride; and silicon for flexible materials [*2*]. The best material is also the most expensive. It is recommended that decisions regarding materials comprising the sampling equipment be based on the parameters to be sampled and the potential interferences involved.

The manner in which the device raises the sample to the surface is also very important. An evaluation of various devices indicated that suction lift, submersible centrifugal, and air lift (air contacts sample) mechanisms significantly altered a number of parameters [*9*].

The literature contains a number of good references on sampling devices and their merits for various parameters and situations. This paper explains what sampling devices were chosen for the CM&E project and why.

Since the principal focus of monitoring in the CM&E project is nitrate and nitrogen species, PVC was chosen as an acceptable material. Pesticide detection is the second most important parameter, so the use of all PVC was scrutinized. But limited contact time may not affect organic compounds [*2*].

The first device purchased was a PVC variable-capacity, double check valve bailer. A bailer is a good sampling instrument for almost any project. It will fit in the well, it is portable, easily maintained and cleaned, and works in all weather conditions. A trifluoroethylene bailer was also purchased in the event that the PVC was interfering with the pesticides.

A bladder (or squeeze) pump was also purchased due to its capacity to purge a well in a reasonable amount of time and its ability to deliver relatively unaltered samples. The bladder pump is one of the best devices available for obtaining ground-water samples and has been used as a control with which other devices are compared [*9,10*].

Lastly, a peristaltic pump was purchased to allow sampling in wells that would not accept the bailer or the bladder pump. The peristaltic pump is limited to 7.62 m (25 ft) of lift, and does alter water chemistries due to degassing of the sample. The peristaltic pump has also been used as a source of pressure for filtering samples.

With these three devices, all of our potential field situations were considered. Whenever possible, the bladder pump is preferred, but the wells must recharge quickly enough for pumping and be accessible to a vehicle that carries a compressor as a gas source for running the pump. Some of the wells are in the middle of fields where, at times, access is denied due to mud or snow. Some of the wells are completed in till and do not recharge quickly enough after purging. In the cold winter weather, the water in the bladder pump tubing will freeze between the well head and the sample bottle.

The bailer is as infallible a device as any, and access is possible as long as the sampling personnel can walk to the well. Bailers do not have the freezing problems of the bladder pump. A bailer can collect high-quality samples by using a bottom-emptying device which minimizes disturbance of the sample [*10*]. Examination of CM&E data to date has revealed no systematic error between bailer and bladder pump samples.

A few wells will not accept the bladder pump or the bailer, but the small-diameter tubing of a peristaltic pump will fit. The peristaltic pump can also be useful when a well has only 0.08 to 0.10 m (3 to 4 in.) of water in the bottom, making it impossible to pump with the bladder pump and extremely difficult to bail. The peristaltic pump is an alternative that can be used in situations where sample collection is not possible by other means and where limited alteration of the parameters is affected by the suction lift delivery. Some parameters can be altered significantly, however, and results should be interpreted carefully.

## Sampling Protocol

Since different sampling techniques can possibly affect the chemistries of samples, an explicit sampling protocol (procedure) should be established and maintained throughout the life of the project.

The first item recorded at the well is the water level. Any pumping and sampling will affect the water level, and a static level is necessary for ground-water flow determinations. The next step is to purge the well of stagnant water, which may have undergone chemical changes during storage in the well casing. Purging continues until a high percentage of fresh formation water is available to the sampling device [*11*]. Overpumping should be avoided as it may introduce ground water from a distant source that may dilute or concentrate certain parameters and result in erratic or misleading data [*11*].

Sampling with the bladder pump requires a certain amount of time for purging the well prior to collecting the sample. The amount of time for purging was established by monitoring temperature and electrical conductivity during pumping until they stabilized. Pumping then continued for an additional two to three times the noted stabilization time to be sure stabilization had been achieved. Thereafter samples would be taken by purging only until the stabilization time had been reached. Periodically the time to stabilization is checked during sampling.

Well purging with the bailer is accomplished by removing three to four volumes of water in the well. To eliminate contact between the water and the rope used for bailing, care is taken to just fill the bailer and then remove it from the well. All-nylon rope is used because it is nonabsorbent and helps to limit carryover between wells. In low-recharge wells completed in till, the well is bailed dry and sampled later (up to three to four days later) as it is recharging [*11*]. The bailer is cleaned prior to sampling a well that has been previously bailed dry.

The peristaltic pump requires a considerable amount of time to adequately purge the well if a substantial amount of water is present. It is advisable to first purge with a higher-volume sampler if possible.

Except when sampling the low-recharge wells that have been bailed dry, sampling devices are considered to be cleaned during the purging operation. This is a legitimate assumption for the low concentrations of very (water) soluble constituents of interest to the CM&E project, but is not an assumption that can be generalized to every project. Data from pesticide sampling are continually being checked for evidence of sample carry-over. To date no carry-over problems have been determined.

A successful CM&E technique is the use of a sampling book that is carried to the field by sampling personnel. The book is organized by site with a map showing well locations within a site. Included for each well is (1) the drilling log for the well, (2) details of the well construction, (3) recharge capabilities of the well and the preferred sampling device, (4) time required for purging with the pump; and (5) an equation to determine the number of bails required for purging based on the water level recorded.

There is a need for sampling personnel to take notes in the field during the entire sampling process. Many apparent anomalies in the data can be explained from observations recorded by the sampler.

Field measurements of dissolved oxygen, water temperature, and specific conductivity in the well immediately after sampling is completed are relatively easy to perform and can contribute invaluable information. Water temperature must be read in the well to be accurate. Dissolved oxygen can be an indicator of spring recharge water, depth within the system, and potential for chemical species to be present. Specific conductivity provides a good overall indicator of water quality that can be used as a quick check of laboratory results.

Proper sample handling, after collecting and tracking the samples from the field through the laboratory, is an important aspect of a ground-water monitoring project. However, the subject is too broad to be adequately addressed in this paper.

*Data Analysis and Evaluation*

Plans for data analysis should be part of the overall monitoring system design. Data analysis should not be performed only when writing the final report. It should be an ongoing procedure to check on laboratory and sampling error by examining analyses as they become available. The project objectives should be reevaluated as data become available, and trends in data should be examined to be sure they are not aberrations due to a procedure being used. Lastly, as major reports are generated and statistical testing is utilized, statistical validity of the test to the data should be assured. Water quality data typically have problems meeting the assumption of normality of the population and equality of variances between populations.

## Conclusions

A ground-water monitoring system cannot be a haphazard plan pieced together without considerable forethought. It is a design conceived and implemented by the joint effort of hydrogeologists, engineers, geologists, chemists, statisticians, and field personnel with experience in drilling and installing wells and sampling of ground water. What is to be monitored and the objectives of the project must be firmly in mind from the start of the system design.

A large land area must be reduced to monitorable-size sites. A preliminary drilling program in the absence of adequate literature sources may be needed to help select potential sites. Monitoring wells, needed to access and assess ground-water systems, are fixed sampling points. Once in place, they are the only means available to collect data. Improper installation, construction, or placement of the wells may make it impossible to meet the objectives of the project or may yield results that are misleading or in error. To properly design and implement the monitoring system, information gained through drilling with detailed logging and preliminary well installations is necessary. Sampling equipment must be chosen based on physical limitations of the wells and sites and chemical requirements of the monitoring objectives. Sampling personnel must be trained and aware of proper techniques and potential pitfalls inherent in the sampling procedure. Proper sampling protocol must be observed throughout the project with the key word being consistency. Data analysis must be an ongoing process starting during determination of the monitoring objectives and continuing until a final report is produced. Periodic reevaluation of goals and objectives, parameters to be monitored, analysis and evaluation plans, and sampling techniques should also be included in a complete monitoring project.

As a summary of the procedures presented in this paper, the flow chart of steps to consider when developing a non-point-source ground-water monitoring project (Fig. 1) is proffered. The single lines represent procedures which for the most part are one-time activities. The double lines represent activities that continue through the life of the project.

The South Dakota Comprehensive Monitoring and Evaluation project has been utilizing the procedures and methodologies presented here. Analyses to date indicate that the data being collected are of good quality and are building the basis to meet project objectives. Failure to follow the proper procedures for monitoring system design or sampling could easily jeopardize the project's chances for success.

*Acknowledgments*

The South Dakota Comprehensive Monitoring and Evaluation project (Oakwood Lakes-Poinsett Project) of the Rural Clean Water Program is a cooperative special study. The following are contributing agencies: Division of Environmental Quality and Division of Geological Survey of the South Dakota Department of Water and Natural Resources; Water Resource Institute, Station Biochemistry, and Microbiology Department of South Dakota State University; U.S. Environmental Protection Agency; Economic Research Service, Cooperative Extension Service, Soil Conservation Service, of the U.S. Department of Agriculture. Special thanks to Dr. Forrest Payne, Dennis Erinakes, and Jeanne Goodman for their review and comments in the preparation of this paper.

## References

[*1*] Everett, L. G., "Monitoring in the Zone of Saturation," *Ground Water Monitoring Review*, Vol. 1, No. 1, Spring 1981.

[2] Barcelona, M. J., Gibb, J. P., and Miller, R. A., "A Guide to the Selection of Materials for Monitoring Well Construction and Ground-Water Sampling," ISWS Contract Report 327, Illinois State Water Survey, 1983.

[3] Perry, C. A., and Hart, R. J., "Installation of Observation Wells on Hazardous Wastes Sites in Kansas Using a Hollow-Stem Auger," *Ground Water Monitoring Review,* Vol. 5, No. 4, Fall 1985.

[4] "1985 Annual RCWP Progress Report," Oakwood Lakes–Poinsett, South Dakota, Project 20 (due for release in 1986), available through the South Dakota Department of Water and Natural Resources, Pierre, SD, 1985.

[5] Kimball, C. G., "The Thermal, Chemical and Physical Effects on the Hydrogeologic System in the Sand Plain of Wisconsin from Water Source Heat Pump Discharge via a Return Well," thesis, University of Wisconsin-Madison, 1983.

[6] Hvorslev, M. J., "Time Lag and Soil Permeability in Ground-Water Observations," Waterways Experimental Station Bulletin 36, U.S. Army Corps of Engineers, 1951.

[7] Freeze, R. A. and Cherry, J. A., *Groundwater,* Prentice-Hall Englewood Cliffs, NJ, 1979, pp. 339–340.

[8] Kimball, C. G., and Bischoff, J. H., "A Method for Bail-Down Test Hydraulic Conductivity Determinations in Rapidly Recharging Wells" (in preparation).

[9] Houghton, R. L. and Berger, M. E., "Effects of Well-Casing Composition and Sampling Method on Apparent Quality of Ground Water" in *Proceedings,* Fourth National Symposium and Exposition on Aquifer Restoration and Ground Water Monitoring, National Water Well Association, May 1984.

[10] Barcelona, M. J., Helfrich, J. A., Garske, E. E., and Gibb, J. P., "A Laboratory Evaluation of Ground Water Sampling Mechanisms," *Ground Water Monitoring Review,* Vol. 4, No. 2, Spring 1984.

[11] Gibb, J. P., Schuller, R. M., and Griffin, R. A., "Procedures for the Collection of Representative Water Quality Data from Monitoring Wells," Cooperative Groundwater Report 7, Illinois State Water Survey, 1981.

*P. Stan Mitchem,*[1] *George R. Hallberg,*[2] *Bernard E. Hoyer,*[2] *and Robert D. Libra*[2]

# Ground-Water Contamination and Land Management in the Karst Area of Northeastern Iowa

**REFERENCE:** Mitchem, P. S., Hallberg, G. R., Hoyer, B. E., and Libra, R. D., **"Ground-Water Contamination and Land Management in the Karst Area of Northeastern Iowa,"** *Ground-Water Contamination: Field Methods, ASTM STP 963,* A. G. Collins and A. I. Johnson, Eds., American Society for Testing and Materials, Philadelphia, 1988, pp. 442–458.

**ABSTRACT:** A statistical analysis of existing northeast Iowa water-quality data showed that systematic nonpoint contamination of ground-water quality was occurring in regional carbonate aquifers. Well-water quality generally improved with increased well depth and with natural geologic protection related to thickness of overlying bedrock or Quaternary aquitards. Nitrate levels were highest in areas where sinkholes were present; areas without sinkholes but with relatively thin protective cover had similar but somewhat lower levels of nitrate.

A detailed analysis of a single karst ground-water basin found the same systematic nitrate contamination along with low but persistent levels of herbicides. Agricultural practices were judged to be the source of the contaminants and infiltration was the principal mechanism. Ninety-five percent of the nitrate and 55 to 85% of the atrazine were delivered by infiltration and percolation.

Various state and Federal agencies, university researchers, local groups, and private organizations are cooperating in a seven-year project in the Big Spring basin to demonstrate and evaluate the effectiveness of best management practices on ground-water quality and on crop production. The project will provide research results to serve as a foundation for future policies and programs that agriculture must address relative to environmental problems.

**KEY WORDS:** ground water, carbonate aquifers, karst, contamination, pesticides (herbicides), nitrates, bacteria (coliform), infiltration/percolation, fertilization, agriculture, soil conservation

Ground water is one of Iowa's most valuable resources. Almost 90% of the water used for municipal supplies and rural domestic and livestock production in Iowa comes from the ground.

Most of Iowa's bedrock aquifers are covered by a thick layer of surficial materials composed of glacial till and loess. These materials serve as an aquitard and protect the aquifers by filtering out disease-causing bacteria and viruses and adsorbing many chemicals. Over large areas of northeastern Iowa, however, this protective layer is quite thin. The bedrock aquifers in this region are largely composed of limestone and dolomite (carbonate rocks). These carbonate aquifers transmit water through fractures, joints, and other secondary openings that have been enlarged by chemical solution. Locally, large caves may develop. Where soils are thin, sinkholes form as a consequence of rock solution and collapse.

Areas with numerous sinkholes are known as karst areas. Sinkholes allow sediment, bacteria,

[1] State geologist, USDA Soil Conservation Service, Casper, WY 82601; formerly geologist, USDA Soil Conservation Service, Des Moines, IA 50309.

[2] Geologists, Iowa Geological Survey, Iowa City, IA 52242.

and other contaminants to run directly into the upper bedrock aquifers. In addition, sinkholes have locally been used as trash dumps, further endangering water quality. Water from wells tapping the upper bedrock aquifers often has objectionable taste or smell. After rains, the water in some wells turns cloudy. For example, whey dumped by a cheese factory in 1963 ran into a sinkhole and into the aquifer, killing all the fish at a trout hatchery supplied by a spring more than 16 km (10 miles) away [1].

Residents have long been concerned about the quality of the ground water in the karst area. Over the past 10 to 15 years, nitrate concentrations in the ground water have risen sharply. Consequently, many residents have had to drill deeper wells in order to obtain water of adequate quality for domestic and livestock purposes.

Because of the concerns about water quality, a number of state and Federal agencies, university researchers, local groups, and private organizations are cooperating in an effort to study and find solutions for the ground-water quality problems. A comprehensive, three-phased evaluation of hydrogeology, water quality, and land management is being conducted to address these concerns.

This case study does not present new ideas for standardizing site-specific sampling. Rather, the purpose is to illustrate the step-by-step procedures used to refine hydrogeologic water-quality observations, from vague regional data sets to more defined local studies, to assess the impacts on ground-water quality due to agricultural nonpoint contamination. This presents a very different scale and scope of problems than sampling needs relative to point-source problems or unusual contaminants. Standards for sampling and analytical work that already exist were followed in this study.

## Phase I

The initial study focused on 22 counties in northeast Iowa that comprise most of the karst area in Iowa (see Fig. 1). The Iowa Geological Survey (IGS), in part supported by a grant from the Iowa Department of Environmental Quality (now called the Iowa Department of Water, Air, and Waste Management, IDWAWM), compiled pertinent geologic, hydrologic, and water quality data [2].

### *Geology*

More than 12700 sinkholes were identified and mapped using published and unpublished modern soil surveys, unpublished field surveys, photo-interpretation of high-altitude color-infrared aerial photography, and unpublished master's theses. The geologic distribution of the sinkholes and factors affecting distribution were determined.

Karst features occur principally in the thicker, relatively pure carbonate rock sequences. The highest concentrations of sinkholes occur in: (1) Silurian age rocks, particularly along an escarpment they form, in southern Clayton, northern Delaware, and eastern Fayette Counties; (2) Galena Group rocks of Ordovician age in southwestern Allamakee and portions of Clayton and Winneshiek Counties; and (3) areas of Middle Devonian outcrops adjacent to the Cedar River in Mitchell, Floyd, Chickasaw, and Bremer Counties. All of these sinkhole concentrations occur in carbonate aquifers which have been major sources of ground-water supplies in northeast Iowa.

### *Hydrology*

Ground-water flow in karst aquifers can be very complex. Unlike water flow in clastic rocks such as sandstone, water in karst aquifers flows along bedding planes and vertical or high-angle fractures and joints. These secondary openings are often enlarged by chemical solution of the carbonate rock. Most carbonate solution takes place below but near the surface of the water table.

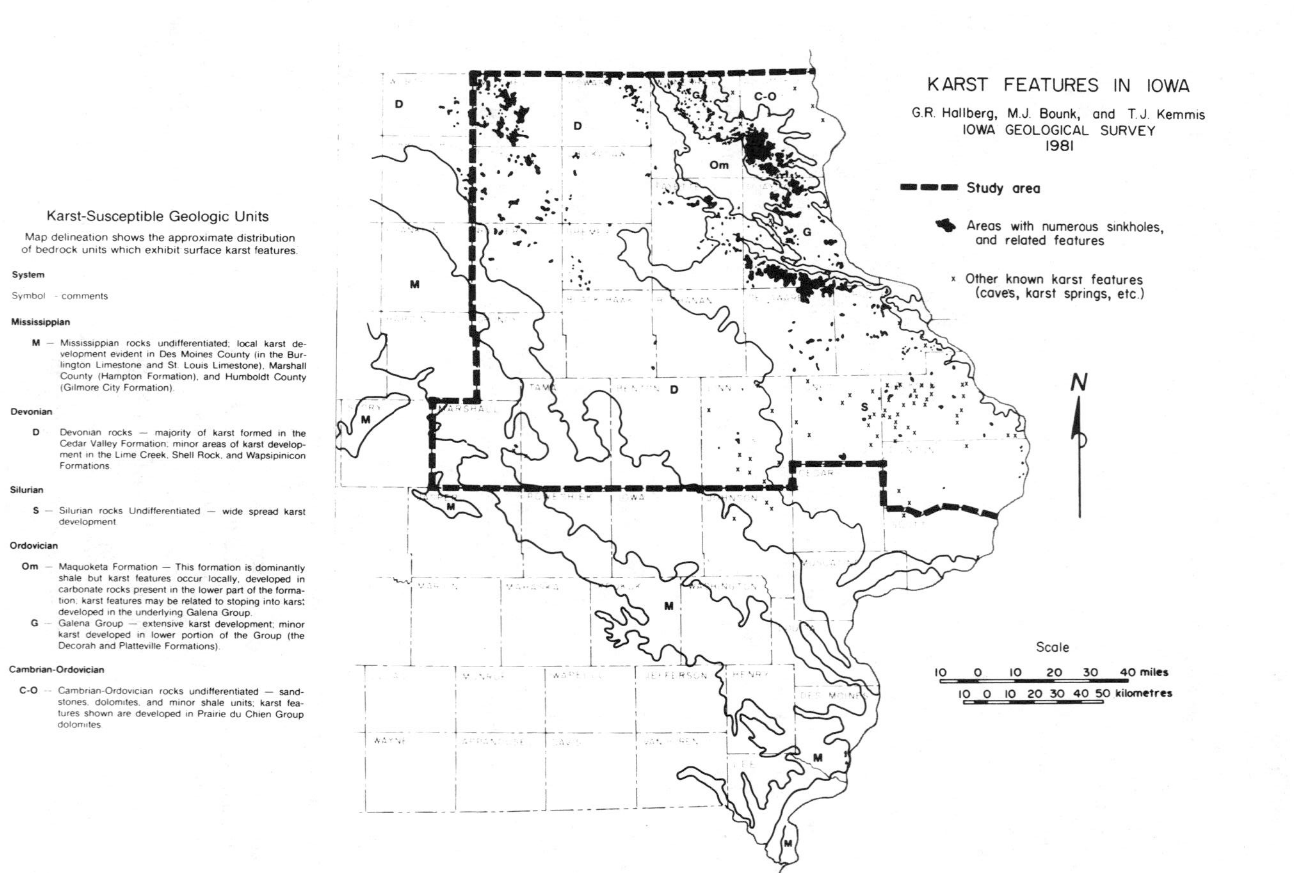


FIG. 1—*Twenty-two county study area and regional geology.*

In places, infiltration from stream beds or sinkholes that "swallow" streams disrupts surface flows. The water levels in the openings may rise and fall rapidly in response to recharge events.

Three dye traces were made in 1981 in the Silurian karst of Fayette County [3] and twelve dye traces were made in the late 1970's in the Galena aquifer of Clayton County [1]. Standard dye-tracing procedures were used [4]. One thousand color-coded polyethylene spheres 1.1 cm (0.4 in.) in diameter with two different specific gravities were also introduced into six of the Galena sinkholes [1]. These traces were made to determine the directions and velocities of ground-water flow as well as the locations of drainage basin divides. Field measurements of joint orientation and mapping of cave passages also aided in analysis of ground-water flow. Regional ground-water divides roughly parallel the major regional surface-water divides; however, local flow systems may cross beneath tributaries of the major streams.

Sinkholes are widespread in rocks of the Galena Group throughout the area in Allamakee, Clayton, and Winneshiek Counties. The deep dissection of the area allows the solutional conduit system to develop to greater depths than in the other geologic settings. The surficial solution-conduit flow systems may therefore be active to proportionally greater depths. If so, contamination from surficial sources may penetrate to greater depths.

*Water Quality*

Several water-quality data sets were used to gain an understanding of the water-quality effects related to the karst aquifers. Records of the University of Iowa Hygienic Laboratory (UHL) were the most extensively used for regional analyses; however, records from the Water Storage and Retrieval System (WATSTORE) data file of the U.S. Geological Survey (USGS) and several other less-extensive data sets were also used.

The UHL data consist of water data analyses performed between January 1977 and December 1980. The data include 6039 analyses of nitrate concentration. Nitrate analyses were run using the U.S. Environmental Protection Agency (EPA) Method 353.2 [5]. Also supplied by UHL were 8130 total coliform bacterial analyses.

All samples were from private sources, and location and geologic control were poor. The only location given was the resident's return postal address. The only information on well construction was total well depth. Comparisons were made with known records. The figures for well depth were accurate about 80% of the time.

Postal addresses were used as sample centers for the well location. Sample centers were then assigned to geologic settings as follows:

1. Deep Bedrock—depth to bedrock more than 15 m (50 ft).
2. Shallow Bedrock—depth to bedrock less than 15 m (50 ft) with no sinkhole concentrations.
3. Karst—depth to bedrock less than 15 m (50 ft) within an area of concentration of sinkholes.

Along with the sample center location, the depth class was entered. Depth classes were 0 to 15 m (0 to 50 ft), 15 to 30 m (50 to 100 ft), 30 to 45 m (100 to 150 ft), 45 to 150 m (150 to 500 ft), >150 m (>500 ft), and unknown.

The WATSTORE file provided additional nitrate concentrations for 1832 water samples from 888 private and public wells in the study area. These data included aquifer tapped, geographic location, and, in many instances, well construction information.

Other water-quality data were available from several sources for the period 1969–1975. Considerable control existed for some of these samples, and they were investigated for supplementary and comparative purposes.

Elevated levels of nitrate and occurrences of coliform bacteria are common throughout the entire data set. Nearly 35% of the total coliform bacterial analyses had unsafe bacterial levels (Most Probable Number of coliform bacteria per 100 mL (0.1 qt) equal to or greater than 2.2), and nearly 18% of the nitrate analyses exceeded the EPA drinking water standard of 45 mg/L (45

TABLE 1—*Summary of median nitrate concentrations (mg/L) from different geologic settings and well depths.*

| Well depth,[a] m | Number of Wells | Median Nitrate Concentrations by Geologic Setting | | |
|---|---|---|---|---|
| | | Karst, mg/L[b] | Shallow Bedrock, mg/L | Deep Bedrock, mg/L |
| 0 to 15 | 593 | 28 | 26 | 33 |
| 15 to 30 | 992 | 34 | 19 | 6 |
| 30 to 45 | 1081 | 23 | 16 | 0 |
| 45 to 150 | 1726 | 10 | 6 | 0 |
| >150 | 125 | 2 | 3 | 0 |
| Unknown | 1522 | 22 | 7 | 0 |
| All Wells | 6039 | 19 | 9 | 0 |

[a] 1 m = 3.28 ft.
[b] 1 mg/L = 1 ppm.

ppm). Statistical analyses of the bacterial data were difficult, and the significance of the results was vague. This vagueness arises from the relatively uniform distribution of elevated bacterial levels throughout all geologic settings and well depths and also from a lack of relationship between bacterial levels and nitrate concentrations. The results generally suggest, however, that bacterial contamination is site-specific and is related to improper well construction or placement, or that it is introduced by the water system rather than caused by aquifer contamination.

Statistical evaluations of the nitrate data sets are much more useful than those of the bacterial data. Median nitrate values generally decrease with increasing well depth and with changes from the karst to shallow and deep bedrock geologic settings (Table 1). Wells less than 15 m (50 ft) deep consistently have the highest levels of nitrates regardless of geologic setting. In well depth categories over 15 m (50 ft), karst areas have significantly higher median nitrate levels than shallow bedrock areas, and shallow bedrock areas have significantly higher levels than deep bedrock areas.

**Phase II**

The Big Spring basin in Clayton County was chosen to provide a controlled assessment of a single, well-defined, karst ground-water basin (see Fig. 2). Work began in November 1981 and continues today [*6–8*]. The work is a cooperative effort of the IGS, Soil Conservation Service (SCS), IDWAWM, and EPA Region 7. Assistance was provided by the Iowa Conservation Commission (ICC), UHL, and USGS.

The primary reason for choosing the Big Spring area was that the ICC Fish Hatchery at Big Spring is located on a spring which accounts for 89% of the discharge from the basin. This unique situation affords the opportunity to quantitatively gage the discharge of ground water. This gaging, coupled with detailed water-quality measurements, allows mass-balance assessments of the chemicals discharged in the ground water.

The Big Spring study included a basin-wide inventory of geology, soils, sinkhole locations, land use, and potentiometric surface of the Galena aquifer. More than 270 rural domestic wells were inventoried, and about 125 of these were sampled for water-quality analysis.

A computerized geographic data base was established for the Big Spring study area to aid in analysis. Earth Resources Laboratory Applications Software (ELAS), developed by the National Aeronautics and Space Administration (NASA), was used to digitize and manipulate the data.

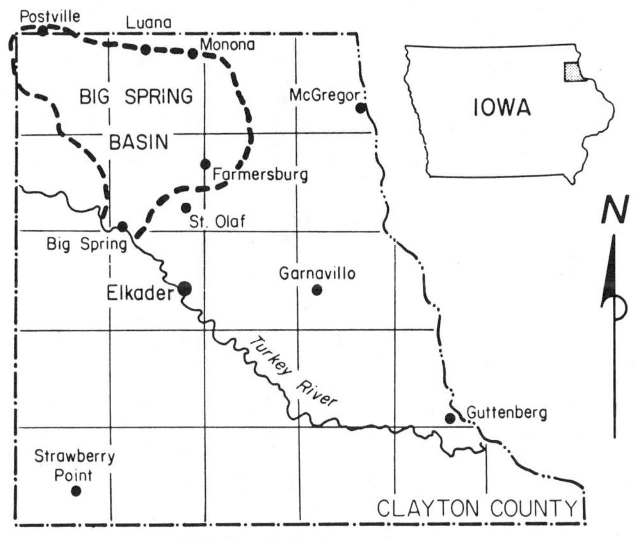


FIG. 2—*Big Spring study area.*

Geographical control was provided by USGS 7½-min topographic maps. The data base consisted of the following sets: soils, land use, bedrock geology, sinkholes, sinkhole basins, nitrates, potentiometric surface, roads, and sample points.

## *Geology*

The Big Spring basin is located within the Paleozoic Plateau Landform Region [*9*]. Topography ranges from moderately rolling in the northern part to steeply sloping in the south along the Turkey River.

Thin isolated patches of Pre-Illinoian glacial till exist, but most of the deposits have been removed by erosion. Loess is the dominant surficial material. Many soil cores were taken during the study to determine the thickness and nature of the surficial materials.

Big Spring draws its water from the Galena aquifer. The Galena carbonates outcrop low in the landscape along the principal stream valleys. Below the Galena aquifer is an aquitard that separates it from the St. Peter aquifer. Overlying the Galena aquifer is the Maquoketa Formation, which can be divided into two hydrologic units. The lower unit is composed of shales and carbonates; the upper unit, the Brainard Member, is a thick clay-shale that is a major aquiclude in northeast Iowa [*6*].

Within the basin, the Galena aquifer averages 67 m (220 ft) thick. Its base dips generally southwest at about 3.5 m/km (18 ft/mile). In the western part of the basin where the thick Brainard shale is present, the Galena aquifer is under confined or artesian conditions, but in the remainder of the basin it is a water-table aquifer.

Sinkhole distribution was mapped from soil-survey maps, field inventory, and review of ICC field-mapping notes. Sinkholes occur in several geologic settings in the Big Spring basin. The areas of greatest extent and concern are in the Galena carbonates and near the contact of the Galena and the overlying lower Maquoketa rocks [*6*]. The karst system and landscape are constantly changing. Numerous "new" sinkholes open and old ones "close" continually.

TABLE 2—*Big Spring land use in 1980.*

| Land Use | Area, $km^2$ | Percent |
|---|---|---|
| Row crop | 160 | 59.9 |
| Cover crop | 81 | 30.3 |
| Forest | 17 | 6.4 |
| Other | 9 | 3.4 |
| Total | 267 | 100.0 |

1 $km^2$ = 0.386 $mile^2$.

### *Land Management*

The basin is predominantly rural with intensive agricultural production. Knowledge of land use and trends is important for evaluating the ground-water resource and predicting responses.

Land use was initially determined by interpretation of 1:80000 scale high-altitude color-infrared photography taken in November 1980 (Table 2). The map was compiled manually and digitized and entered into the computer. Aerial photography provided by the Agricultural Stabilization and Conservation Service (ASCS) from 1970 was used, along with the 1980 photography, to determine the trend in land use for the basin. Digital LANDSAT data were also analyzed to determine trends. Scenes from August 1972 and May 1981 were used. Cultural information, such as roads and urban areas, was obtained entirely from manual digitization of information from USGS topographic maps. Both methods showed similar land use and trends.

The computer data base was used to model soil erosion and runoff to evaluate runoff and erosion effects of alternative management schemes. The Universal Soil Loss Equation [*10*] and SCS Technical Release 55, "Urban Hydrology for Small Watersheds" [*11*], were used for this purpose.

### *Hydrology*

Additional dye traces were done in June and July 1982 in the Big Spring basin to further investigate flow paths and to better define the eastern basin boundary. Fluorescein dye was used in both traces with packets of activated coconut charcoal used to capture the dye. Packets had been placed at the collection points for at least one week immediately before the tracing to test for background levels of fluorescein.

The dye traces indicate flow to Big Spring from the north and northeast, crossing under Roberts Creek, which flows to the southeast. Dye travel times from the north ranged from 24 h under "moderately high-flow conditions" [*1*] for a straight-line distance of 13.9 km (8.6 miles) to 39 to 50 h under lower flows [*6*]. Travel times from the east side of the basin were somewhat longer, ranging up to 134 h for a distance of 10.8 km (6.7 miles). All of these flow rates are rapid compared with flow in clastic rocks such as sandstone.

Previous studies along Roberts Creek indicated a 10-km (6 mile) losing reach. Farther downstream, outside the basin, Roberts Creek is observed to gain water. Portions of Silver Creek have also been observed as losing reaches.

The potentiometric surface (Fig. 3), in conjunction with dye trace data, was used for the final delineation of the Big Spring basin. The ground-water basin as identified includes an area of about 267 $km^2$ (103 $mile^2$). On the north and west sides, the ground-water basin divide is coincident with the surface-drainage divide. On the east side, however, the ground-water divide cuts across several surface-drainage basins.

Within an eight-day period in late June and early July 1983, 280 mm (11 in.) of rain fell within the basin. Peak discharge at Big Spring of about 6.2 $m^3/s$ (220 $ft^3/s$) resulted.

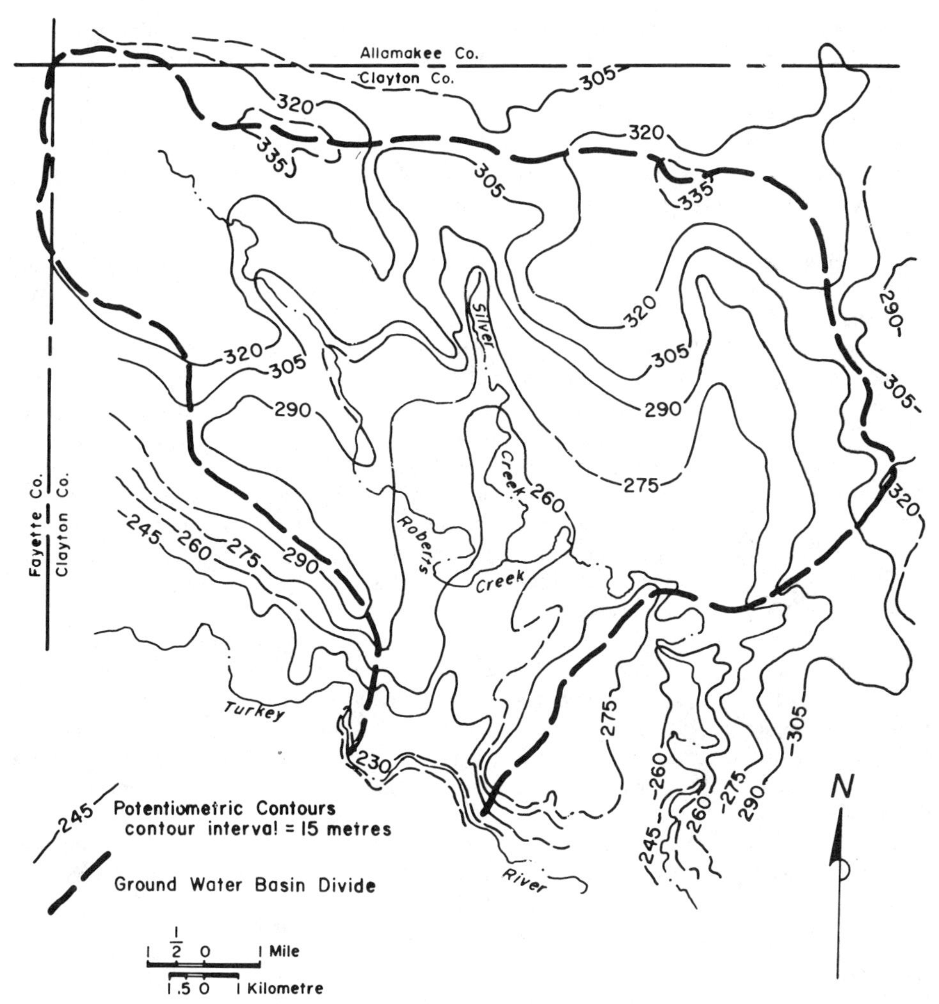


FIG. 3—*Elevation of the water table/potentiometric surface in the Galena aquifer in Big Spring Study area.*

During this discharge, a variety of detailed water-quality data was collected, in addition to two dye traces, to document the effects of a major runoff/run-in event in the Big Spring basin. A different dye was injected at the same time into the water draining into two sinkholes on opposite sides of the basin. Fluorescein (chloric acid yellow 73), a green dye, was used on the west side; amino-G (7-amino 1, 3 Napthalene disulphonic acid), a blue dye, was used on the east side. Both sinkholes had been previously traced in 1982. A Turner Model III filter fluorimeter [*12*] was used to measure the concentration of each of the dyes through time.

These quantitative dye traces substantiate previous qualitative dye traces. Minimum ground-water velocities of 0.2 and 0.1 m/s (0.6 and 0.3 ft/s) were determined for the west and east sides of the basin, respectively. The traces indicate that some water may begin arriving in 18 to 24 h, but the maximum impact of the run-in water may take 36 to 48 h. Dye was detected until 63 h after injection. This finding suggests that at least trace amounts of run-in water and the contaminants it contains may be in the system for at least 2½ days after the run-in event. Great fluctuations in dye concentrations also indicate that the water travels as "slugs" rather than steady flow.

In addition to the major runoff event monitored in June and July, the main snowmelt period in February and several minor runoff-discharge events were also monitored in detail. During these events in 1983, water samples were also collected intermittently from surface water and tile lines. Stream discharges were also measured at selected sites. Water-quality parameters analyzed included suspended sediment, organic nitrogen (N), ammonium-*N*, nitrate-*N*, specific conductance, and pesticides. Separation of the Big Spring hydrograph into the diffuse infiltration-percolation flow and the rapid run-in-conduit flow components was performed using both analytical methods and a method based on measured physical and chemical parameters [*6,7*]. Both methods had very similar results.

Two analytical methods were used [*13*]. The first method utilizes base-flow recession parameters to calculate conduit-flow recession parameters. The parameters describe the recessions of the two flow components as linear responses. Several initial estimates of the base-flow recession parameter are used to generate corresponding conduit, flood-flow parameters until a combination of base-flow and conduit-flow recessions are generated that best fit the observed discharge hydrograph. In the second method, daily discharges are separated into five-day sets, and the minimum flow for each set is identified. The minima are then compared sequentially in groups of three, and "turning points" are plotted on the hydrograph and connected, forming the base-flow hydrograph. While both methods yielded nearly identical results over the span of a water year, the first method was more accurate for single-stage events.

If the concentrations of a particular constituent are known in the ground-water base flow and in the runoff, flood-flow component, and their resultant, mixed concentrations are measured in the outflow during a discharge event, the respective water-discharge components can be separated by resolving the simple proportions of the constituents [*14*]. This was done using specific conductance and atrazine. Atrazine concentrations yielded a better check on the hydrograph separation because it has no natural subsurface source, although some atrazine is likely attached to sediments within the conduit system.

About 77% of the discharge at Big Spring during these events was a result of diffuse infiltration-percolation flow. Weighted values indicate that this diffuse flow delivered 87% of the nitrate and the conduit flow (23% of the total discharge) delivered 80% of the atrazine.

### *Water Quality*

Existing rural domestic wells were used to sample ground-water quality. The water in or near a well casing may not represent the water quality within the aquifer being sampled. To flush the well so that it contains water representing the aquifer, several methods have been recommended [*15–17*]. Temperature was chosen as the factor to monitor because of the ease of measurement and because other recommended parameters, such as pH, will not stabilize until the temperature has stabilized. The temperature was monitored while the well was pumped with the existing pump. In most cases, a stable temperature was reached within 10 min of pumping, after which time samples were collected and other field measurements, such as pH and conductivity were taken.

Water chemistry can also be changed if the water passes through a cistern, storage tank, or water conditioner. To avoid this problem, all samples in this inventory were taken at the wellhead or at the hydrant closest to it.

The median nitrate concentration in wells sampled during the initial inventory was 35 mg/L (35 ppm). Nitrate levels in wells in the western part of the basin, where the Maquoketa provides protection, were below the detection limit [5 mg/L (5 ppm)] and represent the background or natural concentration. In the more "open" parts of the basin, where the Galena outcrops, nitrate levels ranged up to 280 mg/L (280 ppm). Water quality and discharge have been measured at Big

TABLE 3—*Comparison of the total water and chemical discharge from the Big Spring basin for 1982, 1983, and 1984 water years* [6,8].

| Measurement | Units[a] | Water Year 1982 | 1983 | 1984 |
|---|---|---|---|---|
| 1. *Precipitation and discharge* | | | | |
| Precipitation | mm | 864 | 1130 | 833 |
| Total ground-water discharge | million $m^3$ | 58.5 | 63.4 | 52.5 |
| Ground water discharge less change in storage | million $m^3$ | 46.1 | 63.4 | 49.7 |
| Streamflow discharge | million $m^3$ | 42.9 | 85.9 | 45.1 |
| Total water yield | million $m^3$ | 101.4 | 149.3 | 97.6 |
| 2. *Ground-water components* | | | | |
| Run-in-peak conduit flow | million $m^3$ | 4.1 | 5.6 | 2.3 |
| Infiltration-base flow | million $m^3$ | 42.0 | 45.5 | 38.0 |
| 3. *Nitrate-N discharge in ground water* | | | | |
| Flow-weighted mean nitrate concentration | mg/L | 39 | 46 | 43 |
| Nitrate-*N* discharge for basin | kg/$km^2$ | 1787 | 2296 | 1743 |
| 4. *Atraxine discharge in ground water* | | | | |
| Flow-weighted mean atrazine concentration | μg/L | 0.18 | 0.28 | 0.45 |
| Atraxine discharge for basin | kg/$km^2$ | 0.02 | 0.05 | 0.07 |

[a] 1 mm = 0.0393 in.
1 000 000 $m^3$ = 810 acre-ft.
1 mg/L = 1 ppm.
1 kg/$km^2$ = 0.00892 lb/acre.
1 μg/L = 1 ppb.

Spring since November 1981 (Table 3). In addition, selected tile lines and surface sites have been sampled.

Because the volumes of both surface and ground waters are monitored as well as their chemical constituents, the total mass of constituents discharged in the water can be calculated. The amount of nitrate-N lost in water years 1982, 1983, and 1984 ranged from about 33 to 55% of the amount of chemical-*N* fertilizer applied in the basin. The highest concentrations and largest mass of nitrates are delivered through the infiltration component (95%), comprising about 90% of the total ground-water discharge. The total mass of pesticides discharged is estimated to be 5% of that normally applied. Fifteen to 45% of the total mass of pesticides are delivered through the run-in component comprising only 10% of the total ground-water discharge. Thus, most (55 to 85%) of the pesticide in the ground water is delivered by infiltration recharge. Higher than normal annual precipitation, such as in 1983, resulted in higher percentages of the contaminants reaching the ground water.

Soil samples were taken at 21 sites in 1982 and 1983 to gather background information on nitrate and pesticide movement through the soil profile under different land uses and management. Soil cores, taken with a trailer-mounted Giddings hydraulic soil-coring machine, ranged in depth from 1 to 11 m (3 to 35 ft), depending on depth to bedrock.

Analysis of the soil samples was done by UHL. Nitrates were analyzed using the Army Corps of Engineers' elutriate method [*18*]. Pesticide residues were analyzed using EPA methods [*19*].

These soil-core studies agree with numerous other studies of nitrate buildup in soils. Pesticides and large nitrate accumulations are occurring at depths below the root zone. In areas where the aquifer is shallow, these surficially derived chemicals are being delivered to the aquifer.

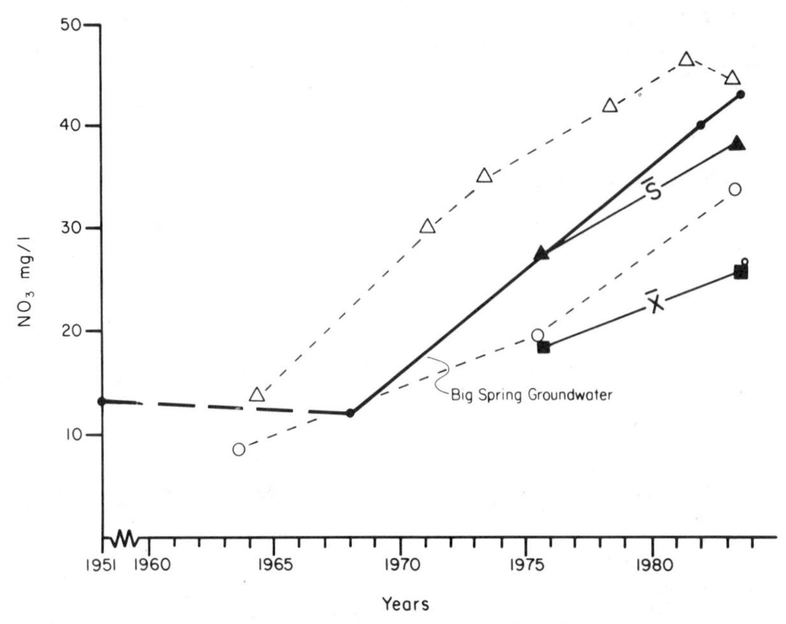


FIG. 4—*Change in mean nitrate concentrations in ground water from Big Spring outlet (*solid circles*), the surficial aquifer data in 50-well network (*$\bar{s}$, solid triangles*), the total 50-well network (*$\bar{x}$, solid squares*), and two individual wells (*open circles *and* triangles*).*

Historic ground-water quality data from Big Spring show an increase in ground-water nitrate concentrations of approximately 230% between 1968 and 1982. Historic data from 50 wells, in surrounding areas, similar to the Big Spring basin, show the same magnitude of change (Fig. 4).

Since the late 1960's, corn acreage increased about 40% and the application rate of fertilizer-*N* increased about 80%, resulting in a total increase of applied fertilizer-*N* of about 250%. Over the same period, manure from cattle and hogs increased only about 30%.

Figure 5 graphically summarizes the change in ground-water nitrate concentrations, fertilizer-*N* application, and manure-*N* contributions. The primary reason for the increase of nitrates in the ground water is clearly the dramatic increase in chemical fertilizer-*N* applied.

## Phase III

The agencies and individuals working on the first two phases of this study became aware of the lack of state or Federal programs to deal with the contamination of the ground water from agricultural nonpoint sources, especially chemicals and chemical fertilizers. Another problem was this: many land-treatment practices were believed to reduce ground-water contamination, but their quantitative effects on ground-water quality and crop production were unknown. Concerned state, Federal, and local agencies and university organizations formed an ad hoc committee in 1983 to take action to reduce the hazards of ground-water contamination. It is jointly chaired by the Cooperative Extension Service (CES) and the Northeast Iowa Conservancy District (NEICD).

The ad hoc committee recommended a pilot project to develop and demonstrate best management practices (BMPs) in the field, evaluate the effects of BMPs on both ground-water quality and crop production, assess the economic impact of these practices, and provide a foundation of information and experience from which policies, programs, and institutional arrangements can be built to help agriculture address Iowa's increasing ground-water problems.

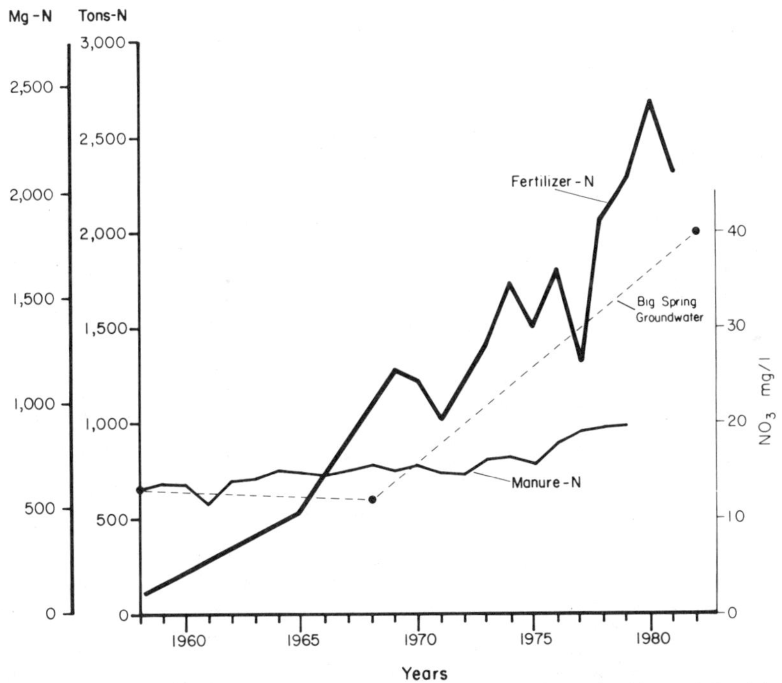


FIG. 5—*Estimated tons of fertilizer and manure nitrogen applied in Big Spring basin (*left axis*) and mean nitrate concentration (*right axis*) in ground water from Big Spring.*

A demonstration project has been developed to begin meeting the challenge of protecting ground water from agricultural chemicals and fertilizers over broad areas of northeast Iowa. The Big Spring basin was chosen as the focus for the project because of the previous work done there, the unique opportunities to monitor ground-water quality in the area, and the excellent cooperation of the residents.

The primary focus of the seven-year cooperative project is to find and demonstrate methods to reduce the delivery of nitrates and pesticides to ground water from agricultural land. Components of the project include educational, research, demonstration, monitoring, and evaluation programs. Figure 6 summarizes the conceptual approach to the overall demonstration project.

Various state and Federal agencies and university researchers will participate with coordination and review of the project conducted by the ad hoc committee. Funding for the project will come from landowners, agency resources, and outside funding sources.

As part of the demonstration project, CES, ASCS, and SCS will provide records of changes in land treatment and cropping practices. An annual survey of all basin residents will ascertain changes in chemical usage, livestock population, and tillage practices. Color-infrared aerial photography will be flown annually, and along with field mapping, will provide annual updates on cropping patterns and other land use changes.

An automated data processing system will be developed on the IGS computer system for storage and analysis of all hydrologic and water-quality data gathered during the project. Data will be summarized on a water/year basis and brief reports will be prepared annually with more extensive reports compiled at the end of the third and seventh years.

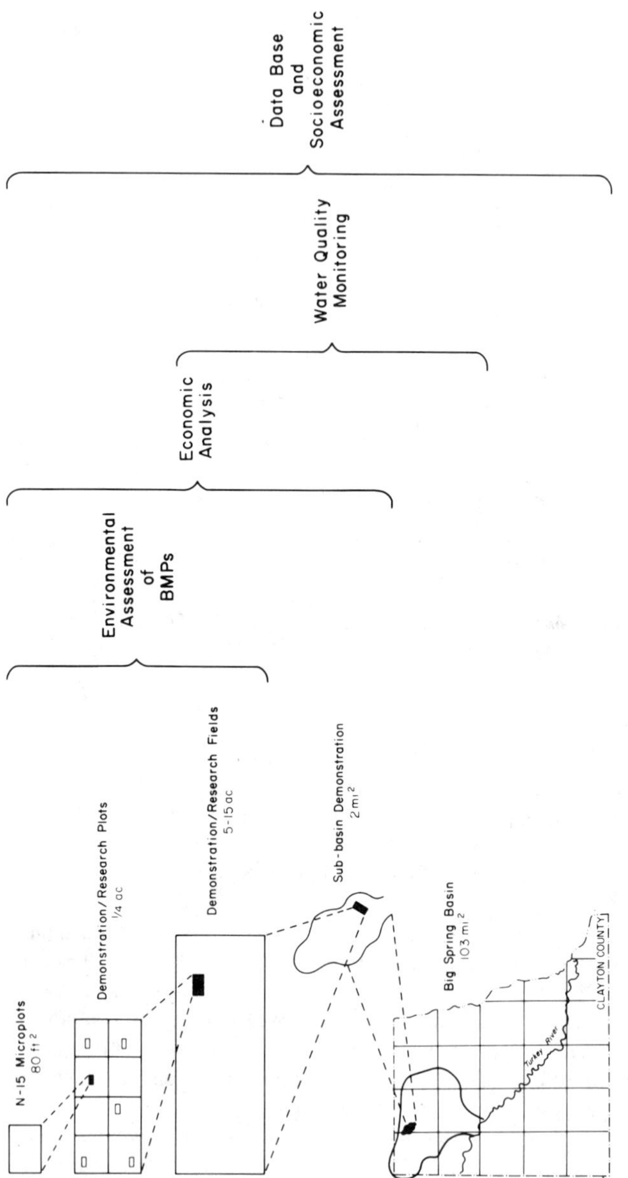


FIG. 6—*Conceptual approach to Big Spring Demonstration Project.*

### *Basin-Wide BMP Implementation, Monitoring, and Evaluation*

An expanded educational program, conducted in conjunction with ongoing soil conservation efforts, will serve as the primary mechanism for achieving accelerated application of BMP's throughout the Big Spring basin. Educational activities may include workshops on the ground-water system and movement of contaminants, sprayer calibration, and agricultural chemical and manure management.

Some activities have already begun. The CES has been publishing a quarterly newsletter for basin residents and is assisting farmers in pest management procedures that minimize the use of excessive amounts of pesticides. They have also had demonstration plots for fertilizer management.

During this portion of the study, various sampling will be done to address potential health related issues. Only a limited number of pesticides in use in Iowa have been found in ground water, and it is not known if metabolites, or breakdown products, are occurring. The Institute of Agricultural Medicine, University of Iowa through the Iowa Pesticide Hazard Assessment Program (IPHAP) will look for the occurrence of these metabolites.

This implementation effort should produce improvements in ground-water quality proportionate with the success at implementing BMP's. Success will be monitored using the land-management data base, including data on cropping, chemical use, and conservation practices. Monitoring will include both the discharge and chemical concentrations of surface water and ground water so that a water budget can be computed and chemical losses can be evaluated in relation to land management. The ultimate proof of BMP effectiveness will be improvement in ground-water quality at Big Spring and in the basin.

### *Subbasin Implementation of BMP's*

Accelerated cost-sharing and one-on-one educational and technical assistance activities will be employed to encourage land users to implement BMP's within a single 4.4-$km^2$ (1100 acre) subbasin. In addition to traditional soil and water conservation practices, fertilizer, pest, and animal waste management practices will be used under long-term contracts with land users.

Installation of land treatment practices in the Big Spring basin has already begun. ASCS has chosen the Big Spring basin as a special project, allowing 75% cost sharing on selected practices. SCS is also providing accelerated cost-sharing assistance for the subbasin. The results in this demonstration subbasin will be used in educational efforts throughout the Big Spring project area.

### *Field Demonstration/Research Projects*

Iowa State University (ISU) and IGS scientists will intensively study the environmental, crop yield, and economic effects of selected BMP's through demonstration and research plot projects. The effects of nitrogen shielding, nitrogen stabilization, multiple nitrogen applications, different nitrogen application rates, and crop rotations will all be evaluated in conjunction with soil conservation practices.

The results of the field demonstration/research projects will help in education efforts and will ultimately provide the technical database which is needed to make sound farm management recommendations and build future programs.

The total cost for the proposed project during the seven years is $6786200. This includes $2321700 from existing programs of state and Federal agencies, $37100 from subbasin landowners as their contribution in cost-sharing programs, and $4427400 from external funds specifically allocated towards the project.

**Discussion**

IGS has been the lead agency in the study of ground-water quality in northeast Iowa. They are also one of the key agencies on the ad hoc committee which is directing the Big Spring Demonstration Project. This has provided a consistency throughout the study, but, more importantly, there has been a strong emphasis on geology, which is necessary for any ground-water investigation.

The regional ground-water quality study in northeast Iowa was initiated because of concerns that surface contaminants were reaching the ground water through sinkholes. While sinkholes turned out to have a relatively small impact on the total problem, the geology of northeast Iowa significantly affects the contamination. The aquifers which produce drinking water in other areas are often more deeply buried than they are in northeastern Iowa, and problems are more localized. Thus, the infiltration of agricultural chemicals has not reached and affected these aquifers to the same scale yet.

Northeast Iowa has drawn considerable attention the last few years relative to ground-water quality. A misconception has developed, because of the intensive study, that water quality in northeast Iowa, and particularly the Big Spring basin, is worse than other areas of Iowa or the Midwest. This is not the case, however. The work that IGS has done elsewhere in the state indicates that the quality of water occurring in the Big Spring basin is typical of that occurring in shallow ground-water acquifers across Iowa and probably the entire Cornbelt.

The Big Spring basin was chosen for further study and demonstration because the spring and the facilities used to develop it for the fish hatchery, allow ground water to be studied in ways not possible elsewhere. Practically the entire discharge of ground water from the basin is through Big Spring. This allows an accurate measurement of discharge and the computation of the total volumes of contaminants.

Considerable effort went into the analysis of the regional water-quality data sets in relation to the geology during Phase I, and defining the geology and geography of the Big Spring basin in Phase II. The following studies were an integral part of the overall Big Spring study:

(*a*) geologic mapping,
(*b*) inventory of 90%, of the basin wells,
(*c*) potentiometric surface mapping,
(*d*) dye tracing,
(*e*) historic water quality and land management inventories,
(*f*) stream gaging,
(*g*) spring gaging,
(*h*) land use mapping, and
(*i*) land treatment and chemical use surveys.

The analyses that were made in the Big Spring study are the direct result of this effort, which allowed the determination of aquifer characteristics and their relation to land utilization in the basin.

In ground-water investigations it is necessary to keep the number of uncontrolled variables to a minimum. This is generally more difficult to do when investigating contamination from nonpoint sources than contamination from point sources. The Big Spring area is predominantly agricultural and has no discharges of municipal wastes. The contamination in the Big Spring aquifer is almost entirely related to agricultural practices—the application of agricultural chemicals in combination with total nutrient management.

Nitrate levels at Big Spring have nearly tripled in the past 20 years in response to increasing intensive agricultural production, and there is no evidence to indicate a change in this trend. The amount of nitrogen discharged annually at Big Spring equals about 5600 to 8400 kg/km$^2$ (50 to 75 lb/acre) for the long-term cropped area of the basin. This is about 33 to 55% of average fertilizer-*N* normally applied.

An area of concern in ground-water studies is quality control during sampling, storage, and transportation of samples to the laboratory. All sampling during the more detailed portions of this study was done under the direction of IGS in cooperation with UHL. An approved EPA quality assurance/quality control program was followed. IGS has custody of the samples until turning them over to the UHL laboratory.

During Phase I of the study, a need was discovered relative to the existing water-quality data sets. Over 16000 analyses of ground water were available, but the control on them varied widely. The UHL data had previously been aggregated by town and county; however, data handling and statistical analysis required a great deal of effort and manual handling. An improved database including more precise information on source of water (aquifer) and geographic location is necessary to better evaluate regional water-quality problems.

This database is also necessary to aid researchers in their study of the health effects of nitrates and herbicides. The flow-weighted mean concentration of nitrates in ground water issuing from Big Spring has fluctuated around the EPA drinking water standard of 45 mg/L (45 ppm). The health effects of elevated nitrate levels are not well documented, and there has been much recent discussion about whether this standard is too restrictive. Recent studies, however, have suggested that nitrate levels below the current standards may contribute to congenital malformations in humans [*20,21*].

The volumes of pesticide losses from the Big Spring basin have been quite small in comparison to nitrate losses. Pesticide losses in ground water were less than 1% of the amount applied. Concentrations are very low, below levels thought to contribute to long-term chronic health problems, but there are many uncertainties, especially regarding combinations of pesticides and nitrates.

## Conclusions

The Big Spring Demonstration Project is the next step in the development of state, Federal, and local programs to integrate profitable agriculture, soil and water conservation, and ground-water protection. The results will be particularly applicable for implementation over substantial portions of northeast Iowa. The detailed evaluations of the practices and hydrologic system will allow application of results through substantial portions of Iowa as well as other cornbelt states.

Cooperating agencies are optimistic about the results of the project. Residents of the Big Spring Project area are well aware of the ground-water quality problems and very concerned about them. Improvement in ground-water quality is expected within one or two years of the application of BMP's and in proportion to their application.

## References

[*1*] Heitmann, N., "Water Source of Big Spring Trout Hatchery, Clayton County, Iowa," *Proceedings, Iowa Academy of Science,* Vol. 87, 1980, pp. 143–147

[*2*] Hallberg, G. R. and Hoyer, B. E., "Sinkholes, Hydrology, and Ground-Water Quality in Northeast Iowa," Iowa Geological Survey, Iowa City, IA, 1982.

[*3*] Bounk, M. J., "Some Factors Influencing Phreatic Cave Development in the Silurian Strata of Iowa," *Proceedings, Iowa Academy of Science,* Vol. 90, 1982, pp. 19–25.

[*4*] Aley, T. and Fletcher, M. W., "The Water Tracer's Cookbook," *Missouri Speleology,* Vol. 16, No. 6, 1976, pp. 1–31.

[*5*] "Methods of Chemical Analysis of Water and Waste," EPA-600/4-79-020, U.S. Environmental Protection Agency, 1979.

[*6*] Hallberg, G. R., Hoyer, B. E., Bettis, E. A., III, and Libra, R. D., "Hydrology, Water Quality, and Land Management in the Big Spring Basin, Clayton County, Iowa," Iowa Geological Survey Open-File Report 83-3, Iowa City, IA, 1983.

[*7*] Hallberg, G. R., Libra, R. D., Ressmeyer, G. G., Bettis, E. A., III, and Hoyer, B. E., "Temporal

Changes in Groundwater Nitrates in Northeast Iowa,'' Iowa Geological Survey Open-File Report 84-1, Iowa City, IA, 1984.

[8] Hallberg, G. R., Libra, R. D., Bettis, E. A., III, and Hoyer, B. E., ''Hydrogeologic and Water Quality Investigations in the Big Spring Basin, Clayton County, Iowa,'' Iowa Geological Survey Open-File Report 84-4, Iowa City, IA, 1984.

[9] Prior, J. C., ''A Regional Guide to Iowa Landforms,'' Iowa Geological Survey Educational Series 3, 1976.

[10] Wischmeier, W. H. and Smith, D. D., ''Predicting Rainfall Erosion Losses—A Guide to Conservation Planning,'' U.S. Department of Agriculture, Agriculture Handbook No. 537, 1978.

[11] ''Urban Hydrology for Small Watersheds,'' U.S. Department of Agriculture Soil Conservation Service, Engineering Division, SCS Technical Release No. 55, 1975.

[12] Smart, P. L. and Laidlow, I. M. S., ''An Evaluation of Some Fluorescent Dyes for Water Tracing,'' *Water Resources Research,* Vol. 13, 1977.

[13] Singh, K. P. and Stall, J. B., ''Derivation of Base Flow Recession Curves and Parameters,'' *Water Resources Research,* Vol. 7, No. 2, 1971, pp. 292–303.

[14] Freeze, R. A. and Cherry, J. A., *Groundwater,* Prentice-Hall, Englewood Cliffs, NJ, 1979.

[15] Scalf, M. R., McNabb, J. F., Dunlap, W. J., Cosby, R. L., and Fryberger, J., *Manual of Ground-Water Sampling Procedures,* NWWA/EPA Series, Ada, OK, 1981.

[16] Gibb, J. P., Schuller, R. M., and Griffin, R. A., ''Procedures for the Collection of Representative Water Quality Data from Monitoring Wells,'' Illinois State Water Survey and Illinois State Geological Survey, Cooperative Groundwater Report 7, Champaign, IL, 1981.

[17] *National Handbook of Recommended Methods for Water-Data Acquisition,* Office of Water Data Acquisition, U.S. Geological Survey, Reston, VA, 1984.

[18] ''Ecological Evaluation of Proposed Discharge of Dredged or Fill Material into Navigable Waters,'' U.S. Army Engineers, Waterways Experiment Station, Miscellaneous Paper D-76-17, 1976.

[19] ''Manual for the Analysis of Pesticides in Humans and Environmental Samples,'' EPA-600/8-800-083, Section 11, U.S. Environmental Protection Agency, June 1980.

[20] Dorsch, M. M., Scragg, R. K., McMichael, A. J., Baghurst, P. A., and Dyer, K. F., ''Congenital Malformations and Maternal Drinking Water Supply in Rural South Australia: A Case Control Study,'' *American Journal of Epidemiology,* Vol. 119, 1984, pp. 473–486.

[21] Fraser, P. and Chilvers, D., ''Health Aspects of Nitrate in Drinking Water: The Science of the Environment, Water Supply and Health,'' *Studies in Environmental Science,* Vol. 12, 1980, pp. 103–116.

*Bradley G. Waller*[1] *and Barbara Howie*[1]

# Determining Nonpoint-Source Contamination by Agricultural Chemicals in an Unconfined Aquifer, Dade County, Florida: Procedures and Preliminary Results

**REFERENCE:** Waller, B. G. and Howie, B., "**Determining Nonpoint-Source Contamination by Agricultural Chemicals in an Unconfined Aquifer, Dade County, Florida: Procedures and Preliminary Results,**" *Ground-Water Contamination: Field Methods, ASTM STP 963*, American Society for Testing and Materials, Philadelphia, 1988, pp. 459–467.

**ABSTRACT:** South Dade County, Florida, is one of the few areas in the continental United States where winter fruits and vegetables are grown. About 85% of the agricultural area is underlain by Rockdale soil composed of crushed limestone and organic debris. The remainder of the agricultural area is underlain by Perrine marl. Both soils must be irrigated and require frequent applications of fertilizer, micronutrients, and pesticides. The agricultural area overlies permeable limestone of the unconfined Biscayne aquifer, which is designated as the sole source of drinking water for southeast Florida. The unsaturated zone in the agricultural area ranges in thickness from 0.3 to 4.6 m.

An evaluation of nonpoint-source contamination from agricultural chemicals began in March 1985. Representative fields from the 30 000 ha under cultivation were selected for ground-water quality monitoring. Agricultural production is comprised of row crops, seasonal tropical vegetables, groves, and commercial wholesale nurseries. Five test plots underlain by both Rockdale and Perrine marl soils were selected for intense evaluation. Wells ranging in depth from 2.4 to 15.7 m were installed in areas hydraulically upgradient and downgradient and near the center of these plots. Additional wells were installed in the shallow mixing zone within and adjacent to the agricultural area to determine background and baseline water-quality conditions.

The ground-water monitoring program was designed to leave in place a permanent well network and sampling stratagem that can be adapted to future water-quality concerns. Wells are drilled by clean water rotary and left open hole in the sampling interval so that organic, inorganic, or biological samples could be taken without contamination by drilling fluids, grouting material, or well casing and screens. A quality assurance program involving 10 to 20% of all samples was used to check field and laboratory procedures and to establish the accuracy of analytical data.

Preliminary results of water-quality sampling indicate that agriculture has had a minimal effect on ground-water quality. Of the constituents sampled, only nitrate and potassium from fertilizers are found at elevated concentrations beneath farmed areas. Elevated concentrations of some nutrients, major ions, and trace metals are found beneath areas used for storage and disposal of agrichemicals.

**KEY WORDS:** ground water, well design, water-quality sampling, quality assurance, agricultural chemicals, nonpoint source

South Dade County, Florida (Fig. 1), is one of the few localities in the continental United States where winter fruits and vegetables are grown. The area provides most of the fresh produce, primarily tomatoes, corn, beans, squash, and potatoes, for the eastern United States. In addition,

[1] Research hydrologist and hydrologist, respectively, U.S. Geological Survey, Miami, FL 33178.

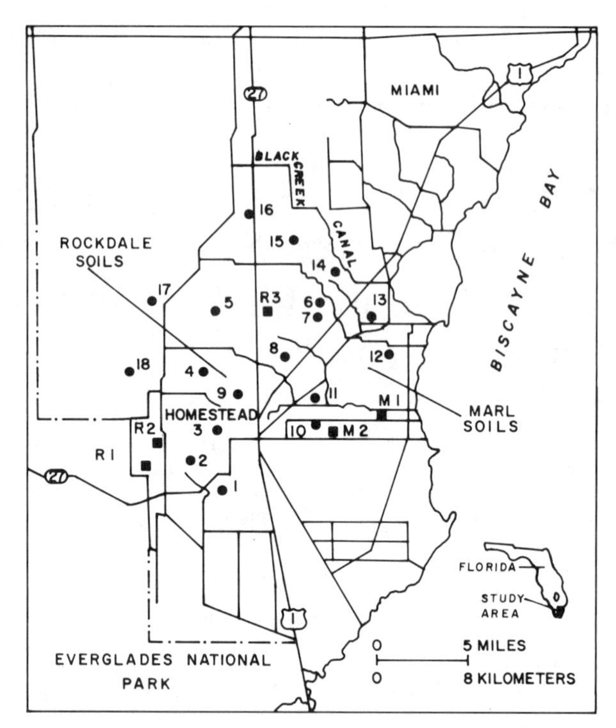


FIG. 1—*South Dade County and locations of sampling sites.*

year-round crops include groves of avocados, limes, mangos, papayas, lychees, and other tropical fruits and vegetables.

Soils in south Dade County are naturally infertile and require large inputs of fertilizers and micronutrients to maintain profitable crop production levels. One method available to increase both nutrient and moisture retention capacity is to apply sludge from treated domestic wastewater to the fields and groves. Concern has been raised by regulatory agencies about the effect of treated wastewater sludge on ground-water quality in the Biscayne aquifer. In 1985, the U.S. Geological Survey, in cooperation with the South Dade Soil and Water Conservation District and the Florida Department of Environmental Regulation, began a 4-year investigation to determine the effects that applications of sludge and selected agricultural chemicals have on ground-water quality in the south Dade County agricultural area.

The source of all potable and irrigation water in the study area (Fig. 1) is the unconfined Biscayne aquifer described by Parker et al [*1*]. Work by Klein and Hull [2] provided an impetus for this aquifer to be designated as the sole source of drinking water for the area. In south Dade County, the Biscayne aquifer is composed of limestone and sandstone with some pockets of sand of Pleistocene and Pliocene age.

Measured hydraulic conductivities in the Biscayne aquifer are greater than 3,000 m/day, and the vertical permeability greatly exceeds the horizontal permeability ([*1*], p. 102). The Biscayne aquifer in the area of investigation ranges in thickness from about 18 to 37 m, and transmissivities exceed 60,000 $m^2$/day throughout the area.[2]

There is little true soil development in south Dade County. Surface deposits comprise thin

[2] Fish, J. E., U.S. Geological Survey, written communication, 1986.

Holocene materials, which include freshwater marl, peat, organic detritus, and quartzose sand. Two types of soils are under cultivation in south Dade County. The most extensively farmed soil in south Dade County is the Rockdale soil association [*3*]; which is found along the Atlantic Coastal Ridge ([*4*], p. 51). Rockdale soils are composed of any surface material, including limestone which is subsequently crushed and worked until a material that can be cultivated is produced. Perrine marl is comprised of freshwater deposits found in the physiographic province of the Southern Coast ([*4*], p. 53) and the Transverse Glades [*5*]. About 26 000 ha of Rockdale soil and 4000 ha of Perrine marl are currently (1986) in agricultural production.

The climate of south Florida has two distinct seasons: a hot and humid June through October when seasonal rainfall averages approximately 1140 mm, and a dry and mild November through April when rainfall is about 380 mm. May is transitional.

This paper presents the methodology used in site selection, monitor well placement, monitoring well installation and construction, sampling procedures, quality assurance measures, and initial results of the evaluation of nonpoint-source contamination from agricultural chemicals to the Biscayne aquifer.

Site selection, installation of the monitoring wells, and collection of water-quality samples began in 1985. Ground-water samples will be collected eight times during 1986 to 1987 under varying climatological conditions, such as the first flush of summer rains, rising water-table, high water-table, and low water-table conditions. Sampling will be coordinated with agricultural chemical applications.

## Methodology and Procedure

Two types of sites were selected: (1) Baseline sites (labeled 1 to 18) where single monitoring wells were installed to determine water-quality conditions in the uppermost part of the saturated zone along interbasin areas—away from canals (Fig. 1); and (2) five test fields (labeled R1, R2, R3, M1, and M2) where multidepth clusters of monitoring wells were installed for intensive water-quality monitoring. Of a total of 18 baseline sites, two (Sites 17 and 18) are considered background because they are hydraulically upgradient of development and two (Sites 6 and 16) are at known sources of agricultural contamination where chemicals and byproducts are stored. The unsaturated zone at the baseline sites ranged in thickness from 0.3 to 4.6 m. The five test fields are summarized in Table 1.

Three multidepth well clusters were installed within or adjacent to each of the five test fields—one upgradient, one near the center, and one downgradient. The downgradient cluster was within

TABLE 1—*Test field summary.*

| Field Label (Fig. 1) | Type of Field | Crop | Field ha | Years Under Cultivation |
|---|---|---|---|---|
| | | ROCKDALE SOIL | | |
| R1 | control | none | 16 | 0 |
| R2 | row crops | tomatoes | 16 | 11 |
| R3 | grove | avocadoes | 12 | 30 |
| | | PERRINE MARL | | |
| M1 | control | none | 4 | 0 |
| M2 | row crops | corn | 8 | 50 |

10 m of the test field because of the generally low water-table gradients (10 cm/km). Changes in flow direction could cause erroneous observed data, but these changes are highly unlikely in south Florida where ground-water gradients are extremely low. Each three-well cluster at the five test fields consisted of the following: a shallow well finished in the uppermost part of the aquifer (2.4 to 4.6 below land surface); the middepth well extends 7.3 to 9.1 m below land surface, but not penetrating an extensive, dense, freshwater limestone layer [6]; and a deeper well drilled to a depth of 12.2 to 15.2 m below land surface in a highly permeable part of the upper Biscayne aquifer for a total of three wells in each cluster.

Wells were drilled by rotary method using ground water for circulation. Shallow wells were drilled to depths 0.3 to 0.6 m below the lowest historic water levels. Casings were set to depths just above the highest recorded water level, enabling a sample to be taken at the water table. The bottom 1.3 m of deep wells were left as an open hole. Wells were cased above the open-hole interval by filling the open-hole interval with clean silica sand, installing 51-mm threaded schedule 40-PVC (polyvinyl chloride) pipe above the sand, and grouting it to land surface (Fig. 2*a*), and then pumping out the sand (Fig. 2*b*). This assures that there is minimal contact between the PVC pipe and ground-water sample. Wells were secured at the surface by grouting a 152-mm PVC casing around the wellhead. No samples were taken before the well had been developed by pumping three times at two-week intervals.

Three different sampling methods are used to collect ground-water samples. The methodology used is dependent on the chemical constituents to be analyzed. First, a reinforced Tygon[3] tube is lowered to the open-hole interval, and ten casing volumes of water are withdrawn with a centrifugal pump. Samples are then collected for physical properties, trace metals, potassium, sodium, chloride, and nutrient analyses. The Tygon tube is then removed. Second, a Teflon tube rinsed with methanol and organic-free water is inserted for sampling neutral and acid-base extractable synthetic organic compounds, insecticides, and herbicides by using a peristaltic pump and vacuum chamber.

Third, a Teflon bailer is used to sample for volatile organic compounds and dissolved organic carbon. Samples for all categories of chemical constituents analyzed can be taken from the same monitoring well by using these sampling techniques.

## Quality Assurance Program

A quality assurance program outlined in Table 2 was included in this investigation to verify sampling methods and analytical procedures. Because background concentrations of many constituents were expected to be close to the limits of detection, even minimal sample contamination or analytical error could be significant. Analytical recoveries that range from 30 to 120% are considered acceptable for some organic analyses; thus, it was imperative that other sources of sample variance be minimized when sampling for organic compounds.

In the field, 10 to 20% of the samples collected were field blanks, duplicates, or reference samples. Trip blanks consisted of triple distilled, organic-free water samples that were treated in the field and transported to the laboratory in the same manner as ground-water samples. Trip blanks were submitted for analysis to determine if samples were contaminated between the time of collection and the time of arrival in the laboratory. Duplicates were samples split from a single source and were used to determine the precision of the analytical procedures used by the laboratory. Reference samples containing known concentrations of a particular constituent were used to determine analytical accuracy and the effects of shipping, holding time, and sample handling on constituent concentrations.

[3] Use of brand names in this report is for identification purposes only and does not constitute endorsement by the U.S. Geological Survey.

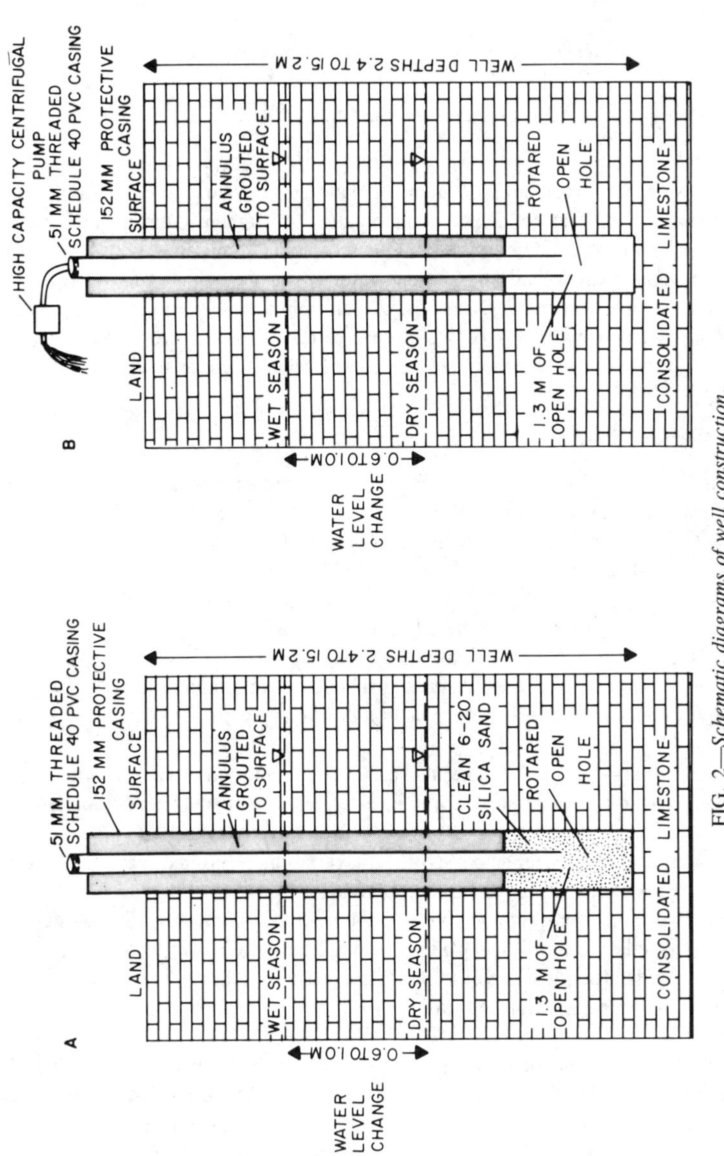


FIG. 2—*Schematic diagrams of well construction.*

TABLE 2—*Type of quality assurance.*

| Field (10 to 20% of samples) | Laboratory |
|---|---|
| Trip blanks | laboratory blanks |
| Duplicate samples | surrogates |
| Reference samples | spikes |
| | internal standards |

The quality assurance procedures used in the laboratory were similar to those in the field. Laboratory blanks were used to detect contamination of samples during analysis. Analytical recovery (accuracy) was determined by adding known concentrations of constituents to a sample to produce a surrogate sample and known concentrations to deionized water to produce a spiked sample. Before a sample was injected into the gas chromatograph for analysis, internal standards consisting of known concentrations of representative constituents were added to the sample to determine the accuracy of the injection and detection phases of the analysis.

**Preliminary Results**

Initial water-quality sampling at completed sites was conducted in March and April 1985, near the end of the winter growing season. Rainfall amounts were well below average, and irrigation of row crops was frequent. The following chemical constituents and physical properties were analyzed from ground-water samples collected from 30 wells:

*Physical properties*—Specific conductance, alkalinity, pH.

*Nutrients*—Nitrogen, phosphorus, dissolved organic carbon.

*Micronutrients*—Copper, iron, magnesium, manganese, potassium, zinc.

*Sludge-derived contaminants*[4]—Arsenic, cadmium, chloride, chromium, lead, mercury, nickel, sodium.

*Organic constituents*—Acid extractable and base-neutral compounds, volatile organic compounds, chlorophenoxy herbicides, organophosphorus insecticides, organochlorine compounds.

The initial sampling results reflected significant variation in ground-water chemistry in south Dade County that is not related to agricultural activities. Specific conductance and concentrations of sodium, chloride, and magnesium (Table 3) were high in ground-water samples collected in an extensive area affected by saline water from a flowing artesian well [*1*], and by saltwater intrusion in the Biscayne aquifer found to the east along the coast [*8*]. Iron concentrations (Table 3) also varied greatly as indicated by concentrations of 170 and 470 μg/L at the two background sites (Sites 17 and 18).

Nutrients and inorganic ions determined in ground water (3) include nitrogen, phosphorus, and potassium derived mainly from fertilizer, magnesium derived from fertilizer or preexisting saline ground water, and sodium and chloride derived mainly from preexisting saline water or from sludge application. Potassium and nitrate concentrations were higher at the agricultural sites (Sites 1–5, 7–9, 11–16, M2, and R2–R3) than at the background sites (Sites 17 and 18) because of agricultural chemical applications. Potassium in agricultural areas averaged 9.4 times background concentrations. Nitrate concentrations were higher in areas planted as groves (1.6 to 11 mg/L) than as row crops (0.01 to 2.4 mg/L). At one lime grove concentrations exceeded the 10-mg/L

[4] Determined by analysis of domestic wastewater sludge in Miami and those commercially available.

TABLE 3—*Summary of nutrient and inorganic ion concentrations in ground water.*[a]

| Constituent | Background Sites (2 wells), Range | Agricultural Sites (26 wells)[b] | | Sources (2 wells), Range |
|---|---|---|---|---|
| | | Range | Average | |
| Specific conductance | 430 to 550 | 285 to 2,200 | 718 | 775 to 935 |
| Nitrogen | | | | |
| Ammonia | 0.02 to 0.34 | <0.01 to 0.69 | 0.125 | 1.5 to 1.7 |
| Organic | 0.66 to 0.96 | 0.06 to 0.94 | 0.366 | 0.2 to 1.5 |
| Nitrate | 0.01 | <0.01 to 11 | 2.04 | 0.01 to 15 |
| Nitrite | <0.01 | <0.01 to 0.06 | 0.006 | <0.01 |
| Phosphorus | <0.01 to 0.01 | <0.01 to 0.04 | 0.010 | 0.01 |
| Alkalinity as $CaCO_3$ | 264 | 224 to 472 | 281 | 360 to 376 |
| Magnesium | 2.6 to 5.7 | 1.1 to 39 | 6.38 | 7 to 9.3 |
| Potassium | 0.50 to 1.1 | 1.7 to 16 | 7.55 | 43 to 47 |
| Chloride | 18 | 4.4 to 480 | 63.9 | 53 to 57 |
| Sodium | 11 | 1.5 to 280 | 37.0 | 15 to 36 |

[a] Concentrations shown in milligrams per liter, except for specific conductance which is in microsiemens per centimeter; <, less than known contaminant.
[b] Includes baseline sites and shallow wells at test field sites.

Primary Drinking Water Standards [*9*]. Concentrations of magnesium, sodium, potassium, and chloride were higher in the agricultural areas than in nonagricultural areas primarily because of inflow from saline artesian water.

The two sites (Sites 6 and 10) where known sources of contamination were present had higher concentrations of most nutrients and major ions than any of the other sites. One site (Site 6), which is an agricultural dump, had concentrations of ammonia, organic nitrogen, and potassium that were one order of magnitude above the background concentrations, and nitrate concentrations that were two orders of magnitude higher than background. Alkalinity and magnesium concentrations were also elevated at this site. The other site (Site 10), where agricultural chemicals are stored, had concentrations of ammonia and potassium one order of magnitude higher than background, and concentrations of alkalinity and magnesium were elevated.

Metal concentrations (Table 4) were generally well below the limits established for drinking water in both background and agricultural sites except for iron and manganese. Iron exceeded the 300-μg/L drinking water regulation [*10*] at eight agricultural sites (Sites 1 and 12, three wells at R1, two wells at R2, and one well at M1) and at one background site (Site 17). Parker et al. [*1*] noted that high concentrations of iron were not unusual for ground water in south Florida. However, iron is a commonly applied mirconutrient, and higher concentrations were found associated with an avocado grove and a tomato field, thus indicating that soluble iron from fertilizers may be reaching the water table. Manganese, which is also a micronutrient, was found to be equal to the 50-μg/L limit [*10*] at one potato field. This high manganese concentration may have been related to recent fertilization and irrigation practices because manganese was usually found at or below the detection limit (10 μg/L). None of the other trace elements (arsenic, cadmium, chromium, copper, lead, mercury, nickel, and zinc) exceeded drinking water limits at background (Sites 17 and 18) or agricultural sites (Sites 1-5, 7-9, 11-16, M1, M2, and R1-R3). In fact, when these constituents were detected at these sites, concentrations generally were close to the detection limits. The two wells at sites (Sites 6 and 10) of known contaminations had higher than background

TABLE 4—*Summary of metal concentrations in ground water.*

| Constituent | Background Sites (2 wells), Range | Agricultural Sites (26 wells)[b] Range | Agricultural Sites (26 wells)[b] Average | Sources (2 wells), Range |
|---|---|---|---|---|
| Arsenic | 1 | <1 to 13 | 1.9 | 7 to 20 |
| Cadmium | <1 | <1 to 1 | 0.1 | <1 |
| Chromium | <1 | <1 to 1 | 0 | <1 |
| Copper | 1 | <1 to 7 | 1.4 | 1 to 4 |
| Iron | 170 to 470 | 40 to 580 | 187 | 60 to 1,000 |
| Lead | <1 to 1 | <1 to 2 | 0.1 | <1 |
| Manganese | <10 to 30 | <10 to 50 | 9.6 | <10 to 50 |
| Mercury | <0.1 | <0.1 to 0.5 | 0.13 | <0.1 to 2 |
| Nickel | <1 | <1 to 8 | 1.3 | <1 to 1 |
| Zinc | <10 | <10 to 30 | 6 | <10 to 20 |

[a] Concentrations shown in micrograms per liter; <, less than known contaminant.
[b] Includes baseline sites and shallow wells at test field sites.

concentrations of arsenic (7 and 20 μg/L), and one site (Site 6) had high concentrations of iron (1000 μg/L), copper (4 μg/L), and manganese (50 μg/L).

There were no acid-extractable compounds, base-neutral extractable compounds, volatile organic compounds, chlorophenoxy herbicides, organophosphorus insecticides, or organochlorine compounds detected at 13 wells sampled. Dissolved organic carbon ranged from less than 0.1 to 34 mg/L in wells sampled in the agricultural areas, from 4.8 to 13 mg/L in the background wells, and from 1 to 12 mg/L in the wells located at known sources of contamination.

## Discussion

Proper design and implementation of a ground-water monitoring network depend upon extensive knowledge of the hydrologic characteristics of the aquifer to be studied and climatological conditions of the area. Hydraulic characteristics, lithologic composition, and sampling requirements dictated well design in the south Dade County agricultural area. Thickness of the unsaturated zone, variations in lithology and water levels, water-table gradients, and flow directions determined well placement and depth. Climatological conditions dictate agricultural practices and, thus, were used to determine sampling frequency.

The constituents and physical properties analyzed were selected on the basis of historic water-quality information, theoretical geochemistry, on-site inspections, county and state agricultural records, and written and oral communication with the agricultural community. Knowledge of both background information and agricultural use helped minimize expenditures on collection and analysis of water samples and maximize the worth of the chemical and physical analyses.

After the design and placement of wells, determination of sampling frequency, and selection of chemical constituents and the physical properties to be measured, selection of the materials for well construction and sampling procedures were chosen to assure that representative samples of ground water were collected. Quality-assurance procedures were formulated as a check on the treatment and analyses of the samples to assure quantitative and representative data.

Preliminary results indicate that 50 years of agricultural activity have had a minimal effect on ground-water quality. Nitrate and potassium from years of fertilizer application are routinely found at elevated concentrations but are usually below health standards. Water-quality problems that occur in the agricultural area are primarily related to the storage and disposal of agricultural chemicals rather than to their application on fields and groves.

*Acknowledgments*

Appreciation is extended to Court Greenfield, District Conservationist, South Dade Soil and Water Conservation District; the U.S. Department of Agriculture Soil Conservation Service; and the Florida Department of Environmental Regulation for continued support of this investigation. Information and data were also provided by the National Park Service, the Dade County Department of Environmental Resources Management, the University of Florida Institute of Food and Agricultural Sciences, and the Dade County Cooperative Extension Service. The authors thank the numerous individuals in the agricultural and horticultural community who allowed the U.S. Geological Survey to install and sample wells on their property.

## References

[*1*] Parker, G. G., Ferguson, G. E., Love, S. K., et al. "Water Resources of Southeastern Florida, with Special Reference to Geology and Ground Water of the Miami Area," U.S. Geological Survey Water-Supply Paper 1255, 1955.

[*2*] Klein, Howard and Hull, J. E., "Biscayne Aquifer, Southeast Florida," U.S. Geological Survey Water-Resources Investigations 78-107, 1978.

[*3*] "Soil Survey, Dade County, Florida," U.S. Department of Agriculture Series 1947, No. 4, 1958.

[*4*] Davis, J. H., Jr., "The Natural Features of Southern Florida, Especially the Vegetation, and the Everglades," Florida Geological Survey Bulletin 25, 1943.

[*5*] Craighead, F. C., "The Trees of South Florida," University of Miami Press, Coral Gables, FL, 1977.

[*6*] Causaras, C. R., "Geology of the Surficial Aquifer System, Dade County, Florida," U.S. Geological Survey Water-Resources Investigations Report 86-4126, 1986.

[*7*] Waller, B. G., "Areal Extent of a Plume of Mineralized Water from a Flowing Artesian Well in Dade County, Florida," U.S. Geological Survey Water-Resources Investigations 82-20, 1982.

[*8*] Klein, Howard and Waller, B. G., "Synopsis of Saltwater intrusion in Dade County, Florida, Through 1984," U.S. Geological Survey Water-Resources Investigations Report 85-4101, 1985.

[*9*] "National Interim Primary Drinking Water Regulations," *Federal Register*, Part IV, U.S. Environmental Protection Agency Water Programs, 27 Aug. 1980.

[*10*] "National Secondary Drinking Water Regulations, EPA-570/9-76-000, U.S. Environmental Protection Agency, 1979.

*James B. Urban*[1] *and William J. Gburek*[1]

# A Geologic and Flow-System-Based Rationale for Ground-Water Sampling

**REFERENCE:** Urban, J. B. and Gburek, W. J., "**A Geologic and Flow-System-Based Rationale for Ground-Water Sampling,**" *Ground-Water Contamination: Field Methods, ASTM STP 963,* A. G. Collins and A. I. Johnson, Eds., American Society for Testing and Materials, Philadelphia, 1988, pp. 468–481.

**ABSTRACT:** Representative measurements of ground water require a sampling protocol based on the physical and hydraulic properties of the geologic materials. Measurements must be made within the context of the site and areal ground-water flow systems. A multizone geologic system in east-central Pennsylvania is used as an example of instrumentation by geologic and hydrologic criteria. Geology is that of a weathered, fractured, sedimentary rock mantle overlying deeper, less fractured, sedimentary rock strata.

The focus of this paper is on describing the domain of and controls on ground-water flow within the shallow (<15 m) and deep (15 to 100 m) fracture systems. A systematic approach, using commonly available information, identifies the areal extent, depth, hydraulic conductivity distribution, and paths of ground-water flow in the two systems. This approach was tested within a 7.4-km$^2$ watershed and at stream-valley cross sections. The findings indicate that ground-water quality distribution is accurately portrayed in both settings if the sampling scheme is based on geologic and flow system controls. Ground-water sampling wells must be located, and their data interpreted, by grouping according to geology and the ground-water flow system elements of recharge, lateral flow, and discharge zones.

**KEY WORDS:** ground water, sampling, fractures, flowlines

Nonpoint source contamination of ground water is an important water resource issue in the Eastern United States. Here, upland watersheds typically have perennial streams with much of their total flow derived from subsurface sources. Thus, there may be a direct link between a low level but persistent degradation of ground-water quality and the low-flow water quality of major rivers and estuaries, such as Chesapeake Bay. However, there are virtually no databases available that would enable investigators to analyze typical rural-area ground-water flow systems, and thereby establish the downstream impacts of this ground-water quality.

Upland watersheds of Pennsylvania typically have shallow depths to ground water. Perennial streamflow is sustained by frequent rainfall, percolation to and drainage from perched water tables, and discharge of the local and regional aquifers. Nonglaciated upland areas are typically underlain by bedrock that is fractured (weathered) at shallow depths. Soil percolation drains to the weathered zone, and areal leakage from this zone discharges deep percolate to sustain the regional aquifer. However, the dominant flow mechanism within the weathered zone is often lateral flow to upland springs, seeps, and streams. Rural residents of the upland areas utilize the shallower aquifer of

[1] Geologist and hydrologist, respectively, Northeast Watershed Research Center, USDA-ARS, University Park, PA 16802.

[2] Trade names and company names are included for the benefit of the reader and imply no endorsement or preferential treatment of the product by the U.S. Department of Agriculture.

the weathered zone for homestead or livestock water supply, and depend upon streamflow for supplemental irrigation.

At the Northeast Watershed Research Center, efforts are underway to characterize shallow (<15 m) and deep (15 to 100 m) fracture layer controls on subsurface flow and contaminant transport in a 7.4-km$^2$ east-central Pennsylvania watershed underlain by sedimentary rock. Characteristics of the two zones are presented here to illustrate progress and problems in defining, sampling, and interpreting subsurface flow pathways and resultant chemical transport at the small-watershed scale.

## Hydrogeology of the Regional Ground-Water Aquifer

To provide insight into the hydrogeologic setting of the 7.4-km$^2$ study area, climate, hydrology, and hydrogeology at the larger-watershed scale are first described. The study area is a subwatershed of Upper Mahantango Creek, a 116-km$^2$ watershed within the Appalachian Ridge and Valley Province approximately 40 km north of Harrisburg, Pennsylvania. Climate is temperate and humid. Average precipitation is approximately 1100 mm/year, while streamflow is about 460 mm, approximately 40% of precipitation. About 60 to 70% of streamflow is estimated to be subsurface return flow, of which ground water is a major component. Most ground-water recharge occurs during the late fall, winter, and spring months; minor recharge occurs during the growing season. Snowmelt is not a major water input [*1*]. Deciduous mature forest covers the ridges; cropland and pasture dominate the rolling hills of the watershed interior. Elevation ranges from about 200 to 460 m. Major farming activities are dairy, poultry, livestock raising, and cash cropping on small (~100 ha) family farms.

Resistant quartzite strata of the Mississippian Age cap the mountain ridges, while underlying strata outcropping in the valleys is a sequence of Devonian Age sandstone, siltstone, sandy shale, and claystone. Locally, a thick layer of clay shale present at the base of the sequence acts as an impermeable barrier to the movement of ground water. Quantification of ground-water flow at this scale requires definition of the domain of flow and determination of the hydraulic conductivity and storativity distributions within the aquifer. Ground-water levels at 160 similarly constructed farm wells within the watershed were first used to develop a ground-water elevation map. The ground water was found to exist at very steep gradients related directly to topography. Aquifer (pumping) tests at 76 of the wells established the conductivity distribution within the watershed as estimated by specific capacity [*2,3*]. Specific capacity, normalized by saturated thickness, was tabulated by frequency [*3,4*], and the distribution exhibited a distinct slope break near the fortieth percentile; Parizek [*5*] attributes this break to wells that are located either on or off fracture lineaments. This frequency distribution analysis of specific capacity indicates two populations of data: wells located in or near fracture concentrations with specific capacities typically $1.0 \times 10^{-3}$ m$^3$ /min/m/m, and wells located in poorly fractured rock with specific capacities on the order of $1.0 \times 10^{-4}$ m$^3$/min/m/m.

The effect of hydrogeologic setting was also evaluated using the frequency distribution of specific capacities. Valleys in the physiographic area considered are developed along structurally controlled concentrations of joints, so bottomland well yields reflect the high conductivity of fracture zones. Bottomland wells were found to represent a unique statistical population, being generally seven or eight times more productive than upland or valley wall wells. Valley bottom wells exhibited specific capacities ranging from $0.1 \times 10^{-3}$ to $10.0 \times 10^{-3}$ m$^3$/min/m/m, while specific capacities of upland and valley wall wells ranged from $0.1 \times 10^{-4}$ to $10.0 \times 10^{-4}$ m$^3$/min/m/m.

Variation in well yield with rock type was also tested. However, the effects of valley wells was so strong that these data had to first be removed. Differences between nonvalley wells in different outcrop zones could then be evaluated. Sandstone and siltstone are only somewhat more productive water-yield units than shale or siltstone shale. The coarser-grained rocks yield about three times as much water as the shale-associated beds. Our overall interpretation of the above findings is that rock fracturing is more significant than rock type, so we consider flow in the deeper aquifer to be controlled predominantly by the fracture system.

These findings infer hydraulic properties of the deeper aquifer as distributed by topography and geologic strata. The subsurface water divides defined by the ground-water table map are found to be coincident with major surface divides, and pumping tests indicate that water yield to wells becomes negligible below 80 to 100 m depth. Thus, we have defined the lateral and bottom boundaries to the regional flow system, as well as the importance of the geologic and topographic factors governing hydraulic conductivity within the watershed. Any sampling scheme intended to evaluate the hydrogeology and ground-water quality of a watershed at this larger scale must be developed in the context of these boundaries and aquifer property distributions.

## Hydrogeology of the 7.4-km$^2$ Study Area

At this point, our focus is scaled down from regional hydrogeology to the 7.4-km$^2$ watershed shown in Fig. 1. This watershed, WE-38, contains all flow elements described previously. Topography and geology of WE-38 are typical of upland watersheds in the unglaciated, intensely folded, and faulted Appalachian Ridge and Valley Province. The watershed is underlain by two formations, Trimmers Rock (Late Devonian) and Catskill (Late Devonian–Early Mississippian). The Trimmers Rock Formation, predominantly shale in the study area, outcrops at the watershed outlet in a near-horizontal position and increases in dip within the watershed to a maximum of 22 deg at its uppermost contact to the north. The overlying Catskill Formation consists of interbedded shales, siltstones, and sandstones, becoming increasingly coarse-grained from south to north. Siltstone outcrops in midwatershed, and a relatively pure quartz-sandstone-conglomerate outcrops to form the north watershed divide. Dip of the Catskill strata increases from 22 deg (midwatershed) to 30 deg at the divide [*6*]. Based on specific capacity data [*7*], the two geologic formations are hydrologically similar. Shallow residual soils (1 to 2 m) cover virtually all of the bedrock. Detailed descriptions of the soils, geology, and hydrology of the WE-38 area can be found in Cline [*3*] and Urban [*7*].

### *Interaction of Land Use and Geology on Ground-Water Quality*

Given the findings of the larger-scale investigations, a ground-water flow system for WE-38 can be constructed as shown in Fig. 2, taken from Pionke and Urban [*1*]. Flow is from left to right, with recharge dominating at the left boundary (X) and ground-water discharge dominating at the upper right boundary (X1). Zone B is characterized primarily by vertical flows, and exhibits the highest recharge rates per unit land area. It is classified as the recharge zone. Flow in Zone C, downgradient from B, is a combination of ground-water recharge and throughflow. In contrast to B, most recharge to C comes from cropland, rather than from forest. Ground-water recharge to C dominates the shallower ground water, whereas the deeper ground water, originating in B, is hypothesized to flow horizontally through C largely unchanged in chemical character. Zone D is predominantly a zone of mixing and discharge, where deep and shallow horizontal flows from C converge and resurface to the land surface and stream. Some recharge occurs from the cropland over D, but it is reduced in importance by the strong upward flow components. The rationale for hypothesizing mixing and resurfacing of the ground water in D is that its shale layer has a very low hydraulic conductivity, particularly near the right lower boundary of the cross section. The combination of geologic structure (decrease in dip), stratigraphy (presence of low permeability shale), and geometry of the flow system favors a return flow to the stream and land surface in this zone.

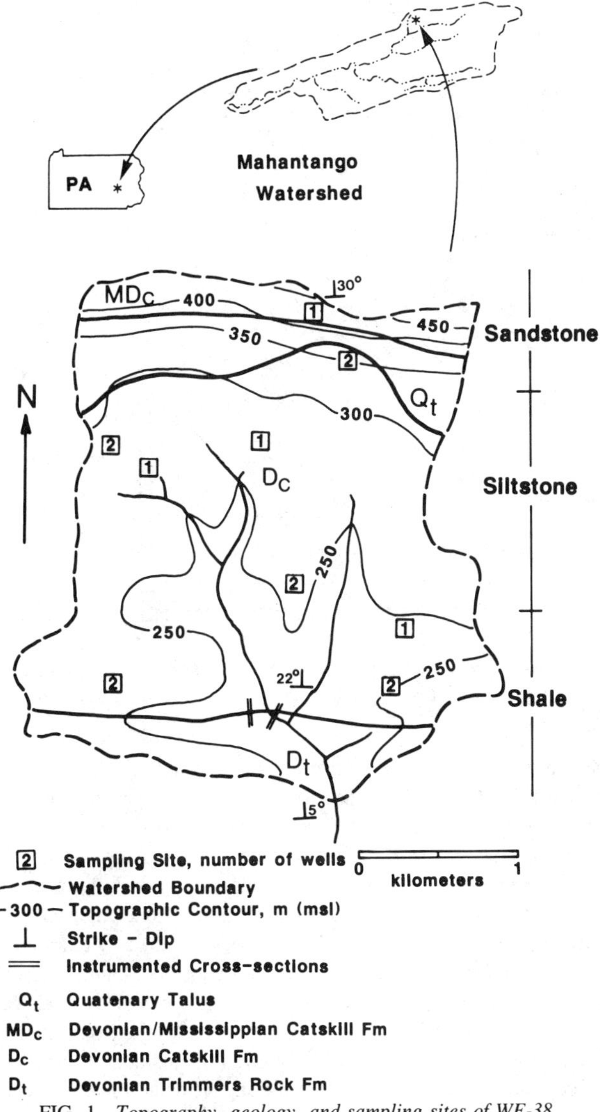


FIG. 1—*Topography, geology, and sampling sites of WE-38.*

The concern is to sample this flow system so as to evaluate the effects of both the ground-water flow system and the overlying agricultural practices. The data collection network consists of nine sites containing 14 wells (Fig. 1). Effects of perched zones are excluded from the data by use of casing and clay grout to depths of 7 m. Boreholes are open to about 60 m and are not screened within the rock formations; the rock is sufficiently competent to resist caving for many years after drilling.

The wells were sampled for inorganic water quality analysis ten times over a ten-year period. Details of sampling and analyses are given in Pionke and Urban [*1*]. A plot of selected chemical constituents from all samples and all wells in a triangle-diagram format, as percent of total

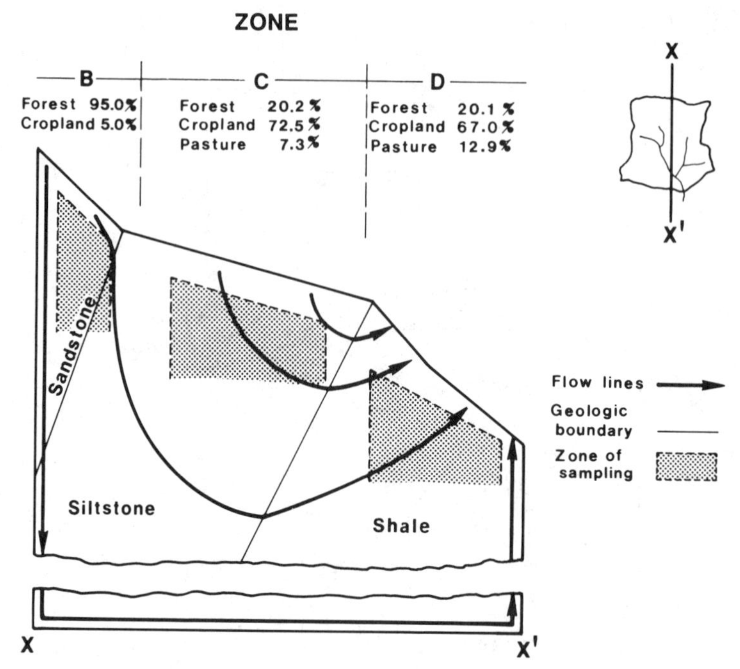


FIG. 2—*Conceptual subsurface flow system of WE-38 from Pionke and Urban* [1].

milliequivalent per litre (meq/L), was found to exhibit a wide, seemingly random, scatter of data points. However, if the data are grouped by flow system elements as shown in Fig. 3, a pattern becomes apparent. Zone B data (recharge-forest) are widely dispersed. This recharge zone ground-water chemistry represents waters from various wetting fronts passing through soils with no history of fertilizer applications and arriving at the water table at the particular times of sampling. Zone C (lateral flow-cropland) also shows a dispersion of points, but anions begin to display some clustering, and cations shift to a calcium-sodium water. Actual concentrations show a dramatic increase in nitrate-nitrogen ($NO_3$-N), chloride (Cl), and phosphate-phosphorus ($PO_4$-P) from Zone B to Zone C, as well as a moderate increase in sodium [*1*]; increases in all cations are reflected by increased electrical conductivity. The major sources of nitrogen, phosphorus, and chloride on the watershed are leaching of fertilized fields and rainfall deposition.

Zone D (discharge-cropland) has a definite clustering of data points, indicating stable ground-water geochemistry of a bicarbonate type. Here, $NO_3$-N, Cl, and $PO_4$-P concentrations have decreased compared with those of Zone C, while sodium has continued to increase. Since Cl decreased in Zone D while sodium increased, we can exclude a deep ground water brine source, giving added evidence that the high $NO_3$-N, Cl, and $PO_4$-P content in Zone C was derived from the cropland area recharge. Thus, the ground-water chemistry substantiates the flow system hypothesized in Fig. 2.

## Hydrogeology of The Shallow Fracture Zone

Based on our findings within the 7.4-km$^2$ watershed and other research within the Appalachian region [*8–11*], we anticipate that the shallow weathered fracture zone has the potential to transmit large quantities of subsurface flow. Since all the deep ground-water observation wells noted in

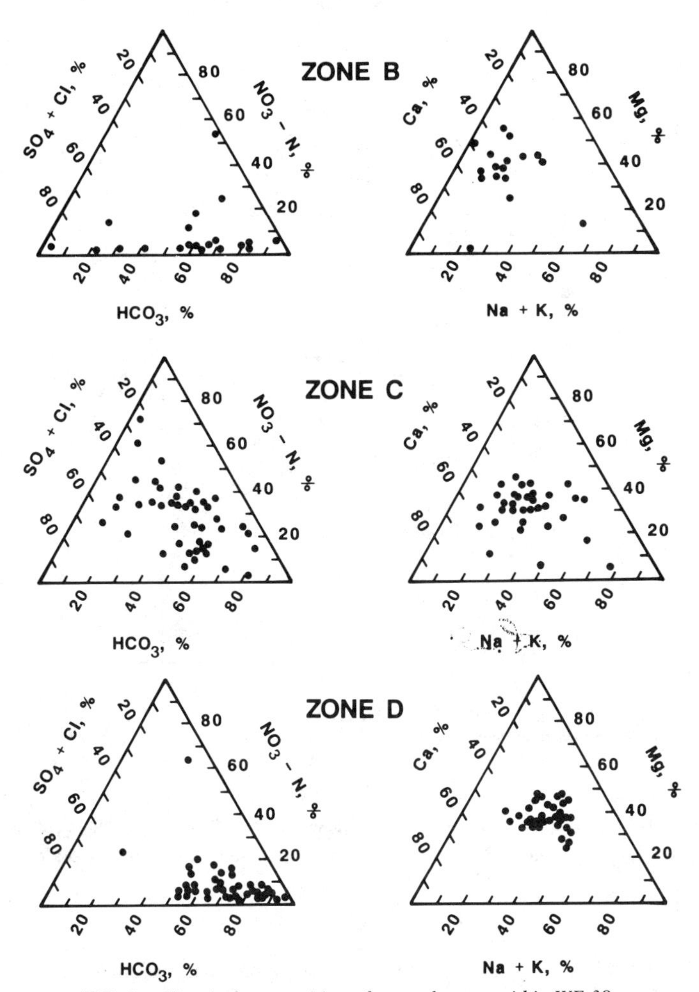


FIG. 3—*Chemical composition of ground water within WE-38.*

Fig. 1 were constructed to eliminate the effects of this zone, their data are useless for its characterization.

Based on past observations, the weathered fracture zone is normally fully drained in the uplands, where potential leakage to deeper ground water from the weathered zone exceeds root zone percolate; measurement of the zone's hydraulic properties in this setting is restricted to unsaturated monitoring techniques (that is, neutron moisture measurement) and special water injection techniques. However, ground-water return flow sustains near-surface saturation in the discharge zone (Zone D) of the 7.4-km$^2$ watershed. Here, leakage to the deeper aquifer is near zero or negative (that is, upward flow), so that lateral saturated flow is permanent within the shallow weathered fracture layer. Thus, methods available to characterize the layer's hydraulic properties in this zone can be those of traditional well and piezometer tests, and tracer observations.

We selected two cross sections within Zone D for detailed study of the shallow weathered fracture layer. Each cross section extends across a stream valley from valley wall to valley wall;

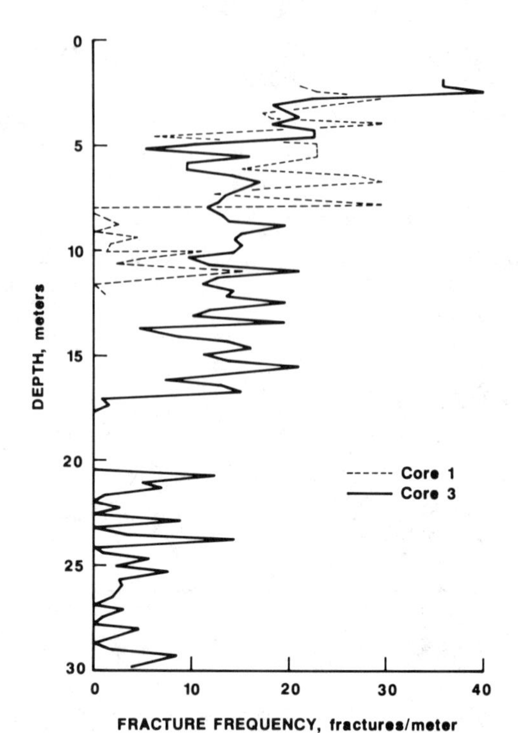


FIG. 4—*Fracture frequency distribution within two cores from the east cross section.*

locations are shown on Fig. 1. The purpose of these investigations is to characterize the physical and hydrologic controls imposed by the shallow fractured layer at the small scale.

*Rock Cores*

Rock cores were taken from each cross section, one directly in the stream channel and two on each side up to lateral distances of 75 m. Cores were obtained from the base of the soil to below the depth of intense fracturing at all sites (approximately 2 to 15 m); one core was drilled to 30-m depth. The coring extracted a high percentage of the rock from most depths, but those zones which were extremely weathered and fractured suffered relatively high sample loss. Fracture frequency for two cores from the east cross section is shown in Fig. 4. Locations of these cores, along with a qualitative indication of fracture frequency based on visual interpretation for all cores in the section, are shown in Fig. 5. Bedrock is seen to be more highly fractured in the upper 5 to 10 m than at lower depths across the entire section. Depth of fracturing appears to be greatest directly under the channel, the same pattern reported by Wyrick and Borchers [*10*]. The west cross section duplicates this pattern.

Within the weathered zone, the cores exhibit oxidized iron and manganese minerals, deposition of amorphous silica, and weathering of shale to clay-sized fragments in the fractures. The weathered zone appears as a dramatic visual feature of the cores; brown, red, and light-colored rock materials and fragments occur in the upper section of the cores, whereas the zone below contains the dark blues and grays of unweathered parent material (clay shale). The contact between the two colors is distinct and consistent from core to core. Fracture width in the weathered zone appears to range from a fraction of a millimetre to about 10 mm, while widths in the unweathered zone appear too

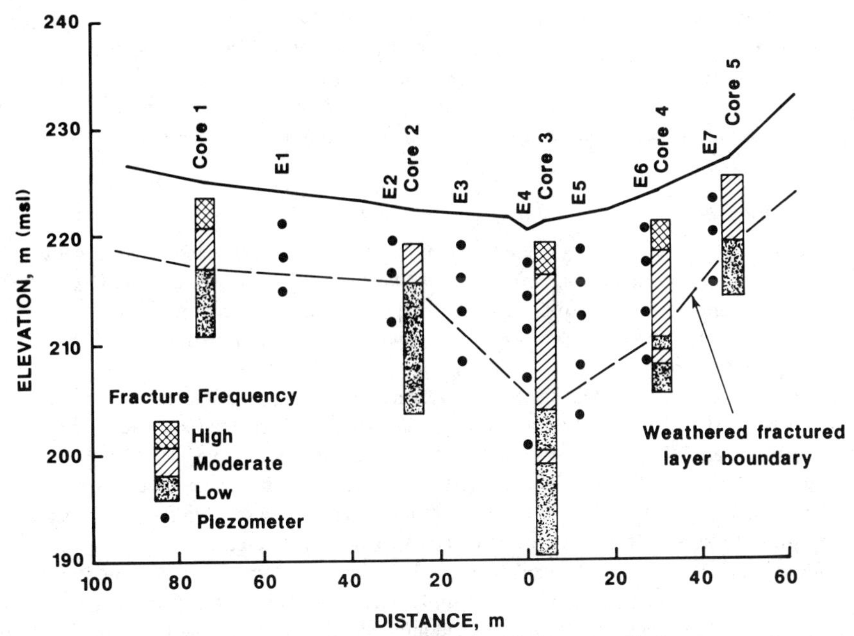


FIG. 5—*Schematic of fracture distribution and location of piezometers in the east cross section.*

small to be measurable. Occasionally, there are single or groups of isolated fractures within the unweathered zone which appear to be weathered and are of the same dimensions as those in the weathered zone.

Matrix characteristics of these cross sections were evaluated by the Mineral Sciences Laboratory of the Pennsylvania State University using samples from the cores; values derived are given in Table 1. Hydraulic conductivity values, determined using an air permeameter, indicate that interconnected large-pore matrix conductivity is so small as to be virtually unmeasurable by the

TABLE 1—*Matrix characteristics of selected core samples.*

| Core | Depth, m | Effective Porosity,[a] % |
|---|---|---|
| 2 | 4.6 | 0.546 |
| 2 | 6.1 | 0.533 |
| 4 | 6.4 | 0.278 |
| 4 | 9.8 | 0.270 |
| 4 | 10.7 | 0.582 |
| 4 | 12.2 | 0.557 |
| 4 | 15.2(t)[b] | 0.166 |
| 4 | 15.2(b)[c] | 0.091 |
| 4 | 18.3 | 0.146 |

[a] Absolute porosity <2% for all samples; hydraulic conductivity <0.0003 m/day for all samples.
[b] (t) = sample from top of core piece.
[c] (b) = sample from bottom of core piece.

equipment at the laboratory. Effective matrix porosity, measured by a kerosene displacement technique, is extremely low in all cases. Total porosity is also low. The shale bedrock matrix does not appear to constitute a significant transport media for ground water, nor to have great potential to store water. Thus, within the shallow weathered fracture layer, the flow system is essentially governed by fracture geometry, with little potential for interaction with the matrix.

### *Geophysical Methods*

Ground-penetrating radar (GPR) and seismic refraction surveys were used in attempts to characterize the soil, weathered bedrock, and unweathered bedrock layers. GPR and seismic refraction are complementary geophysical techniques, both of which can provide a map of subsurface conditions. However, each system responds to different properties of materials and has different capabilities and limitations.

GPR is a relatively new geophysical tool. Numerous reports document its use to characterize rock contacts, faults, bedding planes, and jointing patterns (for example, Refs *12,13*). Our experience using an SIR System-8[2] GPR unit with microprocessor, and 80, 120, 300, and 500 MHz antennas, indicates that significant interference problems are encountered in shale, which has numerous bedding planes. The GPR configurations used were totally ineffective in determining characteristics of the weathered zone near the cross-section study areas.

Seismic refraction surveys have been used widely in hydrogeologic investigations (for example, Refs *14, 15*). The equipment used within the study area was a Nimbus Instruments ES-100 Single Geophone Signal Enhancement Seismograph equipped with a refraction geophone. Results of seismic surveys at the east cross section are given in Table 2. Computed depths to the weathered zone-unweathered zone contact compare favorably with weathered zone depths observed in the rock cores. Since the unweathered zone is well cemented, compact, and of low fracture density, it has a high seismic velocity as indicated in the table. The results of this investigation show that there is a significant difference between the seismic signatures of the weathered and the unweathered layers. The field-determined seismic signatures for the soil, weathered bedrock, and unweathered bedrock zones appear to provide a field technique for mapping hydrogeologic units in the valley study area.

### *Cross-Section Piezometers*

The shallow weathered fracture layer was identified and characterized by both coring and seismic techniques, but the proof of its control on the flow system and associated chemical transport can be characterized only by sampling within the layer. Each cross section was instrumented with a network of piezometers to determine hydraulic head distribution within and immediately below the layer, and to provide access to points throughout the layer for hydraulic testing and water quality sampling. Each piezometer was drilled to depth with a 61-cm-long, 10.2-cm-diameter pilot bar leading a 25-cm three-cone rock bit. Schedule 80 polyvinyl chloride (PVC) casing was installed

TABLE 2—*Summary of seismic data.*

| Layer | Velocity, m/s | Standard Deviation, m/s |
|---|---|---|
| Moist soil | 400 | 80 |
| Weathered shale | 1700 | 600 |
| Unweathered shale | 4000 | 1100 |

from the shelf created by the roller rock bit to 1 m above the land surface, with the annular space between the casing and borehole wall being backfilled with pelletized bentonite. The piezometer point was then completed as an unscreened, 61-cm-long, 10.2-cm-diameter hole within the fractured rock matrix extending beyond the bottom of the casing; based on fracture frequencies discussed previously, this point size samples the fracture-field at a scale that represents fracture-controlled properties of the porous media.

Fracture pattern in both cross sections is similar; consequently, each section was instrumented similarly. Responses were also found to be similar, so only the east section is discussed here. The piezometer net installed is shown in Fig. 5, with piezometer lines being identified. One set of piezometers (E4) samples from 3 m below the land surface to 20 m (below the depth of fracture layer determined from the coring) directly under the channel. Five additional sets sample at distances of up to 55 m on either side of the channel, and at depths from 3 m to below the fracture layer. Each set is aligned parallel to the stream, with individual piezometer casings being spaced 1.5 m apart.

Generally, the piezometers farthest from the stream indicate recharge conditions in the shallow zone; that is, head decreases with depth. Those closer to the stream, where the effects of flowline convergence are more pronounced, indicate ground-water discharge conditions; heads increase with depth. Piezometer responses to a recharge event are similar in timing at all depths. Figure 6 shows head distribution within the cross section under differing hydrologic regimes. The 5/27/82

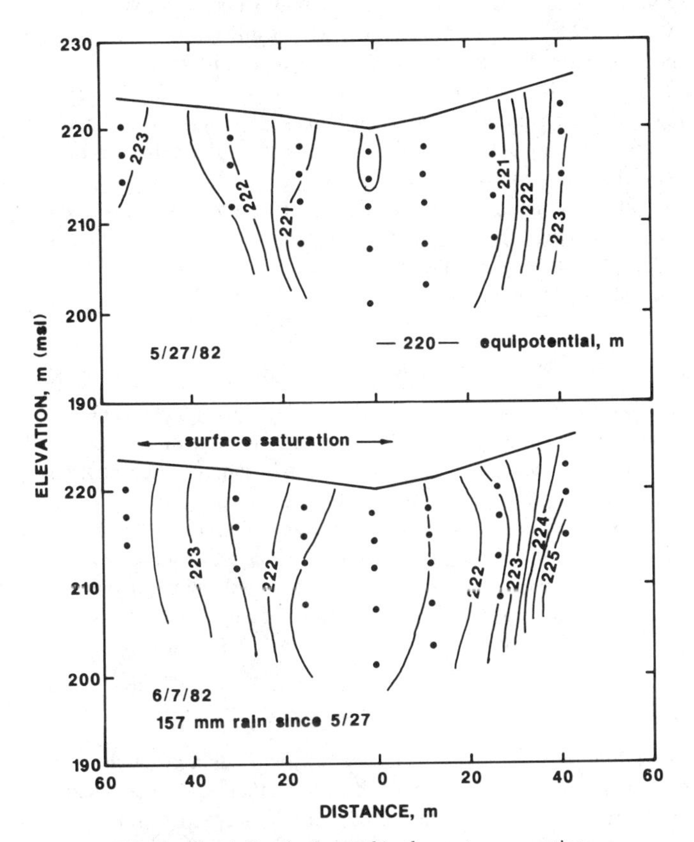


FIG. 6—*Head distribution within the east cross section.*

head distribution was during a relatively dry condition, while the 6/7/82 distribution was after 157 mm of rain over 11 days. The shallow fracture layer exhibited basically lateral flow under the dry conditions, but the flow lines were bent toward the stream at depth. This implies upward flow from the deeper less-fractured zone to the shallow fracture layer that was acting as a drain. After the series of storms, the equipotential lines had migrated toward the channel, and there was more of a lateral flow pattern at all depths, with increasing gradients near the channel discharge point. In each case, lateral flow dominated to within approximately 15 m of the channel, where the flow lines converged rapidly to the stream. At the start of this period, even those 3-m piezometers farthest from the stream had water standing within them; thus, the water table was within 3 m of the land surface over the entire section. Further, after the 157 mm of rain, hydraulic heads at all depths on the north side of the section were greater than the land surface up to approximately 60 m laterally from the channel. An extensive seepage face discharging ground water directly to the land surface had developed as the result of this sequence of storms; its location is also noted in Fig. 6.

### *Hydraulic Conductivity Distribution*

A total of 30 slug tests were performed in the section's piezometers, and results were analyzed using the Hvorslev [*16*] technique. The two injection volumes used, 15 and 45 L, strongly affected the conductivity values determined. Seventeen tests of fractured zone piezometers used 15-L slugs, and three used 45-L slugs. Mean hydraulic conductivity values resulting from each group, $25 \times 10^{-4}$ cm/s (15 L) and $29 \times 10^{-4}$ cm/s (45 L), were similar, but there was more variability in results from the 15-L tests. Within the less-fractured zone, different slug volumes gave different mean conductivities, $8 \times 10^{-4}$ cm/s (15 L) and $36 \times 10^{-4}$ cm/s (45 L). The number of unfractured zone tests using a 45-L slug (three) is too small to make a valid statistical comparison, so we can compare conductivities from only the 15-L slug tests to evaluate the fractured and unfractured zones. The tests yielded a mean hydraulic conductivity of $25 \times 10^{-4}$ cm/s for the shallow fractured weathered zone, but the standard deviation of $25 \times 10^{-4}$ cm/s indicates a wide spread of data. Unfractured zone tests show a mean hydraulic conductivity of $8 \times 10^{-4}$ cm/s with only a $5 \times 10^{-4}$ cm/s standard deviation. Thus, the fractured zone is about three times as permeable as the unfractured zone when both are considered as units.

### *Interaction of Land Use and Geology on Ground-Water Quality*

As in the larger-scale investigation, land use-geology-flow system interactions affect ground-water quality at this smaller scale also. Results of limited sampling of the sections' 9-m piezometers for water quality analyses are given in Table 3. High sodium concentrations result from the bentonite used to seal the piezometer casing. The data indicate a direct effect of overlying land use on ground-water quality within this zone. For example, high nitrate concentrations are observed under cropland (E5-E7), while under meadow (E1-E3), concentrations are almost zero. Only in the vicinity of the stream (E4), where the two shallow lateral flows converge, does the flow field influence the nitrate concentrations. Here, concentrations being input to the stream are intermediate to those of the two fracture-zone-derived lateral flows as a result of their mixing.

## **Summary and Implications for Subsurface Sampling**

Ground water within the study area exists as a layered system of perched and regional ground water. The perched ground water is in a zone of bedrock that is weathered and fractured to depths of 15 m, while the regional ground water exists within a deeper less-fractured zone extending to depths of approximately 100 m. Subsurface waters from the two zones merge near the perennial

TABLE 3—*Water quality within the shallow fracture zone.*

| Date | Ion | [b]Piezometers | | | | | | | [c]Watershed Zone "D" |
|---|---|---|---|---|---|---|---|---|---|
| | | E1 | E2 | E3 | E4 | E5 | E6 | E7 | |
| | | Meadow | | | Stream | Cropland | | | Deep Aquifer |
| 10-10-84 | $NO_3$ (mg/L) | 0.640 | 0.005 | 0.090 | 7.60 | . . . | 9.4 | 12.50 | 1.32 |
| | Cl (mg/L) | 10.1 | 9.3 | 8.6 | 5.3 | . . . | 6.8 | 8.6 | 1.10 |
| | [a]TSP (ppb) | 0.020 | 0.056 | 0.023 | 0.020 | . . . | 0.012 | 0.030 | 0.012 |
| | Ca (mg/L) | 17.65 | 9.20 | 18.95 | 9.88 | . . . | 12.50 | 8.06 | 5.48 |
| | Mg (mg/L) | 9.00 | 3.62 | 5.67 | 5.30 | . . . | 6.63 | 3.94 | 4.38 |
| | Na (mg/L) | 53.1 | 39.3 | 29.4 | 4.95 | . . . | 2.24 | 25.80 | 7.00 |
| | K (mg/L) | 3.14 | 0.52 | 4.84 | 0.57 | . . . | 1.25 | 2.42 | 1.05 |
| | pH | 7.93 | 8.04 | 7.79 | 7.24 | . . . | . . . | 7.36 | 6.96 |
| | Cond (μmhos/cm) | 410 | 250 | 270 | 133 | . . . | 151 | 220 | 93 |
| 1-2-85 | $NO_3$ (mg/L) | <0.1 | <0.1 | 0.4 | 8.4 | 10.4 | 10.5 | 11.3 | 1.32 |

[a]Total soluble phosphorus, ppb.
[b]10-m depth.
[c]Watershed zone "D" average of 10 samplings at 3 wells 1973–1984.

upland streams. The open character of the shallow, weathered, fracture zone marks a horizon of high hydraulic conductivity and a water transfer zone of major importance within upland watersheds. The layer acts as a direct conduit to the stream for ground water discharging upward from the regional aquifer and that flowing laterally within the shallow fracture zone itself. Water quality within this zone is affected directly by overlying land use, and indirectly by the quality of the regional ground-water flow; thus, subsurface flow input to the stream is a composite of both these sources.

Ground-water sampling networks designed for water quantity or quality measurement in such a system require characterization of the geologic controls on water movement. In addition, the sampling scheme requires determination of the general ground-water flow system. Site-specific data are rarely available in rural areas; however, approaching the problem from the large area definition, then proceeding downscale to the site-specific, yields the important parameters needed for flow-system definition. These parameters are subsurface watershed boundaries, aquifer depths, hydraulic conductivity distributions, and ground-water flow-system units.

The research-based approach to definition of subsurface flow and water quality presented here, considering decreasing scale, allows the significance of the general hydrogeology (that is, position in the regional ground-water flow system) to be retained, while bringing into focus the interactions between large- and small-scale phenomona. The change in scale affects the time, as well as the distances, of concern. Investigations within the 116-$km^2$ basin and focus on the 7.4-$km^2$ study area establish guidelines for sampling the regional pattern of geology-hydrology-land use relationships. Important variabilities of aquifer parameters controlling the regional hydrogeologic relationships are at the 10's-of-square-kilometres scale, 1000's-of-metres distance, and 10's-of-metres depth. At a smaller scale, though, important subcomponents of geology and hydrogeology may exert controls, such as shallow, intensively fractured, weathered rock. Variabilities of these systems must be examined at the scale of 100's of square metres, distances of 10's of metres and depths of a few metres, requiring more frequent sampling and more site-specific instrumentation. Yet, there is a commonality to both scales of study which dictates a sampling rationale. The commonality requires that ground-water sampling wells be located, and their data interpreted, by grouping according to geology and the ground water flow system elements of recharge, lateral flow, and discharge to the stream and land surface.

*Acknowledgments*

This work is a contribution from the U.S. Department of Agriculture, Agricultural Research Service, in cooperation with the Pennsylvania Agricultural Experiment Station, The Pennsylvania State University, University Park, Pennsylvania.

## References

[*1*] Pionke, H. B. and Urban, J. B., *Ground Water,* Vol. 23, No. 1, Jan.–Feb. 1985, pp. 68–80.

[2] Theis, C. V., "Estimating Transmissibility from Specific Capacity,"USGS Ground Water Note No. 24, U.S. Geological Survey, 1954.

[*3*] Cline, G. D., "Geologic Factors Influencing Well Yields in a Folded Sandstone-Siltstone-Shale Terrain Within the East Mahantango Creek Watershed," M.S. Thesis, Pennsylvania State University, University Park, PA, 1968.

[*4*] Kimball, B. G., *Transactions,* American Geophysical Union, Vol. 27, 1946, pp. 843–846.

[*5*] Parizek, R. R., "On the Nature and Significance of Fracture Traces and Lineaments in Carbonate and Other Terrains" in *Proceedings,* U.S.-Yugoslavian Symposium on Karst Hydrology and Water Resources, Dubrovnik, Yugoslavia, June 1975.

[*6*] Trexler, J. P., "The Geology of Klingerstown, Valley View and Lykens Quadrangles, Southern Anthracite Field, Pennsylvania," Ph.D. Thesis, The University of Michigan, Ann Arbor, MI, 1964.

[7] Urban, J. B., "The Mahantango Creek Watershed—Evaluating the Shallow Ground Water Regime" in

*Watershed Research in Eastern North America,* D. L. Correll, Ed., Smithsonian Institution, Washington, DC, 1977, pp. 251–275.

[8] Urban, J. B., *Journal of the Soil Conservation Society of America,* Vol. 20, 1965, pp. 178–179.

[9] Ferguson, H. F., "Valley Stress Relief in the Allegheny Plateau," *Association of Engineering Geologists Bulletin,* Vol. 4, 1967, pp. 63–68.

[10] Wyrick, G. G. and Borchers, J. W., "Hydrologic Effects of Stress Relief Fracturing in an Appalachian Valley," USGS Water Supply Paper 2117, U.S. Geological Survey, 1981.

[11] Gerhart, J. M., *Ground Water,* Vol. 22, No. 2, March–April 1984, pp. 168–175.

[12] Morey, R. M., "Continuous Subsurface Profiling by Impulse Radar" in *Proceedings,* Subsurface Exploration for Underground Excavation and Heavy Construction, American Society of Civil Engineers, Henniker, NH, 1974, pp. 213–232.

[13] Shi, S. F. and Doolittle, J. A., *Soil Science of America Journal,* Vol. 48, No. 3, 1984. pp. 651–656.

[14] Tibbetts, B. L., Dunrud, C. R., and Osterwald, F. W., "Seismic Refraction Measurements at Sunnyside, Utah," USGS Professional Paper 550D, U.S. Geological Survey, 1966, pp. D132–D137.

[15] Watkins, J. S. and Spieker, A. M., "Seismic Refraction Survey of Pleistocene Drainage Channels in the Lower Great Miami River Valley, Ohio," USGS Professional Paper 605B, U.S. Geological Survey, 1971, pp. B1–B17.

[16] Hvorslev, M. J., "Time Lag and Soil Permeability in Ground-Water Observations," Bulletin 36, Waterways Experiment Station, U.S. Army Corps of Engineers, Vicksburg, MS, 1951.

# Author Index

# Subject Index